ARTIFICIAL EARTH SATELLITES

ARTIFICIAL EARTH SATELLITES

Volume 3

Volume 4

and

Volume 5

Edited by L. V. Kurnosova

Authorized translation from the Russian

Springer Science+Business Media, LLC 1961

The publishers express their appreciation to the National Aeronautics and Space Administration for providing copies of eight NASA Technical Translations which were used as the basis of the corresponding translations in Volume 3.

The original texts of Volumes 3, 4, and 5 were published by the USSR Academy of Sciences Press, Moscow, in 1959 and 1960.

Library of Congress Catalog Card Number: 59-14596

ISBN 978-1-4899-5931-7   ISBN 978-1-4899-5929-4 (eBook)
DOI 10.1007/978-1-4899-5929-4

Copyright 1961 by Springer Science+Business Media New York
Originally published by Plenum Press, Inc. in 1961.
Softcover reprint of the hardcover 1st edition 1961

All rights reserved. No part of this publication may be reproduced in any form without written permission from the publisher.

NOBLE OFFSET PRINTERS, INC.
NEW YORK 3, N.Y.

CONTENTS

From the Publisher viii

Volume 3

The Capture Problem in the Three-Body Restricted Orbital Problem, V. A. Egorov 3
The Libration of a Satellite, V. V. Beletskii 18
Perturbations of the First Order in the Motion of Artificial Satellites, Caused by the Flattening of the Earth, V. F. Proskurin and Yu. V. Batrakov 46
Perturbations in the Orbits of Artificial Satellites Caused by Air Resistance, Yu. V. Batrakov and V. F. Proskurin 56
Tracking of Artificial Satellites Based on Expected Point of Arrival, V. M. Vakhnin and V. V. Beletskii 67
Secular Variations in Orbit Element as a Function of Air Resistance, P. E. Él'yasberg 76
The Problem of Piercing at Cosmic Velocities, M. A. Lavrent'ev 85
The Determination of the Density of the Atmosphere at an Altitude of 430 Kilometers by the Sodium Vapor Diffusion Method, I. S. Shklovskii and V. G. Kurt ... 92
Methods for the Control of Interfering Currents Originating at the Input of an Electrostatic Fluxmeter During Its Operating in a Conducting Medium, I. M. Imyanitov and Ya. M. Shvarts 108
Some Results of the Determination of the Structural Parameters of the Atmosphere Using the Third Soviet Artificial Earth Satellite, V. V. Mikhnevich, B. S. Danilin, A. I. Repnev, and V. A. Sokolov 119

A Radio-Frequency Mass Spectrometer for Investigations of the Ionic Composition of the Upper Atmosphere, V. G. Istomin 137

The Manometer Error Caused by Small Leaks in the Casing of a Satellite, S. A. Kuchai 161

On the Problem of the Interaction Between a Satellite and the Earth's Magnetic Field, Yu. V. Zonov 169

Volume 4

The Motion of an Artificial Satellite in the Normal Gravitational Field of the Earth, M. D. Kislik, ... 183

Determination of Upper-Atmosphere Density from the Results of Measurements of the Flight of the Third Soviet Artificial Earth Satellite, P. E. Él'yasberg and V. D. Yastrebov 202

Variations in Density of the Upper Atmosphere from Data Obtained from the Measured Period of Rotation of Artificial Earth Satellites, G. A. Kolegov 220

Determination of Illumination Conditions and Periods of Illumination and Darkness for the Artificial Satellite, I. M. Yatsunskii 225

Determination of the Parameters of the Orbit of an Artificial Satellite from the Results of Ground Measurement, T. M. Éneev, A. K. Platonov, and R. K. Kazakova 236

Methods for the Numerical Solution of Finite Difference Equations and Their Application to Computations of Satellite Orbits, G. P. Taratynova 254

Equations of Disturbed Motion in the Kepler Problem, A. I. Lur'e 288

Elements of the Theory of the Impact of Solid Bodies with High (Cosmic) Velocities, K. P. Stanyukovich . . . 292

Meteoric Matter and Some Geophysical Problems of the Upper Atmosphere, B. A. Mirtov 334

Magnetometers in the Third Soviet Earth Satellite, S. Sh. Dolginov, L. N. Zhuzgov, and V. A. Selyutin   358

Method of Determining the Electrical Potential of a Body in a Plasma, Ya. M. Shvarts 397

Study of Meteoric Particles Through Instruments on the Third Soviet Artificial Satellite, T. N. Nazarova . . . 402

Some Results of Measurement of Mass Spectra of Positive Ions on the Third Soviet Artificial Earth Satellite, V. G. Istomin 411

Cosmic Ray Measurements by Geophysical Rockets, Yu. G. Shafer and A. V. Yarygin 430

An Artificial Comet as a Method for Optical Tracking of Cosmic Rockets, I. S. Shklovskii 445

Volume 5

The Orbits of Cosmic Rockets in the Direction of the Moon, L. I. Sedov 469

Magnetic Measurements with the Second Cosmic Rocket, S. Sh. Dolginov, E. G. Eroshenko, L. N. Zhuzgov, N. V. Pushkov, and L. O. Tyurmina 490

Radiation Measurements During the Flight of the Second Lunar Rocket, S. N. Vernov, A. E. Chudakov, P. V. Valukov, Yu. I. Logachev, and A. G. Nickolaev 503

Cosmic Radiation Studies During the Flight of the Second Lunar Rocket, L. V. Kurnosova, V. I. Logachev, L. A. Razorenov, and M. I. Fradkin 512

Results of a Study of Impacting of Meteoric Matter by Means of Instruments Mounted on Space Rockets, T. N. Nazarova 524

Some Direction Control Problems in Interplanetary Space, B. V. Raushenbakh and E. N. Tokar' 528

Determination of the Conditions for Visibility of Space Rockets, O. V. Gurko 547

About the Formation of NO^+ in the Upper Atmosphere, A. D. Danilov 556

Observation of Signals from the Third Soviet Artificial Earth Satellite, at Cape Chelyuskin, L. P. Kuperov 564

Change of the Albedo of the First Artificial Earth Satellite as a Result of the Action of External Factors, I. M. Yatsunskii and O. V. Gurko 573

FROM THE PUBLISHER

Volumes 3, 4, and 5 of the continuing Soviet series, ARTIFICIAL EARTH SATELLITES, which were published separately in the original Russian, are here published in English translation as a single volume, for the convenience of the reader.

These three volumes together comprise a symposium of 37 reports of Soviet research in passive orbit mechanics and satellite-borne space physics experiments. Thirteen of the papers deal with the fields of celestial mechanics (orbit mechanics) and satellite tracking. There are ten papers concerned with satellite instrumentation for a variety of physical experiments. Problems of atmospheric physics, naturally overlapping those of the instrumentation used to study them, account for seven more papers. These do not include the six papers dealing with meteorites. There is, in addition, one paper on satellite guidance. Communications problems are touched upon in a number of the papers.

Subsequent volumes in this series will be made available in English translation as they appear in Russian.

Volume 3

THE CAPTURE PROBLEM IN THE THREE-BODY RESTRICTED ORBITAL PROBLEM

V. A. Egorov

STATEMENT OF THE PROBLEM

By the restricted orbital problem of three bodies in celestial mechanics we mean the problem of the motion of material point m_0 of negligibly small mass under the attraction of two finite point masses m and μ, rotating about a common center of inertia in orbits with constant angular momentum ω. The problem is restricted in the sense that attractions due to the mass m_0 are not included, that is, the accelerations imparted by it to the masses m and μ are neglected.

In this case, in the equations of the 3-point problem, it may simply be assumed that point m_0 is of negligible mass.

The equations of motion of the point m_0 in coordinates that rotate with the straight line joining m and μ, have the well-known Jacobi integral

$$\frac{v^2}{2} - U = h,$$

where

$$U = \frac{fm}{r} + \frac{f\mu}{\rho} + \omega^2(x^2 + y^2). \tag{1}$$

Here h is a constant of integration, which is identified with the energy of the motion, and r and ρ are the distances of the point m_0 from m and μ, respectively. The x-axis is always directed from μ to m, the y-axis passes through the center of mass of the system m, μ in the plane of their orbit, and the z-axis forms a right-handed coordinate system with the x- and y-axes; v is the velocity of the point m_0 in this system. The first two terms in the potential function U are the attractive potentials of the masses m and μ, and the third term is the potential of the centrifugal force.

It follows from the Jacobi integral that the motion with the constant h is always confined to a surface of zero velocity.

$$U = -h. \qquad (2)$$

These limiting surfaces were studied in detail by Hill,[2] and are often called Hill surfaces.

The term "capture" in the general problem of three bodies refers to the phenomenon in which three bodies, initially infinitely far apart, approach each other in such a way that, after the approach, one of the distances between a pair of points remains limited for all time.

In the restricted orbital problem of three bodies, "capture" refers to the phenomenon in which a point of zero mass is made to approach a system of finite masses from infinitely far away and is found not to depart again from the system to an infinite distance but to remain at distances that do not exceed certain finite values.

By considering the motion of a point m_0 in phase space, that is, in a coordinate-velocity space, Hopf[3] showed that for fixed initial values of position in phase space, a point m_0 as time increases without limit will either tend to approach the boundary of the region of possible motion as close as may be required, or repeatly go through a neighborhood around the initial values, which can be taken to be as small as is required. A point of zero mass starting at an infinite distance will, in general, however, in time again become infinitely remote. The expression "in general" is used here in the sense that the numerous initial values for which the motion is constant belong to a set of Lebesgue measure zero.

It is of interest to inquire whether in general such exceptional initial values exist. For among the trajectories with just such initial values would obviously be found trajectories corresponding to capture and also trajectories where a body m_0, being a satellite of one attracting mass, becomes a satellite of one attracting mass, becomes a satellite of another (if, of course, either kind of trajectory actually exists). The latter type of trajectory is obviously of great interest in the development of artificial satellites of the moon and planets without the aid of supplementary motors. Actually, although it would be impossible to realize the exceptional initial values exactly, nontheless, by achieving values sufficiently close to the correct ones, it might be possible to obtain trajectories that would result in a sufficiently large number of revolutions around the moon or the planets before it again moved away from them.

The present article presents the proof of a theorem connected with the question of capture in the restricted orbital three-body problem. It is shown that in the case where one of the attractive masses is sufficiently small with respect to the other, a point of zero mass starting infinitely far away cannot remain forever in the sphere of attraction of the smaller mass, that is, become its satellite.

The sphere of attraction of the smaller mass with respect to the larger refers to the region of space in which the attraction of the smaller mass is greater than that of the larger. We take for the unit of mass the sum of the masses $m + \mu$, and for the unit of length, the distance $m\mu$. Setting equal the attraction of the smaller mass μ to the attraction of the larger m, we get an equation for the limit of the sphere of attraction. In fact it is found to be the equation of a sphere of radius

$$\sqrt{\mu}\sqrt{\frac{1}{1-2\mu} + \frac{\mu}{(1-2\mu)^2}},$$

but with a center displaced from the point μ along the line $m\mu$ in a direction away from m, a distance of $\mu/(1-2\mu)$. Since for small μ the radius of the sphere of attraction is $\sqrt{\mu}$ to an accuracy of $\mu^{3/2}$, and the displacement of its center from the point μ is small with respect to the radius, the difference of the radius from $\sqrt{\mu}$ and the displacement of the center for small μ can be neglected, and one can consider as the sphere of attraction a sphere of radius $\sqrt{\mu}$ with a center at the smaller mass.

To show that a point of zero mass m_0 proceeding from an infinitely remote distance cannot remain forever within the sphere of attraction of the smaller mass, one considers the osculating μ-centered conic section, that is, the ellipse, parabola, or hyperbola, which the point m_0 would describe in a progressively moving coordinate system having its origin at the attractive center μ, if a perturbation from the direction of m were taken into account in the moment considered. The major semi-axis a, the parameter p, the inclination i, and other elements of this conic section are called the osculating elements of the trajectory.

The proof consists of two parts.

In the first part we prove a lemma for sufficiently small values of μ, which shows that if a point of zero mass enters the sphere of attraction $\rho = \rho_T$ of the mass μ, then while it remains in this sphere the major semi-axis of the osculating μ-centered ellipse $a > 2\rho_T$.

In the second part, using the lemma for the case of the oscillating ellipse and without using the lemma for the other cases, we prove that for sufficiently small μ a point of zero mass must inevitably emerge from the sphere of attraction of the smaller mass with respect to the larger one.

PROOF OF THE LEMMA

Turning to the proof of the lemma, we consider, following Hill,[2] surfaces (2) of zero velocity for various values of the constant h.

For large negative values of h the motion is shown to be possible only within the noncontiguous surfaces S' and S", which are nearly small spheres with centers at m and μ, and also outside the surface S, which encloses S' and S" and is nearly an extended circular cylinder having as its axis the z-axis. The surfaces S, S', and S" are of this nature because for large negative h, r, or ρ must be small to satisfy conditions (2), or the sum $x^2 + y^2$ must be large. In the first case the surface (2) must be close to a small sphere r = Constant or ρ = Constant, and in the second, to the extended cylinder $x^2 + y^2$ = Constant. A section of any of the surfaces S, S', and S" in the xy-plane will obviously be nearly a circle.

With an increase in h the surfaces S' and S" expand, and the surface S contracts. For a particular $h = h_1$, there will appear a point L_1 common to the surfaces S' and S" (Fig. 1), and for small $h - h_1 > 0$ the surfaces will be joined by an orifice around the point L_1. Half of the cross-section in the xy-plane of the surfaces S, S', and S", corresponding to $h = h_1$ (the curves S_1, S_1', S_1'') is shown in Fig. 1, which was constructed for the relative mass of the earth and the moon $m/\mu = 81.45$. It is apparent that for $h = h_1$ the penetration of trajectories from regions bounded by S' to a region bounded by S", and vice versa, becomes possible for the first time.

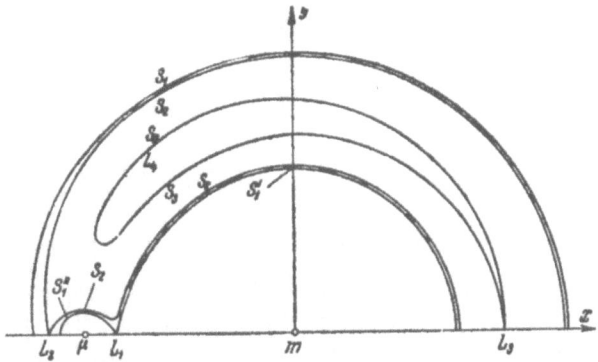

FIG. 1. Section of the surfaces of zero velocity of the surface x,y for the earth-moon mass system.

With a further increase in magnitude, h goes through a critical value h_2, corresponding to the appearance of a point L_2 common to the two surfaces S" and S (the curve S_2 in Fig. 1), so that for $h > h_2$ it becomes possible for a point m_0 to proceed from an infinitely remote distance and to pass through the orifice near the point L_2 into the region S".

In addition to the critical values h_1 and h_2 there are still further critical values $h_3 > h_2$ and $h_4 > h_3$. The value h_3 corresponds to the possibility of penetration of the initially remote point m_0 into the region S" through the orifice near the point L_3. (See the S_3 curve in Fig. 1.) The value $h_4 < 0$ also corresponds to the possibility of penetration into the region S'' by an initially remote point m_0 along an arbitrary direction in the xy-plane. The vanishing of surfaces determined by constraints of motion in space occurs when $h = 0$.

The points L_i [$i = 1, 2, 3$ (the so-called points of libration)] are obviously found as particular points of the surfaces (2); the values of h_i are found from (2) according to the coordinates r_i, ρ_i of the points L_i. In particular, for $i = 1, 2$, on taking ω^{-1} as the time unit and recalling that in this case, because of Kepler's third law for masses μ and m, f will be equal to one in equation (1), we find

$$-h_i = U_i = \frac{[(1-\mu)+(-1)^i \rho_i]^2}{2} + \frac{1-\mu}{1+(-1)^i \rho_i} + \frac{\mu}{\rho_i} \qquad (3)$$

For small μ with $i = 1, 2$ we have [4]

$$\rho_i = \left(\frac{\mu}{3}\right)^{1/3} + (-1)^i \left(\frac{\mu}{3}\right)^{2/3} - \frac{1}{9}\left(\frac{\mu}{3}\right)^{3/3}, \qquad (4)$$

$$-2h_i = 3 + 10\left(\frac{\mu}{3}\right)^{2/3} - \left[3 + (-1)^i \frac{4}{9}\right]\mu^{4/3} \qquad (5)$$

which are accurate to terms in $(\mu/3)^{4/3}$. Since the quantities ρ_i are of the order of $\mu^{1/3}$, and the quantity $\rho_T = \mu^{1/2}$, for sufficiently small μ we have $\rho_T < \rho_i$ ($i = 1, 2$). From an analysis of the evolution of the surfaces of zero velocity with a decrease in energy h, it follows that if point m_0 enters the sphere of attraction from the interior of the region S', $h > h_1$; but if it arrives from infinity, $h > h_2$.

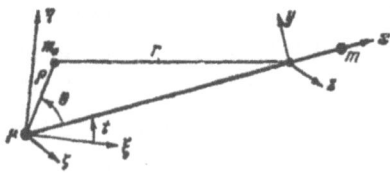

FIG. 2. Transformation from the rotating system of coordinates x, y to a progressively moving system of coordinates.

We shall now carry out the transformation of the Jacobi integral to the μ-centered osculating elements a, ρ, and i. This transformation is similar to that used in the derivation of the well-known Tisserand criterion.[4] For this purpose we transform first from the rotating coordinate system x, y, z to the continuously moving μ-centered system of coordinates ξ, η, ζ, whose ξ, η-plane coincides with the xy-plane and whose ξ-axis coincides at the initial moment t = 0 with the x-axis (Fig. 2). The Jacobi integral on transformation to ξ, η, ζ assumes the form

$$-w^2 + \frac{2\mu}{\rho} + 2\left(\xi\frac{d\eta}{dt} - \eta\frac{d\xi}{dt}\right) +$$
$$+ (1-\mu)^2 - 2(1-\mu)\left[(\xi\cos t + \eta\sin t)\frac{1}{r}\right] = -2h,$$

where w is the velocity in the ξ, η, ζ-system.

Now employing formulas from the theory of finite sections [4]

$$w^2 = \frac{2\mu}{\rho} - \frac{\mu}{a}, \quad \xi\frac{d\eta}{dt} - \eta\frac{d\xi}{dt} = \sqrt{\mu p}\cos i$$

and the formula, which is apparent from Fig. 2,

$$\xi\cos t + \eta\sin t = \rho\cos\theta,$$

where θ is the angle of the vector ρ with the x-axis, we get

$$\frac{\mu}{a} + 2\sqrt{\mu p}\cos i = -2h - (1-\mu)\left[(1-\mu) - 2\rho\cos\theta + \frac{2}{r}\right]. \quad (6)$$

Assuming that ρ becomes less than ρ_T, and employing the development of r^{-1} in powers of ρ, given by the equation

$$r^2 = 1 - 2\rho\cos\theta + \rho^2,$$

and taking into account the inequality $-2h < -2h_1$, we get inequality

$$\frac{\mu}{a} + 2\sqrt{\mu p}\cos i < 10\left(\frac{\mu}{3}\right)^{2/3} + \mu F, \quad (7)$$

where

$$F = 1 - (-1)^i \frac{4}{9} + \frac{\rho^2}{\mu}(1 - 3\cos^2\theta) + \mu^{1/3} F_1,$$

and F_1 is a finite value. Because of the condition $\rho < \sqrt{\mu}$ the function F is also finite.

We shall now introduce proof of the converse. If it is assumed that $a < 2\rho_T$, the inequality (7) can hold, putting

$$\cos i = -1, \quad \rho = a = 2\rho_T;$$

we then get

$$\frac{1}{2}\mu^{1/2} - 2\sqrt{2}\mu^{3/4} < 10\left(\frac{\mu}{3}\right)^{2/3} + \mu F. \tag{8}$$

Since the small power of μ on the left is smaller than that on the right, for sufficiently small values of μ the last inequality becomes a contradiction. This means that for $\rho < \rho_T$ we have

$$a > 2\rho_T.$$

Thus the first part of the theorem (lemma) is proved.

PROOF THAT THE PARTICLE NECESSARILY EMERGES FROM THE SPHERE OF ATTRACTION

We show that if the ratio of the attractive masses is sufficiently small, a point of zero mass approaching from infinity and entering the sphere of attraction $\rho = \rho_T$ of the smaller mass with respect to the larger must inevitably emerge from this sphere. Generally speaking it is necessary to show that if the initial distance ρ_0 of the point m_0 from the smaller mass satisfies the inequality $\rho_0 < \rho_T$, then with time the distance ρ necessarily becomes greater than ρ_T.

This may appear obvious when $a < \rho_T$, but in fact it is required for proof. Actually, in the absence of perturbations, point m_0, after entering the sphere of attraction and approaching the center of attraction, would escape from it, thanks to the condition $a > \rho_T$, to a maximal distance $\rho' > \rho_T$. However, in the presence of perturbations the maximal and minimal distances on the osculating ellipse may occasionally be unattainable. For example, in the motion of a satellite in a circular orbit in the equatorial plane of an oblate terrestial spheroid, the radius of the osculating ellipse at apogee is never reached.[5] In the same way, in the motion of a circular satellite in the equatorial plane of a prolate material ellipsoid, the orbit never reaches as far as the radius at perigee. Although in the problem under consideration perturbations of Kepler motion by the mass m are distinguished from perturbations resulting from nonsphericity in the examples given, nonetheless it is not apparent beforehand that after the point m_0 enters the sphere of attraction of the mass μ the distance $\rho(t)$ will not asymptotically approach some definite distance $\rho_* < \rho_T$, or that it will not oscillate without reaching the value ρ_T. If, however, it can be shown that neither the one nor the other type of change in $\rho(t)$ is possible, the point m_0 must in fact sooner or later move away from the mass μ to distances exceeding the radius of the sphere of attraction.

In order to obtain equations for the changes of ρ with time, we shall use the equations of μ-centered motion,[1] taken as linear for small ρ and given in vector form by

$$\ddot{\boldsymbol{\rho}} = -\frac{\mu}{\rho^3}\boldsymbol{\rho} - (1-\mu)\boldsymbol{\rho} + [3(1-\mu)\rho\cos\theta]\overrightarrow{(\mu m)}.$$

The first term on the right side expresses the attraction of the mass μ, and the remaining terms are the perturbing influences of the mass m. Dots over the letters designate, here and later, differentiation with respect to time.

The equation that defines $\rho(t)$ is obtained from the vector equation by using scalar multiplication of the vector $\boldsymbol{\rho}$ and using the identity

$$\rho\ddot{\rho} = \boldsymbol{\rho}\ddot{\boldsymbol{\rho}} + (w^2 - \dot{\rho}^2),$$

where $w = |\dot{\boldsymbol{\rho}}|$. The equation that determinates ρ assumes the form

$$\rho\ddot{\rho} = -\frac{\mu}{\rho} - (1-\mu)\rho^2 + 3(1-\mu)\rho^2\cos^2\theta + w^2 - \dot{\rho}^2.$$

Since at each instant we have from a determination of the osculating major semi-axis, $a(t)$,

$$w^2 = \frac{2\mu}{\rho} - \frac{\mu}{a(t)},$$

after combining similar terms we get

$$\rho\ddot{\rho} + \dot{\rho}^2 = \frac{\mu}{\rho} - \frac{\mu}{a(t)} - (1-\mu)(1-3\cos^2\theta)\rho^2, \qquad (9)$$

or after multiplying by $2\rho\dot{\rho}$ and integrating

$$(\rho\dot{\rho})^2 - (\rho_0\dot{\rho}_0)^2 = 2\mu(\rho-\rho_0) - \int_{\rho_0^2}^{\rho^2}\frac{\mu}{a(\rho^2)}d(\rho^2) - \frac{1-\mu}{2}\int_{\rho_0^4}^{\rho^4}(1-3\cos^2\theta)d(\rho^4), \qquad (10)$$

where $\rho_0 < \rho_T$ (initial values designated by zeros).

With the help of equation (9), and using the smallness of the perturbations for small μ, it can be shown that for $a > 2\rho_T$ the quantity $\rho(t)$ cannot approach a constant $\rho_* < \rho_T$ indefinitely closely as $t \to \infty$. In fact if equation (9) is multiplied by ρ and ρ is replaced on the right side by $k\rho_T = k\sqrt{\mu}$, where $k < 1$, and one takes the limit as $t \to \infty$, then, since the limit of the left side is equal to zero, we get after combining terms in μ

$$0 = \lim_{t \to \infty}\left[1 - \frac{\rho}{a} - (1-\mu)(1 - 3\cos^2\theta)\, k^3\sqrt{\mu}\right].$$

The difference of the first two terms in square brackets, because of the condition $\rho < \rho_T$ and from the result of the lemma $a > 2\rho_T$, is always greater than 1/2, and the third term may be taken as small as desired for sufficiently small μ. As a result, the equation under consideration is a contradiction, and the distance $\rho(t)$ cannot approach a constant $\rho_* < \rho_T$ indefinitely closely as t increases.

From the fact just shown it follows that the point m_0 having approached the mass μ to some minimum distance begins to move away, just as in the absence of perturbations, while at the same time the distance $\rho(t)$ for sufficiently small μ cannot monotonically approach a constant $\rho_* < \rho_T$ as time increases. Consequently, if it is supposed that the point m_0 does not in time leave the sphere of attraction $\rho = \rho_T$ then it must reach a maximum distance $\rho' < \rho_T$. We shall show that for $a > 2\rho_T$ this is impossible.

Putting $\rho = \rho'$ in (10), employing the condition $\dot\rho = 0$, which holds for $\rho' = \rho_{\max}$, and applying the theorem of the mean, we get an equation for ρ'

$$-(\rho_0 \dot\rho_0)^2 = 2\mu(\rho' - \rho_0) - \frac{\mu}{a_{\mathrm{av}}}(\rho'^2 - \rho_0^2) - \frac{1-\mu}{2}(1 - 3\cos^2\theta_{\mathrm{av}})(\rho'^4 - \rho_0^4). \qquad (11)$$

Since according to the lemma $a(t) > 2\rho_T$, then also $a_{\mathrm{av}} > 2\rho_T$. The root ρ' of this fourth degree equation can be found exactly, but for our purposes it is sufficient to show that $\rho' > \rho_T$.

We suppose the opposite, that is, that $\rho' < \rho_T$. Putting the left side of (11) equal to zero, we obtain an inequality, which division by $\mu(\rho' - \rho_0)$ brings to the form

$$2 < \frac{\rho' + \rho_0}{a_{av}} + \frac{1-\mu}{2\mu}(1 - 3\cos^2\theta_{av})(\rho' + \rho_0)(\rho'^2 + \rho_0^2). \tag{12}$$

With the help of the relations $\rho' < \rho_T$, $\rho_0 < \rho_T$, and $a_{av} > 2\rho_T$ each term of the right side of the inequality (12) can be increased and we get

$$2 < 1 + \frac{1}{2\mu} \cdot 1 \cdot 2\rho_T \cdot 2\rho_T^2. \tag{13}$$

Since $\rho_T^2 = \mu$, for sufficiently small μ the last inequality is a contradiction. This means that $\rho' > \rho_T$ and the theorem for the case where the osculating conic section is an ellipse is proved.

In case the osculating conic section is a parabola or hyperbola, the proof that the point m_0 necessarily emerges from the sphere of attraction is almost trivial. In fact, in the parabolic case $a = \infty$, and in the hyperbolic case $a < 0$, and the proof of the fact that the distance of the point m_0 from the mass μ after it enters the sphere of attraction cannot asymptotically approach a constant $\rho_* < \rho_T$ remains valid. As distinguished from the elliptical case, in the case under consideration after the radial velocity ρ passes through its value at the minimum distance, it cannot return to a value of zero at a finite distance ρ from the center of attraction; and the point m_0 obviously emerges from the sphere of attraction in time. As a result, the theorem is also proved for those cases where the osculating conic section is a parabola or a hyperbola.

Cases are still possible, where the types of osculating conic section change once or more as the point m is removed from the mass μ. The proof in this case, obviously, is just a combination of the proofs of the preceding cases.

Thus, the following theorem has been completely proved.

If, in the restricted orbital problem of three points the ratio of the attractive masses is sufficiently small, a point of zero mass initially at an infinitely remote distance and approaching the sphere of attraction of the smaller mass necessarily leaves this sphere.

REMARKS

1. If one considers the so-called sphere of action of the mass μ,[1] instead of the sphere of attraction, the proof can be carried out in the same way.

The sphere of action of the smaller mass μ with respect to the larger mass m is determined in the following way. One considers the ratio of the force with which the mass m perturbs the μ-centered motion of the point m_0 to the force of attraction of the mass μ. One also considers the ratio of the force with which the mass μ perturbs the m-centered motion of the point m_0 to the force of attraction of the mass m. The region around the mass μ in which the first ratio is smaller than the second is called the sphere of action of the mass μ. In this region, obviously, the error made in the μ-centered equations by neglecting the perturbations of the mass m is less than the made by neglecting the perturbations from the mass μ in the m-centered equations. For small values of μ this region is practically a μ-centered sphere, with a radius given in our units by $\rho_g = \mu^{2/5}$.[1]

The proof of the theorem when ρ_T is substituted for ρ_g proceeds because the smaller powers of μ in the left-hand sides of (8) and (13) are, as before, smaller than those on the right-hand sides. On equating these, a contradiction can be avoided only by using for ρ_T a quantity of the order of $\mu^{1/3}$, that is, of the order of the distance of the point of libration L_1 from μ. If in general the dimensions of the region are determined by a quantity of the order of μ^n, the proof holds for $n > 1/3$. If $n \leq 1/3$ the proof does not hold, and the question remains undecided.

2. If, following V. G. Fesenkov,[6] we call capture of a particle by an attracting mass μ the special case of capture in the restricted orbital three-body problem when the particle starting at infinity always stays in the sphere of attraction of a mass, then the proof of the above theorem shows that capture of the zero-mass point m_0 by the smaller mass μ is impossible. However, this theorem does not indicate that the capture of a point m_0 by a mass system m, is in general

impossible. In fact, it does not follow from the theorem that a point starting at infinity might not always become a satellite of the mass system m, μ, occasionally moving away from the mass μ to a distance that could be larger than the distance from μ to the point of libration L_1.

3. Since the proof of the theorem was done at the same time for the values $h = h_2$ and $h = h_1$, it follows that for sufficiently small μ capture of a point by a mass μ is also impossible if the point reaches the sphere of attraction of the mass μ (or its sphere of action) not from infinity, but from the region bounded by S' that surrounds the mass m. In particular, if the mass m has a satellite, this cannot be captured by a mass μ if the value of μ is sufficiently small.

4. The first part of the proof of the theorem given above has been briefly noted;[7] but the result derived in the second part of the proof was assumed to be obvious.[7] However, as can be seen from the examples given above, it is needed the proof, and the author is extremely grateful to G. A. Merman for having pointed this out. The omission referred to in Ref. 7 also occurs in Ref. 6, where an attempt is made to derive a result close to that of the lemma given above. However, the omissions referred to in Refs. 7 and 6 ruin the rigor of the proof.

5. The largest values of μ, for which equations (8) and (12) are still not contradictory, are of the order of 10^{-4}, as can immediately be checked by noting that the function F_1 in (8) has a magnitude of order unity. Therefore for the majority of planets in the solar system, capture by a planet of particles of small mass is impossible. For Jupiter, the mass of which is about 0.001 solar masses, the question of the possibility of capturing particles of small mass remains undecided.

AN INFORMAL REVIEW OF THE QUESTION OF THE POSSIBILITY OF CAPTURE BY THE MOON OR PLANETS OF MISSILES FROM THE EARTH

The theorem proved above is not applicable to the earth-moon system ($\mu > 0.01$); but the important question of whether the moon can capture missiles from the earth can be approximately answered very simply. In the case where a trajectory starting on earth enters the sphere of action of the moon on its first revolution around the earth (that is, according to the definition of Ref. 7, it is an approach trajectory) it can be shown, neglecting perturbations, that this trajectory must leave the sphere of action on the first revolution around the moon. Capture of missiles by the moon is impossible for such trajectories, because the segments of them that lie within the sphere of action of the moon is always practically a hyperbola referred to continuously moving moon-centered coordinates. At the boundary of the sphere of action the moon-centered velocity exceeds the local parabolic velocity $v_p = \sqrt{2\mu/\rho_g}$ = 383 m/sec by more than a factor of two.

Actually, for initial velocities that are less than the parabolic one for the earth, one can conclude from the earth-centered surface integral (which hold approximately up to the point of entry of the trajectory into the sphere of action of the moon) that the upper limit of the transverse component of the entering earth-centered velocity is no more than 200 m/sec.[7] Since the angle formed by the velocity of the moon and the entering earth-centered radius is not much different from a straight line (not by more than 10°), and the velocity of the moon itself is about 1 km/sec, then one component of the entering moon-centered velocity, which is also orthogonal to the earth-centered radius vector of the point of entry into the sphere of action, is at least 0.8 km/sec, that is, it exceeds $2v_p$.

For initial velocities larger than or equal to the parabolic one for the earth, the radial component of the entering earth-centered velocity is greater than 1 km/sec, and the magnitude of the entering moon-centered velocity, as mentioned, is more than twice v_p.

Perturbations from the earth cannot significantly change such a sharply defined hyperbolic motion within the sphere of action, and missiles entering the sphere of action should leave it on the first revolution around the moon if it does not strike the surface of the moon.[4]

In the same way for a planet-sun system, disregarding perturbations, it can be shown that capture of missiles from the earth by a planet on its first revolution around the sun should not occur. It may be mentioned that the error caused by disregarding perturbations, that is, the influence of a planet beyond its sphere of action and the influence of the sun in the sphere of action of the planet,[1] will be much smaller than in the problem of the flight to the moon. The hyperbolicity of velocities of a missile from the earth in the sphere of action of planets would appear to be much larger than in the sphere of action of the moon. Much smaller entering velocities of a missile from the earth into the sphere of action of the target planet occur for Mars and Venus. However, even these velocities are about twice as large as the planet-centered parabolic velocities at the boundary of the sphere of action of the planet. Thus, capture of a missile from the earth by a planet on its first revolution around the sun is also impossible.

ACKNOWLEDGEMENT

The author is indebted to D. E. Okhotsimskii for his careful attention in reviewing the manuscript, for his helpful criticism, and for a number of valuable observations.

LITERATURE

1. M. F. Subbotin, Kurs nebesnoy mekhaniki (Text in Celestial Mechanics), Vol. II, ONTI, 1937, pp. 108–109, 194.
2. G. W. Hill, Lunar Theory, The American Journal of Mathematics, 1, 23 (1877).
3. E. Hopf, Mathematische Annalen, 103 (1930).
4. F. R. Moulton, An Introduction to Celestial Mechanics, Macmillan, New York, 1935.
5. D. E. Okhotsimskii, T. M. Eneev, and G. P. Taratynova, Uspekhi fizicheskikh nauk (Successes in the Physical Sciences), 63, 1a (1957).
6. V. G. Fesenkov, Astronomicheskii zhurnal (Astronomical Journal), 23, 1 (1946).
7. V. A. Egorov, Uspekhi fizicheskikh nauk, 63, 1a (1957).

THE LIBRATION OF A SATELLITE

V. V. Beletskii

In this paper the methods of Lyapunov-Chetaev are applied to find conditions for the existence and stability of the relative equilibrium of a material body in orbit in a Newtonian central-force field and to consider the oscillations of the body around the position of relative equilibrium. The problem considered is an idealized motion, which actually occurs in the solar system (the motion of the moon relative to the earth and the possible motions of artificial earth satellites).

PRELIMINARY ANALYSIS

In Ref. 1 the theory was considered of the motion of an artificial satellite of the earth relative to the center of mass in the presence of perturbing factors (aerodynamic and gravitational moments; orbital regression). In this work it was shown that if the kinetic energy of rotation around the center of mass is sufficiently large with respect to the external forces acting, as was the case for the first artificial satellites, the motion of the satellite can be resolved into unperturbed Euler-Poinsot motion around the vector of momentum and into a secular precession-nutation motion of the same vector.

But in general it is necessary to consider the case of small kinetic energy as well. Then under the action of external forces it can be shown that motion of another type is possible, namely a librational (oscillatory) motion. The motion of the moon is a typical example of such motion.

The present article is devoted to the investigation of conditions for the existence and stability of the position of relative equilibrium of a satellite, that is, equilibrium in a system of coordinates fixed with the radius vector of the center of mass of the satellite. The librational motion around a position of relative equilibrium is also considered.

Since the existence of librations is caused by the kind of action on the satellite of the Newtonian central-force field, we shall consider only this action and shall neglect perturbations due to the compression of the earth, aerodynamics, etc. From this point of view the theory of librations refers to idealized motion, but one which actually occurs in the solar system (the motion of the moon with respect to the earth) and motions that are possible for artificial earth satellites.

Before proceeding to the exact theory we shall consider the phenomena qualitatively. The existence of a stable relative equilibrium and the libration are caused by the mutual action of the forces of Newtonian attraction of the particles of the body toward the center of gravitation and by centrifugal forces which cause the motion of the center of mass of the body to follow an orbit.

We shall consider first the action of a Newtonian central-force field, neglecting the motion of the center of mass. Let the center of mass of the satellite be found at a distance R from the center of attraction. We put at the center of mass of the satellite a right-handed orthogonal coordinate system x_1, y_1, and z_1 such that the z_1 axis is directed along the radius vector, and the y_1 axis is along the binormal to the trajectory of the center of mass. On a particle of the satellite with a mass dm and coordinates x_1, y_1, and z_1, there acts a Newtonian force F in the direction of the center of attraction

$$F = -\frac{\mu dm}{x_1^2 + y_1^2 + (z_1 + R)^2} \cdot r_0, \qquad (1)$$

where μ is the gravitational constant, and the unit vector r_0 is determined by the direction cosines

$$\cos(\widehat{x_1, r}) = \frac{x_1}{r}, \quad \cos(\widehat{y_1, r}) = \frac{y_1}{r}, \quad \cos(\widehat{z_1, r}) = \frac{z_1 + R}{r}, \\ r = \sqrt{x_1^2 + y_1^2 + (z_1 + R)^2}. \qquad (2)$$

The dimensions of the satellite are small with respect to the distance R to the center of attraction. This means that the actions of the force F on particles of the satellite can be considered for only small (with respect to R) values of x_1, y_1, and z_1. Then apart from differentials of higher order the components of the force F with respect to the x_1, y_1, and z_1 axes have the form

$$F_{x_1} = -\frac{\mu\,dm}{R^2} \cdot \frac{x_1}{R}, \quad F_{y_1} = -\frac{\mu\,dm}{R^2} \cdot \frac{y_1}{R},$$
$$F_{z_1} = -\frac{\mu\,dm}{R^2} + 2\frac{\mu\,dm}{R^2} \cdot \frac{z_1}{R}. \qquad (3)$$

The resultant of all the elementary forces $-\mu dm/R^2$ which enter into F_{z_1} is applied at the center of mass of the satellite and is collinear with the radius vector of the center of mass. Under the action of this resultant, the center of mass moves in a Kepler orbit. The second term in F_{z_1} and F_{x_1}, F_{y_1} by their total action result in a moment which causes the rotation of the satellite around the center of mass. For example, let us consider a plot of the lines of force of this force field in the plane z_1, x_1 of the orbit (in any other given plane containing the z_1 axis the picture is the same). The differential equation of the lines of force

$$\frac{dz_1}{dx_1} = \frac{F_{z_1}}{F_{x_1}} = -\frac{2z_1}{x_1} \qquad (4)$$

after integration gives the equation

$$z_1 = \frac{C}{x_1^2}. \qquad (5)$$

of a family of lines of force (Fig. 1). The direction of the action of forces is shown on Fig. 1 by arrows. We see that a particle moving under the action of the force field will tend to move toward the z_1 axis, i.e., toward the radius vector of the center of mass. We suppose that in the force field shown in Fig. 1 there is a certain liquid mass. Then the particles of liquid will tend in the direction of the z_1 axis. This widely known phenomenon of currents is observed for example on the earth in the ocean tides under the action of the moon and the sun.

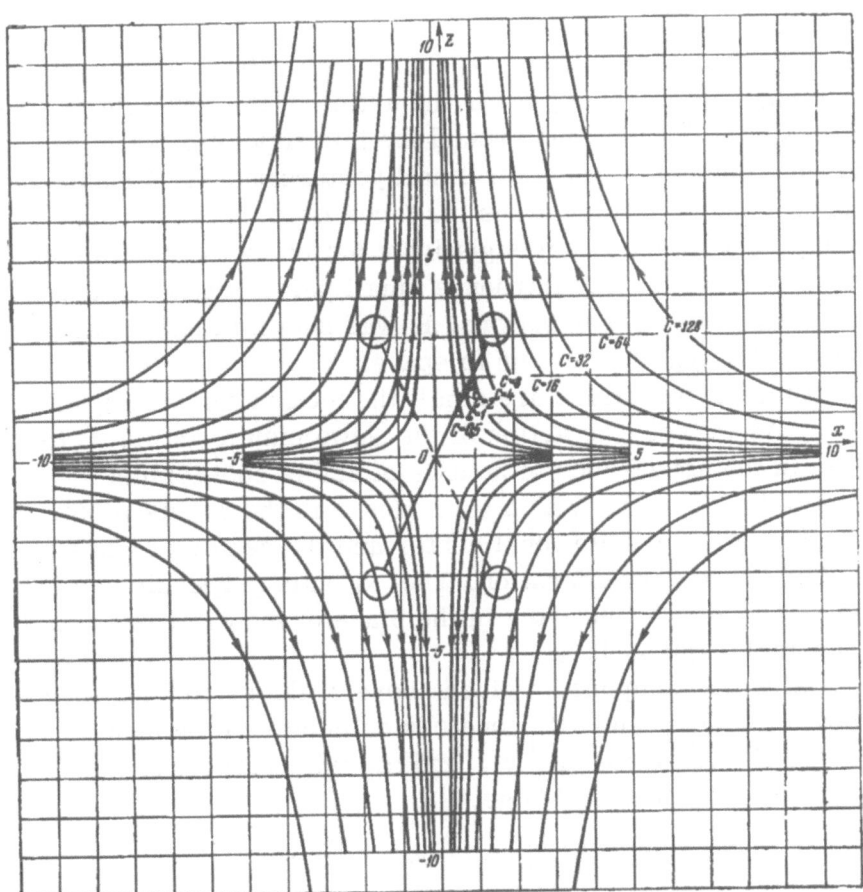

FIG. 1. Field of perturbing forces.

If in the force field considered there is found a completely rigid body the center of mass of which coincides with the center of field, the body will rotate about the center of mass. For purposes of illustration let the body consist of an extended axially symmetric figure, for example like a dumbbell (two equal point masses at the end of a bar of negligible mass). Then it is easy to see that if at the initial instant the axis of the body coincides with the x_1 axis, then for small perturbations of the body under the action of the forces it will turn in order to align its axis with the radius vector of the center of mass (i.e., the z_1 axis). If, however, the axis of the body initially coincides with the z_1 axis, then with small initial perturbations the body, under the action of the force field, will execute oscillations around the initial position (Fig. 1). We note that this occurs for small perturbations. For sufficiently large perturbations the force field, because it is weak, does not exert any significant action on the rotation of the body and will only introduce small perturbations in the rotation. These perturbations, as was shown in Ref. 1, are of the secular type for motion in space.

If the perturbations are small the motion can display librations. As we see, the action of the Newtonian force field results in the appearance of an axis of stable equilibrium, namely the z_1 axis, which coincides with the radius vector connecting the gravitational center and the center of mass of the body under consideration. The x_1 axis perpendicular to the radius vector and lying in the plane of the orbit is an axis of unstable equilibrium.

Using the components given above of the perturbing force field F_{x_1}, F_{y_1}, and F_{z_1} one can calculate the moment of this field at a given point x_1, y_1, and z_1 with respect to the origin of coordinates. We get

$$M_{x_1} = + 3y_1 z_1 \cdot \frac{\mu \, dm}{R^3}, \quad M_{y_1} = - 3x_1 z_1 \frac{\mu \, dm}{R^3},$$
$$M_{z_1} = 0. \tag{6}$$

For a circular orbit $\mu/R^3 = \omega^2$; ω is the angular velocity of the motion of the center of mass in the orbit.

For a dumbbell of half-length l, and half-mass m, located in the plane of the orbit, we get the restoring moment

$$M_{y_1} = - 3ml^2\omega^2 \sin 2\alpha_0,$$

and in a plane normal to the plane of the orbit and passing through the radius vector, the restoring moment, as follows from a consideration of the plot of the lines of force, has the form

$$M_{x_1} = - 3ml^2\omega^2 \sin 2\beta_0,$$

where α_0 and β_0 are the angles between the positive direction of the axis z_1 and the dumbbell, taken counterclockwise in the planes considered, as viewed from the positive directions of the y_1 and x_1 axes, respectively. In the plane perpendicular to the radius vector there is no moment. Thus a Newtonian force field tends to align an elongated body along the radius vector of the orbit.

In this discussion we have neglected the motion of the center of mass of the body. As a result of the motion of the center of mass of a satellite in orbit, there will be superimposed on a Newtonian force field the action of centrifugal forces. The nature of this action is clearly shown by the same simple example of the dumbbell.

For simplicity we consider a circular orbit in which the center of mass of a satellite moves with an angular velocity ω. Then at each point m(x_1, y_1, and z_1) of a system of coordinates x_1, y_1, and z_1 there acts a centrifugal force

$$\mathbf{\Phi} = \omega^2 \mathbf{r}_1 dm,$$

where $\mathbf{r}_1$ is the distance from the point m to the axis of rotation, that is, to an axis passing through the gravitational center normal to the plane of the orbit. Obviously

$$\cos(\widehat{r_1, x_1}) = \frac{x_1}{r_1}, \quad \cos(\widehat{r_1, y_1}) = 0,$$

$$\cos(\widehat{r_1, z_1}) = \frac{R + z_1}{r_1}, \quad |\mathbf{r}_1| = \sqrt{x_1^2 + (R + z_1)^2},$$

and the components of the centrifugal force may be taken as

$$\Phi_{x_1} = x_1 \omega^2 dm, \quad \Phi_{y_1} = 0, \quad \Phi_{z_1} = (R + z_1) \omega^2 dm.$$

One may consider the plot of the lines of force of the force field Φ in each of the planes x_1, y_1; x_1, z_1; y_1, z_1 in the same way as was done in the Newtonian force field. Such a consideration serves to establish the restoring or tumbling moment of the forces in each of these planes. The moment of the force Φ with respect to the center of mass of the satellite, apart from terms which are negligible when one sums over the total volume of the body, will have components

$$M_{x_1} = y_1 z_1 \omega^2 dm, \quad M_{y_1} = 0, \quad M_{z_1} = -y_1 x_1 \omega^2 dm.$$

Thus, in the plane of the orbit the centrifugal force does not produce a moment. In the plane normal to the radius vector the centrifugal force produces a restoring moment M_{z_1} which for the dumbbell will have the form

$$M_{z_1} = -ml^2\omega^2 \sin 2\gamma_0,$$

where γ_0 is the angle between the x_1 axis and the axis of the dumbbell, taken with the same conventions as the angles α_0 and β_0. From a consideration of the lines of force in this plane it follows that this moment will tend to align the dumbbell along the x_1 axis. In a plane perpendicular to the plane of the orbit the centrifugal force produces a restoring moment

$$M_{x_1} = -ml^2\omega^2 \sin 2\gamma_0,$$

which, as follows from a consideration of the lines of force in the plane y_1, z_1, tends to align the axis of the dumbbell along the direction of the radius vector.

Thus, for all three axes there are restoring moments under the joint action of gravitational and centrifugal force: along the normal to the plane of the orbit

$$M_{y_1} = -3ml^2\omega^2 \sin 2\alpha_0;$$

along the tangent to the orbit

$$M_{x_1} = -4ml^2\omega^2 \sin 2\beta_0;$$

along the radius vector of the orbit

$$M_{z_1} = -ml^2\omega^2 \sin 2\gamma_0.$$

One can assume that for a three-dimensional dumbbell the indicated action of the gravitational and centrifugal forces will tend to align the axes of the dumbbell along the axes x_1, y_1, and z_1. Finally, on going to the general case of an arbitrary solid body, from similar considerations one can conclude that Newtonian and centrifugal forces will have a tendency to align the body in such a way that two of the principal axes of inertia, just as the axes of the three-dimensional dumbbell, coincide with the radius vector of the orbit and the perpendicular to it in the plane of the orbit. We shall call such a position of a body in orbit relative equilibrium. The question arises of how the positions of relative equilibrium will be established. One would expect that the largest axis of the inertial ellipsoid should be directed along the radius vector, so that, in the same way as for the dumbbell, the elongation along the radius vector best assists the restoring action of the Newtonian force field. In fact, it has been shown[2] that in a stationary Newtonian field absolute equilibrium is established when and only when the major axis of the inertial ellipsoid coincides with the direction of the center of attraction. But then one would expect that the second axis in the plane of the orbit (which, for a circular orbit is directed along the tangent to the trajectory) should be the intermediate axis of the inertial ellipsoid. Actually, in this case one can best use the remaining dynamic "elongation" of the body for stabilization along the tangent to the orbit under the action of centrifugal forces. Such a position for the intermediate axis can also be deduced from the fact that it cannot be located along the binormal to the orbit, since relative equilibrium of the body amounts to absolute rotation around the direction of the binormal, and the rotation of a free body around the intermediate axis is unstable; Newtonian and centrifugal forces do not remove the instability.

We should note that in Tisserand's text[3] the assignment of the axes of the inertial ellipsoid given above is indicated. Tisserand arrived at this assignment by an approximate analysis of the equations of motion and qualitative considerations.

In the following sections it will be shown by the methods of Lyapunov and Chetaev[4,5] that the preliminary considerations of this section lead to the correct conditions for the stability of relative equilibrium: the assignment of axes of the inertial ellipsoid given above is a sufficient condition for the stability of relative equilibrium.

STABILITY

We now consider the problem of the motion of a solid body in a Newtonian central-force field from the most general point of view.

Let a stationary gravitational center O be located outside the solid body. Fixed at O there is a stationary coordinate system X, Y, and Z and at the center of mass of the body C there is a stationary system x, y, and z the axes of which are directed along the principal axes of inertia of the body. The relative position of these systems is determined from tables of the direction cosines

$$\begin{array}{c|ccc} & X & Y & Z \\ \hline x & \alpha_1 & \alpha_2 & \alpha_3 \\ y & \beta_1 & \beta_2 & \beta_3 \\ z & \gamma_1 & \gamma_2 & \gamma_3 \end{array} \qquad (7)$$

and the coordinates X_C, Y_C, and Z_C of the center of mass of the body. The potential function of the attracting point on the body is

$$U = \iiint \frac{\mu dm}{\rho}, \qquad (8)$$

$$\rho^2 = X_C^2 + Y_C^2 + Z_C^2 + 2X_C(x\alpha_1 + y\alpha_2 + z\alpha_3) + \\ + 2Y_C(x\beta_1 + y\beta_2 + z\beta_3) + 2Z_C(x\gamma_1 + y\gamma_2 + z\gamma_3) + x^2 + y^2 + z^2. \qquad (9)$$

μ is the gravitational constant.

Furthermore

$$\rho^2 = R^2 + 2R(x\gamma + y\gamma' + z\gamma'') + x^2 + y^2 + z^2, \qquad (10)$$

where $R = \sqrt{X_C^2 + Y_C^2 + Z_C^2}$ is the radius vector of the center of mass of the body, and γ, γ', and γ'' are the direction cosines of the axes x, y, and z with R

$$\begin{pmatrix} R\gamma \\ R\gamma' \\ R\gamma'' \end{pmatrix} = \begin{pmatrix} x \\ y \\ z \end{pmatrix} \begin{pmatrix} \alpha_1 & \alpha_2 & \alpha_3 \\ \beta_1 & \beta_2 & \beta_3 \\ \gamma_1 & \gamma_2 & \gamma_3 \end{pmatrix}.$$

If the dimensions of the body are small with respect to the distance to the attracting center, i.e.,

$$l = \sqrt{x^2 + y^2 + z^2} \ll R,$$

the integral (8) is easily evaluated by developing U as a power series in x/R, y/R, and z/R and by disregarding differentials of order higher than two. Then

$$U = \frac{\mu M}{R} - \frac{3}{2} \cdot \frac{\mu}{R^3}(A\gamma^2 + B\gamma'^2 + C\gamma''^2) + \frac{3}{2} \cdot \frac{\mu}{R^3} \cdot \frac{A+B+C}{3}. \quad (11)$$

Here M is the mass of the body; A, B, and C are its principal moments of inertia. In (11), terms of the order $(l_0/R) \times (\mu/R^3)I$ and higher are omitted; $l_0 = \max l$; $I = \max(A, B,$ and $C)$. We note that in actual cases it is quite sufficient to consider the approximate value of U (11), since the terms disregarded are very small. For the moon $l/R = 0.0045$ and for artificial satellites $l/R = 10^{-4} - 10^{-6}$.

Such effects as the dimensions of the regions of stability, the sizes of the perturbations affecting the motion because of oscillations around the stable position, etc., will differ in the exact case from the approximate case by magnitudes of the order l/R, i.e., they will be negligibly small. With this in mind, we shall write in the following the letter U to designate either the general expression (8) or the approximation (11).

Then the equations of motion will have the form:

$$M\frac{d^2 X_C}{dt^2} = \frac{\partial U}{\partial X_C}; \quad M\frac{d^2 Y_C}{dt^2} = \frac{\partial U}{\partial Y_C}; \quad M\frac{d^2 Z_C}{dt^2} = \frac{\partial U}{\partial Z_C}, \quad (12)$$

$$\left. \begin{array}{l} A\frac{dp}{dt} + (C-B)qr = M_x \\ B\frac{dq}{dt} + (A-C)pr = M_y \\ C\frac{dr}{dt} + (B-A)pq = M_z \end{array} \right\} \quad (13)$$

As is well known (see, for example, Ref. 6) if the potential function U depends on all the absolute direction cosines, the components of the moment of forces will have the form:

$$M_x = \frac{\partial U}{\partial \alpha_2}\alpha_3 - \frac{\partial U}{\partial \alpha_3}\alpha_2 + \frac{\partial U}{\partial \beta_2}\beta_3 - \frac{\partial U}{\partial \beta_3}\beta_2 + \frac{\partial U}{\partial \gamma_2}\gamma_3 - \frac{\partial U}{\partial \gamma_3}\gamma_2,$$

$$M_y = \frac{\partial U}{\partial \alpha_3}\alpha_1 - \frac{\partial U}{\partial \alpha_1}\alpha_3 + \frac{\partial U}{\partial \beta_3}\beta_1 - \frac{\partial U}{\partial \beta_1}\beta_3 + \frac{\partial U}{\partial \gamma_3}\gamma_1 - \frac{\partial U}{\partial \gamma_1}\gamma_3,$$

$$M_z = \frac{\partial U}{\partial \alpha_1}\alpha_2 - \frac{\partial U}{\partial \alpha_2}\alpha_1 + \frac{\partial U}{\partial \beta_1}\beta_2 - \frac{\partial U}{\partial \beta_2}\beta_1 + \frac{\partial U}{\partial \gamma_1}\gamma_2 - \frac{\partial U}{\partial \gamma_2}\gamma_1.$$

Furthermore, introducing the relative direction cosines γ, γ', and γ''

$$M_x = \frac{\partial U}{\partial \gamma'}\gamma'' - \frac{\partial U}{\partial \gamma''}\gamma'; \quad M_y = \frac{\partial U}{\partial \gamma''}\gamma - \frac{\partial U}{\partial \gamma}\gamma''; \quad M_z = \frac{\partial U}{\partial \gamma}\gamma' - \frac{\partial U}{\partial \gamma'}\gamma$$

or, on account of (11), approximately

$$M_x = 3\frac{\mu}{R^3}(C-B)\gamma''\gamma', \quad M_y = 3\frac{\mu}{R_3}(A-C)\gamma\gamma'',$$

$$M_z = 3\frac{\mu}{R^3}(B-A)\gamma\gamma'.$$

Equations (12) and (13) are connected by the usual kinematic relations of Poisson for the direction cosines (7), and together with these relations we have the following first integrals

Energy integrals

$$\frac{1}{2}Mv^2 + \frac{1}{2}(Ap^2 + Bq^2 + Cr^2) - U = h, \qquad (14)$$
$$v^2 = \dot{X}_C^2 + \dot{Y}_C^2 + \dot{Z}_C^2;$$

Three angular momentum integrals

$$M(Y_C\dot{Z}_C - Z_C\dot{Y}_C) + Ap\alpha_1 + Bq\alpha_2 + Cr\alpha_3 = L_1, \qquad (15)$$

$$M(Z_C\dot{X}_C - X_C\dot{Z}_C) + Ap\beta_1 + Bq\beta_2 + Cr\beta_3 = L_2, \qquad (16)$$

$$M(X_C\dot{Y}_C - Y_C\dot{X}_C) + Ap\gamma_1 + Bq\gamma_2 + Cr\gamma_3 = L_3 \qquad (17)$$

and trivial integrals relating the direction cosines, of which we note *

$$V_2 = \beta_1^2 + \beta_2^2 + \beta_3^2 - 1 = 0, \qquad (18)$$

$$V_3 = \gamma^2 + \gamma'^2 + \gamma''^2 - 1 = 0. \qquad (19)$$

We shall look for a particular solution of the system [Equations (12) and (13)] and its connected kinematic relations, namely for motion in a circular orbit R = R_0 with a constant angular velocity ω and relative equilibrium of the body, i.e., the assignment of the principal axes of inertia of the body x, y, and z along the radius vector, the tangent, and the binormal of the unperturbed orbit at all times of the motion. If the z axis is directed along the radius vector (the axis with mo-

*The general equations of motion for n bodies exerting gravitational pull on each other in accordance with Newtonian law have been derived by G. N. Duboshin.[7] Several particular cases of this problem have been examined in a series of papers by Kondurar'[8-12] and in another contribution by G. N. Duboshin.[13]

ment of inertia C), and the y axis along the binormal (the axis with moment of inertia B), the motion considered may be described in the form

$$p = r = 0, \quad q = \omega,$$
$$\gamma = \gamma' = \beta_1 = \beta_3 = 0, \quad \gamma'' = \beta_2 = 1. \tag{20}$$

We introduce further the spherical coordinates of the center of mass

$$X_C = R \cos\psi \sin\varphi,$$
$$Y_C = R \sin\psi,$$
$$Z_C = R \cos\psi \cos\varphi.$$

Then in unperturbed motion

$$R = R_0, \quad \psi = 0, \quad \dot\psi = 0, \quad \dot\varphi = 0, \quad \dot R = 0. \tag{21}$$

It is not hard to see that unperturbed motion (20) and (21) will occur if

$$\left(\frac{\partial U}{\partial \gamma}\right)_{\substack{\gamma''=1\\R=R_0}} = \left(\frac{\partial U}{\partial \gamma'}\right)_{\substack{\gamma''=1\\R=R_0}} = 0, \tag{22}$$

$$\left(\frac{\partial U}{\partial R}\right)_{\substack{\gamma''=1\\R=R_0}} < 0. \tag{23}$$

Equation (22) is the condition for a conditional extremum of U in unperturbed motion when γ, γ', and γ'' are changed under the constraints of (19). Equation (23) shows that the force acting is attractive. For the approximate form (11) of the potential function U, (22) is completely identical. If one considers the exact value (8) of the potential function, then (22) means that some of the integrals over the volume of the solid must be equal to zero. One can mention a large number of bodies for which (22) provides a completely exact solution. For instance, these conditions are satisfied exactly for any body having a mass distribution with even just one plane of symmetry.

The condition (23) is also satisfied for actual cases, since it is satisfied by the principal term of the development of U, and the remaining terms are very small with respect to the principal one and cannot cause a violation of inequality (23).

The angular velocity ω of the motion of the center of mass is determined by the equation

$$\omega^2 = -\frac{1}{MR_0}\left(\frac{\partial U}{\partial R}\right)_{\substack{\gamma''=1\\R=R_0}}, \qquad (24)$$

so that the period of revolution

$$T = \frac{2\pi}{\sqrt{-\frac{1}{MR_0}\left(\frac{\partial U}{\partial R}\right)_{\substack{\gamma''=1\\R=R_0}}}} \qquad (24a)$$

(generalization of Kepler's third law.) Substituting in (24a) the expression for U(11) we get

$$T = T_{\kappa}\left(1+\frac{3}{2}\frac{A+B-2C}{MR_0^2}\right)^{-\frac{1}{2}} \approx T_{\kappa}\left(1-\frac{3}{4}\frac{A+B-2C}{MR_0^2}\right),$$

from which it can be seen that the period T is different from the Kepler period T_K by a quantity of order $(l/R)^2$, which is negligibly small, as are all effects of the dependence of the motion of the center of mass on the rotation around the center of mass. But the exact formulation of the problem is useful, thanks to the existence of the integrals of motion, for a rigorous proof of the stability of relative equilibrium.

We shall look for the conditions of stability of the unperturbed motion (20) and (21).

According to well-known theorems,[4,5] it is sufficient for the stability of the unperturbed motion that there exist an integral of definite sign for the equations of perturbed motion. For the perturbed motion

$$R = R_0 + \Delta, \quad \dot{\varphi} = \omega + y, \quad \dot{R}, \quad \dot{\varphi},$$
$$p, r, q = \omega + x,$$
$$\beta_1, \beta_3, \beta_2 = 1 + \eta,$$
$$\gamma, \gamma', \gamma'' = 1 + \zeta,$$

we write out the integrals (14), (16), (18), and (19) by substituting the spherical coordinates of the center of mass, and we seek the Lyapunov function in the form

$$L = 2\frac{h}{M} - 2\omega\frac{L_2}{M} + gV_2 + fV_3 + \frac{m}{M^2R_0^2}L_2^2. \qquad (25)$$

Here m is some constant, and h, L_2, V_2, and V_3 are defined by (14), (16), (18), and (19) after substituting in them the spherical coordinates of the center of mass and changing to perturbed motion. Again for brevity we shall designate A/M, B/M, C/M, and U/M by A, B, C, and U. We define g and f as follows.

$$g = \frac{\partial U}{\partial \gamma''} + \frac{\partial^2 U}{\partial \gamma \, \partial \gamma''} \gamma + \frac{\partial^2 U}{\partial \gamma' \gamma''} \gamma' + \frac{1}{2}\left(\frac{\partial^2 U}{\partial \gamma''^2} - \frac{\partial U}{\partial \gamma''}\right)\zeta + \frac{\partial^2 U}{\partial R \, \partial \gamma''} \Delta,$$

$$f = \omega^2 B + \omega B x - \frac{B \omega m}{R_0^2}(2R_0 \omega \Delta + R_0^2 y + Bx).$$

Both here and later all partial derivatives of U are taken at the point $R = R_0$, $\gamma'' = 1$, $\gamma = \gamma' = 0$, that is, for the unperturbed motion.

Then L assumes the form

$$L = Ap^2 - 2\omega Ap\beta + \omega^2 B\beta^2 + Cr^2 - 2\omega Cr\beta'' + \omega^2 B\beta''^2 + \omega^2 c_{33}\eta^2 + \dot{R}^2 +$$
$$+ \omega^2 R_0^2 \psi^2 + R_0^2 \dot{\psi}^2 + a_1 \gamma^2 + a_2 \gamma'^2 + c_{11} \Delta^2 + c_{22} y^2 + c_{33} x^2 + 2b\gamma\gamma' +$$
$$+ 2a_{14} \Delta \gamma + 2a_{15} \Delta \gamma' + 2c_{12} \Delta y + 2c_{13} \Delta x + 2c_{23} yx + O(3); \quad (26)$$

O(3) contains only terms of higher than second-order infinitesimals with respect to the perturbations, and the coefficients a, b, and c have the values

$$a_1 = \frac{\partial U}{\partial \gamma''} - \frac{\partial^2 U}{\partial \gamma^2}, \quad c_{11} = -\frac{\partial^2 U}{\partial R^2} + \omega^2(4m - 1),$$

$$a_2 = \frac{\partial U}{\partial \gamma''} - \frac{\partial^2 U}{\partial \gamma'^2}, \quad c_{22} = R_0^2(1 + m),$$

$$b = -\frac{\partial^2 U}{\partial \gamma \, \partial \gamma'}, \quad c_{33} = B\left(1 + m\frac{B}{R_0^2}\right);$$

$$a_{14} = -\frac{\partial^2 U}{\partial R \, \partial \gamma}, \quad c_{12} = 2R_0 \omega m,$$

$$a_{15} = -\frac{\partial^2 U}{\partial R \, \partial \gamma'}, \quad c_{13} = 2\frac{\omega B m}{R_0}, \quad c_{23} = Bm.$$

Since as a result of the equations of perturbed motion dL/dt = 0 sufficient conditions for the stability of unperturbed motion (20) and (21) will be the positive definite quadratic form (26). These conditions are

a) $B > A$, $B > C$, b) $a_1 > \frac{a_{14}^2}{\Delta_3}$; $a_1 a_2 - b^2 > \frac{a_2 a_{14}^2 - a_1 a_{15}^2}{\Delta_3}$

(27)

c) $\Delta_3 = BR_0^2\left\{-\frac{\partial^2 U}{\partial R^2} + \omega^2 \frac{3m - 1 - m\delta^2}{1 + m + m\delta^2}\right\} > 0$

$$m > 0, \quad \delta^2 = \frac{B}{R_0^2}.$$

We recall that all the parameters a_i, b, and Δ_3 entering the condition for stability for the exact value U (8) are calculated from certain definite integrals taken over the volume of the body. These integrals can be calculated approximately quite easily by expanding the integrand in powers of X/R, Y/R, and Z/R, and disregarding terms of higher order. This is equivalent to the introduction of the approximate value (11) for U. Evaluation of the integrals in this way will result in values differing from their exact ones by quantities of the order l/R, i.e., in actual cases sufficiently small (the numerical evaluation of the quantity l/R was given above) as to be unimportant. Conditions (27) simplified in this way will be sufficiently exact to determine the conditions of stability, since for them to be satisfied the Lyapunov function L (25) will be of definite sign as a result of the well-known fact that a form with sufficiently small coefficients added to a form of the same order does not change the definiteness of its sign.[4] The simplified conditions for stability will be given below.

Condition (27, a) means that in unperturbed motion the axis of the largest moment of inertia, i.e., the least axis of the inertial ellipsoid, should be directed along the normal to the plane of the orbit. Condition (27, c) is always satisfied in actual cases for sufficiently large positive m. In other words, condition (27, c) is not a condition of stability but only determines the choice of m in the Lyapunov function L. As $m \to \infty$ condition (27, c) is transformed into the well-known[15] condition of the theory of the Newtonian potential. It is easy to confirm that condition (27, c) is satisfied for sufficiently large m for the leading term of U but the remaining terms because of their smallness cannot cause a violation of the inequality (27,c).

For a better inspection of condition (27, b) we disregard, on the basis of what has been said above, the terms containing a_{14} and a_{15}. Then conditions (27, b) take the form

$$\left. \begin{array}{c} \dfrac{\partial U}{\partial \gamma''} - \dfrac{\partial^2 U}{\partial \gamma^2} > 0 \\ \left(\dfrac{\partial U}{\partial \gamma''} - \dfrac{\partial^2 U}{\partial \gamma^2}\right)\left(\dfrac{\partial U}{\partial \gamma''} - \dfrac{\partial^2 U}{\partial \gamma'^2}\right) > \left(\dfrac{\partial^2 U}{\partial \gamma \partial \gamma'}\right)^2 \end{array} \right\} \quad (28)$$

with

$$R = R_0 \quad \text{and} \quad \gamma'' = 1$$

Conditions (28), together with (22), show that the potential function in unperturbed motion has a conditional maximum. Since condition (27, a) is the condition of a maximum of the potential function of centrifugal forces, we conclude that for stability of relative equilibrium it is sufficient that in unperturbed motion the total potential function of Newtonian and centrifugal forces have a maximum with respect to the coordinates of rotational motion of the body.

We should like to emphasize that the formulation of the problem under consideration permits one to find the condition of stability of relative equilibrium both for perturbed motion around the center of mass and also for perturbed orbits.

Let us examine the physical significance of the conditions of stability (27, b). With this in mind we note that for actual cases the dimensions of the body l are very small with respect to the distance R to the center of attraction, and therefore it is natural to consider the approximate value (11) of the potential function U together with its exact value (8). Then

$$a_{14} = a_{15} = b = 0,$$

$$a_1 = 3\frac{\mu}{R^3}(A-C), \quad a_2 = 3\frac{\mu}{R^3}(B-C),$$

and the condition of stability (27, b) can be written in the form

$$A - C > 0, \quad B - C > 0. \tag{29}$$

These conditions together with (27, a) give the following sufficient condition for stability

$$B > A > C, \tag{30}$$

which means that in unperturbed motion the smallest axis of the inertial ellipsoid is directed along the normal to the plane of the unperturbed orbit, and the largest axis of the inertial ellipsoid along the radius vector of this orbit. As was shown above, condition (30) is one of the exact sufficient conditions for stability, since when condition (30) is satisfied the Lyapunov function L (25) remains positive and definite.

The investigation carried out shows that for stability of relative equilibrium of a solid body in a circular orbit in a Newtonian central-force field it is thus sufficient that in unperturbed motion the largest axis of the inertial ellipsoid be directed along the radius vector of the orbit and the smallest axis along the normal to the plane of the orbit. Then unperturbed motion will be stable with respect to small perturbations of the motion of the center of mass and of motion around the center of mass.

LIBRATION

We return to the problem of determining the oscillations around the position of stable relative equilibrium.

Even in the case where the approximate value (11) of the potential function is used the equations of motion of the center of mass (12) and the equations of motion around the center of mass (13) will be mutually dependent, which leads to additional difficulties in the integration of the equations. However for real motions it can be assumed with a high degree of accuracy that the center of mass moves according to a Kepler law, which is equivalent to disregarding terms depending on the rotation of the body in (12) of the motion of the center of mass. Then (12) can be integrated independently of the other equations and determine a Kepler orbit. The problem reduces to a determination of the motion (13) about the center of mass, which moves on a Keplerian orbit. This is precisely how the problem of the rotation of bodies is formulated in classical celestial mechanics.[3]

In this statement of the problem, which might be termed the "restricted problem of the motion of a body about the center of mass in a Newtonian force field," we come up with a problem differing from the general problem (12)-(13). The equations of this problem for the approximate value (11) of the force function may be written in the form (13) and completed with the kinematic equations for the relative direction cosines given in tabular form:

	x	y	z
x_1	α	α'	α''
y_1	β	β'	β''
z_1	γ	γ'	γ''

where the z_1 axis of a right-handed system of coordinates x_1, y_1, and z_1 is directed along the radius vector of the orbit and the y_1 axis along the normal to the plane of the orbit.

In (13) R is a known function of time as is also the angular velocity ω, entering explicitly in the kinematic relationships for the relative direction cosines. These kinematic relations differ from the usual Poisson relations in that there are additional terms $\omega\alpha$, $\omega\alpha'$, and $\omega\alpha''$ on the right-hand sides of the equations for γ, γ', and γ'' and correspondingly $-\omega\gamma$ $-\omega\gamma'$, and $-\omega\gamma''$ are added in the equations for α, α', and α''. The equations for β, β', and β'' agree with the usual Poisson equations.

The equations of the restricted problem are significantly different from the equations of the complete problem and in the general case do not have first integrals, apart from trivial ones. However, in the case of a circular orbit $\omega^2 = \mu/R^3 =$ Const and (8) and (9) admit of a generalized Jacobi energy integral

$$Ap^2 + Bq^2 + Cr^2 + 3\omega^2(A\gamma^2 + B\gamma'^2 + C\gamma''^2) -$$
$$- 2\omega(Ap\beta + Bq\beta' + Cr\beta'') = h, \qquad (31)$$

which can be written in the form

$$Ap_{rel}^2 + Bq_{rel}^2 + Cr_{rel}^2 + 3\omega^2[(A-C)\gamma^2 + (B-C)\gamma'^2] +$$
$$+ \omega^2[(B-A)\beta^2 + (B-C)\beta''^2] = h. \qquad (32)$$

Here p_{rel}, q_{rel}, and r_{rel} denote the projections of the relative angular velocity of the body on the x, y, and z axes

$$p_{rel} = p - \omega\beta, \quad q_{rel} = q - \omega\beta', \quad r_{rel} = r - \omega\beta''.$$

It follows directly from (32) that the relative equilibrium of a solid body in a circular orbit

$$p_{\text{rel}} = q_{\text{rel}} = r_{\text{rel}} = 0, \quad \gamma = \gamma' = \beta = \beta'' = 0, \quad \gamma'' = \beta' = 1$$

is stable if (30)

$$B > A > C.$$

This is just the requirement that is to be satisfied. The equation of perturbed motion will have a positive definite integral (32), which is, according to the known results of Lyapunov and Chetaev, sufficient for the stability of the unperturbed motion. We note that this shows the stability only for motion around the center of mass, and a circular orbit remains unperturbed; in the preceding section it was shown that the complete formulation of the problem is useful in showing the sufficiency of conditions (30) for the stability of unperturbed motion both with respect to perturbations of the motion around the center of mass and also with respect to perturbations of the orbit of the center of mass.

It is apparent that the condition of stability (30) corresponds to a maximum in the total potential function of the Newtonian and centrifugal forces. The condition for stability (30) can be explained as follows: Relative equilibrium of a body is in absolute motion an equilibrium rotation around one of the principal inertial axes, perpendicular to the plane of the orbit. This axis cannot be the intermediate axis of inertia, since rotation of a free body around the intermediate axis is unstable, and the action of the Newtonian field cannot remove this instability.[2] The largest axis of the inertial ellipsoid also cannot be directed normal to the orbit since in this case the field of perturbing forces would act to cause instability. Therefore the smallest axis of the inertial ellipsoid should be directed normal to the plane of the orbit. The largest axis must be directed along the radius vector since in this case the restoring action of the field of perturbing forces is assured.[2] As a result we get the assignment of axes indicated by condition (30).

The essential positiveness of the components on the left side of (32) is used to evaluate the amplitudes of the perturbed motion. Obviously

$$\gamma^2 \leqslant \frac{h}{3\omega^2(A-C)}, \quad \gamma'^2 \leqslant \frac{h}{3\omega^2(B-C)},$$
$$\beta^2 \leqslant \frac{h}{\omega^2(B-A)}, \quad \beta'^2 \leqslant \frac{h}{\omega^2(B-C)}.$$

For the oscillations to be limited it is necessary that the initial perturbations satisfy the condition

$$h < \min\{3\omega^2(A-C),\ \omega^2(B-C)\}.$$

For example, let all the perturbation except q_{rel}^0 equal zero. Then it follows from (32) that for limited oscillations it is necessary that

$$q_{rel}^0 < \sqrt{3}\,\omega\,\sqrt{\frac{A-C}{B}} < \sqrt{3}\,\omega.$$

Since for artificial satellites ω is of the order of 0.1°/sec—0.01°/sec, it is clear that librational motion (oscillations around the position of relative equilibrium) is possible only for extremely small initial perturbations of angular velocity of rotation around the center of mass of the satellite.

In the general case of a noncircular orbit, the integral (31) does not exist, which leads to difficulties in the investigation. In order to obtain a representation of the nature of oscillations in an elliptical orbit we consider the particular case of motion in the plane of the orbit. The equation of such plane motion, on transforming from the time t as an argument to θ, the true anomaly (the angular distance of the radius vector from perigee in the orbit), leads to the form

$$(1 + e\cos\theta)\frac{d^2\delta}{d\theta^2} - 2e\sin\theta\frac{d\delta}{d\theta} + 3a^2\sin\delta = 4e\sin\theta, \qquad (33)$$

where $\delta = 2\vartheta$ is twice the angle of nutation (δ is the angle between the principal inertial axis of the satellite and the radius vector); $a^2 = (A-C)/B$; and e is the eccentricity of the orbit. In a circular orbit $e = 0$, $\theta = \omega t$, and (33) assumes exactly the same form as the equation of plane oscillation of a solid body in a Newtonian force field in the case of a stationary center of mass of the body[16] and is integrable in elliptic functions,[3] so that for an angle of nutation $\vartheta = \delta/2$ we get in a general form

$$\sin\vartheta = k\operatorname{sn}\left\{\sqrt{3\frac{A-C}{B}}\,\omega(t-t_0) + F_0(\varphi_0, k)\right\},$$

$$F_0(\varphi_0, k) = \int_0^{\sin\varphi_0} \frac{dx}{\sqrt{(1-x^2)(1-k^2x^2)}}; \quad \sin\varphi_0 = \frac{1}{k}\sin\vartheta_0;$$

$$k^2 = \frac{\dot\vartheta_0^2}{3\omega^2\frac{A-C}{B}} + \sin^2\vartheta_0.$$

The motion will be periodic if $k^2 < 1$, that is

$$|\dot\vartheta_0| < \omega\sqrt{3\frac{A-C}{B}}\,|\cos\vartheta_0|.$$

If $k^2 = 1$ or $k^2 > 1$, the motion will be limiting (tending to a position of equilibrium) or regressive (continuous rotation in one direction), respectively, which confirms what was said earlier about the nature of the action of a Newtonian force field on the rotation of a satellite: the librational motion is possible only for small angular velocities of the order of the angular velocity of the motion of the center of mass in the orbit; consequently only for small kinetic energy of rotation. If, however, the initial angular velocity exceeds a critical value the Newtonian force field will introduce only small perturbations in the usual unperturbed rotation of the body around the center of mass.

For periodic motion the period of oscillations

$$T = \frac{4}{\sqrt{3\omega^2\frac{A-C}{B}}}\,F\!\left(\frac{\pi}{2}, k\right),$$

where $F(\pi/2, k)$ is the complete elliptic integral of the first kind. Expanding it in a power series in k and introducing the period $T_0 = 2\pi/\omega$ for an orbital revolution of the satellite, it is possible to write

$$T = \frac{T_0}{\sqrt{3\frac{A-C}{B}}}\left\{1 + \left(\frac{1}{2}\right)^2 k^2 + \left(\frac{1\cdot 3}{2\cdot 4}\right)^2 k^4 + \cdots + \left[\frac{(2n+1)!!}{2n!!}\right]^2 k^{2n} + \cdots\right\}.$$

Note: The usual notation for the general term would be

$$\left[\frac{(2n)!}{2^{2n}(n!)^2}\right]^2 k^{2n}.$$

We see that the oscillations will occur very slowly. The period of oscillations has as a lower limit the quantity

$$T_{\min} = \frac{T_0}{\sqrt{3}} = 0.577\, T_0$$

and in the general case the period can only be larger.

Equation (33) for plane motion in an elliptical orbit differs from the corresponding equations for a circular orbit by the periodicity of the coefficients, the presence of a term in the first derivative, and the presence of a term on the right-hand side which brings about an increase in the constrained oscillations over those of the circular orbit and which we shall call the eccentric oscillations. Apart from terms in e^3 the eccentric oscillations ϑ_e are given by the equation

$$\vartheta_e = \frac{2e \sin \theta}{1 + e \cos \theta} \left\{ \frac{1}{3a^2 - 1} + \frac{e}{3a^2 - 4} \cos \theta \right\}, \tag{34}$$

which is valid for values of the parameter a^2 that are not close to the resonant value $a^2 = 1/3$ (on physical grounds $a^2 < 1$).*

Small oscillations in an elliptical orbit are described by an equation of the Hill type, which can be obtained by using a linear expansion in (33) and substituting

$$\vartheta = \frac{z}{1 + e \cos \theta}.$$

*The difference in motion on an elliptic orbit from motion on a circular orbit is illustrated by the following curious example: when $a^2 = 2e$, equation (33) has the particular solution $\vartheta = \theta/2$, i.e., the body rotates continuously in one direction so that it reassumes its initial position in two revolutions of the center of mass of the satellite along the orbit. This solution is valid, however, only in the restricted statement of the problem as considered above, and not in the general statement of the problem (see section on Stability).

Then

$$\frac{d^2z}{d\theta^2} + \left\{\frac{3a^2 + e\cos\theta}{1 + e\cos\theta}\right\}z = 2e\sin\theta. \qquad (35)$$

With the exception of the resonant case $3a^2 = 1$, (35) is not integrable when $e \neq 1$; however, for small e an analysis of the solution can be carried out without difficulty, since the particular solution of the nonhomogeneous equation gives (34), and the homogeneous equation apart from terms of order e^2 reduces to the well-known Mathieu equation. In the general case of small oscillations in space in an elliptical orbit, the direction cosines γ, γ', and α' correspond to small angles of pitch, roll, and yaw and the equations for these angles have the form

a) $\quad \dfrac{d^2\gamma}{dt^2} + 3\zeta\,\dfrac{A-C}{B}\,\gamma = -\dot{\omega},$

b) $\quad \dfrac{d^2\gamma'}{dt^2} + (3\zeta + \omega^2)\dfrac{B-C}{A}\gamma' - \omega\left(1 + \dfrac{C-B}{A}\right)\dfrac{d\alpha'}{dt} - \dot{\omega}\alpha' = 0,\qquad (36)$

c) $\quad \dfrac{d^2\alpha'}{dt^2} + \omega^2\,\dfrac{B-A}{C}\,\alpha' - \omega\left(-1 + \dfrac{B-A}{C}\right)\dfrac{d\gamma'}{dt} + \dot{\omega}\gamma' = 0,$

where the variable coefficients ω, ζ, and $\dot{\omega}$ are given by the relations

$$\zeta = \frac{g_0 R_0^3}{\rho^3}, \qquad \omega = \frac{R_0\sqrt{g_0 p_0}}{\rho^2},$$

$$\rho = \frac{p_0}{1 + e\cos\theta}, \qquad \frac{d\theta}{dt} = \omega.$$

(R_0 is the radius of the earth, g_0 is the acceleration of gravity at the surface of the earth, p_0 is the focal parameter of the orbit, and e is the eccentricity of the orbit.) We see that the roll and yaw motions are interdependent, but do not depend on pitch.

The particular effect of ellipticity of the orbit is displayed, as we shall see, in the presence of constrained oscillations of the pitch angle. These oscillations have been considered above. The angles γ' and α' have no constrained oscillations and their behavior is close to that in a circular orbit. For a circular orbit $\varphi = \omega = \text{Const}$; $\dot{\omega} = 0$ and (36) are easily integrated

$$\gamma = A_\gamma \sin(\lambda_\gamma t + \delta_\gamma)$$
$$\gamma' = A_1 \sin(\lambda_1 t + \delta_1) + A_2 \sin(\lambda_2 t + \delta_2)$$
$$\alpha' = \frac{1}{Q}\left(\lambda_1 - 4\frac{B-C}{A}\omega^2 \cdot \frac{1}{\lambda_1}\right) A_1 \cos(\lambda_1 t + \delta_1) \qquad (37)$$
$$+ \frac{1}{Q}\left(\lambda_2 - 4\frac{B-C}{A}\omega^2 \cdot \frac{1}{\lambda_2}\right) A_2 \cos(\lambda_2 t + \delta_2).$$

Here A_γ, A_1, A_2, δ_γ, δ_1, and δ_2 are constants determined by the initial conditions

$$\lambda_\gamma = \sqrt{3\frac{A-C}{B}}\,\omega$$
$$\lambda_1 = \sqrt{\frac{\Omega^2 + P^2}{2} + \sqrt{\frac{(\Omega^2 + P^2)^2}{4} - 4\omega^2\Omega^2}}$$
$$\lambda_2 = \sqrt{\frac{\Omega^2 + P^2}{2} - \sqrt{\frac{(\Omega^2 + P^2)^2}{4} - 4\omega^2\Omega^2}} \qquad (38)$$
$$\Omega^2 = \frac{(B-A)(B-C)}{AB}\omega^2, \qquad P^2 = \left[1 + \frac{3(B-C)}{A}\right]\omega$$
$$Q = \left(1 - \frac{B-C}{A}\right)\omega.$$

The solution (37) is given by the roots of the characteristic equation of the system of the first approximation

$$\left[\lambda^2 + 3\omega^2\frac{A-C}{B}\right]\left\{\lambda^4 + \lambda^2\omega^2\left[1 - 3\frac{C-B}{A} - \frac{(C-B)(B-A)}{AC}\right]\right.$$
$$\left. - 4\omega^2\frac{(C-B)(B-A)}{AC}\right\} = 0.$$

The absence roots of the characteristic equation having a negative real part is a necessary condition for the stability of unperturbed motion. These conditions are the following [3]

$$(B-A)(B-C) > 0,$$
$$AC + 3C(B-C) + (B-C)(B-A) > 0,$$
$$[AB + 3C(B-C) + (B-C)(B-A)]^2 + 16AC(B-A)(B-C) > 0,$$
$$A - C > 0.$$

If even one of these conditions is not satisfied the motion is unstable. In Fig. 2 are shown regions of unstable motion (the abbreviations $\epsilon = C/A$, $\delta = B/A$ are used). However, even if all the necessary conditions are satisfied, the question of whether the motion will be stable or not remains undecided. In order to answer this question it is necessary to investigate the sufficient conditions for stability. It was shown that such sufficient conditions for stability are $B > A > C$ (cross hatched on Fig. 2). As can be seen in Fig. 2, besides in this region, the necessary conditions for stability are satisfied in still another small section of the region of values of the parameters (heavy diagonal stripes). From qualitative considerations it would be expected that for parameters connected with this region the motion will be unstable. This is confirmed by the fact that in this region the potential function has no maximum (it has negative coefficients of Poincaré stability) which shows[5] the instability of the relative equilibrium. Thus conditions (30) are necessary and sufficient conditions of stability of relative equilibrium in the restricted formulation of the problem.

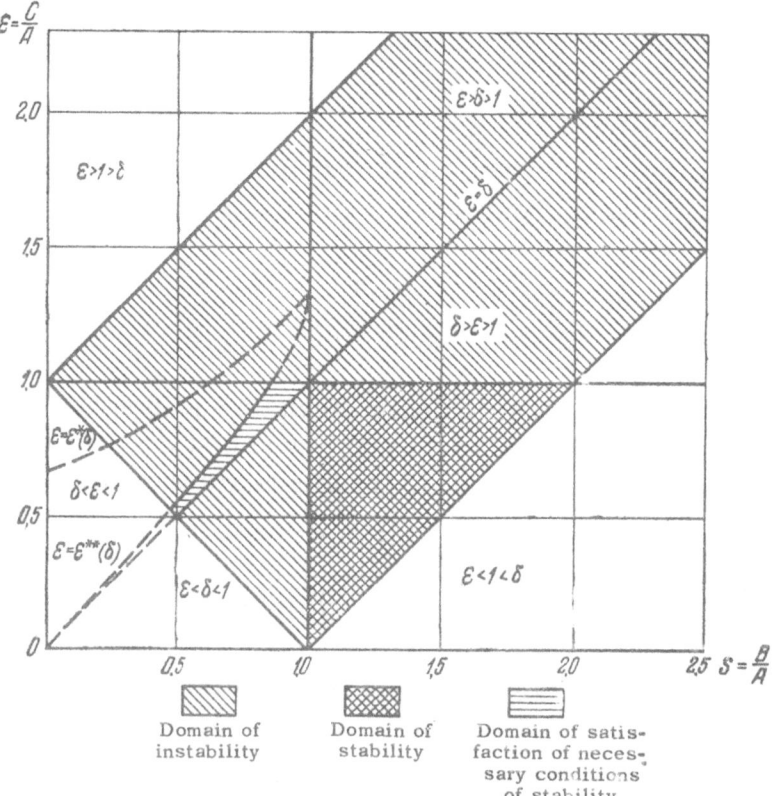

FIG. 2. Domains of stability and instability.

We note that in Fig. 2 only physically real values of the parameters are given, namely those obeying the conditions

$$A + B > C, \ B + C > A, \ C + A > B.$$

The solution (37) of the equations of small oscillations holds when the sufficient conditions for stability B > A > C are satisfied.

It can be seen from this solution that the pitch angle has the period

$$T_\gamma = \frac{T_0}{\lambda_\gamma / \omega} > 0.577 \, T_0,$$

and the roll and yaw angles are represented as sums of two periodic functions with periods

$$T_1 = \frac{T_0}{\lambda_1 / \omega}, \quad T_2 = \frac{T_0}{\lambda_2 / \omega},$$

where T_0 is the period of revolution of a satellite in its orbit.

It can be inferred from (38) that

$$T_2 > T_0, \ T_1 > 0.5 \, T_0$$

in view of the condition of stability B > A > C and the condition C + A > B which holds because of the physical meaning of the quantities A, B, and C. A similar evaluation of the periods of pitch, roll, and yaw is obtained from the examples of the dumbbell.

Thus the oscillations occur slowly with periods comparable to the period of revolution of a satellite in its orbit.

Thus when the conditions of stability are satisfied and when the initial angular velocities of rotation around the center of mass are very small and are comparable with the angular velocity of the motion of the center of mass, the motion of the satellite around the center of mass will have a librational character. If, however, the initial angular velocities are relatively large, as was the case for the first Soviet satellites, the motion will have a different character, considered in Ref. 1.

In conclusion we note that in the actual libration of an artificial satellite, if it occurs, a number of small perturbing factors have an influence: the moments of the aerodynamic forces, perturbing moments caused by deviations of the attractive field of the earth from a central-force field, regression of the orbit as a result of the compression of the earth and aerodynamic resistance, moments of the electromagnetic fields, etc. Investigation shows that if the basic conditions of stability (30) are satisfied as well as certain other natural conditions, libration in the presence of the perturbations indicated will differ only slightly from unperturbed libration.

A number of additional factors also influence the libration of the moon. But a consideration of the perturbations on the libration is beyond the scope of this work.

ACKNOWLEDGEMENT

The author is indebted to D. E. Okhotsimskii and V. A. Sarychev for reviews and discussions.

LITERATURE

1. V. V. Beletskii, "The Motion of an Artificial Earth Satellite About Its Center of Mass," in Artificial Earth Satellites, Plenum Press, New York, 1960, Vol. 1, p. 30.
2. V. V. Beletskii, Some Problems of the Motion of a Solid Body in a Newtonian Force Field, PMM, 21, 6 (1957).
3. F. Tisserand, Traité de Mécanique Céleste, Paris, 1891.
4. A. M. Lyapunov, Obshchaya Zadacha ob ustoychivosti dvizheniya (The General Problem of Stability of Motion), Gostekhizdat, 1950.
5. N. G. Chetaev, Ustoychivost' dvizheniya (The Stability of Motion), Gostekhizdat, 1955.
6. D. N. Goryachev, Nekotoryye obshchiye integraly v zadache o dvizhenii tverdogo tela (Some General Integrals in the Problem of the Motion of a Solid Body), Warsaw, 1910.
7. G. M. Duboshin, The Differential Equations of Progressively Rotating Motion of Mutually Attracting Solid Bodies, Astr. zh. (Astronomical Journal), 35, 2 (1958).
8. V. T. Kondurar', The Problem of the Motion of Two Ellipsoids Under the Action of Mutual Attraction, Part I, Astr. zh., 13, 6 (1936).
9. V. T. Kondurar', The Problem of the Motion of Two Ellipsoids Under the Action of Mutual Attraction, Part II, Trudy GAISh, 9, 2 (1949).

10. V. T. Kondurar', The Problem of the Motion of Two Ellipsoids Under the Action of Mutual Attraction, Part III, Trudy GAISh, 21, (1952).
11. V. T. Kondurar', The Problem of the Motion of Two Ellipsoids Under the Action of Mutual Attraction, Part IV, Trudy GAISh, 21 (1952).
12. V. T. Kondurar', The Problem of the Motion of Two Ellipsoids Under the Action of Mutual Attraction, Part V, Trudy GAISh, 24 (1954).
13. G. M. Duboshin, A Particular Case of the Problem of Progressively Rotating Motion of Two Bodies, Astr. zh., 36, 1 (1959).
14. I. G. Malkin, Teoriya ustoychivosti dvizheniya (Theory of the Stability of Motion), Gostekhizdat, 1952.
15. G. M. Duboshin, The Expansion of the Potential Function in the Theory of the Satellites of Saturn, Trudy GAISh, 15 (1945).
16. V. V. Beletskii, The Integrability of the Equations of Motion of a Solid Body Around a Stationary Point of Attraction Under the Action of a Newtonian Central-Force Field, DAN (Reports of the Academy of Sciences), 113, 2 (1957).

PERTURBATIONS OF THE FIRST ORDER IN THE MOTION OF ARTIFICIAL SATELLITES, CAUSED BY THE FLATTENING OF THE EARTH

V. F. Proskurin and Yu. V. Batrakov

A consequence of the fact that the earth is not strictly spherical is the appreciable deviation of the orbits of artificial satellites from unperturbed Keplerian ellipses. Particularly noticeable perturbations in the motion of satellites are caused by the flattening of the earth.

The problem of the motion of satellites in the gravitational field of a flattened planet is not new. It has already been considered to a certain extent in connection with the theory of motion of the satellites of large planets, and also in connection with lunar theory. The orbits of artificial satellites possess certain characteristics, however, that differ considerably from those of natural satellites. These differences are primarily the greater declination of the orbit, and the much closer approach of the artificial satellite to the surface of the earth. Thus the former theories of motion, derived to be applicable to natural satellites in orbits inclined only slightly to the equatorial plane and at relatively great distances from the surface of the earth, can not, generally speaking, be used for artificial earth satellites.

The necessity therefore arises of deriving a new analytical theory, applicable to artificial satellites with orbits having arbitrary inclinations to the equatorial plane, that will be sufficiently accurate even in the case when the satellite passes very close to the surface of the earth.

In the present work, we will consider the motion of a satellite in the gravitational field of a flattened planet, and will assume that the inclination of the orbital plane of the satellite

FIRST-ORDER PERTURBATIONS

to the equatorial plane is arbitrary. We will also assume that the planet has the form of a regular ellipsoid of revolution, and that the flattening of the planet is small enough that the expansion of the perturbation function in powers of the contraction can be limited to terms containing only the first power of the contraction. The part of the perturbation function that is taken into account is expressed in a series of powers of the eccentricity; the coefficients of the series can be expressed very simply in terms of trigonometric functions of the angle of declination. Integration of the ordinary Lagrange equations yields analytical expressions for the perturbations of the first order relative to the contraction in all the elements of the orbit, with an accuracy up to and including the fourth power of the eccentricity.

1. THE PROBLEM. THE EXPANSION OF THE PERTURBATION FUNCTION

We will consider the following problem. A satellite of zero mass (S in Fig. 1) is moving in the gravitational field of the

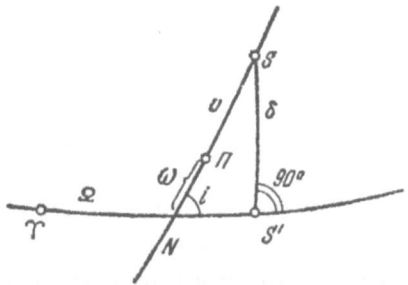

FIG. 1. Part of the geocentric celestial sphere and the projection on it of the earth's equator (γ, S'), the satellite's orbit (NS), and some elements of the orbit.

earth, with the latter assumed to have the form of a regular ellipsoid of revolution. The flattening of the earth, and its angular rate of rotation are assumed to be small.

The potential of a regular ellipsoid of revolution at an ex-

external point, with an accuracy of the first power of the contraction, is known to be

$$V = \frac{fm}{r} + \frac{1}{3} J \frac{fma'^2}{r^3} (3\sin^2 \delta - 1),\qquad(1)$$

where f is the gravitational constant, m the mass of the ellipsoid, r the radius vector of the point, a' the equatorial radius of the ellipsoid, and δ the angular distance of the point from the equatorial plane. The coefficient J is given by the formula

$$J = \varepsilon - \frac{1}{2} \cdot \frac{\omega^2 a'^3}{fm},$$

where ε is the compression of the ellipsoid and ω is the angular velocity of rotation of the ellipsoid.

The first term in (1) corresponds to the undisturbed motion of the satellite on an elliptical Keplerian orbit, while the second term is the disturbing potential or the so-called perturbation function, which we will denote by R.

In the spherical triangle SNS', we have

$$\sin \delta = \sin i \sin(v + \omega),\qquad(2)$$

where i is the declination of the orbit, v the real anomaly, and ω the angular distance of the perigee from the node (the argument of the perigee). After some transformations, we obtain the expression

$$R = \frac{1}{6} Jfm \frac{a'^2}{a^3} \Big[(2 - 3\lambda^2)\Big(\frac{a}{r}\Big)^3 + 3\lambda^2 \cos 2\omega \Big(\frac{a}{r}\Big)^3 \cos 2v - 3\lambda^2 \sin 2\omega \Big(\frac{a}{r}\Big)^3 \sin 2v\Big],\qquad(3)$$

where a is the semimajor axis of the orbit, and where for brevity we have used the notation sin i = λ.

Expansions already exist for the combinations $(a/r)^3$, $(a/r)^3 \cos 2v$, and $(a/r)^3 \sin 2v$ in series of multiples of the mean anomaly (for example, in the book by M. F. Subbotin[1]). We substitute these expansions in (3), and obtain an expression for the perturbation function of the form

$$R = JfM \frac{a'^2}{a^3} \sum F_{j,j'}(e, \lambda) \cos(jM + j'\omega)\qquad(4)$$

where the coefficients $F_{j,j'}$ are series in e multiplied by certain finite expressions in λ = sin i.

2. PERTURBATIONS OF THE FIRST ORDER

The Lagrange equations for the determination of the osculating elliptic elements are[1]

$$\frac{da}{dt} = \frac{2}{na} \cdot \frac{\partial R}{\partial \varepsilon}$$

$$\frac{de}{dt} = -\frac{\sqrt{1-e^2}}{na^2 e} \cdot \frac{\partial R}{\partial \pi} - \frac{e\sqrt{1-e^2}}{1+\sqrt{1-e^2}} \cdot \frac{1}{na^2} \cdot \frac{\partial R}{\partial \varepsilon}$$

$$\frac{di}{dt} = -\frac{\operatorname{cosec} i}{na^2 \sqrt{1-e^2}} \cdot \frac{\partial R}{\partial \Omega} - \frac{\operatorname{tg} \frac{i}{2}}{na^2 \sqrt{1-e^2}} \left(\frac{\partial R}{\partial \pi} + \frac{\partial R}{\partial \varepsilon} \right)$$

$$\frac{d\Omega}{dt} = \frac{\operatorname{cosec} i}{na^2 \sqrt{1-e^2}} \cdot \frac{\partial R}{\partial i} \qquad (5)$$

$$\frac{d\pi}{dt} = \frac{\operatorname{tg} \frac{i}{2}}{na^2 \sqrt{1-e^2}} \cdot \frac{\partial R}{\partial i} + \frac{\sqrt{1-e^2}}{na^2 e} \cdot \frac{\partial R}{\partial e}$$

$$\frac{d\varepsilon}{dt} = -\frac{2}{na} \cdot \frac{\partial R}{\partial a} + \frac{\operatorname{tg} \frac{i}{2}}{na^2 \sqrt{1-e^2}} \cdot \frac{\partial R}{\partial i} + \frac{e\sqrt{1-e^2}}{1+\sqrt{1-e^2}} \cdot \frac{1}{na^2} \cdot \frac{\partial R}{\partial e},$$

where a, e, i, Ω, $\pi = \omega + \Omega$, and ε are the elements of the satellite's orbit, n is the mean daily motion, and R the perturbation function.

We note that in the equation for π, the derivative $\partial R/\partial e$ is multiplied by a negative power of e. This leads to a decrease of accuracy in the expression for the perturbations of the first order in π.

We expand the coefficients of the first terms on the right-hand sides of the equations in (5) in powers of the eccentricity, substitute R from (4), integrate, and obtain the following expressions for the perturbations of the first order in the elliptic elements of the orbit

$$\frac{\delta a}{a} = 2J\left(\frac{a'}{a}\right)^2 \left(1 - \frac{3}{2}\lambda^2\right)\left\{\left(e + \frac{9}{8}e^3\right)\cos M + \frac{3}{2}e^2 \cos 2M + \right.$$
$$+ \frac{53}{24}e^2 \cos 3M\Big\} + J\left(\frac{a'}{a}\right)^2 \lambda^2 \Big\{-\frac{1}{2}\left(e - \frac{1}{8}e^3\right)\cos(M + 2\omega) +$$
$$+ \frac{1}{48}e^3 \cos(M - 2\omega) + \left(1 - \frac{5}{2}e^2\right)\cos(2M + 2\omega) +$$
$$+ \frac{7}{2}\left(e - \frac{123}{56}e^3\right)\cos(3M + 2\omega) + \frac{17}{2}e^2 \cos(4M + 2\omega) +$$
$$\left. + \frac{845}{48}e^3 \cos(5M + 2\omega)\right\}, \qquad (6)$$

$$\delta e = J\left(\frac{a'}{a}\right)^2 \left(1 - \frac{3}{2}\lambda^2\right)\left\{\left(1 + \frac{1}{8}e^2\right)\cos M + \frac{3}{2}\left(e - \frac{2}{9}e^3\right)\cos 2M + \right.$$
$$+ \frac{53}{24}e^2\cos 3M + \frac{77}{24}e^3\cos 4M\bigg\} +$$
$$+ \frac{1}{2}J\left(\frac{a'}{a}\right)^2 \lambda^2 \left\{\frac{1}{2}\left(1 - \frac{1}{8}e^2\right)\cos(M + 2\omega) + \frac{1}{16}e^2\cos(M - 2\omega) + \right.$$
$$+ \frac{1}{12}e^3\cos(2M - 2\omega) - \frac{1}{2}\left(e - \frac{11}{4}e^3\right)\cos(2M + 2\omega) +$$
$$+ \frac{7}{6}\left(1 - \frac{235}{56}e^2\right)\cos(3M + 2\omega) + \frac{17}{4}\left(e - \frac{383}{102}e^3\right)\cos(4M + 2\omega) +$$
$$+ \frac{169}{16}e^2\cos(5M + 2\omega) + \frac{533}{24}e^3\cos(6M + 2\omega)\bigg\}; \qquad (7)$$

$$\delta i = \frac{1}{2}J\left(\frac{a'}{a}\right)^2 \lambda\sqrt{1-\lambda^2}\left\{-\left(e + \frac{3}{8}e^3\right)\cos(M + 2\omega) - \frac{1}{24}e^3\cos(M - 2\omega) + \right.$$
$$+ (1 - 2e^2)\cos(2M + 2\omega) + \frac{7}{3}\left(e - \frac{95}{56}e^3\right)\cos(3M + 2\omega) +$$
$$+ \frac{17}{4}e^2\cos(4M + 2\omega) + \frac{169}{24}e^3\cos(5M + 2\omega)\bigg\}; \qquad (8)$$

$$\delta\Omega = -J\left(\frac{a'}{a}\right)^2 \sqrt{1-\lambda^2}\left\{(1 + 2e^2)nt + 3\left(e + \frac{13}{8}e^3\right)\sin M + \right.$$
$$+ \frac{9}{4}e^2\sin 2M + \frac{53}{12}e^3\sin 3M + \frac{1}{2}\left(e + \frac{3}{8}e^3\right)\sin(M + 2\omega) -$$
$$- \frac{1}{48}e^3\sin(M - 2\omega) - \frac{1}{2}(1 - 2e^2)\sin(2M + 2\omega) - \qquad (9)$$
$$- \frac{7}{6}\left(e - \frac{95}{56}e^3\right)\sin(3M + 2\omega) - \frac{17}{8}e^2\sin(4M + 2\omega) - \frac{169}{48}e^3\sin(5M + 2\omega)\bigg\};$$

$$e\delta\pi = -J\left(\frac{a'}{a}\right)^2 (\sqrt{1-\lambda^2} - 1 + \lambda^2)\left\{(e + 2e^3)nt + 3e^2\sin M + \right.$$
$$+ \frac{9}{4}e^3\sin 2M + \frac{1}{2}e^2\sin(M + 2\omega) - \frac{1}{2}(e - 2e^3)\sin(2M + 2\omega) -$$
$$- \frac{7}{6}e^2\sin(3M + 2\omega) - \frac{17}{8}e^3\sin(4M + 2\omega)\bigg\} +$$
$$+ J\left(\frac{a'}{a}\right)^2 \left(1 - \frac{3}{2}\lambda^2\right)\left\{(e + 2e^3)nt + \left(1 + \frac{23}{8}e^2\right)\sin M + \right.$$
$$+ \frac{3}{2}\left(e + \frac{19}{18}e^3\right)\sin 2M + \frac{53}{24}e^2\sin 3M + \frac{77}{24}e^2\sin 4M\bigg\} -$$
$$- \frac{1}{4}J\left(\frac{a'}{a}\right)^2 \lambda^2 \left\{\left(1 - \frac{7}{8}e^2\right)\sin(M + 2\omega) - \frac{1}{8}e^2\sin(M - 2\omega) + \right.$$
$$+ 5\left(e - \frac{23}{20}e^3\right)\sin(2M + 2\omega) - \frac{1}{6}e^3\sin(2M - 2\omega) -$$
$$- \frac{7}{3}\left(1 - \frac{397}{56}e^2\right)\sin(3M + 2\omega) - \frac{17}{2}\left(e - \frac{511}{102}e^3\right)\sin(4M + 2\omega) -$$
$$- \frac{169}{8}e^2\sin(5M + 2\omega) - \frac{533}{12}e^3\sin(6M + 2\omega)\bigg\}; \qquad (10)$$

FIRST-ORDER PERTURBATIONS

$$\delta\varepsilon = 2J\left(\frac{a'}{a}\right)^2 \left(1 - \frac{3}{2}\lambda^2\right)\left\{\left(1 + \frac{7}{4}e^2\right)nt + \frac{13}{4}\left(e + \frac{133}{104}e^3\right)\sin M \right.$$
$$+ \frac{21}{8}e^2 \sin 2M + \frac{265}{96}e^3 \sin 3M\Big\} + 3J\left(\frac{a'}{a}\right)^2 \lambda^2 \left\{-\frac{13}{24}\left(e - \frac{17}{104}e^3\right)\sin(M + \right.$$
$$+ 2\omega) + \frac{5}{192}e^3 \sin(M - 2\omega) + \frac{1}{2}\left(1 - \frac{35}{12}e^2\right)\sin(2M + 2\omega) + \frac{91}{72}\left(e - \right.$$
$$\left. - \frac{143}{56}e^3\right)\sin(3M + 2\omega) + \frac{119}{48}e^3 \sin(4M + 2\omega) + \frac{845}{192}e^3 \sin(5M + 2\omega)\Big\} -$$
$$- J\left(\frac{a'}{a}\right)^2 (\sqrt{1-\lambda^2} - 1 + \lambda^2)\left\{(1 + 2e^2)nt + 3\left(e + \frac{13}{8}e^3\right)\sin M \right.$$
$$+ \frac{9}{4}e^2 \sin 2M + \frac{53}{24}e^3 \sin 3M + \frac{1}{2}\left(e + \frac{3}{8}e^3\right)\sin(M + 2\omega) -$$
$$- \frac{1}{48}e^3 \sin(M - 2\omega) - \frac{1}{2}(1 - 2e^2)\sin(2M + 2\omega) -$$
$$- \frac{7}{6}\left(e - \frac{95}{56}e^3\right)\sin(3M + 2\omega) - \frac{17}{8}e^2 \sin(4M + 2\omega) -$$
$$- \frac{169}{48}e^3 \sin(5M + 2\omega)\Big\}. \tag{11}$$

These expressions can also be used for calculating the perturbed positions of the artificial satellite. For this we apply the usual formulas

$$x = r(\cos u \cos \Omega - \sin u \sin \Omega \cos i),$$
$$y = r(\cos u \sin \Omega + \sin u \cos \Omega \cos i),$$
$$z = r \sin u \sin i,$$
$$u = v + \omega,$$
$$\text{tg}\frac{v}{2} = \sqrt{\frac{1+e}{1-e}}\,\text{tg}\frac{E}{2},$$
$$r = a(1 - e \cos E),$$
$$E - e \sin E = M,$$
$$M = M_0 + n_0(t - t_0) + \delta\varepsilon - \delta\pi - \frac{3}{2} \cdot \frac{n_0}{a_0}\int_{t_0}^{t} \delta a \, dt,$$

where symbols with zero subscripts refer to the undisturbed values of the elements. It should be noted that secular disturbances occur only in the elements Ω, π, and ε. They are absent in the remaining elements.

3. SECULAR PERTURBATIONS OF THE FIRST ORDER

Among perturbations of the first order, the most important are the secular perturbations, since these are the ones that determine the evolution of the orbit with time. It is therefore useful to know their values with greater accuracy than the values of the periodic perturbations.

Because of the special form of the perturbation function, the secular perturbations of the first order can be obtained as expressions in finite form, independent of the expansions in powers of the eccentricity. Instead of using the elements π and ϵ in the calculation of the secular perturbations, it is more convenient to use the quantities ω and M_0 defined by the relations

$$\omega = \pi - \Omega, \qquad M_0 = \epsilon - \pi.$$

The Lagrange equations for ω and M_0 are

$$\left. \begin{aligned} \frac{d\omega}{dt} &= -\frac{\operatorname{ctg} i}{na^2 \sqrt{1-e^2}} \frac{\partial R}{\partial i} + \frac{\sqrt{1-e^2}}{na^2 e} \cdot \frac{\partial R}{\partial e}, \\ \frac{dM_0}{dt} &= -\frac{1-e^2}{na^2 e} \cdot \frac{\partial R}{\partial e} - \frac{2}{na} \cdot \frac{\partial R}{\partial a}. \end{aligned} \right\} \quad (12)$$

We will denote the coefficients of the elements Ω, ω, and M_0 in the secular perturbations by Ω', ω', and M_0', respectively. In addition, we will use the real anomaly v as the independent variable, where v is related to the time t by the formula

$$r^2 dv = \sqrt{fm} \sqrt{a(1-e^2)}\, dt.$$

Then

$$\Omega' = \frac{1}{2\pi \sqrt{fm} \sqrt{a(1-e^2)}} \int_0^{2\pi} r^2 \frac{d\Omega}{dt}\, dv,$$

$$\left. \begin{aligned} \omega' &= \frac{1}{2\pi \sqrt{fm} \sqrt{a(1-e^2)}} \int_0^{2\pi} r^2 \frac{d\omega}{dt}\, dv, \\ M_0' &= \frac{1}{2\pi \sqrt{fm} \sqrt{a(1-e^2)}} \int_0^{2\pi} r^2 \frac{dM_0}{dt}\, dv. \end{aligned} \right\} \quad (13)$$

The formulas in (13) give the coefficients Ω', ω', and M_0' under the assumption that the time is measured in terms of the unit $T/2\pi$, where T is the period of rotation of the satellite about the earth. If it is desirable to use a day as the unit of time, then the right-hand sides of the equations in (13) must be multiplied by the mean daily travel n of the satellite. It

FIRST-ORDER PERTURBATIONS

should be noted that when n is expressed in degrees or radians, then Ω', ω', and M_0' will be obtained in the same units.

According to Ref. 2, calculations starting from the formulas in (13), with the right-hand sides multiplied by n, yield the following equalities[1]

$$\left.\begin{aligned} \Omega' &= -J\left(\frac{a'}{a}\right)^2 \frac{\cos i}{(1-e^2)^2}\, n, \\ \omega' &= \frac{1}{2} J\left(\frac{a'}{a}\right)^2 \frac{5\cos^2 i - 1}{(1-e^2)^2}\, n, \\ M_0' &= \frac{1}{2} J\left(\frac{a'}{a}\right)^2 \frac{3\cos^2 i - 1}{(1-e^2)^{3/2}}\, n. \end{aligned}\right\} \quad (14)$$

From this it is easy to obtain expressions for the coefficients for the secular perturbations π and ϵ

$$\left.\begin{aligned} \pi' &= \Omega' + \omega' = \frac{1}{2} J\left(\frac{a'}{a}\right)^2 \frac{5\cos^2 i - 2\cos i - 1}{(1-e^2)^2} \cdot n, \\ \varepsilon' &= \pi' + M_0' = \frac{1}{2} J\left(\frac{a'}{a}\right)^2 \frac{(5+3\sqrt{1-e^2})\cos^2 i - 2\cos i - 1 - \sqrt{1-e^2}}{(1-e^2)^2} \cdot n. \end{aligned}\right\} \quad (15)$$

The expressions in (14) are valid for any value of the eccentricity less than one.

We see from (14) that the secular motion of the node attains its maximum for i = 0° or 180°, i.e., when the orbit lies in the equatorial plane. For i = 90°, i.e., when the plane of the orbit passes through the pole of the earth, $\Omega' = 0°$. The secular motion of the perigee attains its maximum for i = 0° or 180° and becomes zero for i ~ 63° 40'. The secular motion of the element M_0 reaches its maximum for i = 0° or 180°, and is zero for i = 54° 44'.

4. A SAMPLE CALCULATION OF PERTURBATIONS OF THE FIRST ORDER

As an example of perturbation calculations, we will take an orbit with the following elements:

$$a = 7286.88 \text{ km},$$
$$e = 0.099\,493,$$
$$i = 65°.4900.$$

We will assume the following values for the constants a' and J

$$a' = 6378.39 \text{ km}, \quad J = 0.0016\,4147.$$

When these values of the orbit elements and the values of a' and J are substituted in formulas (6)-(11), we obtain the following final expressions for the perturbations of the first order caused by the flattening of the earth

$\delta a = -0.45 \cos M - 0.07 \cos 2M - 0.01 \cos 3M - 0.38 \cos(M + 2\omega) +$
$\quad + 7.40 \cos(2M + 2\omega) + 2.58 \cos(3M + 2\omega) + 0.52 \cos(4M + 2\omega) +$
$\quad + 0.13 \cos(5M + 2\omega)$

$\delta e = -0.00030 \cos M - 0.00005 \cos 2M - 0.00001 \cos 3M +$
$\quad + 0.00026 \cos(M + 2\omega) - 0.00003 \cos(2M + 2\omega) +$
$\quad + 0.00058 \cos(3M + 2\omega) + 0.00021 \cos(4M + 2\omega) +$
$\quad + 0.00005 \cos(5M + 2\omega) + 0.00001 \cos(6M + 2\omega),$

$\delta i = -0°.001 \cos(M + 2\omega) + 0°.013 \cos(2M + 2\omega) +$
$\quad + 0°.003 \cos(3M + 2\omega) + 0°.001 \cos(4M + 2\omega),$

$\delta\Omega = -2°.674 t - 0°.009 \sin M - 0°.001 \sin 2M -$
$\quad - 0°.001 \sin(M + 2\omega) + 0°.015 \sin(2M + 2\omega) +$
$\quad + 0°.003 \sin(3M + 2\omega) + 0°.001 \sin(4M + 2\omega),$

$\delta\pi = -0°.311 t - 0°.018 \sin M - 0°.003 \sin 2M - 0°.015 \sin(M + 2\omega) -$
$\quad -0°.006 \sin(2M + 2\omega) + 0°.033 \sin(3M + 2\omega) + 0°.012 \sin(4M + 2\omega) +$
$\quad + 0°.003 \sin(5M + 2\omega) + 0°.001 \sin(6M + 2\omega),$

$\delta\varepsilon = -4°.780 t - 0°.016 \sin M - 0°.002 \sin 2M - 0°.009 \sin(M + 2\omega) +$
$\quad + 0°.095 \sin(2M + 2\omega) + 0°.024 \sin(3M + 2\omega) +$
$\quad + 0°.005 \sin(4M + 2\omega) + 0°.001 \sin(5M + 2\omega).$

Here the coefficients of the periodic perturbations of the semimajor axis are given in kilometers, and in the secular perturbations of the elements Ω, π, and ε, the time must be measured in mean solar days.

The numerical values of the perturbations given above clearly show the relative influence of such perturbations. Thus, in the semimajor axis the greatest influence comes from the periodic perturbations with arguments $(2M + 2\omega)$ and $(3M + 2\omega)$; the amplitude of the resulting variations are, respectively, about 7.4 and 2.6 km. The perturbation having the greatest influence on the eccentricity has the argument $(3M + 2\omega)$ and this can produce a deviation in the height of the perigee of approximately 4.2 km. The greatest periodic perturbations in the remaining elements are of the order of a few minutes of arc.

Rather large perturbations thus occur in all the elements, and these must be taken into consideration for the best possible processing of the results obtained.

LITERATURE

1. M. F. Subbotin, A Course in Celestial Mechanics (in Russian), Vol. 2 (ONTI, 1937).
2. D. E. Okhotsimskii, T. M. Éneev, and G. P. Taratynova, Usp. fiz. nauk, 63, 1a, 33 (1957).

PERTURBATIONS IN THE ORBITS OF ARTIFICIAL SATELLITES CAUSED BY AIR RESISTANCE

Yu. V. Batrakov and V. F. Proskurin

If the orbit of an artificial satellite passes relatively close to the surface of the earth, then in this portion of its orbit the satellite experiences a significant braking action from the air through which it is moving. The result of this periodic braking process is a decrease in the total mechanical energy of the satellite and a consequent rapid increase in the secular variations in the form and dimensions of the orbit, which cause the satellite to pass ever closer to the earth, and lead to its eventual destruction in the denser layers of the atmosphere.

The properties of the motion of an artificial satellite in an air medium have, up to the present, been insufficiently investigated. In works dedicated to this problem, only the most essential of these properties have been considered — secular perturbations[1] and perturbations with periods that are significantly longer than the period of a single orbit of the satellite.

In the present work, we will attempt to obtain the general form of perturbations of the first order in the elements of the elliptic orbit of a satellite that are produced by atmospheric resistance only. It will be assumed that the earth's atmosphere has a strictly spherical density distribution, and that the earth's attraction can be replaced by that of a material point located at its center with the same mass as the earth. With these assumptions, in addition to the secular perturbations, there are short-period perturbations with periods not exceeding the period of a single revolution of the satellite. As far as we know, these short-period perturbations caused by air resistance have not yet been studied.

We will also give a numerical example, showing the relative magnitude of the perturbation of the first order caused by air resistance.

1. THE EQUATIONS OF MOTION, AND THE GENERAL FORM OF PERTURBATIONS OF THE FIRST ORDER

The resistance force acting on a body moving with a velocity V in a rarefied-air medium of density ρ is given by the formula

$$F = \frac{1}{2} C_x S \rho V^2, \qquad (1)$$

where S is the area of an (arbitrary) cross section of the body and C_x is the aerodynamic resistance. The force F always acts in the direction opposite to that of the motion of the body. We will assume that the air density is a decreasing function of the distance from the surface of the earth, denoted by h.

If the mass of the moving body is m, then the disturbing acceleration produced by the force F is $\alpha \rho V^2$, where

$$\alpha = \frac{1}{2} C_x \frac{S}{m}.$$

The coefficient α depends only on the shape, the dimensions, and the weight of the moving body.

The Lagrange equations for the perturbations of the elliptical orbit are[2]

$$\begin{aligned}
\frac{da}{dt} &= 2a^2 e \sin \theta \cdot S' + 2a^2 p r^{-1} T', \\
\frac{de}{dt} &= p \sin \theta \cdot S' + p(\cos \theta + \cos E) T', \\
\sin i \frac{d\Omega}{dt} &= r \sin(\theta + \Pi - \Omega) W', \\
\frac{di}{dt} &= r \cos(\theta + \Pi - \Omega) \cdot W', \\
e \frac{d\Pi}{dt} &= 2e \sin^2 \frac{i}{2} \frac{d\Omega}{dt} - p \cos \theta \cdot S' + (r+p) \sin \theta \cdot T', \\
\frac{d\varepsilon}{dt} &= 2 \sin^2 \frac{i}{2} \frac{d\Omega}{dt} - 2r \sqrt{1-e^2} \cdot S' + \\
&\quad + \frac{e}{1+\sqrt{1-e^2}} [-p \cos \theta \cdot S' + (r+p) \sin \theta \cdot T'],
\end{aligned} \qquad (2)$$

where a is the semimajor axis, e the eccentricity, i the declination, Ω the longitude of the node, Π the longitude of the perigee, ϵ the mean longitude of the epoch, r the radius vector, θ the real anomaly, and E the eccentric anomaly. $p = a(1-e^2)$ is the orbit parameter. If S, T, and W denote the components of the perturbing acceleration in the direction of the radius vector, in the direction lying in the osculating orbit perpendicular to the radius vector, and in the direction normal to the plane of the orbit respectively, then we obtain the expressions

$$S' = \frac{1}{k\sqrt{mp}} S, \quad T' = \frac{1}{k\sqrt{mp}} T, \quad W' = \frac{1}{k\sqrt{mp}} W,$$

where k is the Gauss constant.

Since the disturbing force acts in the direction opposite to that of the velocity of the body, and since the unperturbed motion lies in a plane, the perturbed motion also remains in a plane, and the component W is zero.

We note that the projection of the velocity vector on the radius vector, and on a direction in the plane of the orbit perpendicular to the radius vector, are respectively $\dot{r}$ and $r\dot{u}$, where $u = \theta + \omega$ is the distance of the satellite from the node of the orbit (the latitude argument) and the dot indicates differentiation with respect to time. Then, if we use the relations

$$\dot{r} = \frac{nae}{\sqrt{1-e^2}} \sin\theta,$$

$$r\dot{u} = \frac{na^2\sqrt{1-e^2}}{r},$$

which are valid for unperturbed motion, we obtain the relations

$$\left.\begin{array}{l} S' = -\alpha\rho \dfrac{ne}{(1-e^2)^{3/2}} \sqrt{1 + 2e\cos\theta + e^2} \cdot \sin\theta, \\[6pt] T' = -\alpha\rho \dfrac{n}{(1-e^2)^{3/2}} \sqrt{1 + 2e\cos\theta + e^2}(1 + e\cos\theta). \end{array}\right\} \quad (3)$$

which are accurate up to small quantities of higher order than the disturbing forces.

When (3) is substituted in (2) and certain simplifications are made, we obtain the formulas

$$\begin{aligned}
\frac{da}{dt} &= -\frac{2\alpha p n a^2}{(1-e^2)^{3/2}}(1+2e\cos\theta+e^2)^{3/2} \\
\frac{de}{dt} &= -\frac{2\alpha p n a}{(1-e^2)^{1/2}}(1+2e\cos\theta+e^2)^{1/2}(e+\cos\theta), \\
\frac{d\Omega}{dt} &= \frac{di}{dt} = 0, \\
e\frac{d\Pi}{dt} &= -\frac{2\alpha p n a}{(1-e^2)^{1/2}}(1+2e\cos\theta+e^2)^{1/2}\cdot\sin\theta, \\
\frac{d\varepsilon}{dt} &= \frac{2\alpha p n a e}{(1-e^2)^{3/2}}(1+2e\cos\theta+e^2)^{1/2}\left(\frac{\sqrt{1-e^2}}{1+e\cos\theta}-\frac{1}{1+\sqrt{1-e^2}}\right)\sin\theta.
\end{aligned} \qquad (4)$$

In order to obtain the perturbations of the first order in the elements of the satellite's orbit, we must expand the right-hand sides of (4) in trigonometric series in the mean anomaly M, or, what amounts to the same thing, in trigonometric series in the time. According to our assumptions, the air density ρ is a function of only the distance from the surface of the earth, which is denoted by the symbol h. The quantity h is an even periodic function of the mean anomaly with period 2π, and can consequently be expanded in a Fourier series containing only cosines of integral multiples of the mean anomaly. The functions $\cos\theta$ and $\sin\theta$ can also be expanded in Fourier series containing only cosines or sines of the mean anomaly respectively.

It thus follows that after expanding the right-hand sides of the equations in (4) in Fourier series in the mean anomaly, we will have the equations of the form

$$\begin{aligned}
\frac{da}{dt} &= a' + \sum_{j=1}^{\infty} a_j \cos jM, \\
\frac{de}{dt} &= e' + \sum_{j=1}^{\infty} e_j \cos jM, \\
\frac{d\Omega}{dt} &= \frac{di}{dt} = 0, \\
\frac{d\Pi}{dt} &= \sum_{j=1}^{\infty} \Pi_j \sin jM,
\end{aligned} \qquad (5)$$

$$\frac{d\varepsilon}{dt} = \sum_{j=1}^{\infty} \varepsilon_j \sin jM, \quad \Bigg\} \tag{5}$$

where all the coefficients of the sines and cosines are functions of the semimajor axis and the eccentricity, and also depend on the parameters occurring in the distribution law for the air density in terms of the distance from the earth's surface.

In the first approximation, the eccentricity and the semimajor axis are assumed constant; integration of (5) then yields the equations

$$\begin{aligned} a &= a_0 + a'(t-t_0) + \sum \frac{a_j}{jn} \sin jM, \\ e &= e_0 + e'(t-t_0) + \sum \frac{e_j}{jn} \sin jM, \\ \Omega &= \Omega_0, \quad i = i_0, \\ \Pi &= \Pi_0 - \sum \frac{\Pi_j}{jn} \cos jM, \\ \varepsilon &= \varepsilon_0 - \sum \frac{\varepsilon_j}{jn} \cos jM, \end{aligned} \Bigg\} \tag{6}$$

where a_0, e_0, ... are constants of integration, t_0 is the time of passage through the perigee, and $M = n(t - t_0)$ where n is the mean motion.

Thus, in the semimajor axis and in the eccentricity, in addition to periodic perturbations, there are secular terms, while in the elements π and ϵ the secular terms are absent. The elements i and Ω are constant, since the motion lies in a plane. It should be noted that the periodic perturbations in a and e are odd functions of M and that they are zero for M = 0, i.e., at the perigee of the orbit. The periodic perturbations of Π and ϵ are even functions, and reach their extreme values at the perigee and apogee.

The coefficients a' and e', and also a_j, e_j, Π_j, and ϵ_j can be calculated from the usual Euler formulas

$$a' = \frac{1}{\pi} \int_0^{\pi} \frac{da}{dt} dM, \quad e' = \frac{1}{\pi} \int_0^{\pi} \frac{de}{dt} dM, \Bigg\} \tag{7}$$

$$a_j = \frac{2}{\pi} \int_0^\pi \frac{da}{dt} \cos jM \, dM, \qquad e_j = \frac{2}{\pi} \int_0^\pi \frac{de}{dt} \cos jM \, dM,$$
$$\Pi_j = \frac{2}{\pi} \int_0^\pi \frac{d\Pi}{dt} \sin jM \, dM, \qquad \varepsilon_j = \frac{2}{\pi} \int_0^\pi \frac{d\varepsilon}{dt} \sin jM \, dM,$$
(7)

either numerically, or, if possible, in the form of explicit analytic expressions in a_0, e_0, and the parameters occurring in the law used to give the relation of the density to the height.

The signs of the coefficients a' and e', which determine the character of the satellite's orbit, can be determined without any preliminary calculation. In fact it follows from (4) that da/dt is always negative, so that a' will always have a negative sign. The same results holds for the coefficient e', where this follows from the fact that the derivative de/dt is negative in the interval $0 < \theta < \pi/2$ and in this interval its absolute value exceeds any values of de/dt in $\pi/2 < \theta < \pi$. It is assumed here that the density ρ is a decreasing function of the satellite's height above the surface of the earth, and is consequently a decreasing function of θ in the interval $(0, \pi)$. Since a' and e' are negative, both the semimajor axis and the eccentricity decrease with time, and the orbit becomes smaller and approaches a circular form.

The formulas (5) give a complete representation of the character of perturbations of the first order that are caused by air resistance. If higher-order perturbations must be taken into account, the elements a and e must be expressed in the form

$$a = a_0 + a'(t - t_0) + a''(t - t_0) + \cdots,$$
$$e = e_0 + e'(t - t_0) + e''(t - t_0) + \cdots,$$
(8)

where the theoretical determination of the coefficients a", e", ... is more difficult than that of a', e', ...

2. AN EXAMPLE OF THE CALCULATION OF PERTURBATIONS OF THE FIRST ORDER IN THE ELEMENTS OF A SATELLITE'S ORBIT CAUSED BY AIR RESISTANCE

For the calculation of real perturbations, we must first of all have a definite expression for the air density as a function of the height. To obtain such an expression, we use the model of the atmosphere introduced in Ref. 3, which is defined by the data in the first four columns of the table. The first column gives the height above the earth's surface, while the second, third, and fourth give the concentrations of N_2, O_2, and O, which are the main constituents of and which determine the properties of air at great heights. The density of the air, given in the fifth column, is calculated from the formula

$$\rho = m(N_2) n(N_2) + m(O_2) n(O_2) + m(O) n(O),$$

where $m(N_2) = 4.688 \cdot 10^{-23}$ g, $m(O_2) = 5.354 \cdot 10^{-23}$ g, and $m(O) = 2.677 \cdot 10^{-23}$ g are the particle masses, and $n(N_2)$, $n(O_2)$, and $n(O)$ are the number of particles per unit volume. In the sixth column the values of the density are given in technical units.

The dependence of density on height given in the table is now approximated by a formula of the type

$$\rho = A \left(\log \frac{h}{10} \right)^B, \tag{9}$$

where A and B are parameters, and the logarithm in the parentheses is to the base ten. Taking logarithms of (9), we obtain the relation

$$\log \rho = \log A + B \log \log \frac{h}{10}, \tag{10}$$

which gives a linear relation between log A and B.

If we substitute in (10) the values of h (in kilometers) and the corresponding densities from the sixth column, we obtain 15 equations for the determination of log A and B. An application of the method of mean-squares yields

$$\log A = -6.98787, \quad B = -26.8132. \tag{11}$$

The values of the density calculated by using formula (9) with the parameters A and B from (11) are given in the last column of the table.

h	Number of particles per cm³			ρ, g/cm³	ρ, kg·sec²/m⁴	ρ, kg·sec²/m⁴
	$n(N_2)$	$n(O_2)$	$n(O)$			
100	$2.82 \cdot 10^{13}$	$7.56 \cdot 10^{12}$	$1.09 \cdot 10^{12}$	$1.742 \cdot 10^{-9}$	$1.775 \cdot 10^{-7}$	$1.03 \cdot 10^{-7}$
110	$8.58 \cdot 10^{12}$	$8.25 \cdot 10^{10}$	$3.59 \cdot 10^{12}$	$4.989 \cdot 10^{-10}$	$5.084 \cdot 10^{-8}$	$3.47 \cdot 10^{-8}$
120	$2.54 \cdot 10^{12}$	$5.29 \cdot 10^{8}$	$1.06 \cdot 10^{12}$	$1.463 \cdot 10^{-10}$	$1.491 \cdot 10^{-8}$	$1.33 \cdot 10^{-8}$
130	$8.60 \cdot 10^{11}$	$1.72 \cdot 10^{7}$	$3.45 \cdot 10^{11}$	$4.917 \cdot 10^{-11}$	$5.010 \cdot 10^{-9}$	$5.70 \cdot 10^{-9}$
150	$1.47 \cdot 10^{11}$		$1.08 \cdot 10^{11}$	$9.707 \cdot 10^{-12}$	$9.891 \cdot 10^{-10}$	$1.33 \cdot 10^{-9}$
175	$2.73 \cdot 10^{10}$		$3.28 \cdot 10^{10}$	$2.164 \cdot 10^{-12}$	$2.205 \cdot 10^{-10}$	$3.01 \cdot 10^{-10}$
200	$4.31 \cdot 10^{9}$		$1.75 \cdot 10^{10}$	$6.653 \cdot 10^{-13}$	$6.779 \cdot 10^{-11}$	$8.87 \cdot 10^{-11}$
250	$3.28 \cdot 10^{8}$		$2.55 \cdot 10^{9}$	$8.299 \cdot 10^{-14}$	$8.457 \cdot 10^{-12}$	$1.29 \cdot 10^{-11}$
300	$4.71 \cdot 10^{7}$		$7.15 \cdot 10^{8}$	$2.118 \cdot 10^{-14}$	$2.158 \cdot 10^{-12}$	$2.95 \cdot 10^{-12}$
400	$1.93 \cdot 10^{6}$		$1.08 \cdot 10^{8}$	$2.958 \cdot 10^{-15}$	$3.014 \cdot 10^{-13}$	$3.33 \cdot 10^{-13}$
500			$2.63 \cdot 10^{7}$	$6.985 \cdot 10^{-16}$	$7.118 \cdot 10^{-14}$	$6.93 \cdot 10^{-14}$
600			$8.49 \cdot 10^{6}$	$2.255 \cdot 10^{-16}$	$2.298 \cdot 10^{-14}$	$3.04 \cdot 10^{-14}$
700			$3.41 \cdot 10^{6}$	$9.057 \cdot 10^{-17}$	$9.229 \cdot 10^{-15}$	$7.57 \cdot 10^{-15}$
800			$1.38 \cdot 10^{6}$	$3.665 \cdot 10^{-17}$	$3.735 \cdot 10^{-15}$	$3.30 \cdot 10^{-15}$
900			$7.33 \cdot 10^{5}$	$1.947 \cdot 10^{-17}$	$1.984 \cdot 10^{-15}$	$1.62 \cdot 10^{-15}$

When we compare the original values of the density with those calculated from formula (9), we see that the difference between them reaches 30% of the value of the density in some cases. When we consider, however, that the density of air at great heights is only known within limits that may exceed the density itself by several times, it can be seen that such differences can be tolerated.

For the actual calculation of the perturbations, we simplify the equations (4) somewhat, and with this aim in view we set the eccentricity equal to zero everywhere in the right-hand sides of these equations (except in the case of the dependence of ρ on e). We also introduce the notation

$$\overset{*}{\rho} = \left(\log \frac{h}{10}\right)^B. \tag{12}$$

The equations for the elements a, e, Π, and ε then take the form

$$\frac{da}{dt} = -2\alpha A n a^2 \rho^*, \qquad e\frac{d\Pi}{dt} = -2\alpha A n a \rho^* \sin M,$$
$$\frac{de}{dt} = -2\alpha A n a \rho^* \cos M, \qquad \frac{d\varepsilon}{dt} = 0. \qquad (13)$$

The height h occurring in the formula for ρ^* is expressed in terms of M according to the relations

$$h = r - R, \quad r = a(1 - e\cos E), \quad E - e\sin E = M, \qquad (14)$$

where R is the mean radius of the earth. For the example we are considering, we take the elements

$$\left.\begin{array}{l} n_0 = 5024°.40 \\ n' = 1°.191 \end{array}\right\} \text{per day} \qquad (15)$$
$$e = 0.099493$$

where n_0 and n' are related to a_0 and a' by the formulas

$$a_0^3 n_0^2 = k^2 m, \quad n' = -\frac{3}{4}\cdot\frac{n_0}{a_0} a'. \qquad (16)$$

We take the value of $k^2 m$ from Ref. 4, and use a system of units that is convenient for the problem

$$k^2 m = 9.75787 \cdot 10^{15} \text{ km}^3 \cdot \text{deg}^2/\text{day}^2$$

Then from (16) we obtain

$$a_0 = 7286.88 \text{ km}, \quad a' = -2.303 \text{ km/day} \qquad (17)$$

For the value a_0 = 7286.88 km with R = 6367.55 km, and when the usual formulas of harmonic analysis are applied for the case when the period is divided into 24 parts, formula (14) yields

$$10^4 \rho^* = 0.7433 + 1.4633 \cos M + 1.3968 \cos 2M + 1.2953 \cos 3M +$$
$$+ 1.1702 \cos 4M + 1.0337 \cos 5M + 0.8971 \cos 6M + 0.7696 \cos 7M +$$
$$+ 0.6582 \cos 8M + 0.5679 \cos 9M + 0.5017 \cos 10M +$$
$$+ 0.4614 \cos 11M + 0.2239 \cos 12M. \qquad (18)$$

Formula (18) clearly shows how slowly the Fourier series for the function $\rho = \rho(M)$ converges.

If we substitute (18) in (13) and integrate, we obtain both the secular and the periodic perturbations of the elements.

All the resulting formulas contain the factor αA, which is known only very approximately for a variety of reasons. First of all, the aerodynamic coefficient C_x is not known very accurately; the instantaneous cross-sectional area S can not be determined with sufficient accuracy; finally, the coefficient A is known only very approximately. It is therefore more convenient not to calculate the factor αA theoretically, but to determine this factor by starting from the quantity a′, introduced in (17). In fact, if we use the relation

$$a' = -2\alpha A n a^2 [\rho^*],$$

where the symbol $[\rho^*]$ denotes the free term in the Fourier expansion, we obtain

$$2\alpha A n a^2 = -\frac{a'}{[\rho^*]}. \tag{19}$$

If we substitute (19) in (13) and integrate, we obtain the following formulas

$$\left. \begin{array}{l} a = a_0 + \frac{a'}{[\rho^*]} \int \rho^* dt, \qquad e\Pi = e_0 \Pi_0 + \frac{a'}{a[\rho^*]} \int \rho^* \sin M \, dt, \\ e = e_0 + \frac{a'}{[\rho^*]a} \int \rho^* \cos M \, dt, \quad \varepsilon = \varepsilon_0. \end{array} \right\} \tag{20}$$

If we use (20), (16), (17), and (18), we finally arrive at the following formulas for the perturbations in an orbit with the elements (15)

$$\begin{aligned}
a = a_0 &- 2.303\, t \quad -0.052 \sin M \quad -0.004 \sin 7M \\
&-0.025 \sin 2M \quad -0.003 \sin 8M \\
&-0.015 \sin 3M \quad -0.003 \sin 9M \\
&-0.010 \sin 4M \quad -0.002 \sin 10M \\
&-0.007 \sin 5M \quad -0.002 \sin 11M \\
&-0.006 \sin 6M \quad -0.001 \sin 12M;
\end{aligned} \tag{21a}$$

$$\begin{aligned}
e = e_0 &- 0.0002584\, t - 0.0000070 \sin M \quad -0.0000005 \sin 7M \\
& - 33 \sin 2M \quad\quad -4 \sin 8M \\
& - 20 \sin 3M \quad\quad -4 \sin 9M \\
& - 14 \sin 4M \quad\quad -2 \sin 10M \\
& - 10 \sin 5M \quad\quad -2 \sin 11M \\
& - 7 \sin 6M \quad\quad -1 \sin 12M;
\end{aligned} \tag{21b}$$

$$\begin{aligned}\Pi = \Pi_0 &+ 0°.00013 \cos M & &+ 0°.00006 \cos 6M \\ &+ 12 \cos 2M &+& 5 \cos 7M \\ &+ 11 \cos 3M &+& 3 \cos 8M \\ &+ 09 \cos 4M &+& 2 \cos 9M \\ &+ 08 \cos 5M &+& 1 \cos 10M, \end{aligned} \quad (21c)$$

where the coefficients of the perturbations in the semimajor axis are given in kilometers. It is evident from (21) that the amplitude of the greatest periodic term in the semimajor axis is approximately equal to 50 m, and that the periodic perturbations in the longitude of the perigee attain a value of half a second of arc. The perturbations in e are still smaller than the perturbations in the longitude of the perigee, since they contain the eccentricity of the orbit as a factor.

The periodic perturbations caused by air resistance are thus rather small, and they need not be taken into account in the processing of visual or rough photographic observations of artificial earth satellites.

LITERATURE CITED

1. D. E. Okhotsimskii, T. M. Eneev, and G. P. Taratynova, Usp. fiz. nauk, 63, 1a (1957).
2. M. F. Subbotin, A Course in Celestial Mechanics (in Russian) Vol. 2 (ONTI, 1937).
3. S. K. Mitre, The Upper Atmosphere (Russian translation) (IL, 1955).
4. I. D. Zhongolovich, Byull. ITA, 6, 8 (1957).

TRACKING OF ARTIFICIAL SATELLITES BASED ON EXPECTED POINT OF ARRIVAL

V. M. Vakhnin and V. V. Beletskii

The problem of predicting the appearance of an earth satellite within the zone accessible to tracking equipment is related to the need to measure all the elements of the orbit with high precision. However, on account of the spread in initial data and certain perturbations which constantly take their toll, the period of revolution of a satellite, its eccentricity, and other orbit parameters cannot be known beforehand with sufficient accuracy. It is therefore expedient to work out some searching method in satellite tracking wherein an inexact knowledge of several orbit parameters would not present a serious hindrance to predicting the time and place of appearance.

In this paper the authors suggest a method which should make possible repeated successful tracking of a satellite already observed once, in the case where the satellite's period of revolution is not known.

We begin by assuming that the inclination of the orbit, the longitude of the node, and the position of perigee remain unaltered. This assumption is reasonable for orbits approximating circular polar orbits. In that case, we may formulate the following "local time rule":

If the inclination of the satellite orbit is not zero, then the traversal by the satellite of any given latitude will always take place at the same local sidereal time.

This rule continues in force even when the period of revolution, the eccentricity, or the position of perigee undergo change. For example, a change in period of revolution is attended by a shift in the point of latitude intersection, but the time of intersection, stated as local time at the new point, will have the same value that it had at any other point where the satellite intersected that same latitude previously.

Figure 1 illustrates the operation of the above rule. Let the satellite intersect a specified parallel (point A) on the

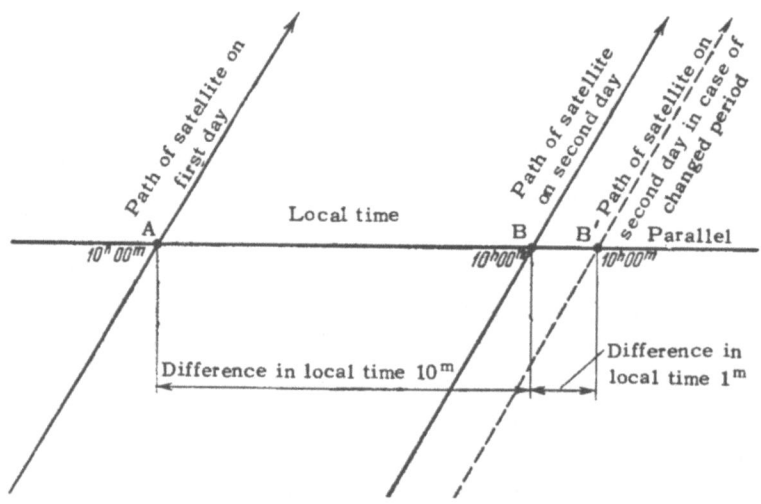

FIG. 1. Example illustrating operation of the "local time rule." At points A, B, B', 10^h00^m local time (referred to time at each point). At point B 9^h50^m with respect to the time at point A, and at point B', 9^h49^m with respect to the time at point A.

first day at $10^h 00^m$ local sidereal time (LST). If the period of revolution is not some multiple of the sidereal day, then the satellite will intersect the same parallel at point B on subsequent days; but the time based on the positions of the stars at point B will be the same as the local time earlier at point A.

If the period of revolution undergoes a change and the satellite manages to intersect the parallel at some point B', then the time of intersection in local time referable to point B' will again be the same as it was at point A. If we now define all of these times in a single system of time reckoning, as for instance the local time at point A, we shall find that the difference between times of intersection of a given latitude on the first and second days is equal to the LST difference of the points of intersection. The existence of this close relation between the position of the point of intersection and the time of intersection enables us to plot a "prediction graph," relating each direction of possible satellite appearance with a completely specified instant of time.

Figure 2 illustrates the simplest example of such a graph. The plot shows that the satellite came into observation on the first day at point 0, say, so that on subsequent days it could

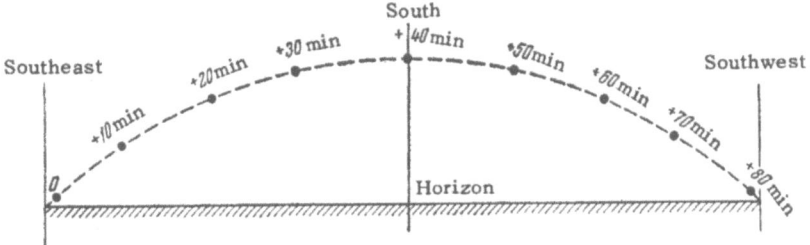

FIG. 2. Prediction graph.

be searched for at the same point at the same LST. If the satellite fails to appear at the specified point, it then becomes necessary to effect a smooth transit of the tracking equipment so as to home in, 10 min later, on the point "+10 min," or, after 20 min, on the point "+20 min," etc., until the satellite is finally detected.

For real nonidealized orbits, in which precession of the orbital plane and drift of perigee over the orbital plane occur, the simple rule outlined above becomes inexact and is applicable only for short time intervals and for tracking equipment scanning a broad angle of vision. Nevertheless, the prediction plot proves useful even for orbits of varying orientation.

Let us consider this case in greater detail. We shall make use of the equatorial system of coordinates on the celestial sphere: right ascension read off from the vernal equinox, and declination read off from the celestial equator. Since the satellite is orbiting relatively close to the earth, its apparent equatorial coordinates differ from the true coordinates. Let the observer be stationed at a point A (Fig. 3) on the earth's surface at the time of the first observation, and let the satellite appear at point B in space. The true equatorial coordinates of the satellite are right ascension α_0 and declination δ_0, defined by the position in space of the vector OB produced

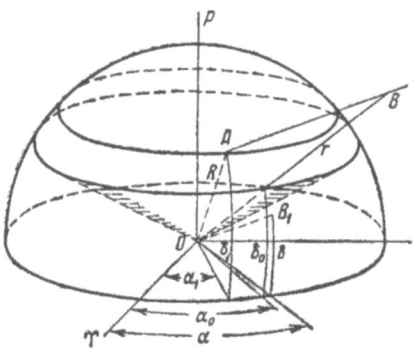

FIG. 3. Apparent and true coordinates of the satellite. Hatching indicates the cone on which the expected point of arrival will be found.

from the center of the earth O to the point B. The apparent coordinates α and δ of the satellite are defined by the position of vector AB or, restated, the position of vector OB_1, produced from the center of the earth parallel to AB. Let us trace a cone whose apex is at the center of the earth and which cuts the parallel δ_0 (Fig. 3). Point B will be found to lie on this surface, and the plane of the satellite's orbit will cut it along the straight line OB.

The method proposed for spotting the satellite is based on the fact that the satellite will again pass through point B in subsequent passes, so that point B may be viewed as an "expected point of arrival." If the position of the tracking equipment located at point A is controlled in such a way that the equipment is consistently directed at the expected point of arrival throughout the search, it is inevitable that the satellite will be detected at some moment provided the search time is adequate (longer than the period of revolution of the satellite). It is obvious that the direction of the optical axis of the instrument in space will be described by δ, α the apparent coordinates of the expected point of arrival B. The locus defined by coordinates δ, α on the celestial sphere as these coordinates change will be termed the prediction curve.

The coordinates of the prediction curve at each instant of

time are completely defined in terms of the coordinates of the point of observation A and expected point of arrival B, projected onto the celestial sphere. Point A executes a diurnal rotation: the declination δ_1 and the distance to the center of the earth R of point A remain constant, with the right ascension α_1 varying uniformly with time. The position of expected point of arrival B on the cone is defined by the declination δ_0, the right ascension α_0, and the distance from the center of the earth r. In consequence of the deviation of the earth's gravitational field from a central field, and because of the effect of the earth's atmosphere, the plane of the satellite's orbit precesses, resulting in a variation, generally speaking, in the longitude of the point B. Because of these same perturbing factors, the satellite orbit turns in its plane (drift of orbit perigee), so that the distance r separating point B from the center of the earth is also free to vary. The variables r(t) and $\alpha_0(t)$ describe the motion of the expected point of arrival over the latitude cone.

When the orbit is close to a polar orbit, the rate at which the orbit precesses is very small, and we may assume that $\alpha_0 \approx$ const (this is precisely the case for a polar orbit). If we may assume further that r ≈ const, which will be true when the orbit is circular or close to circular, we will then have the ideal case where the expected point of arrival B remains motionless on the latitude cone. This case was considered at the beginning of our discussion. In this case, the prediction curves will be closed curves, and the task of tracking the satellite becomes very simple.

In the general case, however, point B will not remain fixed. But even so it is rather simple to compute the time and the coordinates of the origin for tracking, once the time and coordinates of the first detection of the satellite are known.

Let us proceed further, first to derive the parametric equation of the prediction curve.

Coordinates δ_1, α_1, and R of point A, and δ_0, α_0, r of point B uniquely and completely specify the direction of the vector AB in space. Expressed in equatorial coordinates α, δ, this direction is

$$\sin \delta = \frac{\sin \delta_0 - \frac{R}{r} \sin \delta_1}{\sqrt{1 + \left(\frac{R}{r}\right)^2 - 2\left(\frac{R}{r}\right) \cos \varkappa}}, \tag{1}$$

$$\operatorname{tg} \alpha = \frac{\sin \alpha_0 \cos \delta_0 - \frac{R}{r} \cos \delta_1 \sin \alpha_1}{\cos \alpha_0 \cos \delta_0 - \frac{R}{r} \cos \delta_1 \cos \alpha_1}, \tag{2}$$

where the following notation is used

$$\cos \varkappa = \sin \delta_0 \sin \delta_1 + \cos \delta_1 \cos \delta_0 \cos (\alpha_0 - \alpha_1). \tag{3}$$

The right-hand members of these equations are dependent on time via the quantities α_1, α_0, and r. Consequently, (1) and (2) yield a parametric representation of the prediction curve in terms of the time parameter. The explicit expression for α_1 in terms of the time t is given by

$$\alpha_1 = \alpha_{10} + 2\pi \frac{(t - t_0)}{T_1}, \tag{4}$$

where T_1 denotes a sidereal day.

For the right ascension α_0 of the projection of point B, the following approximate equation[1] is valid:

$$\alpha_0 = \alpha_{00} - 2\pi K_0 \frac{(t - t_0)}{T}, \tag{5}$$

$$K_0 = \beta \left(\frac{R}{p}\right)^2 \cos i,$$

where i is the angle of inclination of the orbit to the equator, p is the focal parameter of the orbit (orbit parameter, $\beta = 0.00163$, and T_0 is the period of revolution of the satellite in orbit. If the rate at which the perigee shifts in the orbital plane is known, the function of time can also be expressed for the parameter r.

First let us suppose that the variation in r over a 24 hr period can be neglected, so that we have r ≈ const for one whole day. Let the satellite be detected at some time t_H by an observer located at point α_{10}, δ_1, and let the apparent coordinates of the satellite be δ_H and α_H. From these data, we may compute the real satellite coordinates at the time of observation α_{00}, δ_0. When the satellite's height above the

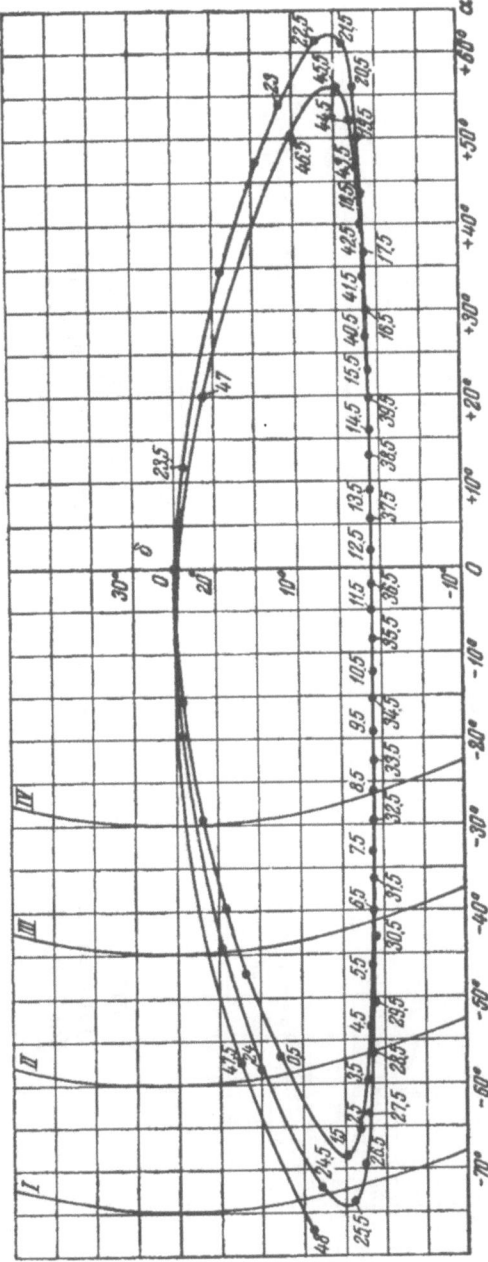

FIG. 4. Prediction graph: $\delta_0 = \delta_1 = 25°$, $\alpha_{00} = \alpha_{10} = 0°$, $h = 500$ km. I) Horizon line for 1 and 25 hr from time of first observation; II) horizon for 2 and 26 hr of tracking; III) horizon for 3 and 27 hr of tracking; IV) horizon for 4 and 28 hr of tracking.

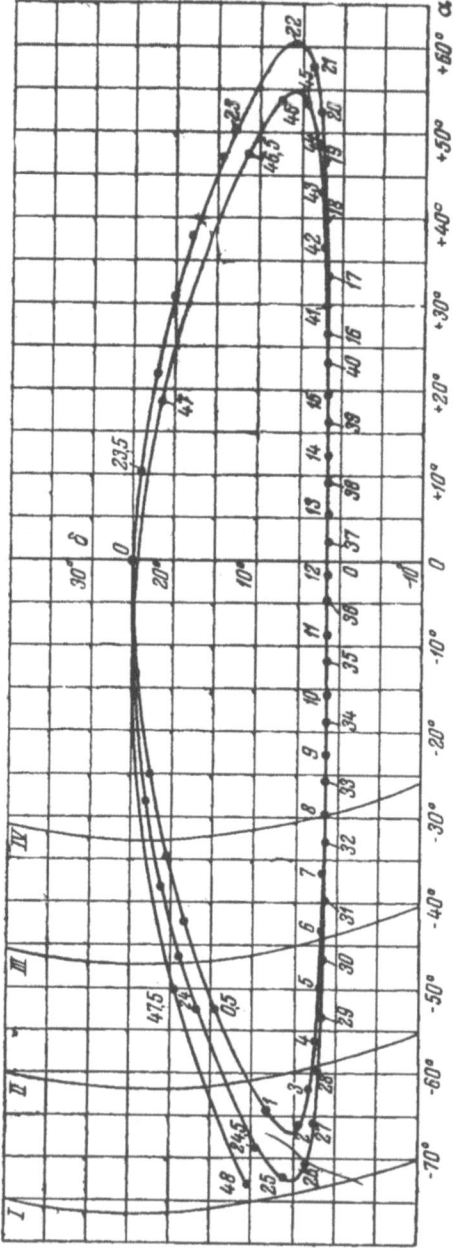

FIG. 5. Prediction graph: $\delta_0 = \delta_1 = 25°$, $\alpha_{00} = \alpha_{10} = 0°$; $h = 600$ km.
I) Horizon line for 1 and 25 hr from time of first observation; II) horizon for 2 and 26 hr; III) horizon for 3 and 27 hr; IV) horizon for 4 and 28 hr.

earth is known, and with r, α_{00}, δ_0, α_1, and δ_1 known, the prediction curve may be plotted without ambiguity by recourse to equations (1)-(5). An example of a prediction graph is shown in Fig. 4, where we have the graph computed for constant r corresponding to a satellite height of 500 km above the earth. The curve was plotted from the following data

$$\alpha_{00} = \alpha_{10} = 0; \quad \delta_0 = \delta_1 = 25°; \quad i = 40°.$$

An analysis of the equations of the prediction curves demonstrates that the motion of the node of the orbit is greatly affected by the slope of the prediction plot. As an example, let us take the longitude drift of the prediction curve seen in Fig. 4, which, over the course of a day, amounts to ~50° at a precessional speed of ~5° for the orbit in a single day.

The effect of the motion of the orbit perigee is relatively small; for our example (Fig. 5), with a 100-km change in the height of the satellite above the earth (h = 600 km), the prediction curve shows almost perfect fit everywhere with the curve computed for h = 500 km. The greatest spread in longitude is roughly 2°, which is of no consequence for reasonably wide-angle equipment.

The successful use of the method proposed above for satellite tracking thus essentially requires that we take due account of one variable factor, the rate of precession of the orbit.

LITERATURE

1. D. E. Okhotsimskii, T. M. Eneev, and G. P. Taratynova. Uspekhi fiz. nauk, 63, No. 1a (1957).

SECULAR VARIATIONS IN ORBIT ELEMENT AS A FUNCTION OF AIR RESISTANCE

P. E. Él'yasberg

Several papers dealing with the effect of air resistance on the secular changes of the elements of artificial earth satellite orbits have already appeared in the literature.[1-3] This present contribution constitutes a further development of the idea, suggested by Yatsunskii,[1] of expanding the secular variations of the orbit elements into a series based on Bessel functions of imaginary argument. This resulted in some straightforward and illustrative formulas useful in solving a number of problems (e.g., determination of the dependence of secular variations of orbit elements on the values of those elements, estimating the accuracy achieved in determining air density from measured secular variations of the orbit elements, etc.)

If we assume the atmosphere to be stationary with respect to the earth (in an inertial system of coordinates), the equations of the osculating elements (Lagrange equations) for the case of motion in a medium presenting drag and in the absence of other perturbations may be stated as

$$\begin{aligned}\frac{dp}{d\vartheta} &= -b\rho p^2 \frac{\sqrt{1+2e\cos\vartheta + e^2}}{(1+e\cos\vartheta)^2}, \\ \frac{de}{d\vartheta} &= -b\rho p \frac{(e+\cos\vartheta)\sqrt{1+2e\cos\vartheta+e^2}}{(1+e\cos\vartheta)^2},\end{aligned} \quad (1)$$

where

$$b = \frac{C_x F_M}{m}; \quad (2)$$

C_x is the drag coefficient of air, F_M is the area of the frontal section presented by the satellite, m is the mass of

SECULAR VARIATIONS AS FUNCTION OF AIR RESISTANCE

the satellite, ρ is the air density, p is the orbit parameter, e is the eccentricity, and ϑ is the true anomaly.

Let us now convert from the true anomaly ϑ to the eccentric anomaly E in (1). Along with the variations in the values of p and e, we here consider the variations in semimajor axis a, period of revolution T, and distance from perigee to the earth's center r. Then, making use of the familiar equations relating the above quantities, we obtain, after some straightforward transformations:

$$\Delta p = -b \int_0^{2\pi} \rho p^2 \frac{\sqrt{1-e^2\cos^2 E}}{1-e^2} \, dE,$$

$$\Delta e = -b \int_0^{2\pi} \rho p \frac{\cos E \sqrt{1-e^2\cos^2 E}}{1-e\cos E} \, dE,$$

$$\Delta a = -b \int_0^{2\pi} \rho a^2 \frac{1+e\cos E}{1-e\cos E} \sqrt{1-e^2\cos^2 E} \, dE, \qquad (3)$$

$$\Delta T = -\frac{3\pi}{\sqrt{\mu}} b \int_0^{2\pi} \rho a^{5/2} \frac{1+e\cos E}{1-e\cos E} \sqrt{1-e^2\cos^2 E} \, dE,$$

$$\Delta r_p = -b \int_0^{2\pi} \rho a^2 (1-e) \frac{1-\cos E}{1-e\cos E} \sqrt{1-e^2\cos^2 E} \, dE,$$

where Δp, Δe, ΔT, Δa, and Δr_p are variations in the corresponding quantities for a single orbit pass, and μ is the coefficient in the expression for acceleration due to gravity $g = \mu/r^2$ (with r being the distance to the earth's center).

We now expand the integrands in the right-hand members of the above equations into series in powers of e. Note that for the case $e < 0.1$ all terms of order higher than third may be discarded in these series. This operation yields the following equations

$$\Delta p = -b \int_0^{2\pi} \frac{\rho p^2}{1-e^2} \left(1 - \frac{1}{2} e^2 \cos^2 E - \cdots \right) dE,$$

$$\Delta e = -b \int_0^{2\pi} \rho p \left(\cos E + e \cos^2 E + \frac{e^2}{2} \cos^3 E + \frac{e^3}{2} \cos^4 E + \cdots \right) dE, \qquad (4)$$

$$\Delta a = -b \int_0^{2\pi} \rho a^2 \left(1 + 2e\cos E + \frac{3}{2} e^2 \cos^2 E + e^3 \cos^3 E + \cdots\right) dE,$$

$$\Delta T = -\frac{3\pi}{V\overline{\mu}} b \int_0^{2\pi} \rho a^{7/2} \left(1 + 2e\cos E + \frac{3}{2} e^2 \cos^2 E + e^3 \cos^3 E + \cdots\right) dE,$$

$$\Delta r_p = -b \int_0^{2\pi} \rho a^2 (1-e) \Big[(1-\cos E) + e(\cos E - \cos^2 E) +$$

$$+ \frac{e^2}{2}(\cos^2 E - \cos^3 E) + \frac{e^3}{2}(\cos^3 E - \cos^4 E) + \cdots\Big] dE.$$

(4)

The air density ρ included in the right-hand members of these equations depends, generally speaking, not only on orbit height h, but also on several other factors. However, we shall neglect these other factors here and assume $\rho = \rho(h)$. Since the greatest perturbing effect on the orbit elements is that of air drag in the region of perigee, the function $\rho(h)$ should be chosen to allow maximum accuracy in computation for the perigee height h_p. With these considerations in mind, we assume that

$$\rho(h) = \rho_p \exp\{-k(h - h_p)\}, \qquad (5)$$

where $\rho_p = \rho(h_p)$ is the air density in the region of perigee and k is a coefficient representing the rate at which ρ decreases with height.

These quite familiar equations may be used to transform (5) to the form

$$\rho(h) = \rho_p \exp(-\nu + \nu \cos E), \qquad (6)$$

where

$$\nu = aek. \qquad (7)$$

We then substitute (6) in the right-hand members of (4). After taking into account the smallness of variations in element orbits during a single satellite pass, we may feel free to neglect the variability of quantities e, p, and a, upon integrating the right-hand members of these equations. The following formulas are derived

$$\Delta p = -\frac{b_{p_p}p^2}{1-e^2}\exp(-\nu)\left(F_0 - \frac{1}{2}e^2F_2 + \cdots\right),$$

$$\Delta e = -b_{p_p}p\exp(-\nu)\left(F_1 + eF_2 + \frac{e^2}{2}F_3 + \frac{e^3}{2}F_4 + \cdots\right),$$

$$\Delta a = -b_{p_p}a^2\exp(-\nu)\left(F_0 + 2eF_1 + \frac{3}{2}e^2F_2 + e^3F_3 + \cdots\right),$$

$$\Delta T = -\frac{3\pi}{V\mu}b_{p_p}a^{1/2}\exp(-\nu)\left(F_0 + 2eF_1 + \frac{3}{2}e^2F_2 + e^3F_3 + \cdots\right),$$

$$\Delta r_p = -b_{p_p}a^2(1-e)\exp(-\nu)\left[(F_0 - E_1) + e(F_1 - F_2) + \frac{e^2}{2}(F_2 - F_3) + \frac{e^3}{2}(F_3 - F_4) + \cdots\right],$$

(8)

where

$$F_n = \int_0^{2\pi}\exp(\nu\cos E)\cos^n E\, dE, \quad n = 0, 1, 2\ldots \tag{9}$$

It is obvious that $F_n = F_n(\nu)$. These terms may be expressed with ease in terms of Bessel functions of imaginary argument $I_n(\nu)$. This requires use of the well-known relations[4]

$$I_n(\nu) = \frac{1}{2\pi}\int_0^{2\pi}\exp(\nu\cos E)\cdot\cos n E\, dE,$$

$$I_{n-1}(\nu) - I_{n+1}(\nu) = \frac{2nI_n(\nu)}{\nu}.$$

(10)

As a result, we obtain

$$F_0(\nu) = 2\pi I_0(\nu),$$
$$F_1(\nu) = 2\pi I_1(\nu),$$
$$F_2(\nu) = 2\pi\left[I_0(\nu) - \frac{I_1(\nu)}{\nu}\right],$$
$$F_3(\nu) = 2\pi\left[I_1(\nu)\left(1 + \frac{2}{\nu^2}\right) - \frac{1}{\nu}I_0(\nu)\right],$$
$$F_4(\nu) = 2\pi\left(1 + \frac{3}{\nu^2}\right)\left[I_0(\nu) - \frac{2}{\nu}I_1(\nu)\right],$$

(11)

The values of the increments in the orbit elements may be

obtained from (7), (8), and (11), with the aid of tables of Bessel functions. Of primary interest to us here is however not the working out of methods of approximate computations (since we can always run the problem on a computer), but the derivation of straightforward equations defining the increments in the orbit elements as functions of the values of those elements and as functions of the basic parameters of the atmosphere. It is fortunate that the value of ν is comparatively large over the main portions of the orbits considered, thus allowing us to utilize existing asymptotic expansions[4] in deriving the desired relations

$$\left.\begin{aligned} I_0 &= \frac{\exp(\nu)}{\sqrt{2\pi\nu}} \left(1 + \frac{1}{8} \cdot \frac{1}{\nu} + \frac{9}{128} \cdot \frac{1}{\nu^2} + \frac{75}{1024} \cdot \frac{1}{\nu^3} + \cdots \right) \\ I_1 &= \frac{\exp(\nu)}{\sqrt{2\pi\nu}} \left(1 - \frac{3}{8} \cdot \frac{1}{\nu} - \frac{15}{128} \cdot \frac{1}{\nu^2} - \frac{105}{1024} \cdot \frac{1}{\nu^3} + \cdots \right) \end{aligned}\right\} \quad (12)$$

As we know, these expansions are fruitful at $\nu \gg 1$. In practice, the expansions may be used at

$$\nu > 4 \qquad (13)$$

(working with the number of terms indicated above) to successfully determine functions $I_0(\nu)$ and $I_1(\nu)$ to an accuracy of 0.1%, which is entirely satisfactory for our purposes.

Assuming $k = 0.02$ km^{-1} at a height $h = 220$-230 km, and that $a \approx 7000$ km, the condition (13) may be replaced by the inequality

$$e > 0.028. \qquad (14)$$

As we know, much larger eccentricities (of the order of 0.05-0.15) have been observed over the main portions of the orbits of existing earth satellites.

It follows directly from (11) and (12) that

$$F_n = \sqrt{\frac{2\pi}{\nu}} \exp(\nu) f_n, \qquad (15)$$

where

$$\left.\begin{aligned} f_0 &= 1 + \frac{1}{8} \cdot \frac{1}{\nu} + \frac{9}{128} \cdot \frac{1}{\nu^2} + \frac{75}{1024} \cdot \frac{1}{\nu^3} + \cdots, \\ f_1 &= 1 - \frac{3}{8} \cdot \frac{1}{\nu} - \frac{15}{128} \cdot \frac{1}{\nu^2} - \frac{105}{1024} \cdot \frac{1}{\nu^3} + \cdots, \end{aligned}\right\} \quad (16)$$

$$f_2 = 1 - \frac{7}{8} \cdot \frac{1}{v} + \frac{57}{128} \cdot \frac{1}{v^2} + \frac{195}{1024} \cdot \frac{1}{v^3} + \cdots,$$
$$f_3 = 1 - \frac{11}{8} \cdot \frac{1}{v} + \frac{225}{128} \cdot \frac{1}{v^2} - \frac{945}{1024} \cdot \frac{1}{v^3} + \cdots, \quad (16)$$
$$f_4 = 1 - \frac{15}{8} \cdot \frac{1}{v} + \frac{489}{128} \cdot \frac{1}{v^2} - \frac{5445}{1024} \cdot \frac{1}{v^3} + \cdots$$

After substitution of (15) into (8), we arrive at the following set of equations

$$\Delta p = -\frac{b_{pp} p^2}{1-e^2} \sqrt{\frac{2\pi}{v}} \left(f_0 - \frac{1}{2} e^2 f_2 + \cdots \right),$$
$$\Delta e = - b_{pp} p \sqrt{\frac{2\pi}{v}} \left(f_1 + e f_2 + \frac{e^2}{2} f_3 + \frac{e^3}{2} f_4 + \cdots \right),$$
$$\Delta a = - b_{pp} a^2 \sqrt{\frac{2\pi}{v}} \left(f_0 + 2 e f_1 + \frac{3}{2} e^2 f_2 + e^3 f_3 + \cdots \right),$$
$$\Delta T = - \frac{3\pi}{\sqrt{\mu}} b_{pp} a^{1/2} \sqrt{\frac{2\pi}{v}} \left(f_0 + 2 e f_1 + \frac{3}{2} e^2 f_2 + e^3 f_3 + \cdots \right), \quad (17)$$
$$\Delta r_p = - b_{pp} a^2 (1-e) \sqrt{\frac{2\pi}{v}} \Big[(f_0 - f_1) + e(f_1 - f_2) + \frac{e^2}{2}(f_2 - f_3) + \frac{e^3}{2}(f_3 - f_4) + \cdots \Big].$$

It is well to carry out a further simplification of the equations derived. As an example, consider the first equation of (17). We may use (16) to state that

$$f_0 - \frac{1}{2} e^2 f_2 + \cdots = f_0 \left(1 - \frac{1}{2} e^2 + \cdots \right) + \frac{1}{2} e^2 (f_0 - f_2) + \cdots =$$
$$= f_0 \left(1 - \frac{1}{2} e^2 + \cdots \right) + \frac{1}{2} e^2 \left(\frac{1}{v} - \frac{3}{8} \cdot \frac{1}{v^2} - \frac{15}{128} \cdot \frac{1}{v^3} + \cdots \right) + \cdots \quad (18)$$

Now note that the series enclosed within parentheses following the term f_0 corresponds to the series used in the first formula of (4) in the case where $\cos E = 1$. Hence, comparing this with the first formula in (3), we obtain

$$1 - \frac{1}{2} e^2 + \cdots = \sqrt{1 - e^2}.$$

We can demonstrate further that, when condition (13) is fulfilled, the remaining terms in the series in (18) yield a correction considerably smaller than 1% of the amount determined. We may state, consequently, that

$$f_0 - \frac{1}{2}e^2 f_2 + \cdots \approx f_0 \sqrt{1-e^2}.$$

Similar simplifications are possible in the right-hand members of the remaining formulas in (17). As a result, we derive the following expressions

$$\left.\begin{aligned}
\Delta p &\approx -\frac{b_{\rho_p} p^2}{\sqrt{1-e^2}} f_0 \sqrt{\frac{2\pi}{\nu}}, \\
\Delta e &\approx -b_{\rho_p} p \sqrt{\frac{1+e}{1-e}} f_1 \sqrt{\frac{2\pi}{\nu}}, \\
\Delta a &\approx -b_{\rho_p} a^2 \frac{1+e}{1-e} \sqrt{1-e^2} f_0 \sqrt{\frac{2\pi}{\nu}}, \\
\Delta T &\approx -\frac{3\pi}{\sqrt{\mu}} b_{\rho_p} a^{3/2} \frac{1+e}{1-e} \sqrt{1-e^2} f_0 \sqrt{\frac{2\pi}{\nu}}, \\
\Delta r_p &\approx -b_{\rho_p} a^2 \sqrt{1-e^2} \frac{1}{2\nu} f' \sqrt{\frac{2\pi}{\nu}};
\end{aligned}\right\} \quad (19)$$

In these formulas

$$f' = 2\nu(f_0 - f_1) = 1 + \frac{3}{8} \cdot \frac{1}{\nu} + \frac{45}{128} \cdot \frac{1}{\nu^2} + \cdots. \quad (20)$$

Further simplification is achieved by assuming, in the above formulas, that

$$f_0 \approx f_1 \approx f' \approx 1. \quad (21)$$

Note that (17) and (19) are suitable only at moderately low eccentricities, where (13) is fulfilled. When (13) is not obeyed, we must take recourse to (8). In particular, at e = 0 (circular orbit), it follows directly from (9) that

$$F_i(0) = \begin{cases} 2\pi & \text{when } i = 0 \\ 0 & \text{when } i > 0 \end{cases},$$

so that

$$\left.\begin{aligned}
\Delta p &= \Delta a = \Delta r_p = -2\pi b_{\rho_p} a^2, \\
\Delta e &= 0, \\
\Delta T &= -\frac{6\pi^2}{\sqrt{\mu}} b_{\rho_p} a^{3/2}.
\end{aligned}\right\} \quad (22)$$

It is obvious that (22) could not be derived from (19) by simply passing to the limit at $e \to 0$.

In conclusion, let us note that (19), in contrast to (17), remains valid even at large eccentricities (it being sufficient to observe the restraint $e < 1$).

In deriving the foregoing equations, we chose to neglect the effect of the earth's departure from spherical shape and the fact that the earth's gravitational field is noncentral. But this effect can be taken into account partially by finding ρ_p in (8), (17), and (19) as the air density at a point on a real, nonideal orbit situated at a minimum height h_{min} above the earth's surface (rather than at perigee in an osculating orbit).

To test the accuracy of the approximate equations (17) and (19) which we derived, we carried out a comparison with the results of numerical integration of the exact equations of motion. The calculations were performed for the orbits of the first and second Soviet earth satellites. As a result, we found that at $e > 0.04$, the error in computing increments Δa, Δe, Δp, and ΔT from the approximate formulas was not in excess of 1-2% of the amount determined. These errors built up as eccentricity declined, reaching 6% at $e = 0.023$, and 15-20% at $e = 0.012$. The error in computing the increment Δr_p (which in and of itself is comparatively small) was 10-40%, somewhat larger.

The proposed approximate formulas are thus entirely suitable for qualitative analysis of the dependencies considered in this paper. In particular, the formulas clearly indicate that the secular variations Δa, Δe, Δp, and ΔT are proportional to the quantity $b\rho_p \nu^{-1/2}$, while the variation Δr_p is proportional to the quantity $b\rho_p \nu^{-3/2}$. Hence, we find from (7) that upon determining air density from measurements of the secular variations ΔT and Δa, we actually find not ρ_p, but the ratio $\rho_p/\sqrt{k}$. To determine the magnitude of coefficient k, we must determine the lowering of perigee Δr_p and the variation in period of revolution ΔT to fairly high accuracy. In fact, (19) will be found useful in formulating

$$\frac{\Delta r_p}{\Delta T} \approx \frac{1}{\nu} \cdot \frac{\sqrt{\mu}}{6\pi \sqrt{a}} \cdot \frac{1-e}{1+e} \cdot \frac{f'}{f_0} .\qquad (23)$$

The ratio $\Delta r_p/\Delta T$ is thus found to be proportional to $1/k$ and independent of the values of b and ρ_p. This in turn shows that the results from measuring this ratio serve in computing the coefficient k.

LITERATURE CITED

1. I. M. Yatsunskii, Uspekhi fiz. nauk, 63, 1a (1957).
2. D. E. Okhotsimskii, T. M. Eneev, and G. P. Taratynova, Uspekhi fiz. nauk, 63, 1a (1957).
3. G. P. Taratynova, Uspekhi fiz. nauk, 63, 1a (1957).
4. G. N. Watson, A Treatise on the Theory of Bessel Functions, Part I, Cambridge University Press, New York, 1944.

THE PROBLEM OF PIERCING AT COSMIC VELOCITIES

M. A. Lavrent'ev

It is known from the theory of cumulative charges that the mechanism of puncturing of metallic plates by cylinders or by balls differs greatly at 3-10 km/sec from the mechanism at velocities up to 1000 m/sec. Two stages take place at high velocities: a) the ball or cylinder, penetrating into the obstacle, flows over the surface of the punched-out crater; b) inertia expansion of the cavity takes place after the "bullet has been annihilated." The first part can be calculated with sufficient accuracy by using an ideal incompressible liquid as a model, but the calculation of the second part is more difficult, although the results available indicate that the principal difficulties can be overcome. The computational schemes yield good agreement with experiment.

Much less investigated is the puncturing problem at 50-100 km/sec; the experimental difficulties at such velocities require particular caution as regards the main hypothesis made in the theory of each phenomenon.

As far as I know, the literature contains only one paper devoted to this problem—that of K. P. Stanyukovich on the formation of the moon's craters and on the determination of the momentum imparted by a falling meteorite. Stanyukovich advances, on the basis of his computation, the hypothesis that as a ball drops its kinetic energy is converted into potential energy of the gas into which it is converted by the impact; assuming further a few less clearly defined hypotheses, he arrives at the following conclusions:

1) the momentum acquired by the struck body is proportional to the energy of the incident body;

2) the volume of the crater is also proportional to the energy.

In the present paper I propose an incompressible-medium model for which the calculations can be carried out to conclusion. My deductions do not agree with those of Stanyukovich.

One-Dimensional Case

We start with a consideration of the one-dimensional case, where the model is particularly simple. We consider the impact of a plate of thickness a, flying at a velocity v_0, against the end of a cylinder (rod) of length l. The thickness a will be assumed much smaller than l. The problem consists of determining the momentum which the rod acquires as a result of the impact.

We assume the striking plate to be incompressible and absolutely hard. The rod is considered as a limiting case of a set of absolutely hard infinitesimally thin platelets, placed infinitely close to each other. During an inelastic impact between the striking plate and the first platelet of the rod, the momentum is conserved and kinetic energy will be lost as a result of the increase of the mass. This will occur as each succeeding platelet goes into motion. Let us now calculate, for the limiting case, the distribution of the kinetic-energy loss in the system along the rod.

Thus, let a rod of length x be set into motion and let v be the velocity of this piece; as a plate of thickness dx is set into motion, the change in velocity dv should satisfy the relation

$$x dv + v dx = 0;$$

from which we obtain after integration, noting that $v = v_0$ when $x = a$,

$$v = v_0 \frac{a}{x}.$$

Let us calculate now the unknown kinetic-energy loss

$$E = \frac{1}{2} x v^2.$$

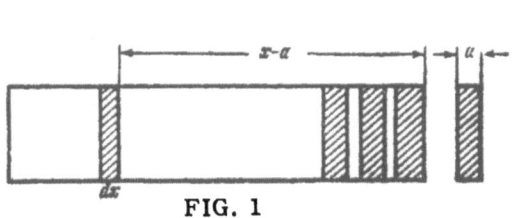
FIG. 1

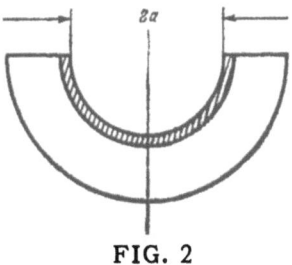
FIG. 2

We have
$$dE = \tfrac{1}{2}v^2 dx + xv\,dv = -\tfrac{1}{2}v^2 dx = -\tfrac{1}{2}v_0^2\left(\frac{a}{x}\right)^2 dx = -E_0 \frac{a}{x^2} dx,$$
where
$$E_0 = \tfrac{1}{2}av_0^2$$
is the initial energy of the striking plate. Assuming that the entire energy lost is converted into heat, we obtain the following distribution of heat along the rod
$$T = -\frac{dE}{dx} = E_0 \frac{a}{x^2}.$$
Let us now denote by T_0 the minimum heat density necessary to vaporize the rod material. From this we find that the impact vaporizes the portion of the rod 0, x, x_0, where the unknown x_0 is found from the relation
$$T_0 = E_0 \frac{a}{x_0^2}.$$
As the vapor expands it separates from the rod, while the remaining portion of the rod acquires a momentum I. Let us calculate I for the case of greatest interest, when
$$a \ll x_0 \ll l$$
and the energy going into vaporization is small compared with the initial energy of the striking plate. We can assume here that the entire energy of the gas goes into kinetic energy. We obtain for the velocity of the gas cloud
$$x_0 V^2 = av_0^2, \quad V = \sqrt{\frac{a}{x_0}}\, v_0.$$
Our problem has reduced to a purely gas-dynamic problem. We confine ourselves now to only very rough calculations in two cases.

a) We make an additional assumption that all the scattered particles of the gas cloud acquire an identical velocity V, and then

$$x_0 V^2 = a v_0^2, \quad V = \sqrt{\frac{a}{x_0}} v_0,$$

obtaining finally for the momentum

$$I = x_0 V = \sqrt{a x_0}\, v_0 = a v_0^{3/2}.$$

b) The second extreme case we consider is one in which each layer dx of the gas cylinder scatters (along the x direction) independently of the other elements. Then the gas element dx, located at a distance x from the end, yields a momentum

$$V dx = \sqrt{2T}\, dx = \sqrt{2E_0 a\, \frac{dx}{x}}.$$

From this we obtain for the unknown momentum

$$I = \int_a^{x_0} V dx = \sqrt{2E_0 a}\, \lg \frac{x_0}{a} = a v_0 \lg \frac{v_0}{\sqrt{2T_0}}.$$

Three-Dimensional Case

Let us now perform a calculation analogous to the preceding one for the case when a ball strikes a hemispherical plate of radius R. For this computation we must make further assumptions. We assume that the action of the ball on the plate reduces to an impact of a spherical layer of the ball against a spherical cavity in the plate. All the velocities of the points of the striking layer are directed along the radii, and the velocity distribution coincides with the corresponding distribution in an incompressible liquid: the velocity v at a point located at a distance r from the center of the layer is determined from the formula

$$v = v_0 \left(\frac{a}{r}\right)^2,$$

where a is the radius of the internal cavity of the striker; v_0 the velocity of the points of this cavity. Let ka be the radius of the external surface of the striker. The medium is represented by a model similar to that considered in the linear formulation, namely that the medium is a set of infinitesimally thin spherical layers of an ideal incompressible liquid, located infinitesimally close to each other. As a result of the impact of the striker, its mass will start increasing as called for by the inelastic-impact scheme; as the mass increases, the momentum is conserved, and the kinetic energy of the system is converted into heat.

Let us perform the calculation. We determine first the radial momentum during the instant when the zone of motion in the medium is a hemisphere of radius r; let in this case the radial velocity of the internal surface of the striker be w = w(r), then

$$\omega(ka) = v_0. \tag{1}$$

Since the medium is incompressible, the velocity v at a point located at a distance x from the center will be

$$v = \omega \left(\frac{a}{x}\right)^2.$$

From this we obtain for the momentum

$$I = \int_a^r 4\pi x^2 \omega \left(\frac{a}{x}\right)^2 dx = 4\pi \omega a^2 (r-a).$$

The condition that the momentum be constant yields

$$(r-a)\,d\omega + \omega\,dr = 0,$$

$$\omega = \frac{C}{r-a},$$

where the constant C is determined from (1). Thus, finally

$$\omega = \frac{(k-1)a}{r-a} v_0, \quad d\omega = -\frac{(k-1)a}{(r-a)^2} v_0\,dr. \tag{2}$$

We now find the kinetic energy E under the same conditions

$$E = \frac{1}{2} \int_a^r 4\pi x^2 \omega^2 \left(\frac{a}{x}\right)^4 dx = 2\pi a^3 \omega^2 \left(1 - \frac{a}{r}\right). \tag{3}$$

From this we can obtain the distribution of heat T over the body. Then

$$T = -\frac{1}{4\pi r^2}\frac{dE}{dr} = -4\pi a^3\omega\left(1-\frac{a}{r}\right)\frac{d\omega}{dr} - 2\pi a^4\frac{\omega^2}{r^2}$$

or, taking (2) into account

$$T = 4\pi a^3\omega\frac{r-a}{r}\cdot\frac{\omega}{r-a} - 2\pi a^4\frac{\omega^2}{r^2} = 4\pi a^5 v_0{}^2\frac{(k-1)^2}{r(r-a)^2}\left(1-\frac{1}{2}\cdot\frac{a}{r}\right).$$

Denoting by E_0 the initial energy of the striker, we obtain

$$T = \frac{2(k-1)k}{r(r-a)^2}\left(1-\frac{1}{2}\cdot\frac{a}{r}\right)a^2 E_0 \approx \frac{2k(k-1)}{r^3}a^2 E_0.$$

If we now assume that an amount of heat T_0 per unit volume is needed to vaporize the material, then as the ball strikes the body, there is formed in it a gas cavity of radius

$$\rho^3 = \frac{2k(k-1)}{T_0}a^2 E_0. \tag{4}$$

This cavity, defined by expression (4), will characterize the lower limit of the size.

Let us now calculate the momentum acquired by the body. Again we consider the case when

$$a \ll \rho \ll R,$$

and assume that a small part of the striker energy is consumed in vaporization. Denoting by V the velocity of the gas cloud, we obtain

$$\frac{2}{3}\pi\rho^3 V^2 = E_0, \quad V = \sqrt{\frac{3E_0}{2\pi\rho^3}}.$$

From this we obtain for the sought momentum I

$$I = \frac{2}{3}\pi\rho^3 V = AaE_0, \quad A = \sqrt{\frac{4\pi}{3}\cdot\frac{k(k-1)}{T_0}}.$$

It is interesting to note that in this scheme, the momentum depends greatly on the dimensions of the striking ball.

In analogy with two-dimensional case, we can carry out the calculations in accordance with the second extreme case. Let us assume that each element dr of the spherical layer of gas scatters independently of the other elements and that the momentum imparted is directed along the normal to the spherical cavity.

The normal velocity V of the gas cloud due to an element

dr of the spherical layer is determined from the energy balance.
$$T = 2\pi r^2 V^2.$$

If we now denote by α the angle between the symmetry axis and a line joining the center of the sphere with the element of the gas, then the component $J = dI/dr$ of the momentum along the symmetry axis due to the scattering of a ring of width $rd\alpha$, $\alpha = $ Const, will be

$$dJ = 2\pi r^2 \sin\alpha \cos\alpha V d\alpha,$$

hence

$$J = \pi r^2 V = \pi r^2 \sqrt{\frac{T}{2\pi} \cdot \frac{1}{r}} = \sqrt{\frac{\pi T}{2}} r \approx \pi \sqrt{2}(k-1)\frac{a^{5/2} v_0}{r^{1/2}}.$$

Integrating the latter expression with respect to r, along the thickness of the gaseous portion of the body, we obtain

$$I = B(k-1)^{4/5} T_0^{-1/5} a^{20/5} v_0^{4/5},$$

where B is a numerical constant.

ERRATUM

An error was inadvertently introduced into the formula for the energy loss in the spherical case. It should be

$$T = -\frac{1}{4\pi r^2}\frac{dE}{dr}, \text{ and not } T = -\frac{dE}{dr},{}^*$$

as in the one-dimensional case. Thus, the radius of the spherical cavity is given by

$$\rho^5 = \frac{2k(k-1)}{4\pi T_0} a^2 E_0.$$

Therefore the momentum according to the first scheme is correctly

$$I = \frac{4\pi}{\sqrt{3}} a^3 v_0 \frac{(k-1)^{11/10}}{\sqrt{k}} \left(\frac{v_0^2}{4\pi T_0}\right)^{3/10},$$

and according to the second scheme

$$I = B(k-1)^{4/5} a^3 v_0 \left(\frac{v_0^2}{4\pi T_0}\right)^{1/10}$$

*This correction has been incorporated into the first formula on p. 90 (Publisher's note).

THE DETERMINATION OF THE DENSITY OF THE ATMOSPHERE AT AN ALTITUDE OF 430 KILOMETERS BY THE SODIUM VAPOR DIFFUSION METHOD

I. S. Shklovskii and V. G. Kurt

In recent times our ideas about the basic physical characteristics, such as the density and temperature, of the upper layers of the atmosphere have undergone significant changes. Observations of the drag on Soviet and American satellites have made it possible to determine the density of the atmosphere at heights from 220 to 750 km.[1-4] It was found that the density at the above-mentioned heights is many times larger than had been concluded earlier on the basis of rocket data. Together with this, the calibration of heights appeared to be significantly greater than had been assumed earlier, which indicates a higher temperature for the upper atmosphere.

In view of the extreme importance of these results it would be very desirable to obtain an independent confirmation of them, since data on atmospheric density obtained from the analysis of the drag of satellites may involve systematic errors. For example, the problem of the so-called "electrical drag"[3] is not yet entirely clear. It is not easy to take into account the influence of the shape of the satellite with sufficient accuracy, and this may also have an effect on the errors. Finally, and perhaps most importantly, from observations of satellite drag one obtains only average values of the density. Meanwhile, at the present time there are important indications that the density and temperature of the upper atmosphere are subject to local variations.[5] As a result of this, one must expect systematic differences in the basic characteristics of the upper atmosphere in the polar and the equatorial regions.

Thus it becomes necessary to develop a method which, at a given point and over a rather short period of time, would make it possible to obtain from a single experiment the basic characteristics of the upper atmosphere.

As the most reasonable of such methods we consider the analysis of the diffusion of sodium vapors released at a given height in the upper atmosphere by a rocket. The idea of investigating the upper atmosphere by means of introducing a small quantity of sodium vapors into it is due to Bates[6] and to several articles along this line[7-9] concerning work already carried out in the U.S.A. These articles are devoted to the study of elementary processes responsible for emission of light from sodium in the upper atmosphere. Sodium vapor is released at relatively low heights of 75-140 km as the rockets are in motion, so that long, narrow, luminous bands are observed in the cloud. The problem of determining the basic characteristics of the atmosphere, the density and temperature, had not yet been solved at the time of these experiments. As will be seen below, the problem was incapable of solution, since the heights of the flights were entirely inadequate for this purpose.

The first experiment for the determination of the density of the upper atmosphere by this method was made September 19, 1958, with a high-altitude rocket.

Descriptions of a few other experiments made with similar rockets were published at the time.[10] In our experiment the height reached by the rocket was 430 km. Two sodium evaporators, each containing 2 kg of metallic sodium and a corresponding quantity of thermite were housed in the nose cone. The thermite was ignited at a predetermined moment when the rocket was near the top of its trajectory. The sodium vapor was ejected into the atmosphere through a nozzle in a direction perpendicular to the axis of the rocket, which was stabilized. The evaporation process lasted 10-20 sec. For all of this time the rocket was not far from the top of its trajectory.

At the time of the experiment described here, which was carried out before sunrise, the height of the earth's shadow at the observation point was within the limits 300-250 km.

Therefore, the cloud of sodium vapor formed as a result of the evaporation process was illuminated by the sun's rays, and, thanks to resonant fluorescence, there was intensive scattering in the D_1 and D_2 lines. A succession of photographs of this cloud was made with a camera of f = 58 mm, d/f = 1:2, provided with an OS-14 light filter. All the photographs were calibrated. In addition, to get absolute values of the intensity of emission of the cloud, a standardization was carried out by relating an out-of-focus exposure of the stars to it. The development of the photographs obtained was made according to standard astrophysical practice.

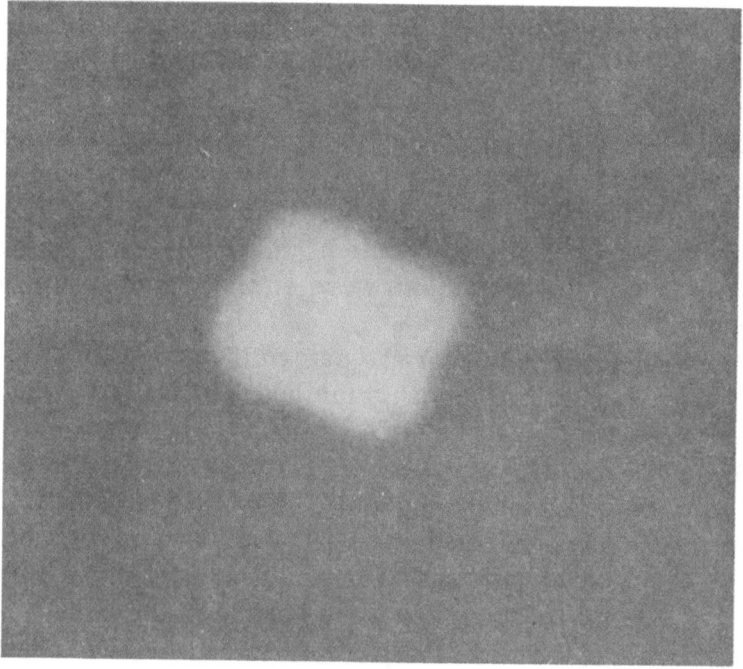

(a) 5 seconds.

FIG. 1. Stages in the development of the sodium cloud.

In Fig. 1 are shown successive photographs of different stages of growth of the cloud formed. The scales of the frames are always the same (36 × 24 deg). The central time of the exposure is given under each photograph.

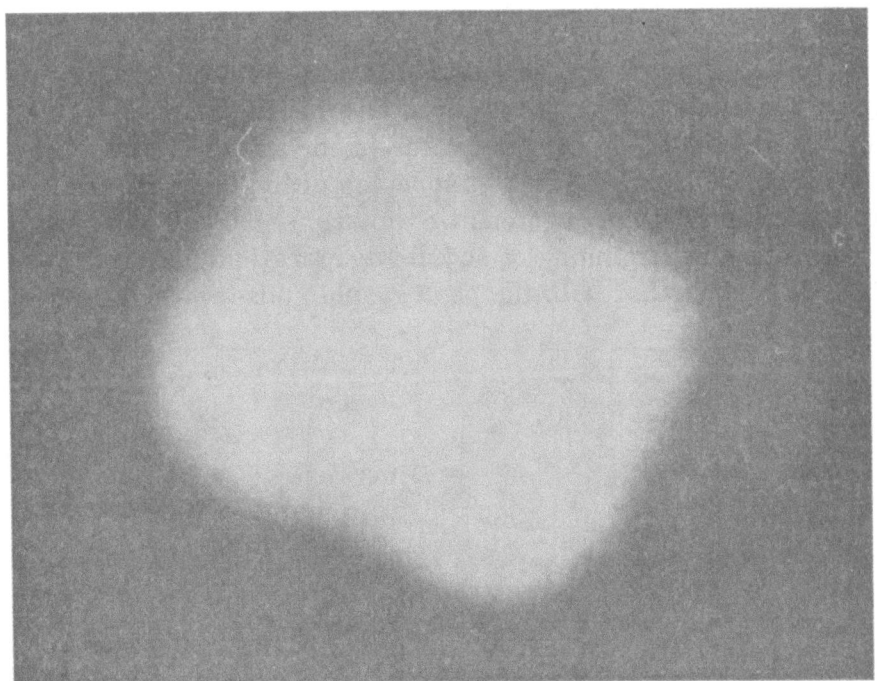

(b) 20 seconds.

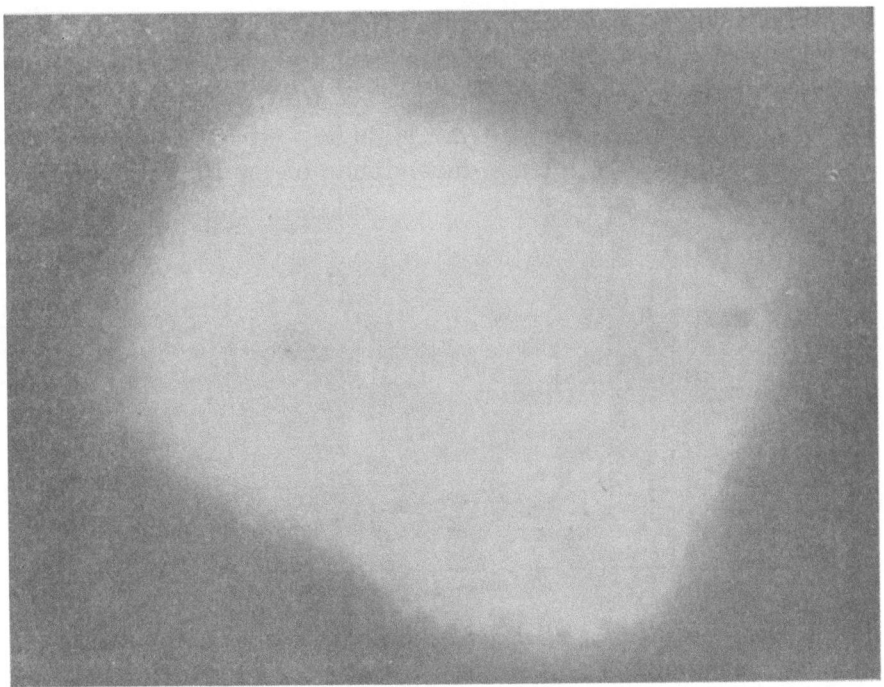

(c) 35 seconds.

FIG. 1. Concluded.

It may be noted that in the first moments after evaporation the shape of the cloud was rather irregular, but even at approximately 100 seconds the cloud assumed a circular outline and this tendency continuously increased with time. This is especially clearly seen from isophotes of the cloud, which give the distribution of intensities for times 90 and 270 seconds (after the beginning of sodium evaporation) as is shown in Fig. 2. Together with the photographic observations, photo-

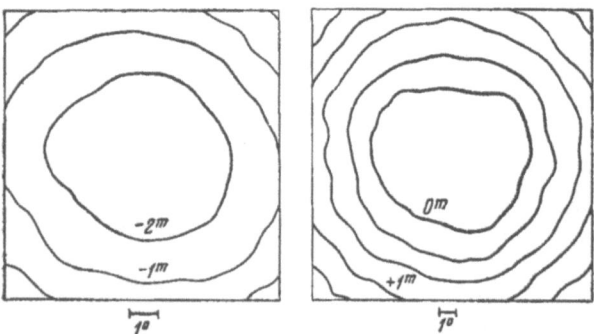

FIG. 2. Isophotes of the sodium cloud for 90 and 270 seconds.

electric observations were made of the brightness of the center of the cloud. For this purpose a photoelectric photometer was used having an objective of f = 100 mm, d/f = 1:2, and an interference light filter with $\Delta \lambda = 30$ A.

Figure 3 shows the time dependence of the brightness of

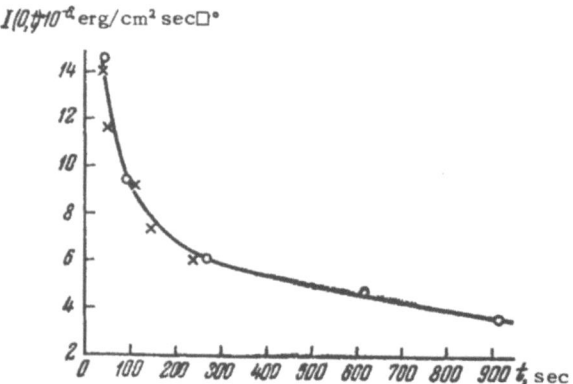

FIG. 3. Dependence of the brightness of the center of the sodium cloud on time. ○ – photographic observations; × – photoelectric observations.

the center of the cloud obtained from photographic (circles) and photoelectric (crosses) observations. By integrating the isophotes, the time dependence of the total light flux was obtained (expressed in visual stellar magnitudes), and it is given in Fig. 4. As can be seen from this figure, the total light flux increases rather quickly initially, and then starting approximately at 100 seconds for the next 15 minutes (during

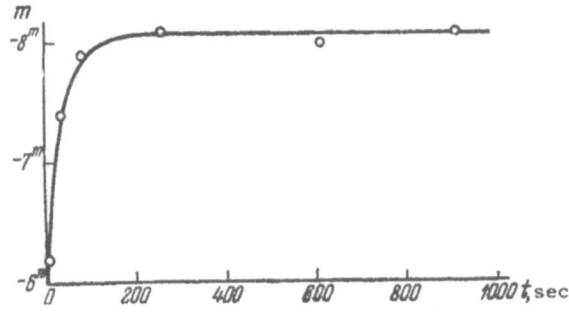

FIG. 4. Dependence of the integral visual stellar magnitude on time.

which observations were made) remains practically constant. The relative smallness of the total flux in the initial stages of dispersion of the cloud are explained by significant self-quenching. When the optical thickness of the dispersed cloud becomes significantly less than unity (which happens at approximately 100 seconds) the total flux becomes constant. At this point the visual stellar magnitude of the cloud was about $8^m.1$ in agreement with theoretical expectations.

It is important to emphasize that there are a large number of independent observations of the sodium cloud that forms made from different points sufficiently far away from the rocket launch point. All observers emphasize the fact that the cloud spread out in all directions in the same way and had, at least during the first 10 minutes, a circular form. During this the center did not change its position with respect to the horizon of the place of observation. These independent observations show that the spatial structure of the cloud was close to spherical and that the center of this sphere remained approximately at a constant height.

The sodium cloud that was formed spread out initially rather quickly and then quite slowly. In Fig. 5 we have the dependence of the square of the effective radius of the cloud S (determined from the isophote having a surface brightness e times less than the central value) on time. As can be seen from this figure, all the points fit a straight line very well. Consequently, the relation

$$S \sim t^{1/2}, \tag{1}$$

holds, which, as is well known, is a characteristic property of diffusion processes. From this empirically obtained dependence it would not be difficult to evaluate the concentration of atmospheric atoms in the region where the sodium diffusion took place.

As is known, the mean displacement of a particle resulting from diffusion in some definite direction from the center (where the particle is located at the initial moment) after a time t will be

$$s = \left(\frac{n}{3}\right)^{1/2} \bar{l}, \tag{2}$$

where $\bar{l}$ is the mean free path length, n is the number of collisions in a time t. Since $n = (vt/\bar{l})$ (where v is the mean

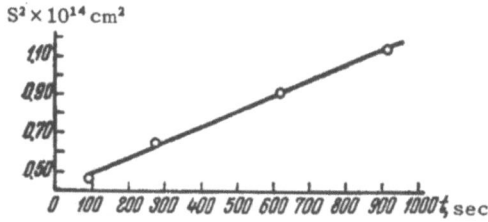

FIG. 5. Dependence of the square of the effective radius of the cloud on time.

velocity of the particle)

$$s = \sqrt{\frac{1}{3} v \bar{l} t}. \tag{3}$$

The quantity $(1/3) v \bar{l}$ (which is nearly the coefficient of diffusion D) can be found directly from the graph made from the observations given in Fig. 5. It is equal to 0.85×10^{11} cm^2 × sec^{-1}. The mean velocity of the sodium atoms is

$$v = \sqrt{\frac{8kT}{\pi m_{Na}}} = 1.5 \cdot 10^5 \text{ cm/sec (for } T = 1600°\text{).}$$

Hence $\bar{l} = 1.7 \times 10^6$ cm = $1/n_1 \bar{Q}_d$, where n_1 is the concentration of atoms in the atmosphere, and $\bar{Q}_d$ is the effective collision cross section, which we take equal to 3.85×10^{-15} g/cm^3 (see below). Hence

$$n_1 = 1.6 \cdot 10^8 \text{ cm}^{-3}. \qquad (4)$$

If the atoms in the atmosphere at a height of 430 km are essentially nitrogen and oxygen, then the density

$$\delta = 4.7 \cdot 10^{-15} \text{ g/cm}^3. \qquad (5)$$

This value agrees well with the value for the density of the atmosphere obtained from an analysis of the drag on satellites.

We shall now solve the problem of determining the density of the atmosphere from observations of the diffusion of a sodium cloud more rigorously. In doing this in the first approximation we shall not consider the influence of the force of gravity on the process of diffusion. According to the calculations of Mitra and Rakshit,[11] at a height of 350 km (which was taken by them in accordance with the old model for the concentration of atoms in the atmosphere and which was close to the value expected for our experiment) the time of effective separation of N_2 and O is about 3 hours. We note further that in our case (diffusion of sodium through atomic nitrogen and oxygen) the time of diffusional separation, on account of the relative closeness of the atomic weights of the diffusing components, should be even greater. Since the time of observations (about 15 min) was less than the time of effective diffusional separation, the effect of the force of gravity may be safely neglected.

We write the differential equation of diffusion for the spherically symmetric problem

$$\frac{\partial N}{\partial t} = D \left(\frac{\partial^2 N}{\partial R^2} + \frac{2}{R} \frac{\partial N}{\partial R} \right), \qquad (6)$$

where N is the concentration of sodium atoms and D is the diffusion coefficient.

The solution of this equation is (see, for example, Ref. 12)

$$N = \frac{1}{2\sqrt{2\pi Dt}} \frac{1}{R} \int_0^\infty \Phi(\rho) \{e^{-\left(\frac{R-\rho}{2\sqrt{Dt}}\right)^2} - e^{-\left(\frac{R+\rho}{2\sqrt{Dt}}\right)^2}\} \rho d\rho, \qquad (7)$$

where $\Phi(\rho)$ represents the distribution of diffusing sodium atoms at the time $t = 0$. We assume that at the initial instant $\Phi = N_0$ within a sphere of radius R_0 and is equal to zero outside this sphere. Then the solution of (6) is

$$N(R, t) = \frac{N_0}{2} \{\psi\left(\frac{R_0 - R}{2\sqrt{Dt}}\right) + \psi\left(\frac{R + R_0}{2\sqrt{Dt}}\right) + \frac{\sqrt{Dt}}{R\sqrt{\pi}}[e^{-\left(\frac{R_0+R}{2\sqrt{Dt}}\right)^2} - e^{-\left(\frac{R_0-R}{2\sqrt{Dt}}\right)^2}]\}, \qquad (8)$$

where

$$\psi(z) = \frac{2}{\sqrt{\pi}} \int_0^z e^{-\xi^2} d\xi.$$

Supposing that $R \to 0$, we find the following law for the change of concentration at the center of the diffusing cloud

$$\frac{N(0, t)}{N_0} = \psi\left(\frac{R_0}{2\sqrt{Dt}}\right) - \frac{R_0}{\sqrt{4\pi Dt}} e^{-\frac{R_0^2}{4Dt}}. \qquad (9)$$

From (9) one may determine the diffusion coefficient (and consequently also the concentration of the atoms in the atmosphere), if it can be determined from the observations how $N(0, t)/N_0$ varies with time.

Therefore, for the solution of the problem, we must first of all determine, from observations at different times, values of $N(0, t)/N_0$ in relative units. The magnitude of $N(R, t)$ can be obtained for a given instant of time from the distribution of intensities in the sodium cloud, which is given by the isophotes through the use of procedures usually applied to similar problems in astrophysics. $N(R, t)$ is proportional to the quantity of energy $F(R, t)$ scattered by a unit volume of the sodium cloud. The surface brightness $I(\rho, t)$ is connected with the quantity $F(R, t)$ through the Abel integral equation

$$I(\rho, t) = \int_{-\infty}^{\infty} F(R, t) dy, \qquad (10)$$

where the integration is carried out over the field of illumination. The solution of (10), as is well known, is

$$F(R, t) = -\frac{1}{\pi} \int_r^\infty \frac{\frac{dI(\rho, t)}{d\rho}}{\sqrt{\rho^2 - R^2}} d\rho. \tag{11}$$

The integration of (11) can be carried out in closed form if the observed intensity is approximated, after correction for errors in the field of the photometer, by the function

$$I = I_0 e^{-k^2 \rho^2}, \tag{12}$$

where I_0 is the value of the intensity at the center of the cloud. Then we have

$$N(R, t) \sim F(R, t) = \frac{I_0(t) k(t)}{\sqrt{\pi}} e^{-k^2 R^2}. \tag{13}$$

We note that in this calculation we have assumed the scattering to be spherical. It would appear that in fact this is not the case. A similar situation occurs in the determination of the electron concentration in the solar corona from the surface brightness of the sun; there the dispersion of scattering angles is significantly greater than in the case at hand. A calculation of anisotropic scattering in the corona made in Ref. 13 leads to a change in the absolute values of N (relative to spherical scattering) of 20-30%. In our case these changes will be even smaller. Furthermore, we are concerned with the relative changes in N, which depend to an even smaller degree on the type of scattering.

The observed distribution of intensity $I(\rho, t)$ can be approximated to a sufficient degree of accuracy by (12).

We obtain values of $1/k(t)$ for the following conditions: 92 sec, 66.5 km; 270 sec, 80.5 km; 620 sec, 95.5 km; 920 sec, 107.5 km.

Since for later stages of growth of the cloud the light flux became constant (see Fig. 4), the following relationship must hold

$$\frac{\pi I_0(t)}{k^2(t)} = L = \text{const.} \tag{14}$$

Furthermore, from (13) for R = 0 we have

$$\frac{N(0,t)}{N_0} = \frac{k(t)I_0(t)}{k(0)I_0(0)} = \left\{\frac{I_0(t)}{I_0(0)}\right\}^{3/2}. \tag{15}$$

In the solution of (6) we assume that $N(R, 0) = N_0$ inside a sphere of radius R_0, where R_0 is determined from the relation

$$\frac{4}{3}\pi R_0^3 N_0 = \frac{N_0 \pi^{3/2}}{[k(0)]^3}. \tag{16}$$

In other words, we shall approximate the initial distribution by a homogeneous sphere with a concentration of sodium atoms equal to the actual concentration at the center at the initial time. In this case

$$R_0 = \frac{\pi^{1/6}}{k(0)}\sqrt[3]{\frac{3}{4}} = \frac{1,10}{k(0)}. \tag{17}$$

The initial time is taken at t = 270 sec. The isophotes for the cloud became practically circular at this time. We may note that almost the same result is obtained if t = 90 sec is taken as the initial time. This shows the slight dependence of the solution of (6) on the choice of the initial time.

In Fig. 6 are shown the time dependence of $N(0,t)/N_0$ obtained directly from observations (solid curve) and a family of curves showing the dependence of $N(0,t)/N_0$ at given initial conditions for various values of the diffusion coefficient.

It can be seen from this figure that the concentration obtained from observations corresponds best to a value of the diffusion coefficient

$$D = 0.59 \cdot 10^{11} \text{ cm}^2 \text{ sec}^{-1}.$$

It is known from Ref. 14 that the diffusion coefficient is connected with the concentration of atoms through which diffusion occurs by the expression

$$D = \frac{3\pi}{32} \cdot \frac{V}{n_1 \overline{Q_d}}, \tag{18}$$

where

$$V = \sqrt{\frac{8kT}{\pi}} \cdot \sqrt{\frac{M_1 + M_2}{M_1 M_2}}$$

is the average relative velocity; M_1 and M_2 are the masses

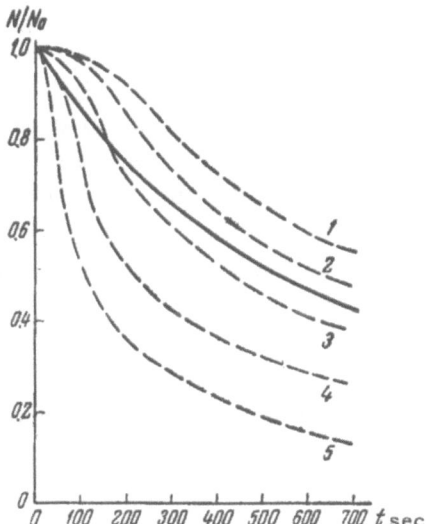

FIG. 6. Relative concentration of sodium atoms as a function of time for the center of the cloud. Solid line, experimental curve; dashed lines, family of curves for various values of the diffusion coefficient D. Curve $1 - D = 0.37 \times 10^{11}$, $n_1 = 4 \times 10^{-8}$; curve $2 - D = 0.45 \times 10^{-11}$, $n_1 = 3 \times 10^{-8}$; curve $3 - D = 0.74 \times 10^{-11}$, $n_1 = 2 \times 10^{-8}$; curve $4 - D = 1.47 \times 10^{-11}$, $n_1 = 10^{-8}$; curve $5 - D = 2.95 \times 10^{-11}$, $n_1 = 0.5 \times 10^{-8}$.

of atoms of the atmosphere; the sodium Q_d is the effective diffusion cross section equal to πa^2, with a equal to the diameter of the atoms considered as solid spheres.

As is well known, for diffusion processes such a representation of collisions leads to very small errors if only the temperature is not too low. The magnitudes of a do not vary greatly for different gases.[15] For our purposes it can be assumed with sufficient accuracy that $a = 3.5 \times 10^{-15}$ cm, whence $\overline{Q}_d = 3.85 \times 10^{-15}$ cm^2.

The temperature of the sodium atoms must be rather close to the temperature of the atoms of the atmosphere, since each atom of sodium will have time to collide several dozen times with atmospheric atoms and will, as a result, assume their temperature. Judging from the fact that the cloud spread out to about a hundred kilometers while maintaining its circular form, the height of the homogeneous atmosphere must be not less than 100 km.*

It is quite natural to assume that at such heights the basic constituent of the atmosphere is atomic oxygen. From this one can conclude that the temperature of the atmosphere must be more than 1600°. We shall assume this magnitude in calculating n_1 from Ref. 18. It is apparent that even relatively large changes in the assumed temperature of the atmosphere have only a slight effect on the value obtained for the concentration of oxygen atoms, which is equal to

$$n_1 = 2.5 \cdot 10^8 \text{cm}^{-3}, \tag{19}$$

from which, on the hypothesis that the atmosphere consists principally of atomic oxygen and nitrogen, we obtain the following value for the density

$$\delta = 6.7 \cdot 10^{-15} \text{ g/cm}^3. \tag{20}$$

Our determination of the concentration of atoms in the atmosphere includes, as must any determination, a certain probable error. The diffusion coefficient D is obtained directly from the observations. To obtain n_1 from D knowledge of the effective diffusion cross section $\overline{Q}_d$ for sodium in oxygen and nitrogen is required. The experimental value of $\overline{Q}_d$ is unknown. The spread of values of a for various vapors of interacting atoms and molecules is not large. However an error of 20-30% in the value of $\overline{Q}_d$ taken by us (3.85×10^{-15} cm^2) is entirely possible. Other possible errors in the determination of n_1 from observations (for example, an inexact approximation in taking the expression $I = I_0 e^{-k^2 \rho^2}$, irregular density of the atmosphere at the edges of the cloud, wrongly neglecting the force of gravity) are obviously less essential. Thus we can estimate the probable error of our determination of n_1 as 30%, i.e.,

---

*We may remark, however, that the exact determination of the height of the homogeneous atmosphere requires specially made observations.

$$n_1 = (2.5 \pm 0.75) \cdot 10^8 \text{ cm}^{-3}. \tag{19a}$$

The value of atmosphere density at a height of 430 km obtained by the authors from an analysis of the diffusion of a sodium cloud can now be compared with data obtained from the analysis of the drag on artificial satellites. According to Ref. 1, at a height of 450 km the density deduced from an analysis of the drag on the satellite 1958 (Explorer I), is equal to $(9 \pm 6) \times 10^{-15}$ g/cm^3. According to the data of Ref. 2, obtained from an analysis of the drag on the same satellite, at a height of 450 km the density is equal to 3×10^{-15} g/cm^3. Our value $(6.7 \pm 2) \times 10^{-15}$ g/cm^3 is evidently in very good agreement with the data obtained from satellites.

We note further that on the basis of data on the electron concentration in the outer regions of the ionosphere, obtained from observations on the radio signals from the first Soviet satellite in Ref. 16, starting with the assumption that ionization is in equilibrium at a height of 400 km, the value of the concentration of neutral atoms is found to be equal to 5×10^8 cm^{-3}. Extrapolating this value to a height of 430 km, we obtain a concentration equal to $(3.2\text{-}3.5) \times 10^8$ cm^{-3}, i.e., 30-40% greater than our value.

Thus we can conclude that methods based on completely different physical principles give (within the limits of observational errors) the same values for the density of the atmosphere at heights of 400-450 km.

In Fig. 7 are shown the results available in the literature at the present time for the density of the upper atmosphere as determined from observations of the drag on Soviet and American satellites, as well as the result obtained from the method described above. The curve is obtained from the condition that best represents the results of observations. Naturally, this curve gives only an average picture. Obviously, at such great heights there must actually be significant changes in density with time and position. Further systematic observations of the diffusion of artificial sodium-vapor clouds in the upper atmosphere would enable us to discover such variations.

The method described for the determination of the density of the atmosphere can be applied over a large range of alti-

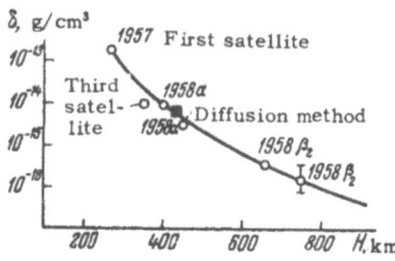

FIG. 7. Density of the atmosphere according to data from the drag on artificial earth satellites.

tudes. The lower limit of this range is determined by the condition that during the time of observation (about 10 minutes) the atoms of sodium should not "vanish" from the earth's atmosphere on account of chemical reactions. This height can apparently be taken at around 200 km. We may remark that if an experiment is carried out at such relatively low altitudes, the linear dimensions of the cloud will be approximately ten times smaller at corresponding times than in our experiment. Thus only a small amount of sodium need be evaporated— 20 or 30 g—since otherwise the optical density of the cloud will be significantly greater than unity, which will make photometric analysis impossible.

It can be assumed that the upper limiting region in the atmosphere where the method of determining the density is applicable is between 500 and 600 km. On the other hand, it is not impossible that our method would be applicable at even higher altitudes if the hydrogen content there is significantly greater than has usually be assumed. This latter possibility can by no means be ruled out. We emphasize once more that the release of sodium vapors in the atmosphere must be carried out near the top of the rocket trajectory, and not during its upward motion as described in Refs. 7-9.

Simultaneously with the determination of density by the diffusion method, it is possible to determine the temperature of the atmosphere, which is of great interest. The solution of this problem requires a special technique. The question will be considered in another paper.

LITERATURE

1. H. E. Lagow, R. H. Horowitz, and I. Ainsworth, IGY World Data Center A, Rockets and Satellites, 145 (1958).
2. Siry, U. S. Naval Res. Lab., Washington, Fifth General Assembly of SC IGY 30, June 1958.
3. Harris and Jastrow, Naval Nucl. Res. Lab., Washington, Fifth General Assembly of SC IGY 30, June 1958.
4. M. L. Lidov, Artificial Earth Satellites, Volume 1, Plenum Press, New York, 1960.
5. V. I. Krasovskii, Report on the Fifth IGY Assembly, Moscow, 1958.
6. D. R. Bates, J. Geophys. Res., 55 (1950), p. 347.
7. H. D. Edwards, J. F. Bedinger, E. R. Manring, and C. D. Cooper, The Airglow and Aurorae, London, 1956, p. 122.
8. J. F. Bedinger, S. N. Ghosh, and E. R. Manring: Trudy konferentsii po kosmicheskomu prostranstvu (Works of the Conference on Cosmic Space), 1958, p. 225.
9. J. F. Bedinger, E. R. Manring, and S. N. Ghosh, J. Geophys. Res., 63, 1 (1958), p. 19.
10. K. I. Gringauz, DAN (Reports of the Academy of Sciences USSR), 120, 6 (1958), p. 1234.
11. S. K. Mitra and H. Rakshit, Ind. J. Phys., 12 (1938), p. 47.
12. Frank and von Mises, Differential and Integral Equations of Mathematical Physics, ONTI, 1937, p. 637.
13. A. F. Bogorodskii and N. A. Khinkulova, Byull. komissii po issledovaniyu Solntsa (Bull. of the Commission on Solar Investigation), 5 (1950), p. 6.
14. H. Massey and E. Burhop, Electronic and Ionic Impact Phenomena, Oxford, London, 1952.
15. J. Jeans, The Dynamical Theory of Gases, Cambridge University Press, 1921, p. 323.
16. Ya. L. Al'pert, F. F. Dobrokhotov, E. F. Chudesenko, and B. S. Shapiro, DAN (Reports of the Academy of Sciences USSR), 120 (1958), p. 743.

METHODS FOR THE CONTROL OF INTERFERING CURRENTS ORIGINATING AT THE INPUT OF AN ELECTROSTATIC FLUXMETER DURING ITS OPERATING IN A CONDUCTING MEDIUM

I. M. Imyanitov and Ya. M. Shvarts

In Ref. 3 the use of an electrostatic fluxmeter of the rotational type has been suggested for the measurement of the electrical charge on a satellite acquired by various processes, from diffusion transfer from various thermal ions and electrons in the plasma of the ionosphere to charging under the action of ultraviolet rays.

Briefly the operation of an electrostatic fluxmeter placed on m, an isolated body located in the plasma is as follows. The measuring plate of the fluxmeter described in detail in a monograph,[1] is a part of the surface of the body; but electrical contact with the remainder of the surface (the main part of the body) is made through a resistance R, which is the input resistance of the apparatus. It is clear that with a fixed screen under stationary conditions the surface of the measuring plate has the same potential and the same charge density as the surface of the body itself would have if located in the position of the measuring plate. As a result of periodically screening the surface of the measuring plate with a rotating screen located rather close to it from the space charge that always surrounds a charged surface in the plasma, there occurs an efflux or influx of surface charge on the measuring plate. The current, passing through the resistance R, causes a decrease in voltage which is measured by a vacuum tube circuit.

The following causes for the appearance of interference are possible:

1. Motion of the satellite with a velocity comparable to

the thermal velocity of ions and electrons, leading, as a result of the charge on the satellite, to the appearance of currents on its surface.

2. A difference of the potențial of the satellite at the position of the electrostatic fluxmeter from the equilibrium value, i.e., from the value of the potential that a stationary isolated body located in the plasma would have. This is caused by the motion of the satellite in a magnetic field, by the operating current in the input resistance, or by constrained variation of the potential in some part of the surface of the satellite.

3. Illumination of the surface of the satellite by ultraviolet, X-ray, or other forms of illumination capable of causing photoemission.

4. The direction of motion in the ionosphere.

An essential property of the interference currents is that their instantaneous value is proportional to the area of the exposed part of the measuring plate. Thanks to this, in the input resistance of the fluxmeter, when operating in the range of resistance given in Ref. 1, there is a phase shift of 90 deg between the operating and the interference currents. An estimate of the order of magnitude of interference currents shows that for a concentration of charged particles $N = 10^{-6}$ cm^{-3}, the limiting value of the current density is no greater than 10^{-7} amp/cm^2. At the same time the value of the current density of the photocurrent is no more than 10^{-8} amp/cm^2.

The appearance of interference currents 3 at the input of the measuring circuit of the electrostatic fluxmeter leads to raising the lower limit of the electrostatic field measureable by the fluxmeter, and consequently to an increase in the lower

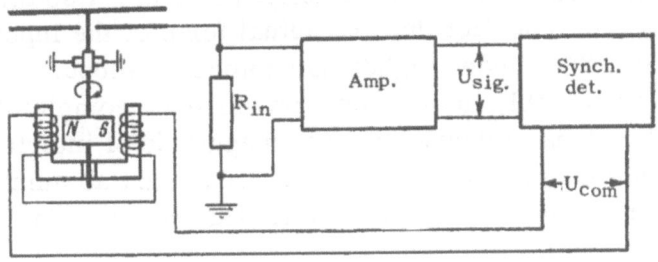

FIG. 1. Measuring circuit of the fluxmeter.

limit of the measurable charge. Therefore it is necessary to find methods of preventing interference currents.

The fact that the interference current is shifted 90 deg with respect to the input operating current when the fluxmeter is operating in the resistance range of Ref. 1 makes it possible to strongly decrease the effect of interference currents by using a synchronous detector in the output of the measuring circuit (Fig. 1).*

Actually,
$$\left. \begin{array}{l} U_o = U_i \cdot k \cdot \cos \varphi, \\ U'_o = U'_i \, k \cdot \cos(\varphi + 90°) = |U'_i \cdot k \cdot \sin \varphi| \end{array} \right\} \quad (1)$$

U_i is the effective value of the input operating voltage, U_o is the value of the operating voltage at the output of the synchronous detector, U_i' is the effective value of the input interference voltage, U_o' is the value of the interference voltage at the output of the synchronous detector, k is the amplification factor of the measuring circuit, and φ is the phase shift between the reference (commutating) voltage of the synchronous detector and the operating (signal) voltage of the detector, caused by the measuring circuit.

In the ideal case with $\varphi = 0$ deg the use of a synchronous detector would make it possible to completely eliminate spurious effects. However, in an actual circuit it is not possible, in the first place, to achieve a zero phase shift in the measuring circuit for the whole range of operating frequencies; in the second place, the change of parameters of the circuit components with changes of temperature also means that it is impossible to maintain $\varphi = 0$ deg over the whole range of operating temperatures. Third, and finally, there are also difficulties arising from the fact that the actual form of the input voltages (Fig. 2) includes higher harmonics on account of which it is impossible to suppress interference completely.

Another means of decreasing the magnitude of interference currents may be noted here. We recall that ac amplifiers are employed in the usual electrostatic fluxmeters. Therefore,

---

*The question of using synchronous detectors in the circuits of electrostatic fluxmeters is considered in detail in Ref. 1.

interference currents are amplified only to the extent that they are modulated during the rotation of the screening plate. Hence, the interference can be cut down by constructing the measuring and screening plates so as to decrease, on the one hand, the modulation of the stream of charged particles reaching the surface and, on the other hand, to decrease the absolute magnitude of the stream colliding with the measuring plate. This is possible when the measuring and screening plates are constructed in the form of metallic grids with a definite electrical and optical transmittance.

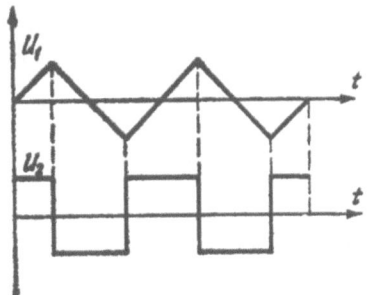

FIG. 2. Form of the input voltages.

To check the ability of grids to pass streams of charged particles without trapping, we conducted experiments to determine the ratio of the electrical currents flowing to the grid and the collector I_g/I_c (Fig. 3) in a gaseous discharge in mercury. A general diagram of the experiment is shown in Fig. 4.

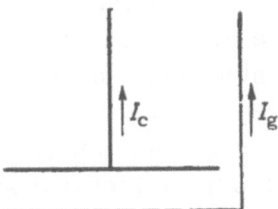

FIG. 3. Experimental plan for determination of the ratio of currents flowing to the grid and collector.

The parameters of the plasma were determined by a probe and were the following for the conditions of the experiment:

temperature of the electrons, 20,000°K; concentration of electrons, 10^9-10^{10} cm^{-3} at a pressure of the mercury vapors of 1.3×10^{-3} to 5×10^{-4} mm Hg.

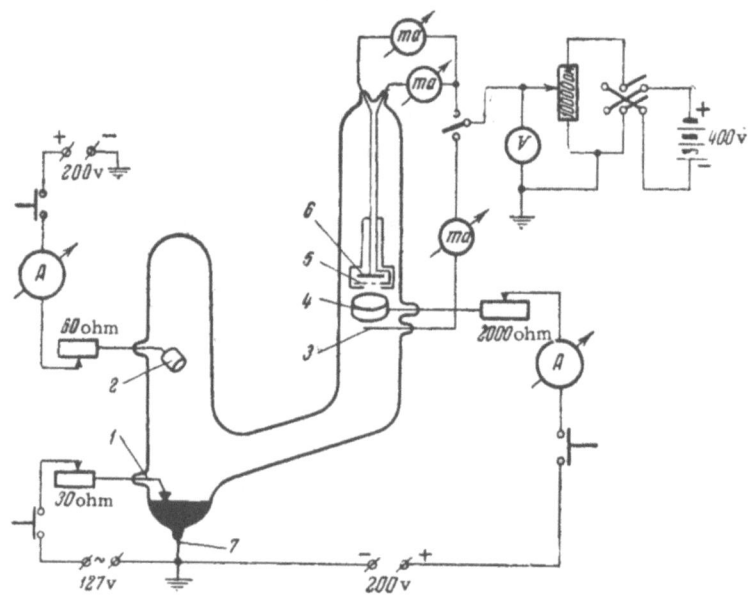

FIG. 4. Overall diagram of the experiment. 1 — igniter; 2 — main anode; 3 — probe; 4 — auxiliary anode; 5 — grid; 6 — collector; and 7 — cathode.

The graph in Fig. 5 shows the dependence of I_g/I_c on the thickness d of a layer of mm volume charge near the grid (calculated theoretically as a function of the potential produced at the grid and the collector).

A metallic grid with a mesh measuring 0.5 × 0.5 mm made of 30 micron wire was used in the experiment. As is seen from Fig. 5, for thicknesses of the layer of volume charge significantly greater than the mesh spacing of the grid, the grid current is several times smaller than the collector current. This results from the fact that streams of charged particles pass through the meshes of the grid and are not held back by it. This occurs as a result of the fact that the distribution of the electric field at the grid, when the thickness of the volume-charge layer is significantly greater than the dimensions of the individual meshes, is of approximately the same type as in a plate condenser, so that particles reaching the plane of the grid have a component of velocity directed perpendicular to the plane of the grid, thanks to which they freely penetrate into the region beyond the grid and are caught by the collector.

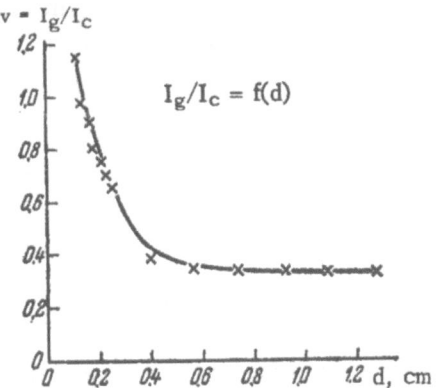

FIG. 5. Dependence of I_g/I_c on the thickness of the volume-change layer near the grid.

Making the screening plate out of wire grid decreases the modulation of the stream of charged particles, while making the measuring plate from wire grid decreases the absolute magnitude of the stream. However the parameters of the grids, the dimensions of the mesh, and the diameter of the wire must be chosen so that the electric field resulting from the volume charges at the surface does not extend beyond the grid; in other words, that the distribution of the electric field at the surface of the grid should be equivalent to that of a metallic surface. Formulas are known for the calculation of the electrical permeability of grids in a vacuum 2, and to a first approximation these can be used to calculate requirements for the grids of the measuring and screening plates for the case where the thickness of the volume-charge layer is significantly greater than the dimensions of the mesh of the grid.

The construction of the collecting part of the electrostatic fluxmeter is shown in Fig. 6, where 1 is the rotating screen, 2 is the measuring plate, and 3 is the collector. 1 and 2 are made following the usual practice in the form of plates with many sectors. The collector serves to trap the stream of charged particles passing through the grids of the measuring and screening plates.

The screen 1 made of wire grid must have a small electrical permeability in comparison with the measuring plate 2, which in its turn must have a small permeability in comparison with the collector 3. We denote the distance between the

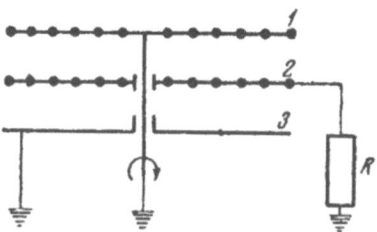

FIG. 6. Diagram of the measuring portion of the electrostatic fluxmeter.

measuring and screening plates by x_{12}, and between the measuring plate and the collector by x_{23}. (Naturally, the size of the mesh should be chosen smaller than x_{12}, x_{23}.)

To calculate the electrical permeability of the screen (considering the measuring plate to be a metallic surface) the following formula was used

$$a_1 = \frac{C_{a_2}}{C_{a_1}} = \frac{\ln \coth \pi \sigma}{2\pi x_{12} Lg - \ln \cosh \pi \sigma}, \quad (2)$$

where σ is the filling coefficient of the grid and for a square grid lattice $Lg = (2/p)[1 - (dg/p)]$.

To calculate the electrical permeability of the measuring plate the following formula was used

$$a_2 = \frac{C_{a_1}}{C_{a_2}} = \frac{\ln \coth \pi \sigma}{2\pi x_{23} Lg - \ln \cosh \pi \sigma} \quad (3)$$

The values a_1 and a_2 indicate what part of the field from the source penetrates the grid.

We note that the formulas give rather low values of a_1 and a_2, since the formulas can only be applied when the field source is at a greater distance than the size of the meshes of the grid.[2] The field resulting from the volume charge near the grid at distances smaller than the spacing of the grid will penetrate significantly more than would be calculated from the above formulas. However, in view of the fact that the thickness of the layers of volume charge for negatively charged bodies at those concentrations and temperatures of ions as are observed in the ionosphere will have dimensions of the order of a centimeter, it is possible to choose a grid spacing so that the error introduced by neglecting the closeness of the volume charge to the grid will be small—less than 10%.

As an example we shall carry out the calculations of (2) and (3) for $x_{12} = 0.3$ cm, $x_{23} = 1$ cm and for grids with square lattices having the following parameters: spacing of winding 0.5 mm, 1 mm, wire diameter 30 microns. The results of the calculations are given below

Spacing, mm	σ	a_1	a_2
0.5	0.12	0.01	0.002
1	0.06	0.05	0.01

The calculations made indicate that grids almost completely screening the electric field pass a stream of charged particles, trapping only a small portion of them. Having the specifications of the circuit used in practice and determining from them permissible values of the coefficients a_1, a_2, and b (see below), one may make the required matching of the parameters of the grids and the collector.

The proposed méthods of suppressing streams of charged particles can be applied equally well to the suppression of photocurrents, since, thanks to the construction of the screening plate in the form of a grid, the modulation of the light flux causing photocurrents is sharply reduced. The coefficient of attenuation is equal in a given case to σ, the filling coefficient. In addition to this, the construction of the measuring plate in the form of a grid sharply reduces the area subjected to the effect of illumination, so that the magnitude of the photocurrent is cut down by a factor of $1/\sigma_{eff}$. However, thanks to the closeness of the collector to the measuring plate, it is possible that photoelectrons may strike the latter, as a result of which σ_{eff} may be increased. However, in general, the magnitude of the quantity J_{phot} effective at the input, which is determined by the photocurrent J_{phot}, is given by $J'_{phot} = J_{phot} \, \sigma \, \sigma_{eff}$.

We now consider yet another method of preventing interference currents, the introduction of negative feedback from the interference signal.

Actually, by using two detectors connected in parallel at the output of the measuring circuit, one of which is turned to the operating voltage, and the other to the interference voltage, and by using the detected interference voltage to introduce negative feedback to a constant voltage at the input,[4] it is possible to sharply decrease the amplitude of the interfering alternating voltage at the input of the operating synchronous detector. The effectiveness of using negative feedback in this way will depend to a large degree on the extent to which the form of the input alternating voltage produced by negative feedback agrees with the form of the input interference voltage; i.e., it depends on the form of the frequency of the input interference voltage.

Concluding the review of methods applicable to the prevention of interference currents at the input of a fluxmeter, we evaluate the minimum value of the intensity of the electrostatic field determined by the charge density on the satellite surface that can be measured by an electrostatic fluxmeter without requiring additional data on its orientation.

We introduce the notation

$$n = \frac{u'}{u''}, \qquad (4)$$

where u' is the value of the minimum measurable operating voltage at the input; u'' is the value of the interference voltage at the output

$$\left. \begin{array}{l} u' = u'_i \cdot k \cdot \cos \varphi, \\ u'' = u''_i \cdot k \cdot \sin \varphi, \end{array} \right\} \qquad (5)$$

where u_i' is the effective value of the minimum measurable operating voltage at the input and u_i'' is the effective value of the input interference. The other notation is that previously used.

From the theory of the electrostatic fluxmeter it follows that

$$u'_i = \frac{E_{min} \cdot f \cdot R}{2\pi}; \quad u''_i = \frac{j_s \cdot bR}{2}, \quad (6)$$

where E_{min} is the measurable value of the intensity of the electrostatic field; f is the commutation frequency; R is the input resistance; J_s is the maximum current density of the interference current on a unit surface; b is the attenuation coefficient, equal to the product $b_1 b_2$, where b_1 is the attenuation coefficient for the modulation of the stream of charged particles by the grid of the screening plate and b_2 is the attenuation coefficient for the stream of charged particles by the grid of the measuring plate. In (6) the area of the measuring plate S is taken to be equal to unity, and the entire calculation is carried out for the first harmonic of the voltage.

Substituting the values of u_i' and u_i'' in (4) we get

$$n = \frac{E_{min} f}{\pi j_s b_1 \cdot b_2 \cdot \tan \varphi}. \quad (7)$$

Setting n = 1, we get for the value of the minimum measurable field E_{min}

$$E_{min} = \frac{\pi j_s \cdot b_1 \cdot b_2 \cdot \tan \varphi}{f}. \quad (8)$$

Taking f = 1500 cycles, $j_s = 10^{-7} A/cm^2$, $\tan \varphi = 0.03$, i.e., $\varphi = 2°$, we obtain a series of values of E_{min} as a function of the product $b_1 b_2$

$b_1 \cdot b_2$	0.1	0.04	0.01
E_{min} v/cm	0.54	0.22	0.05

We emphasize that E_{min} is the minimum measurable magnitude of the intensity of the electrostatic field on the surface of the satellite resulting from its own charge and determined from the conditions for the maximum possible electrical interference current on the surface of the satellite. Since the magnitude of the interference depends strongly on the orientation of a given part of the surface with respect to the motion, the quantity E_{min} also depends on these same parameters.

In order to make a comparison with E_{min} we find the value of the intensity of the electrical field on the surface of a stationary sphere resulting from its own charge under the following parameters of the plasma: $N = 10^6$ and $10^5 \, cm^{-3}$, $T_u = 1000°K$, $T_e = 5000$ and $10,000°K$. (We assume that the radius of the sphere is significantly greater than the thickness of the layer of space charge around it.)

	$T_e = 5000°K$	$T_e = 10\,000°K$
$N = 10^6$	1.0	1.2
$N = 10^5$	0.32	0.38

LITERATURE

1. I. M. Imyanitov, Pribory i metody dlya izucheniya atmosfernogo elektrichestva (Devices and Methods for Studying Atmospheric Electricity), Moscow, Gostekhizdat, 1957.
2. Yu. A. Katsman, Osnovy rascheta radiolamp (Fundamentals of Vacuum Tube Calculation).
3. I. M. Imyanitov, Usp. fiz. nauk (Advances in the Physical Sciences), 63, 1b, 267 (1957).
4. I. M. Imyanitov and V. V. Mikhailovskaya, Trudy GGO (Glavnaya geofizicheskaya observatoriya—Main Geophysical Observatory), No. 018 (1949), p. 64.

SOME RESULTS OF THE DETERMINATION OF THE STRUCTURAL PARAMETERS OF THE ATMOSPHERE USING THE THIRD SOVIET ARTIFICIAL EARTH SATELLITE

V. V. Mikhnevich, B. S. Danilin, A. I. Repnev, and V. A. Sokolov

In Reference 1 the problem of measuring the pressure and density of high layers of the atmosphere using an artificial earth satellite was discussed. An analysis is given in this work of the state of knowledge available up to 1957 of the structural parameters of the upper atmosphere. By that time the mean distributions of pressure, density and temperature up to a height of 100 km[2-4] were known, and it had been established that at heights up to 90 km the atmosphere is intermixed and that oxygen dissociates above 90 km.[5] Up to 1956, direct measurements of the density, pressure, and composition of the atmosphere above 100 km were very few in number, and therefore existing concepts of the atmosphere at these heights were not well defined. In recent years there have been carried out experiments on the determination of the density of the atmosphere at great heights.[6-8] A particularly large contribution to the study of the upper layers of the atmosphere resulted from the investigations made with Soviet artificial earth satellites, which made it possible to determine the density of the atmosphere both from measurements of manometers installed on the third Soviet earth satellite[9] and also from the drag on the satellites.[10,11]

The present article is concerned with a consideration of a part of the results of the determination of the density of the atmosphere obtained from manometer measurements on the third Soviet artificial earth satellite.

APPARATUS

For the determination of the density of the atmosphere on an artificial earth satellite an apparatus (Fig. 1) was used that included two ionization (M_1 and M_2) and one magnetic electrical discharge manometers (M_3).* The command for switching on the apparatus was provided by the program timing mechanism (PTM) to the switching-connecting device (SCD). The current

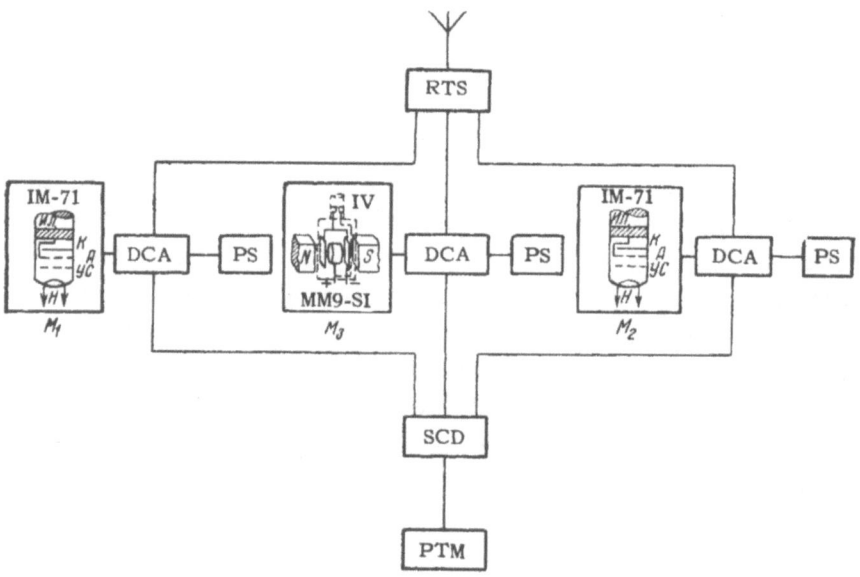

FIG. 1. Block diagram of the manometer apparatus in the satellite. M_1 and M_2 — ionization manometers; M_3 — magnetic electrical discharge manometer; DCA — direct-current amplifier; SCD — switching-connecting device; PS — power supply; PTM — program timing mechanism; RTS — radio telemetry system.

between the electrodes of the manometer, proportional to the number of particles in a unit volume, was fed to the input of a direct-current amplifier (DCA). The output voltage of the amplifier, which depends on the size of the input signal, was transmitted to earth by a radiotelemetry system (RTS).

*The construction and operating principle of the manometers are described in detail in References 9, 12, and 13.

The position of the manometers on the satellite is shown in Fig. 2. The manometers were installed in conical recesses in the nose section of the satellite, in so-called sockets. Fastening the manometers in the sockets was accomplished with the aid of panels. The panels contained hermetic contacts, through which the electrodes of the manometers were connected with the amplifiers and power supplies inside the satellite.

In order to eliminate the effect of atmospheric ions and electrons on the manometer readings, special traps were installed at their inputs.[1,9] In addition, a screen in electrical contact with the body of the satellite was placed at the mouth of the socket, and this protected the collecting outlet of the manometer from incident charged particles.

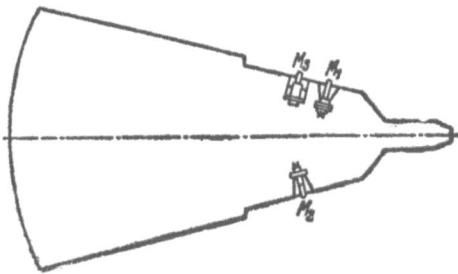

FIG. 2. Location of the manometers on the satellite. M_1 and M_2 — ionization manometers; M_3 — magnetic manometer.

To decrease the effect of desorbed molecules on the manometer readings, the outer satellite skin was made of materials having a maximum rate of desorption and a minimum vapor pressure. In addition the manometers installed in the satellite were first outgassed and evacuated. The manometers were uncapped with the aid of the breaking device during the separation of the protective cone from the satellite after it was in orbit.

METHOD

As a result of the great velocity of a satellite, the temperature difference of the gas in the satellite and in the free atmosphere, as well as changes of the orientation of the inlet aperture of the satellite with respect to the velocity vector of the satellite, the density and pressure within the manometer volume are not equal to their values in the atmosphere.

At the heights of interest to us the length of the mean free path is very much greater than a characteristic dimension of the apparatus (the satellite). Under these conditions there are no collisions between particles in the inflowing stream and and particles remaining in the manometer. If the velocity of a body u is significantly larger than the mean thermal velocity of particles in the medium, in a unit time $NSu \sin \theta$ particles will pass through a manometer opening of area S, where N is the number of particles in a unit volume of the atmosphere, θ is the angle between the velocity vector and the plane of the aperture. Undergoing a series of collisions with the walls of the manometer, the particles attain a most probable velocity v_1, which is determined by the wall temperature T_1. In unit time $(N_1 v_1 / 2\sqrt{\pi})S$ particles flow out of the manometer aperture. At equilibrium both streams of particles are equal. From the condition that the streams are equal the concentration of particles N at a given point in the atmosphere is given by the formula

$$N = \frac{N_{Av} \cdot P_1}{\sqrt{2\pi M R^* T_1} \, u \sin \theta}, \quad (\theta \geqslant 7-8°)$$

if the manometer is graduated not according to the concentration of particles N, but according to the pressure $P_1 = N_1 k T_1$. Here M is the molecular weight of the gas, R^* is the universal gas constant, and N_{Av} is Avogadro's number.

The velocity u, which is needed for the calculations of N, is known with great accuracy from observations of the orbit, T_1 is measured directly on the satellite and is telemetered together with the manometer reading P_1.

The molecular weight was taken equal to that given in Ref. 14 (Table 1). Above 600 km the value of M was obtained by extrapolation.

As is seen from the formula presented, the possible change in the molecular weight does not lead to a significant error in the determination of N. For example, for a change of M from 18 to 24 g/mole, N is decreased less than 20%.

TABLE 1
Changes of Molecular Weight with Height

Height km	Molecular weight g/mole	Height km	Molecular weight g/mole	Height km	Molecular weight g/mole	Height km	Molecular weight g/mole
225	21.28	295	18.63	365	17.25	435	16.53
230	21.04	300	18.50	370	17.18	440	16.50
235	20.80	305	18.37	375	17.12	445	16.46
240	20.57	310	18.25	380	17.06	450	16.43
245	20.36	315	18.14	385	17.00	455	16.40
250	20.15	320	18.03	390	16.94	460	16.37
255	19.94	325	17.92	395	16.89	465	16.34
260	19.75	330	17.82	400	16.84	470	16.31
265	19.56	335	17.73	405	16.79	475	16.30
270	19.40	340	17.64	410	16.74	480	16.27
275	19.22	345	17.55	415	16.70	485	16.24
280	19.07	350	17.47	420	16.65	490	16.22
285	18.91	355	17.39	425	16.61	495	16.19
290	18.77	360	17.32	430	16.57	500	16.16

The orientation of the inlet aperture of the manometer with respect to the velocity vector of the satellite ($\sin \theta$) was determined from a method established in Ref. 15

$$\sin \theta = m \{\nu [A_1 a + B_1 b + C_1 c] + \mu [A_2 a + B_2 b + C_2 c] + \\ + \xi [A_3 a + B_3 b + C_3 c]\} + n \{\nu [- A_1 a' + B_1 b' + C_1 c'] + \\ + \mu [- A_2 a' + B_2 b' + C_2 c'] + \xi [- A_3 a' + B_3 b' + C_3 c']\} + \\ + k \{\nu [A_1 a'' - B_1 b'' + C_1 c''] + \mu [A_2 a'' - B_2 b'' + C_2 c''] + \xi [A_3 a'' - B_3 b'' + C_3 c'']\};$$

$$a = \cos \psi \cos \varphi - \sin \psi \sin \varphi \cos \theta_0,$$
$$b = \sin \psi \cos \varphi + \cos \psi \sin \varphi \cos \theta_0,$$
$$c = \sin \varphi \sin \theta_0,$$
$$a' = \cos \psi \sin \varphi + \sin \psi \cos \varphi \cos \theta_0,$$
$$b' = - \sin \psi \sin \varphi + \cos \psi \cos \varphi \cos \theta_0,$$
$$c' = \cos \varphi \sin \theta_0,$$
$$a'' = \sin \psi \sin \theta_0,$$
$$b'' = \cos \psi \sin \theta_0,$$
$$c'' = \cos \theta_0.$$

Here ψ, φ, θ_0 are the Eulerian angles; ν, μ, ξ are the direction cosines of the velocity vector in an absolute system of coordinates (the Z axis is directed toward the pole star, the X axis toward the vernal equinox, and the Y axis completes a right-handed coordinate system)(Fig. 3); A_i, B_i, C_i (i = 1, 2, 3) are constants for a given loop

$$A_1 = \cos\rho_0 \cos\gamma_0, \quad A_2 = -\sin\rho_0, \quad A_3 = \cos\rho_0 \sin\gamma_0,$$
$$B_1 = \sin\gamma_0, \quad B_2 = 0, \quad B_3 = -\cos\gamma_0,$$
$$C_1 = \sin\rho_0 \cos\gamma_0, \quad C_2 = \cos\rho_0, \quad C_3 = \sin\rho_0 \sin\gamma_0.$$

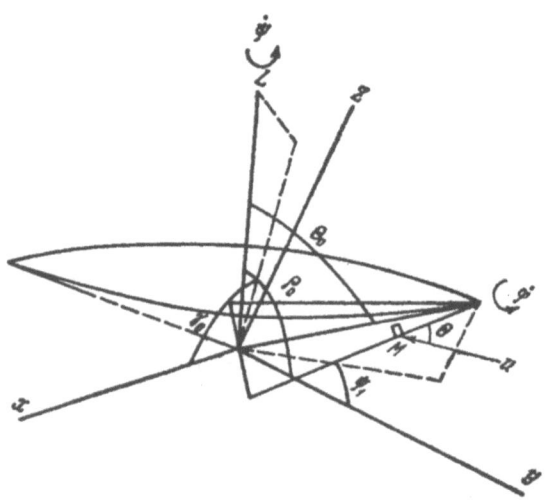

FIG. 3. Diagram explaining the spatial orientation of the manometer. X, Y, Z – absolute coordinate system; L – angular momentum vector of the satellite; ρ_0, γ_0 – its angular coordinates; θ_0 – precession angle; $\dot{\psi}, \dot{\varphi}$ – angular velocities of precession and proper rotation; M – manometer; u – velocity vector; θ – angle between the velocity vector and the cross-section plane of the inlet of the manometer.

ρ_0, γ_0 are angular coordinates of the angular momentum vector, and m, n, and k are the direction cosines of the manometer axes with respect to the principal axes of the satellite.

After determining N (the number of particles in a unit volume) the density ρ was calculated and also the height of the homogeneous atmosphere H*

---

*The determination of H with the formula is made with an error of about 30% with respect to the calculation made by the formula

$$H = -\frac{\Delta h}{2.3(\lg N''/N' - \lg T''/T')}$$

$$H = -\frac{\Delta h}{2.3\,(\lg N'' - \lg N')}.$$

Here N'' and N' are the number of particles in a unit volume of the atmosphere at two points differing in height by $\Delta h = 10$ km.

Next were calculated the temperature

$$T = \frac{MgH}{R^*}$$

and the pressure of the atmosphere

$$P = kNT,$$

where k is the Boltzmann constant.

It should be noted that for the determination of the density of the atmosphere the hypothesis does not enter that there is a local Maxwellian distribution of the velocity of the particles at great heights. The height of the homogeneous atmosphere, the temperature, and pressure were calculated under the assumption that the usual barometric formula is applicable to the rarefied gas in the presence of temperature gradients.

RESULTS

The third Soviet earth satellite was launched May 15, 1958. In accordance with the plan of the experiment, from the moment that the manometers were uncovered their discharge currents (corresponding to density measurements) were continuously recorded, periodic calibration of the amplifiers was made, as were checks of the emission current and the wall temperature of the manometers.

The results of reducing the data obtained on the fifteenth revolution of the satellite are given in Table 2.

The results obtained provide a description of the atmosphere at heights of 225-500 km for May 16, 1958. Measurements at various heights refer to different times of day (1300-1900 hr local time) and different geographical latitudes (57°N-65°N).

The values of the density obtained in the present experiment practically agree in the region of perigee of the Soviet

TABLE 2
Structural Parameters of the Atmosphere at Heights of 225-500 km

Height, km	N, cm^{-3}	ρ, g/cm^3	H, km	T, °K	P, dyne/cm^2	P, mm Hg
225	6.01·10^9	2.12·10^{-13}	40.0	936	7.76·10^{-4}	6.25·10^{-7}
230	5.31	1.79	40.6	938	6.88	5.54
235	4.7	1.7	41.3	941	6.1	4.92
240	4.17	1.42	42.0	946	5.44	4.4
245	3.71	1.25	42.8	952	4.88	3.94
250	3.3	1.1	43.5	958	4.36	3.54
255	2.94	9.73·10^{-14}	44.3	964	3.91	3.17
260	2.64	8.66	45.2	971	3.54	2.88
265	2.36	7.77	46.0	979	3.19	2.6
270	2.12	6.83	47.0	987	2.89	2.35
275	1.91	6.1	47.9	996	2.63	2.14
280	1.72	5.44	48.8	1005	2.39	1.95
285	1.55	4.87	49.7	1015	2.17	1.78
290	1.4	4.36	50.7	1026	1.98	1.62
295	1.27	3.93	51.7	1037	1.82	1.49
300	1.15	3.53	52.7	1048	1.66	1.37
305	1.07	3.26	53.7	1059	1.56	1.29
310	9.57·10^8	2.9	54.5	1072	1.42	1.17
315	8.73	2.63	55.9	1084	1.31	1.008
320	7.98	2.39	57.0	1097	1.21	1.0
325	7.31	2.17	58.1	1110	1.12	9.28·10^{-8}
330	6.7	1.98	59.2	1124	1.04	8.62
335	6.17	1.82	60.3	1138	9.69·10^{-5}	8.06
340	5.68	1.66	61.5	1153	9.04	7.52
345	5.22	1.52	62.8	1169	8.96	7.46
350	4.82	1.4	64.8	1185	7.88	6.58
355	4.46	1.29	65.2	1200	7.39	6.18
360	4.13	1.19	66.7	1219	6.95	5.82
365	3.86	1.1	68.1	1237	6.59	5.53
370	3.56	1.02	69.5	1257	6.18	5.19
375	3.31	9.41·10^{-15}	70.9	1276	5.83	4.9
380	3.08	8.72	72.4	1295	5.51	4.64
385	2.92	8.24	73.9	1305	5.26	4.44
390	2.69	7.56	75.2	1335	4.96	4.19
395	2.52	7.07	76.7	1353	4.71	3.98
400	2.36	6.6	78.9	1373	4.47	3.79
405	2.21	6.16	79.7	1393	4.25	3.6
410	2.08	5.78	81.2	1417	4.07	3.46
415	1.95	5.41	82.9	1440	3.88	3.3
420	1.84	5.09	84.6	1465	3.72	3.17
425	1.73	4.79	86.3	1489	3.56	3.04
430	1.64	4.51	88.1	1514	3.43	2.93
435	1.55	4.25	90.0	1539	3.29	2.82
440	1.47	4.03	91.7	1563	3.17	2.72
445	1.39	3.8	93.6	1589	3.05	2.62
450	1.32	3.6	95.5	1614	2.94	2.53
455	1.25	3.4	98.6	1643	2.84	2.44
460	1.19	3.23	99.9	1675	2.75	2.37
465	1.13	3.06	102.0	1709	2.66	2.3
470	1.08	2.92	104.5	1745	2.6	2.25
475	1.03	2.79	107.0	1781	2.53	2.19
480	9.82·10^7	2.65	109.3	1810	2.45	2.13
485	9.4	2.53	111.5	1845	2.39	2.08
490	8.97	2.42	113.9	1880	2.33	2.02
495	8.61	2.31	116.5	1917	2.28	1.98
500	8.24	2.21	119.0	1953	2.22	1.94

satellites (225-230 km) with values of the density found from observation of the change of orbits[10, 11] (cf. Fig. 4a and Table 3).

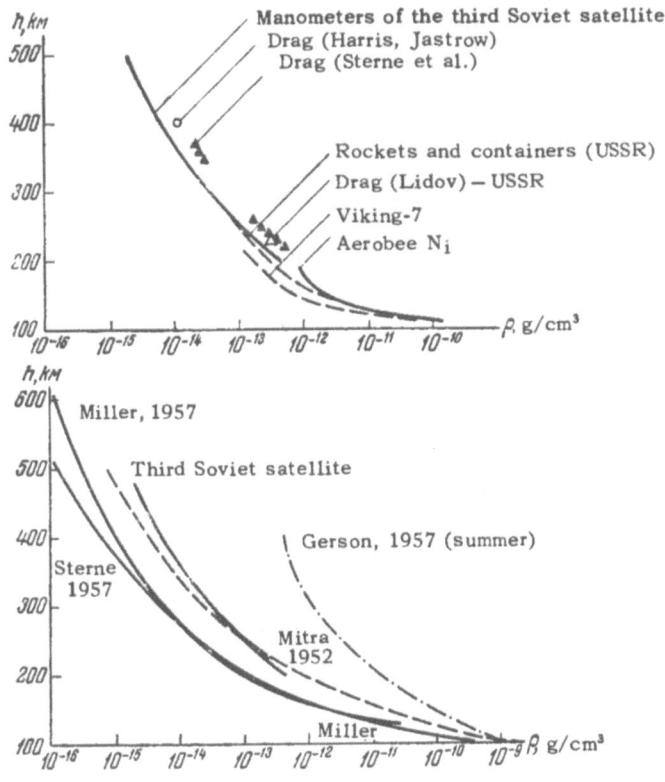

FIG. 4. Change of density with height. (a) From experimental data; (b) from various models.

At the same time the values of density and pressure obtained are about 1.5-10 times greater than the values proposed by numerous models[14, 16, 17] with the exception of a model introduced in Ref. 18 (Fig. 4b and 5). In the majority of cases the models were built on the basis of generalizations of a very few observations on the upper layers of the atmosphere, obtained by various means, at different times of day and season, at different latitudes, etc. Experimental investigations of the upper layers of the atmosphere were isolated. In the construction of various models the assumptions used were

TABLE 3
Values of the Density (in g-cm^{-3}) at Various Heights from Manometer Measurements on Rockets and from the Drag on Satellites

Height, km	Rocket data			Drag data	
	Containers and rockets. Mean latitude of European USSR Ref. 9	Viking-7 33° N lat. Ref. 7	Aerobee-Ni 59° N lat. Ref. 8	1957 α_1 1957 α_2 1957 β_1 Refs. 10, 11	1957 α_1 1957 α_2 1957 β_1 1958 α 1958 β_2 1958 γ Refs. 19-25
100	$4 \cdot 10^{-10}$	$2.5 \cdot 10^{-10}$	$7 \cdot 10^{-10}$		
110	$9.8 \cdot 10^{-11}$	$5 \cdot 10^{-11}$	$1.5 \cdot 10^{-10}$		
120	$2.2 \cdot 10^{-11}$	$1.2 \cdot 10^{-11}$	$2 \cdot 10^{-11}$		$6.9 \cdot 10^{-11}$
130	$7.4 \cdot 10^{-12}$	$3.3 \cdot 10^{-12}$	$6 \cdot 10^{-12}$		$3.01 \cdot 10^{-11}$
140	$3.2 \cdot 10^{-12}$	$1.2 \cdot 10^{-12}$	$3 \cdot 10^{-12}$		$1.49 \cdot 10^{-11}$
150	$1.6 \cdot 10^{-12}$	$6.6 \cdot 10^{-13}$	$2 \cdot 10^{-12}$		$8.07 \cdot 10^{-12}$
160	$9.5 \cdot 10^{-13}$	$4.3 \cdot 10^{-13}$	$1.3 \cdot 10^{-12}$		$4.70 \cdot 10^{-12}$
170	$6.4 \cdot 10^{-13}$	$3.0 \cdot 10^{-13}$	$1 \cdot 10^{-12}$		$2.89 \cdot 10^{-12}$
180	$4.4 \cdot 10^{-13}$	$2.3 \cdot 10^{-13}$	$8 \cdot 10^{-13}$		$1.87 \cdot 10^{-12}$
186					$6.7 \cdot 10^{-13}$
190	$3.3 \cdot 10^{-13}$	$1.8 \cdot 10^{-13}$	$7.5 \cdot 10^{-13}$		$1.25 \cdot 10^{-12}$
197±1					$7.0 \cdot 10^{-13}$
200	$2.7 \cdot 10^{-13}$	$1.4 \cdot 10^{-13}$	$7 \cdot 10^{-13}$		$(3-8.63) \cdot 10^{-13}$
201±4					$6.7 \cdot 10^{-13}$
202±4					$7.37 \cdot 10^{-13}$
206±4					$5.4 \cdot 10^{-13}$
210	$2.0 \cdot 10^{-13}$	$1.1 \cdot 10^{-13}$	$6.0 \cdot 10^{-13}$		$6.04 \cdot 10^{-13}$
211+4					$4.6 \cdot 10^{-13}$
212					$(4.4-4.8) \cdot 10^{-13}$
215					$4.7 \cdot 10^{-13}$
220	$1.6 \cdot 10^{-13}$	$9.0 \cdot 10^{-14}$			$(3.5-5.7) \cdot 10^{-12}$
225				$(2.9-4.1) \cdot 10^{-13}$	
228				$(2.4-3.2) \cdot 10^{-13}$	
230	$1.25 \cdot 10^{-13}$				$3.32 \cdot 10^{-13}$
232					$1.5 \cdot 10^{-13}$
233					$2.2 \cdot 10^{-13}$
240	$1.1 \cdot 10^{-13}$				$2.5i \cdot 10^{-13}$
241					$2.5 \cdot 10^{-13}$
250	$9 \cdot 10^{-14}$			$(1.5-1.6) \cdot 10^{-13}$	$(1.1-1.9) \cdot 10^{-13}$
260	$6.9 \cdot 10^{-14}$				$1.51 \cdot 10^{-13}$
270				$(9.4-10) \cdot 10^{-14}$	$1.19 \cdot 10^{-13}$
275					$8.5 \cdot 10^{-14}$
280					$9.51 \cdot 10^{-14}$
290				$(5.8-7.0) \cdot 10^{-14}$	$7.68 \cdot 10^{-14}$
300					$(5-6.27) \cdot 10^{-14}$
310				$(3.8-4.7) \cdot 10^{-14}$	$5.16 \cdot 10^{-14}$
320					$4.29 \cdot 10^{-14}$
330				$(2.6-3.2) \cdot 10^{-14}$	$3.58 \cdot 10^{-14}$
340					$3.02 \cdot 10^{-14}$
350				$(1.8-2.2) \cdot 10^{-14}$	$(2.1-3) \cdot 10^{-14}$
360					$2.18 \cdot 10^{-14}$
368				$(1.4-1.5) \cdot 10^{-14}$	$(1.4-1.5) \cdot 10^{-14}$

TABLE 3 (continued)

Height km	Rocket data			Drag data	
	Containers and rockets. Mean latitude of European USSR	Viking-7 33° N lat.	Aerobee-Ni 59° N lat.	1957 α_1 1957 α_2 1957 β_1,	1957 α_1 1957 α_2 1957 β_1 1958 α 1958 β_2 1958 γ
370					$1.87 \cdot 10^{-14}$
400					$1.5 \cdot 10^{-14}$
					$9.3 \cdot 10^{-15}$
405					$(9^{+6}_{-4}) \cdot 10^{-15}$
450					$(1.0\text{---}4.5) \cdot 10^{-15}$
500					$(2.3\text{---}6) \cdot 10^{-15}$
550					$(2.2\text{---}4) \cdot 10^{-15}$
600					$2 \cdot 10^{-15}$ $6.8 \cdot 10^{-16}$
650					$1 \cdot 10^{-15}$ $3.8 \cdot 10^{-16}$
700					$7 \cdot 10^{-16}$
720					$(1.2 \pm 0.3) \cdot 10^{-16}$

Conventional Designations of the Satellites:

1957 α_1 — Carrier rocket of the first Soviet satellite (launched 4 Oct. 1957)
1957 α_2 — First Soviet artificial satellite (launched 4 Oct. 1957)
1957 α_3 — Nose cone of first Soviet satellite (launched 4 Oct. 1957)
1957 β_1 — Second Soviet artificial satellite (launched 3 Nov. 1957).
1957 β_2 — Nose cone of second Soviet satellite (launched 3 Nov. 1957)
1958 α — Explorer satellite (launched 1 Feb. 1957)
1958 β_1 — Carrier rocket of Vanguard (launched 17 Mar. 1958)
1958 β_2 — Vanguard I satellite (launched 17 Mar. 1958)
1958 γ — Explorer III satellite (launched 26 Mar. 1958)

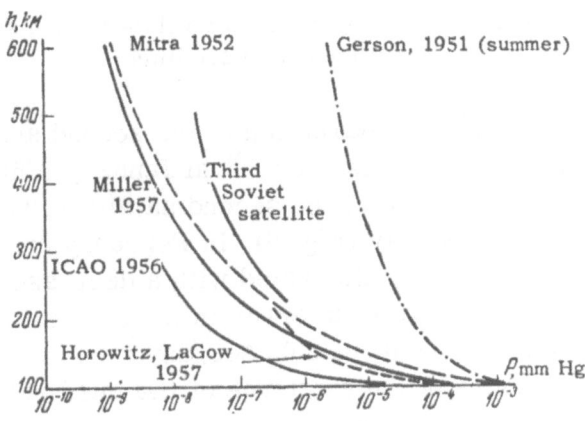

FIG. 5. Change of pressure with height.

distinct and often contradictory, for instance, the use of diffusion separation in some and intermixing in others. All this leads to the fact that the models are noticeably different from each other and do not characterize either the actual state of the atmosphere (since recent investigations indicate the presence of seasonal, daily, latitude, and further variations and indicate that the combination of differently obtained data is incorrect) or statistical average values of the parameters, since the amount of data is insufficient for this.

Some of the differences of data obtained from manometer readings and from satellite drag can be explained, obviously, by the presence of both regular daily, seasonal, and latitude variations as well as variations associated with solar disturbances.

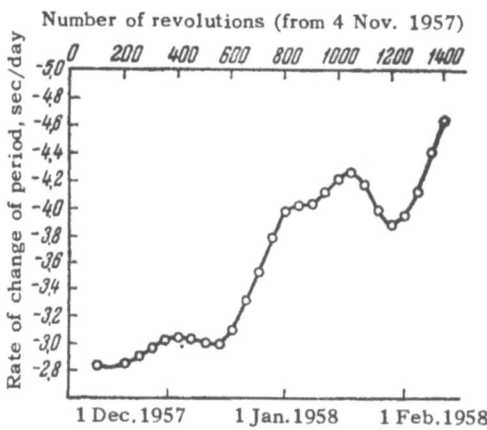

FIG. 6. Rate of change of the period of revolution of the second Soviet artificial earth satellite.

It is known that the retardation of the second and third Soviet satellites was not constant.[27-30] In January 1958 the rate of decrease of the period of the second satellite fluctuated between 4.4 and 3.9 sec/day (Fig. 6). In February an increase of from 3.9 to 5 sec/day alternated with a decrease of as much as 4.5 sec/day. The probable error in the determination of the period was found to be about 0.03 sec/day.[27]

In the first days of the existence of the third satellite the decrease in the period of rotation amounted to 0.97 sec/day, and after 450 revolutions (June 17) to only 0.59 sec/day.[30] A retardation in the decrease of the period is also noticeable in a curve of the change of period with time (Fig. 7). Each point represents an average value over 54 revolutions. In

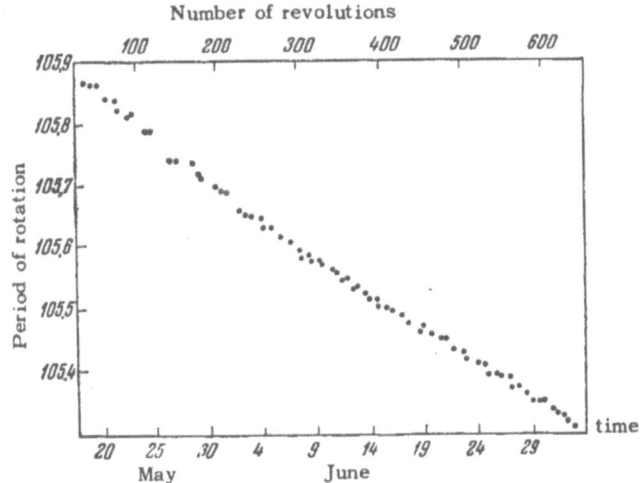

FIG. 7. Change of the period of revolution of the third Soviet artificial earth satellite.

Ref. 30 it is noted that in the period from May 15 to July 17 the atmosphere seemed to contract by an amount equal to 25 km.

In addition, during observations of solar activity and the change of the period of rotation of the second Soviet satellite, it was observed that the strong magnetic storm in the middle of February 1958 and the extension of zones of polarized radiation by 10° toward the equator was accompanied by an increase in the change of the period of the satellite from 3.9 to 5 sec/day.[31] Changes in the intensity of flares on the sun (i.e., the product of the area of a flare by its duration) also correlate with the rate of change of the period of the satellite[32] (Fig. 8).

The results of rocket investigations point to the existence of daily, seasonal, and latitude variations.[33, 8] Thus, according to the data of these articles, at a height of 200 km (Fort Churchill, 59°N) the summer daytime value of the density

$(6.7 \pm 2.0) \times 10^{-13}$ g/cm^3 is approximately twice as high as the corresponding winter value $(3.6 \pm ^{3.0}_{1.5}) \times 10^{-13}$ g/cm^3 and five times larger than the winter night time value of the density $(1.3 \pm 0.6) \times 10^{-13}$ g/cm^3. At the same height the density at latitudes 33 and 59°N on a summer day proved to be equal to $(1.4 \pm 0.5) \times 10^{-13}$ and $(6.7 \pm 2.0) \times 10^{-13}$ g/cm^3, respectively. The presence of a latitude effect is also shown by the results of investigations made with meteorological rockets,[34] by the method of falling spheres,[35] and in other ways.

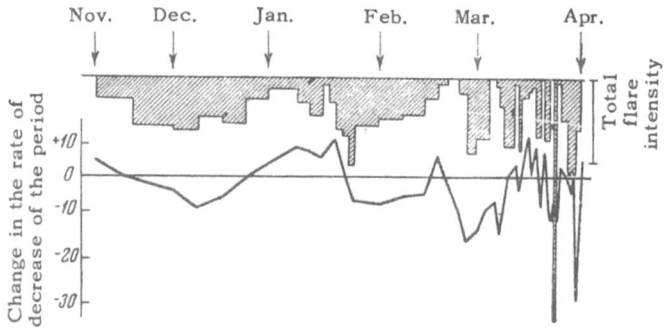

FIG. 8. Correlation of the rate of change of the period of revolution of the second Soviet artificial earth satellite with the intensity of solar flares.

The available experimental data are insufficient to provide the hoped-for determination of the functional dependence of density on latitude in the upper layers of the atmosphere (Fig. 9). However all the results of rocket measurements of density using manometers, relating to a summer day (curve 1), display an increase in density with latitude in the range 33-59°N. For a winter day, however, there is observed a somewhat less pronounced increase of density with latitude (cf. Ref. 33 and Table 3, column 2).

The latitude effect on density in the fall and winter season from observations of the change in orbit of the first and second Soviet satellites is much weaker (see curve 2 in Fig. 9),[36,37] possibly because the same value for the height of the homogeneous atmosphere, H = 30 km, was used for all latitudes in the extrapolation of the data to 206 km. In addition, it is possible that these data refer to different times of day.

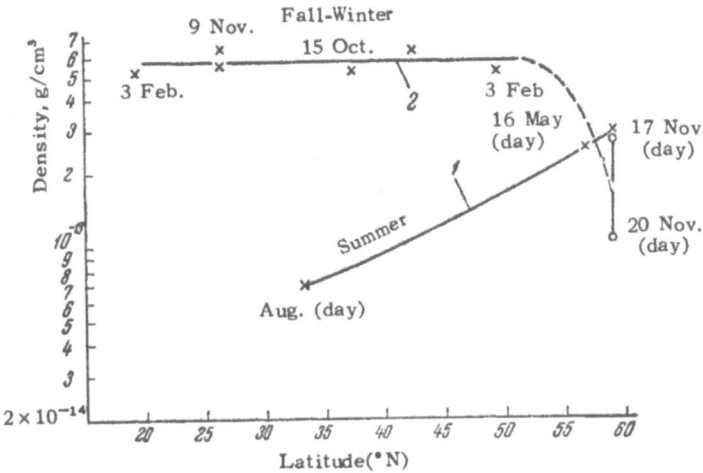

FIG. 9. Change of the density of the atmosphere with latitude. 1 – Summer: Manometer measurements. 33°N – Viking-7 rocket[7]; 57°N – third Soviet satellite; 59°N – Aerobee.[8,27] The data of Ref. 7 and Refs. 8, 27 are extrapolated to a height of ≈230 km. 2 – Fall and winter (October-February): Results obtained from observation of the drag on satellites,[30,31] with the exception of 59°N.[8,27] The data are converted to a height of 206 km with $H = 30$ km.

The observed break in the curves of the change of the periods of rotation of the second and third satellites and the carrier rocket of the third satellite as a function of time as the latitude 30° is approached probably indicates an increase in density on approaching the equator, thanks to the equatorial bulging of the earth.

The facts enumerated attest to the fact that the state of the upper atmosphere may change markedly and that it depends both on external effects on the atmosphere and also on the season, time of day, and geographical latitude.

Values obtained from measurements on the third Soviet artificial earth satellite, which give the density, height of the homogeneous atmosphere, temperature, and pressure, characterize the state of the atmosphere at height of 225-500 km for the daytime hours on May 16, 1958, at latitudes 57-65°N. In particular, on May 16, 1958, the temperature of the atmosphere appeared to be relatively high (Fig. 11).

As a result of further reduction of all the experimental data, further material will be obtained on the density, pressure, and other parameters of the atmosphere, which will enable a clearer explanation of the dependence of the structural parameters on the latitude and time of day to be made.

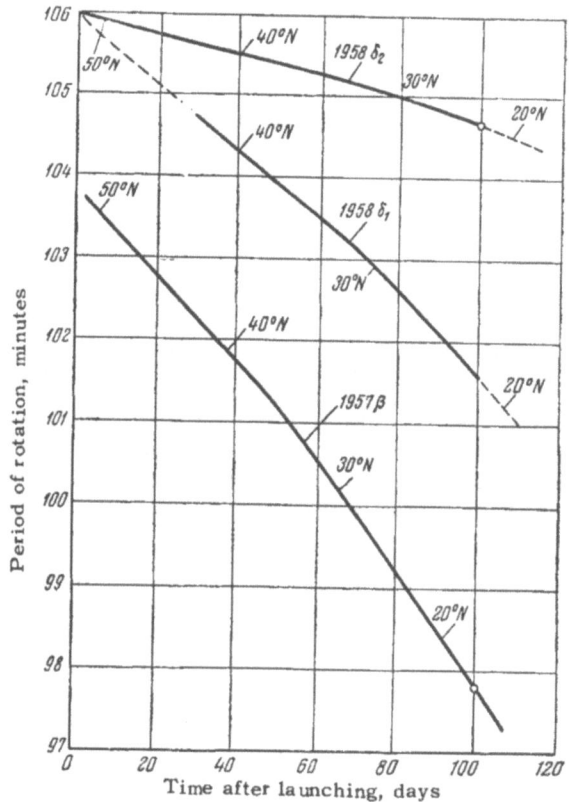

FIG. 10. Change in the period of rotation of the second and third satellites and the carrier rocket of the third satellite with time. Near latitude 30°N there is noted an increase in the rate of change of the period, associated with a change in the resistance to the motion of the satellites. Since the form, dimensions, and weight of the satellites are different, the most probable explanations of this fact is the assumption of a greater density of the atmosphere in the equatorial reqions.[2]

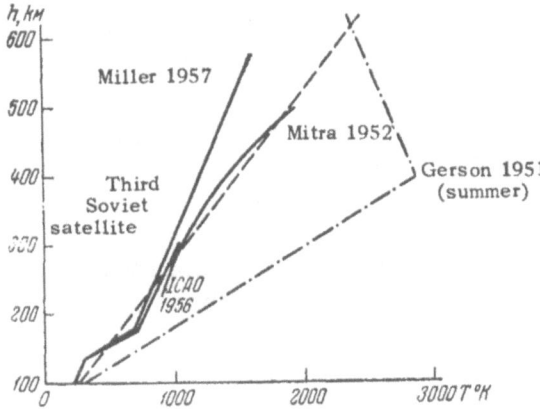

FIG. 11. Temperature of the atmosphere.

LITERATURE

1. B. S. Danilin, V. V. Mikhnevich, A. I. Repnev, and E. G. Shvidkovskii, Usp. fiz. nauk (Advances in the Physical Sciences), 63, 1b (1957).
2. V. V. Mikhnevich and I. A. Khvostikov, Izv. AN SSSR (Reports of the Academy of Sciences USSR), Geophysical Series, 11 (1957).
3. V. V. Mikhnevich, Usp. fiz. nauk (Advances in the Physical Sciences), 63, 1b (1957).
4. P. P. Alekseev, E. A. Besyadovskii, G. I. Golyshev, M. N. Izakov, A. M. Kasatkin, G. A. Kokin, N. S. Livshits, N. D. Masanova, and E. G. Shvidlovskii, Meteorologiya i gidrologiya (Meteorology and Hydrology), 6 (1957).
5. B. A. Mirtov, Usp. fiz. nauk (Advances in the Physical Sciences), 63, 1b (1957).
6. V. V. Mikhnevich, Priroda (Nature), 5 (1958).
7. R. Horowitz and H. E. La Gow, Geophysical Research, 62, 1 (1957).
8. H. E. La Gow, R. Horowitz, and J. Ainsworth, Ann. de Geoph., 14, 2 (1958).
9. V. V. Mikhnevich, Artificial Earth Satellites, Volume 2, Plenum Press, New York, 1960.
10. M. L. Lidov, Artificial Earth Satellites, Volume 1, Plenum Press, New York, 1960.
11. P. E. Él'yasberg, Artificial Earth Satellites, Volume 1, Plenum Press, New York, 1960.
12. E. A. Pen'chko and L. P. Khavkin, Pribory i tekhnika eksperimenta (Instruments and Experimental Techniques), 1 (1959).

13. A. M. Grigor'ev and G. P. Volkov, Elektronika (Electronics), 12 (1958).
14. J. Miller, Geophysical Research, 62, 3 (1957).
15. V. V. Beletskii and Yu. V. Zonov, Article being published.
16. S. K. Mitra, Verkhnyaya atmosfera (The Upper Atmosphere), Izd-vo inostr. lit., 1955.
17. T. E. Sterne, Astron. J., 63, 3 (1958).
18. N. C. Gerson, Adv. in geoph., 1 (1952).
19. H. K. Paetzold, Raketentechn. u. Raumfahrtforsch., 2, 2 (1952).
20. T. E. Sterne, Science, 127, 3308 (1958).
21. F. Whipple, Science, 128, 3316 (1958).
22. L. G. Jacchia, IGY Satellite Report, ser. no. 4, 1958.
23. G. F. Schilling and T. E. Sterne, IGY Satellite Report, ser. no. 4, 1958.
24. I. Harris and R. Jastrow, Science, 128, 3321 (1958).
25. I. Harris and R. Jastrow, Report to the 5th Assembly of the SC of the IGY (Moscow), 1958.
26. T. Sterne, Science, 128, 3321 (1958).
27. L. G. Jacchia, Sky and Telescope, 17, 6 (1958).
28. D. G. King-Hele and R. H. Merson, Journal of British Interplanetary Society, 16, 8 (1958).
29. D. G. King-Hele and D. M. Walter, Nature, 182, 4639 (1958).
30. G. Hergenhahn, Naturwiss., 45, 18 (1958).
31. J. Bartels, Naturwiss., 45, 8 (1958).
32. T. Nonweiler, Nature, 182, 4633 (1958).
33. H. E. La Gow, R. Horowitz, and J. Ainsworth, IGY World Data Center A, IGY Rocket Report Series No. 1, Nat. Acad. Sci. (Washington), July 30, 1958.
34. E. G. Shvidkovskii, Artificial Earth Satellites, Volume 2, Plenum Press, New York, 1960.
35. L. M. Jones, F. F. Fischbach, and J. W. Peterson, IGY World Data Center A, IGY Rocket Report Series No. 1, Nat. Acad. Sci. (Washington), July 30, 1958.
36. G. V. Groves, Nature, 181, 4615 (1958).
37. W. Priester, Sterne, 34, 3-4 (1958).

A RADIO-FREQUENCY MASS SPECTROMETER FOR INVESTIGATIONS OF THE IONIC COMPOSITION OF THE UPPER ATMOSPHERE

V. G. Istomin

The determination of the composition of the upper atmosphere, in particular the determination of the composition of its ionized regions, is one of the most important tasks in studying the physical parameters of the earth's atmosphere. Without knowing its composition, a general theory of the atmosphere cannot be formulated, and the atmosphere and ionosphere cannot be studied. Data on the chemical composition of the ionosphere at different altitudes over different regions of the globe and data on composition changes from daytime to nighttime and over 24-hour periods are extremely important in studying the interaction between solar ultraviolet and corpuscular radiation and the earth's atmosphere and in finding an approach for the solution of such an important geophysical and astrophysical problem as the sun-earth problem. Data on the composition of ionized layers are necessary for a correct interpretation of the results obtained in other experiments involving the ionosphere* and also for the solution of some purely practical problems.

The mass-spectrometry analysis method is the only presently known method for the direct determination of the ionic composition of the upper atmosphere. By installing an ion mass analyzer in a rocket or an earth satellite which moves inside the ionosphere and by transmitting the readings by radio to earth, it is possible to obtain information on the ion

---

*Such experiments include, for instance, the measurement of positive ion concentration[1] and the measurement of electrostatic fields in the ionosphere.[2]

mass spectrum, on the basis of which conclusions concerning the chemical composition of the ionosphere can be drawn.

A mass spectrometer which is designed for studying the composition of upper atmospheric layers will be basically different from the well-known laboratory and industrial models. Like the rest of the equipment built into a rocket or an earth satellite, such a mass spectrometer must be absolutely reliable, simple in handling, and it must operate automatically over a long period of time without any additional regulation or adjustment. It is necessary that the equipment be able to withstand considerable temperature fluctuations and large vibration and static overloads and that its operating ability be preserved throughout the action of these factors. The requirements for small weight and size, which must be satisfied as far as possible, to a certain extent contradict the above conditions. Moreover, the mass spectrometer must operate on low power, it must have a small inertia, and it must satisfy other specifications.

The radio-frequency mass spectrometer,[3] in particular one of its variants, the Bennett radio-frequency mass spectrometer,[4] satisfies these specifications. This mass spectrometer was first used in 1952 by John Townsend, Jr., in investigations with rockets.[5] Although the atmospheric and ionospheric compositions have been studied in the United States over a period of seven years,[6-9] reliable results have been obtained only recently. Thus, as a result of three rocket launchings at Fort Churchill (Canada), which were scheduled in the IGY program, data on the ionic composition of the Arctic winter atmosphere up to an altitude of 251 km were obtained.[10,11]

In the Soviet Union, investigations of the ionic composition of upper atmospheric layers began in 1957. As a result of the investigations performed as a part of the IGY program by means of high-altitude geophysical rockets and the third artificial earth satellite, data on the mass spectrum of positive ions in the ionosphere at altitudes of up to 885 km were obtained.[12] The radio-frequency mass spectrometer described in the present article was used in these investigations.

The theory of mass spectrometers of different types, where the principle of ion separation with respect to velocities in high-frequency electric fields is used, has been repeatedly explained in the literature.[3,4,13-15] It is assumed that the operating principle of the device is known, and, therefore, it will not be presented here.

THE CONSTRUCTION OF THE DEVICE

The RMS-1 radio-frequency mass spectrometer consists of three basic components, which are built as three separate units: the mass spectrometer tube (radio-frequency mass analyzer) with a preamplifier, the electronic unit of the device, and the feed unit (Fig. 1).

FIG. 1. RMS-1 radio-frequency mass spectrometer. 1) Mass spectrometer tube with preamplifier; 2) electronic unit; 3) feed unit (Sputnik variant).

1. The mass spectrometer tube consists of a three-step seven-and-five cycle analyzer. On the basis of data available in the literature,[5,15,16] one can conclude that the seven-and-five-cycle variant is apparently one of the optimum devices with respect to the harmonic level and sensitivity, which explains our choice.*

*A laboratory mock-up of a seven- and five-cycle Bennett mass spectrometer, which was, however, intended for entirely different purposes, was constructed for the first time in the Soviet Union in the West-Siberian Branch, Academy of Sciences, USSR, by a team headed by A. N. Vorsin.[17] The author is deeply grateful to the members of this Branch, M. Ya. Shcherbakova, E. F. Doil'nitsyn, and A. V. Trubetskoi, for their help in the present work.

The mechanically strong structure of the tube was developed on the basis of specified geometric and electric parameters in one of the enterprises of the State Committee for Radio Electronics at the Ministerial Council of the USSR (chief engineer — Z. G. Petrenko). All the mass spectrometer tubes were made in the same place. The specifications for the tube design and the tube's geometric and electric parameters were determined on the basis of laboratory experiments which we performed beforehand. The diagram showing the position of analyzer grids and of the ion source with the distances between the grids and the assigned tolerances is given in Fig. 2. The external view of the analyzer is shown in Fig. 3.

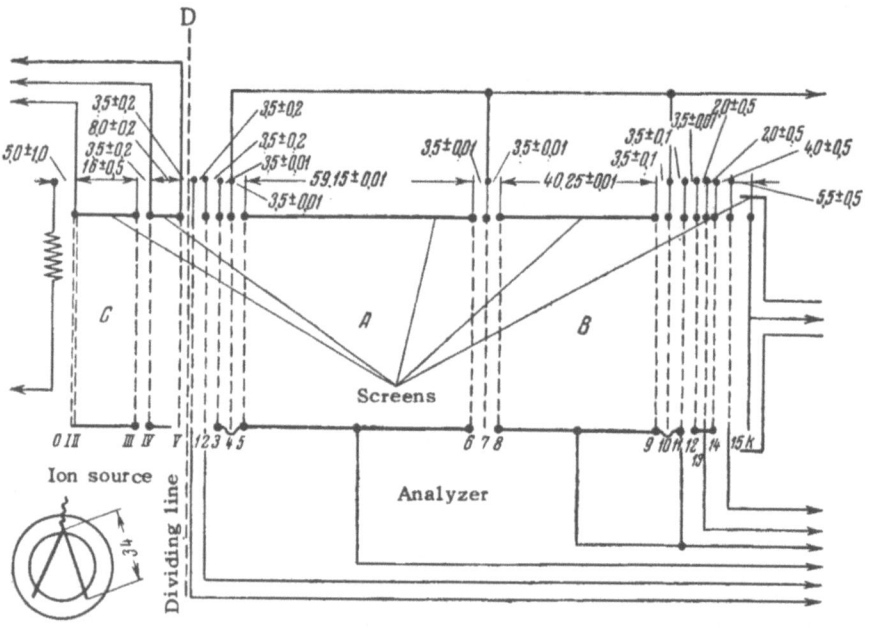

FIG. 2. Arrangement of the analyzer grids and of the ion source of the radio-frequency mass spectrometer. O) V-shaped cathode (VT-10 tungsten, 100 μ); I-V) grids of the ion source; C) ionization space; D) dividing line between the analyzer and the ion source; 1) and 2) analyzer pulling grids; 3), 4), and 5) first step; A) ion drift space (seven cycles); 6), 7), and 8) second step; B) ion drift space (five cycles); 9), 10), and 11) third step; 12), 13), and 14) retarding grids; 15) protective screen; K) ion collector.

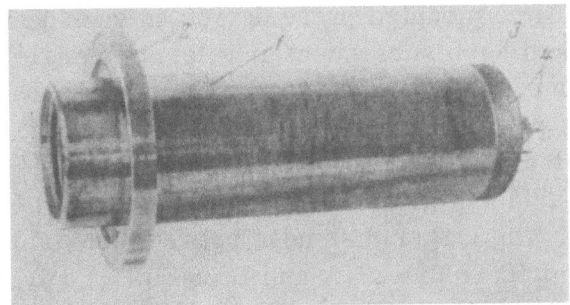

FIG. 3. The mass spectrometer analyzer. 1) Analyzer shell; 2) flange for fastening the analyzer in a rocket or a satellite; 3) threaded part for fastening the preamplifier; 4) lead-outs of the analyzer electrodes.

One of the characteristics of our device is the fact that, in contrast to the radio-frequency mass spectrometers described earlier,[3,4,16] single-row grids, prepared by winding a tungsten wire on Kovar rings, were used in the analyzer and the ion source.[17] For fastening the grid turns on the main ring, another thin (0.01 mm) Kovar ring was placed on top of the wound wire and fixed by spotwelding. Figure 4 shows the photograph of such a grid. The grid operating diameter is 30 mm, the

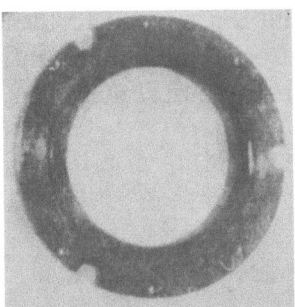

FIG. 4. Grid of the mass spectrometer analyzer.

thickness of a turn is 20 μ, and the winding pitch is 0.5 mm. The transparency of a grid prepared in this manner is 96%. The high transparency of the grid is very important, since the analyzer consists of 15 such grids, and the ion currents obtained from the tube collector increase with an increase in the over-all transparency of all grids, all other conditions being equal. The application of single-row grids makes the construction simpler and less expensive without impairing its efficiency.

The grids are mounted in the device in such a manner that the turns of each subsequent grid form an angle of 120° with the turns of the preceding grid. The plane-parallel arrangement of grids with respect to each other at the required distances with an accuracy of ±0.01 mm and the electrical insulation of adjacent grids is ensured by ground metal-glass ring spacers. The larger distances between the grids which limit the ion-drift volumes (A and B in Fig. 2) are set by ground copper cylinders, which screen the drift volumes at the same time. The package of grids and glass-metal ring-shaped spacers together with the screens and the collector unit is installed between ground ceramic rods 4 mm in diameter and is placed in the analyzer shell, which is afterwards rolled in on the side of the first grid. The Kovar glass-metal adapters in the tube mount serve as lead-outs from the analyzer grids and collector. The Kovar mount header is welded to the shell.

The shell of the mass spectrometer analyzer is made of stainless steel. The shell is provided with a flange which serves for mounting the analyzer in a rocket or a satellite as well as for fastening the ion source to the analyzer. The entire mass spectrometer tube is a demountable glass-metal assembly (Fig. 5). The all metal analyzer is joined to the ion

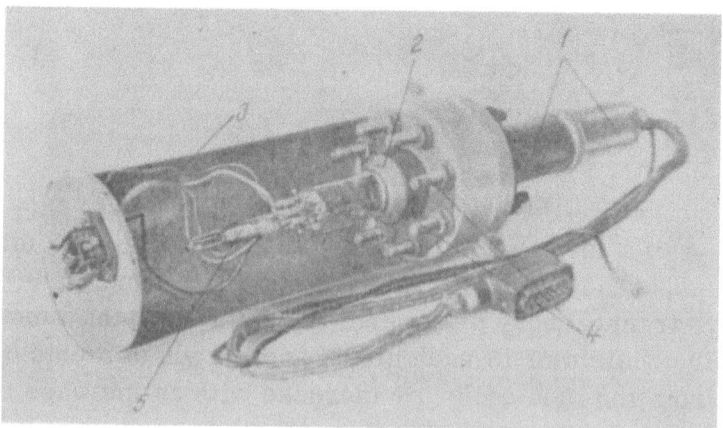

FIG. 5. Mass spectrometer tube. 1) Analyzer with preamplifier; 2) ion source; 3) protective casing (one half removed); 4) mounting bolts; 5) exhaust tip with flashed getter.

source by means of a vacuum packing, which consists of a copper gasket. The ion source has a V-shaped cathode (thoriated VT-10 tungsten 0.1 mm in diameter) and five grids of the type used in the analyzer. The ion source serves for adjusting the device in the laboratory, for checking its operation in all subsequent tests, and for plotting calibration diagrams. When the device is used for analyzing the ionic composition of the atmosphere, the ion source is of course unnecessary, since its function is taken over by the atmosphere itself.

The mass spectrometer tube with the connected ion source is thoroughly degassed by heating to 400° over a period of several hours, after which it is filled with a mixture of argon and neon at a pressure of approximately $3 \cdot 10^{-5}$ mm Hg and then disconnected from the vacuum device and sealed. Two to three hours after sealing, the tube operation is checked in order to make sure that no leaks exceeding the maximum allowable value are present. After this, a getter is flashed in the tube, which is necessary for absorbing the gases which evolve from the ion source cathode during its operation and leak into the tube through the packing. Experiments show that such tubes can be stored for months and that they can operate for a long time without evacuation by means of a vacuum device.

2. The mass spectrometer electronic components are mounted in its electronic unit (Fig. 6). The electronic unit consists of: a) a dc amplifier, b) a high-frequency oscillator; c) "bias" and "stop" voltage rectifiers; d) a saw-tooth voltage generator for scanning with respect to masses; e) a 600-v dc voltage converter; f) a commutation system for the radiotelemetering lead-outs; and g) a relay system for switching on and off the device and for checking.

The mass spectrometer block diagram, in which the above-enumerated basic electronic units are shown, is given in Fig. 7.

The dc amplifier* amplifies the ion current from the mass spectrometer tube collector and supplies pulsed signals (mass

*The circuit recommended in Ref. 18 was used as the basis in developing the amplifier.

FIG. 6. Electronic unit of the radio-frequency mass spectrometer (protective casing removed). 1) Screen of the ion current amplifier; 2) relay system; 3) quartz; 4) oscillator circuit; 5) high-frequency oscillator tube; 6) "bias" and "stop" voltage rectifiers; 7) step-by-step selector; 8) tube of the scanning saw-tooth voltage generator; 9) compensating battery (B_s).

"peaks") to the radiotelemetering line in the range from 0 to +6 v. The complete amplifier circuit is shown in Fig. 8.

The first stage — the electrometer stage — is based on the 6Zh1Zh tube, which operates under normal electrometer conditions. The input resistance is equal to 10^{10} ohm. The 6Zh1Zh tube and the megohm resistor are mounted outside the electronic unit in the preamplifier frame, which is fitted directly on the tail end of the mass spectrometer tube. The wiring diagrams of the analyzer and the preamplifier are shown in Fig. 9. The preamplifier is connected to the electronic unit by means of a multiconductor screened cable and an RK-19 coaxial cable with suitable terminals at their ends. The cable lengths are slightly over one meter.

The remaining amplifier tubes are subminiature tubes of the "fraction" series. The second stage — the 6Zh1B (L_{11})

tube — operates under the so-called "microcurrent conditions", and the anode load resistance is equal to 10^6 ohm.

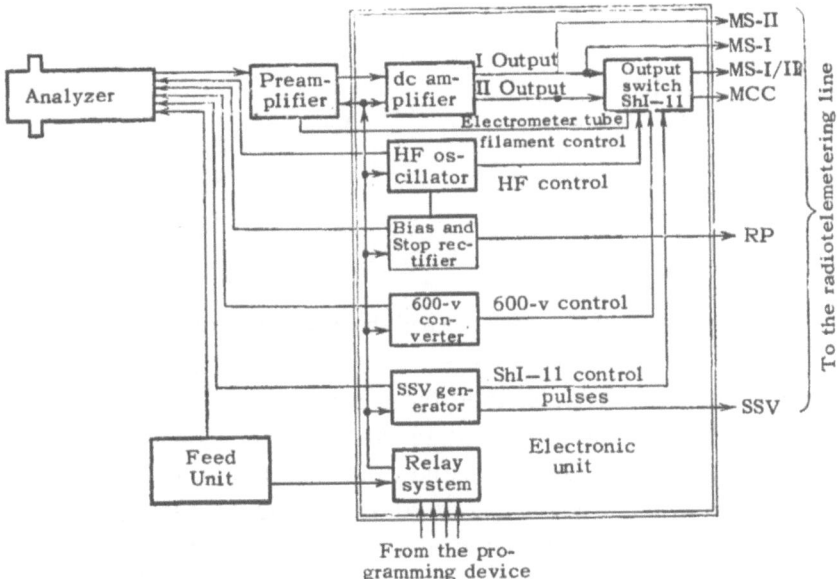

FIG. 7. Block diagram of the radio-frequency mass spectrometer.

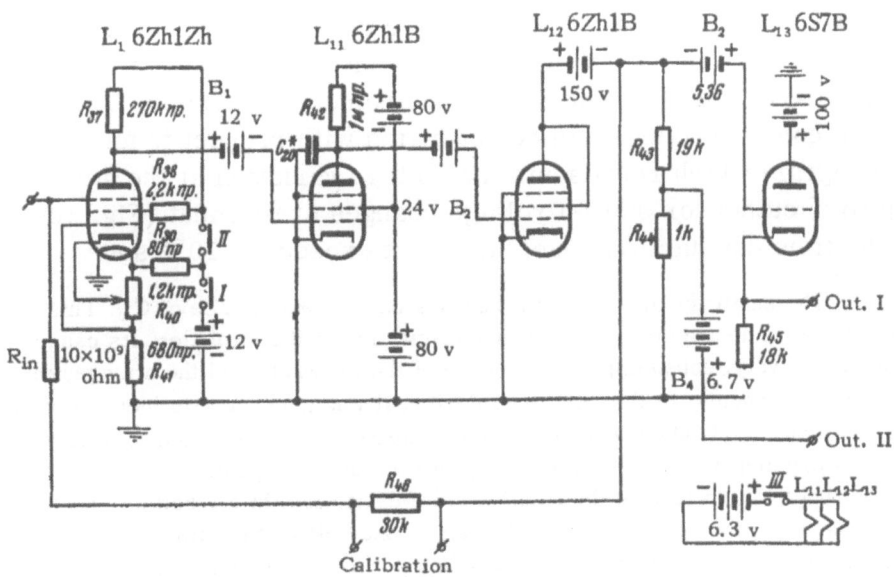

FIG. 8. Wiring diagram of the dc amplifier.

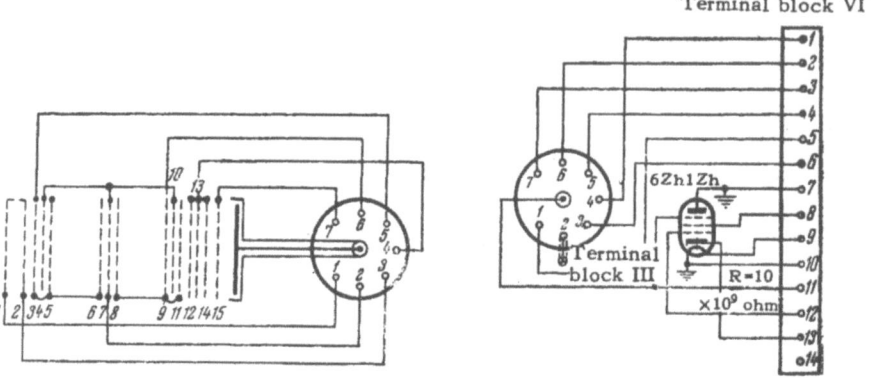

FIG. 9. Wiring diagram of the analyzer and the mass spectrometer preamplifiers.

Such a regime is characterized by a large amplification factor for negligible anode current in the tube. A 6Zh1B tube, which is connected as a triode and operates under ordinary conditions, is also used in the third amplifier stage (L_{12}). The first three stages are encompassed by 100% negative voltage feedback, due to which the voltage amplification factor of the first stages is equal to unity. The voltage amplification factor without feedback is equal to 1000. The application of negative feedback reduces the amplifier time constant and increases its stability. The amplifier time constant secures an undistorted amplification of ion current pulses (mass "peaks"), which are tapped off the mass spectrometer collector. The time constant values can be estimated with respect to an oscillogram which represents the voltage variation at the amplifier output for sudden voltage changes at its input (Fig. 10). The time constant obviously does not exceed $5 \cdot 10^{-3}$ sec*.

*The amplifier time constant depends on the capacitance of C_{20}. This capacitor is necessary to prevent self-excitation of the amplifier; its capacitance can vary, depending on the nature of the assembly and the arrangement of the components. In the described device, it was necessary to increase this capacitance to 3300 $\mu\mu$f because of the extremely closely-packed assembly, which apparently provided many opportunities for the appearance of positive feedbacks, which lead to the amplifier self-excitation. In a laboratory mock-up assembly, the capacitance of C_{20} can be one-half to one-third of that used in the actual device, which, correspondingly, reduces the time constant.

This value is acceptable, since the duration of ion current pulses for the mass spectrometer operating conditions is equal to $3\text{-}5 \cdot 10^{-2}$ sec.

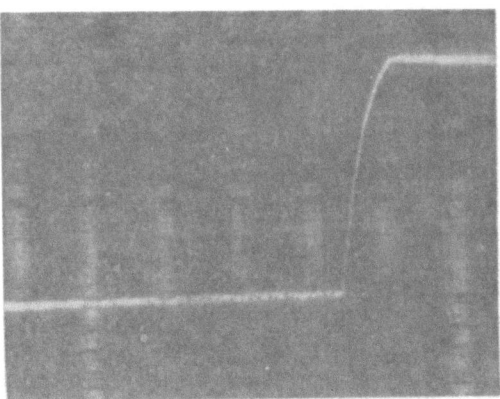

FIG. 10. Oscillogram of the voltage at the dc amplifier output for sudden voltage changes at its input ($R_{in} = 10^{10}$ ohm). The short time marks follow every $2 \cdot 10^{-2}$ sec.

The final amplifier stage (L_{13}) — the cathode follower, which is based on the 6S7B tube — is necessary for matching the sensitive output with the input of the radiotelemetering system. All amplifier stages are galvanically coupled. The coupling batteries (B_1, B_2, and B_3) are mounted together with the amplifier. The B_4 battery, which is also mounted together with the amplifier, provides a positive voltage at the low-sensitivity output. The B_1-B_4 are batteries composed of miniature OR-1K mercuric oxide cells.

The sensitive amplifier output (cathode follower) is linear up to an input voltage of ~5.5 v, and the voltage amplification factor for this output is equal to ~0.8. The low-sensitivity output (divider R_{43} and R_{44}) is linear up to an input voltage of ~110 v for a voltage amplification factor of ~0.05. The amplifier amplitude characteristics are given in Fig. 11.

The amplifier filament circuits are fed from silver-zinc storage batteries, and the anode circuits are fed from a battery of mercuric oxide cells, which are mounted inside the mass spectrometer feed unit. In order to improve the amplifier stability, the filament, the anode, and the screen grid of the electrometer stage (first tube) are fed from a separate storage battery.

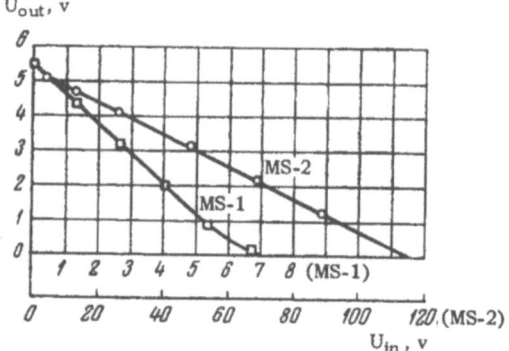

FIG. 11. Amplitude characteristics of the mass spectrometer amplifier.

The circuit diagram of the electronic unit is shown in Fig. 12. The circuit described by Townsend[5] was used as the basis in developing the feed circuit for the mass spectrometer tube. The high-frequency oscillator, which is stabilized with quartz, is based on the Pierce circuit, where a 6P9 tube (L_5) is used. In contrast to the Townsend circuit, the mass spectrometer tube is supplied with high frequency through a capacitive divider, which consists of the C_{10} capacitor and the capacitances of the analyzer grids 4, 7, and 10. Such a feed circuit considerably reduces the power consumed by the oscillator. The high-frequency voltage amplitude is regulated by varying the screen grid voltage in the 6P9 tube by means of resistor R_{17}. The "bias" and "stop" voltages (retarding potential) are provided by rectification and doubling the high-frequency voltage by means of 6D6A tubes (L_6-L_9). Such a feed circuit assures steady operating conditions for the tube if the high-frequency voltage fluctuations are small, because the "bias" and "stop" voltages will be proportional in this case. The small portion of the high-frequency voltages which is rectified by the 6D6A tube (L_{20}) serves for controlling the oscillator operation. The "bias" and "stop" voltages are regulated by means of potentiometers R_{24} and R_{23}. The high-

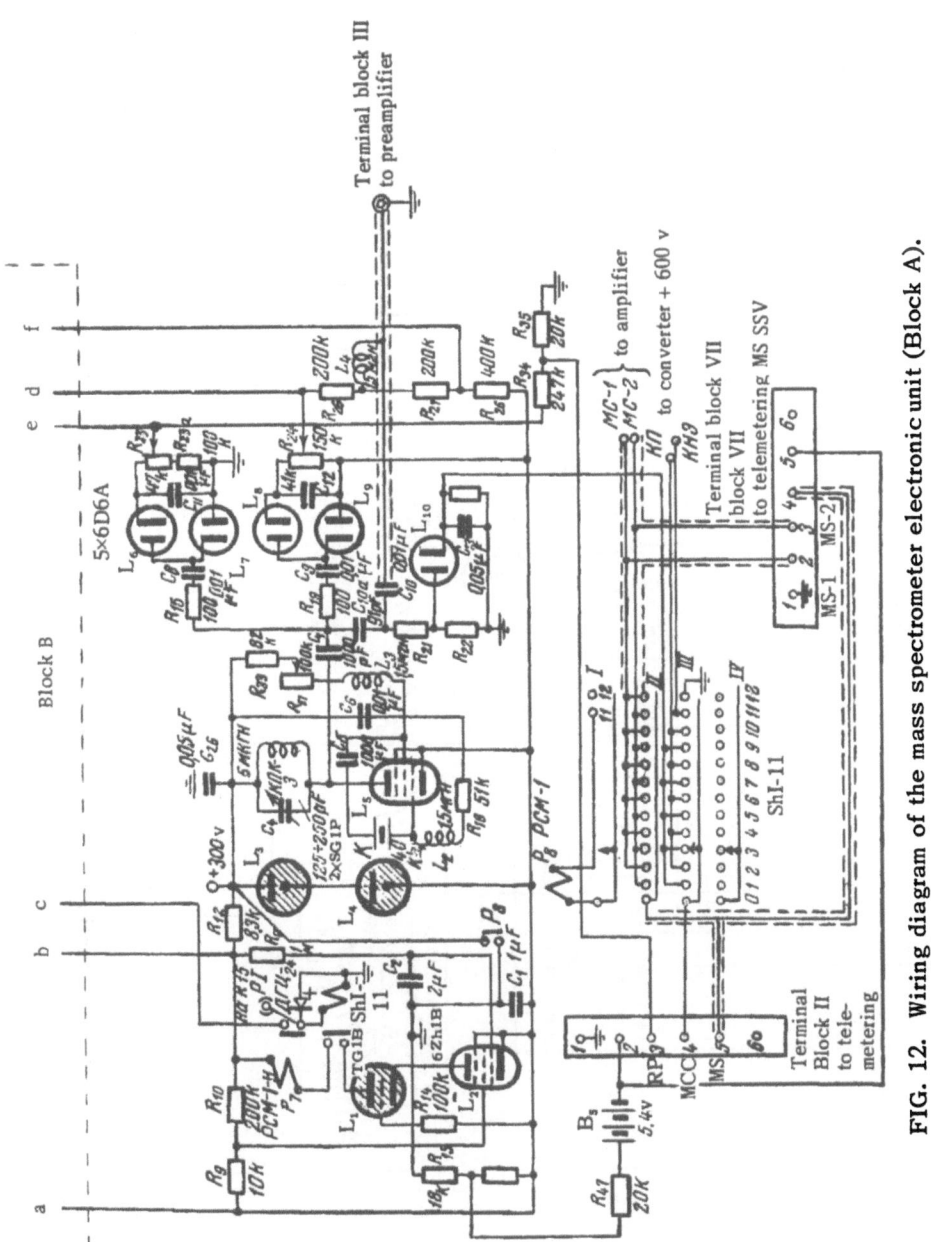

FIG. 12. Wiring diagram of the mass spectrometer electronic unit (Block A).
(Note: $\mu F = \mu f$; $pF = \mu\mu f$; мгн $= mh$; мкгн $= \mu h$; $k =$ kilohm; $M =$ megohm.)

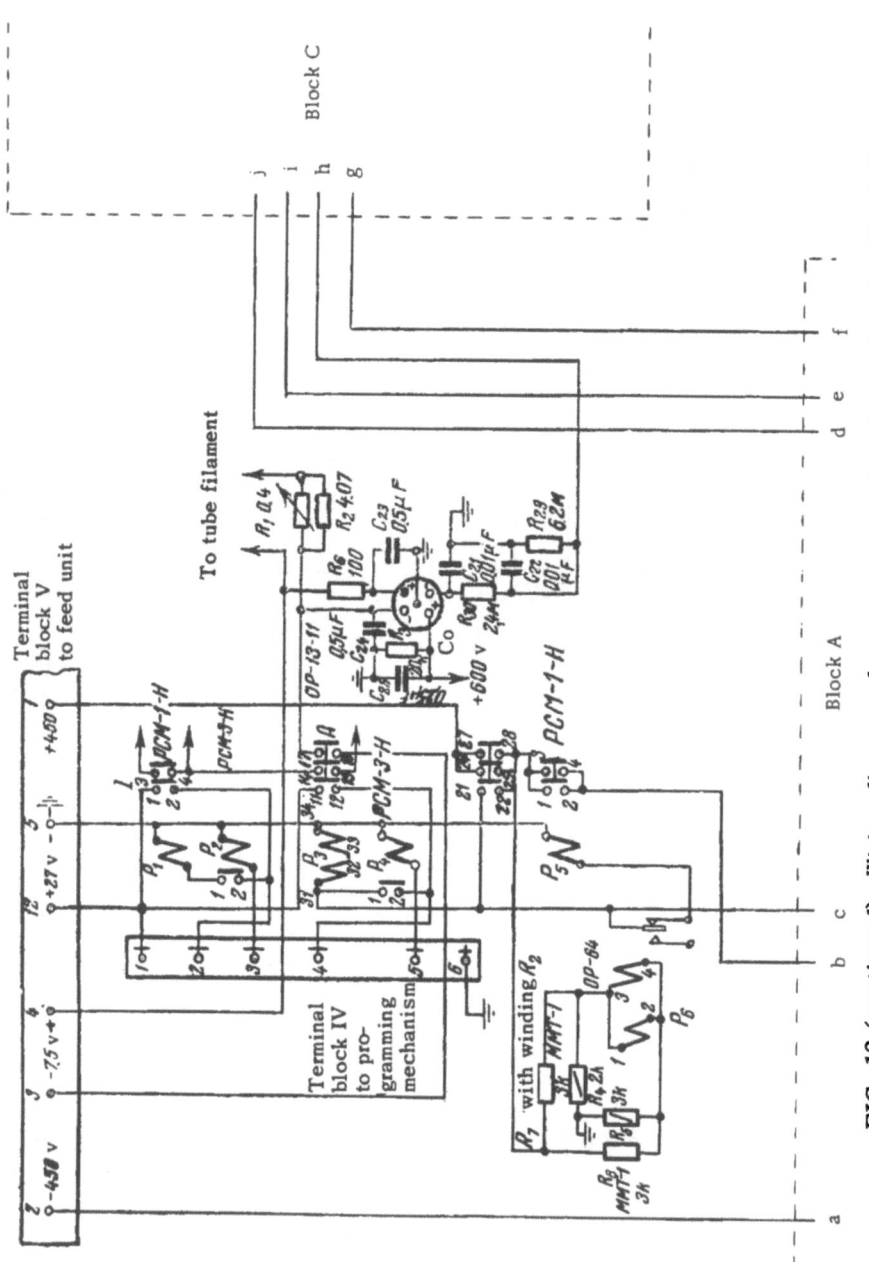

FIG. 12 (continued). Wiring diagram of mass spectrometer electronic unit (Block B).

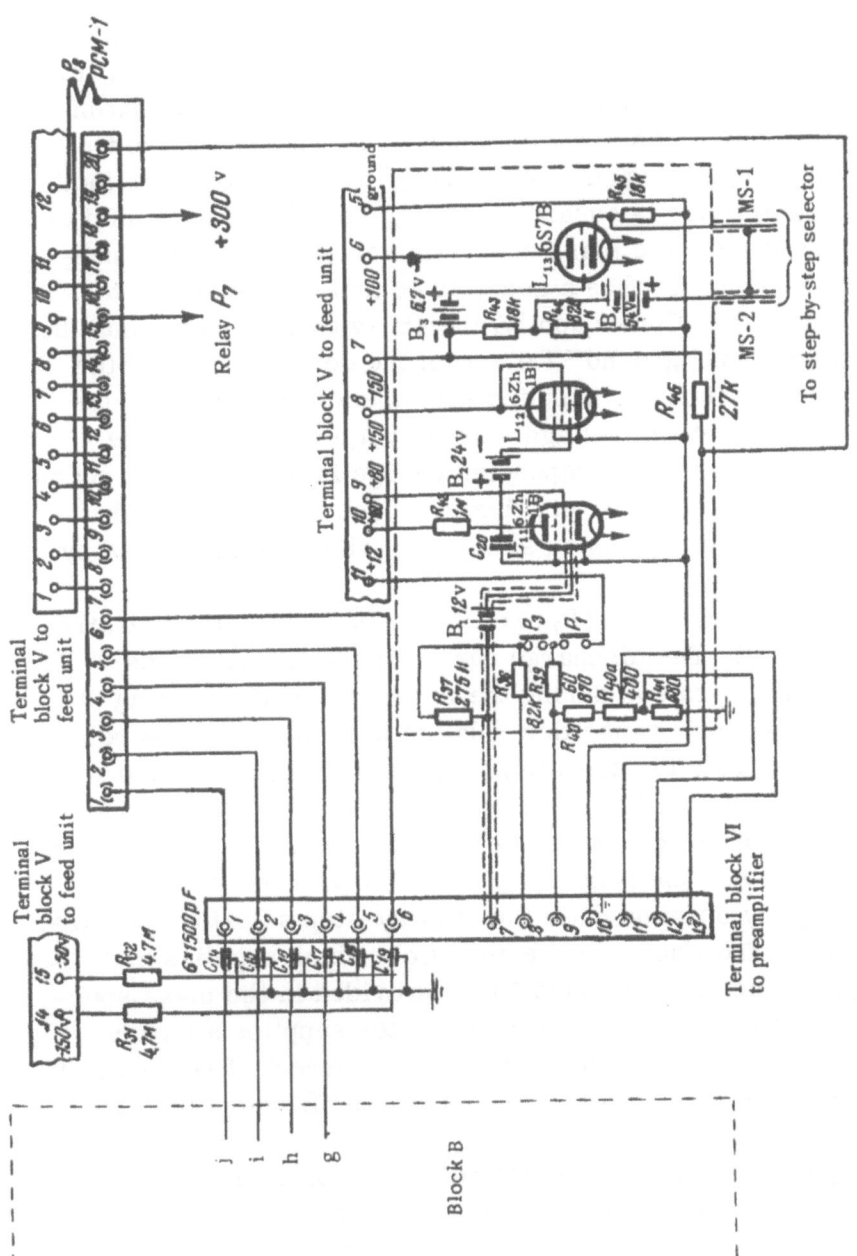

FIG. 12 (continued). Wiring diagram of mass spectrometer electronic unit (Block C).

frequency oscillator anode voltage is stabilized by means of two SG-1P stabilivolts (L_3 and L_4).

The saw-tooth voltage for scanning with respect to masses is generated by a circuit based on a TG-1B thyratron (L_1) with a 6ZH1B linearity-control pentode (L_2). The thyratron operating conditions are somewhat forced, since the anode battery voltage is equal to 450 ± 23 v. For this, the saw-tooth voltage amplitude is equal to ~ 400 v, the voltage being markedly linear. For the mass spectrometer operation, it is necessary that the saw has a negative polarity with respect to the tube frame. Consequently, the 450-v battery poles must be insulated from "ground" i.e., from the device frame. If the scanning oscillator is supplied with unstabilized voltage, the maximum saw-tooth voltage values systematically change, which results in a corresponding change in the maximum mass to be registered. However, this fact does not cause any difficulties, since the scanning saw-tooth voltage is controlled by radiotelemetering.

The winding of the (P_7) relay is connected to the thyratron anode circuit. The discharge current pulses, which pass through the thyratron during the reverse sweep (charging of the C_1 capacitor), cause this relay to operate, while the relay, in turn, controls the operation of the ShI-11 step-by-step selector. The normally closed contacts of the step-by-step selector, which are connected to the break in the thyratron anode circuit, prevent the possibility of an uncontrolled thyratron discharge by opening while the step-by-step selector operates.

A semiconductor converter (Co), which supplies high voltage to the analyzer grid 15, is provided in the mass spectrometer electronic unit. The converter supplies a steady voltage of the order of 600 v for a load current of ~ 50 μa and uses a current of ~ 8 ma at 6.3 v from the filament battery. A part of the converter voltage, which is tapped off the R_{20} and R_3 divider, is used for controlling its operation. This voltage (~ 1 v) is supplied to the even segments of section III of the step-by-step selector. The high-frequency oscillator control voltage and the filament voltage of the first amplifier tube are supplied to the odd segments of this section. Thus, three control parameters are commutated by the step-by-step selector with the mass scanning frequency and are fed to a single telemetering channel (the MCC parameter represents the mass spectrometer circuit control). This was done in order to reduce the number of telemetering channels required for the operation of the device. Beside the enumerated parameters, also a part of the "stop" voltage, which is tapped off the R_{34} and R_{35} divider (the retarding potential parameter RP), and a part of the mass scanning saw-tooth voltage (the SSV parameter) are telemetered. The SSV parameter voltages are tapped off the R_{13} and R_{15} divider; however, since the saw polarity is negative with respect to "ground," a compensating battery (B_5) is connected to its circuit. Like the B_1-B_4 batteries, it consists of miniature OR-1K mercuric oxide cells.

Beside the control parameters, the step-by-step selector also commutates the dc amplifier outputs (Section II). This was done in order to make it possible to use only one telemetering channel with high interrogation capacity for the mass spectrometer operation. However, as can be seen from the electronic unit circuit diagram (see Fig. 12), both amplifier outputs are also brought out without commutation, so that, in normal operation, the sensitive (the MS-1 parameter — the mass spectrum, first range) as well as the low-sensitivity (the MS-2 parameter — the mass spectrum, second range) outputs can be independently telemetered. Section IV of the step-by-step selector can be used in different ways depending on the various tasks to be performed in individual experiments. In particular, by means of Section IV, the analyzer operating conditions can be changed with respect to the retarding potential,* to which 12 different values can be assigned.

The relay system P_1-P_4 serves for connecting and disconnecting two circuits of the apparatus with respect to the

---

*Thus, in the experiment performed on September 9, 1957,[12] the retarding potential was periodically increased for the purpose of eliminating spurious ("harmonic") peaks which could appear in mass spectra due to the intrinsic charge effect.[19]

programming device signals. These signals are supplied as short-duration closures of contacts in terminal bolck IV. The closure of contacts 1 and 2 in terminal block IV causes the operation of relay P_1, while relay contacts 3 and 4 are used for connecting the filament circuit of the amplifier 6Zh1Zh electrometer tube; contacts 1 and 2 block contacts 1 and 2 in terminal block IV, and relay P_1 remains closed until the disconnecting command is supplied. The electrometer tube filament is connected 15 minutes before the start of operation, in order to stabilize the amplifier zero. The connection of the remaining circuits is effected by a short-duration closure of contacts 1-4 of terminal block IV. In this, contacts 11 and 12 of relay P_3 also block contacts 1-4 of terminal block IV, and relay P_3 remains connected until the disconnection command is supplied. The disconnection command is also supplied by a short-duration closure of contacts 1-5 of terminal block IV. In this case, relay P_4 operates, and its normally closed contacts 1 and 2 break the feed circuit of the relay P_3 winding. The disconnection of the electrometer tube filament is effected in a similar manner by short-duration closure of contacts 1-3 of terminal block IV.

Due to the presence of a thyratron in the saw-tooth voltage generator circuit, the anode voltage connection cannot take place simultaneously with the connection of tube filaments. Therefore, the anode voltage is supplied 20 to 40 seconds after the tube filaments are connected. The lag is ensured by a thermorelay circuit. A polarized OR-64 relay (P_6) is included in the diagonal of the bridge which is formed by two MMT-1 thermoresistors (R_7 and R_8) and two MLT resistors (R_4 and R_5). One of the thermoresistors is heated by the tube filament current. For this purpose, a winding (R_2), through which a portion of the tube filament current passes, is provided on the thermoresistor. This disturbs the bridge balance, and the polarized relay operates and connects the P_5 relay, which closes the circuit for the anode voltage that is supplied to the mass scanning oscillator and the high-frequency oscillator.

3. The mass spectrometer feed unit serves for feeding the

electronic components and the pulling grids 1 and 2 of the analyzer. The feed unit is provided with silver-zinc storage batteries, which have a high capacity for a small weight and size, and batteries of mercuric oxide cells, which are the lightest and most compact of all electrochemical systems available at the present time.

WORK WITH THE MASS SPECTROMETER

In preparing experiments for determining the ionic composition of the ionosphere, it was found that for a reliable interpretaion of the data obtained it was necessary to determine the SSV parameter voltage (mass scanning saw-tooth voltage*) with great accuracy. An error in determining SSV of ~2% would result in an error of 1 amu in determining the mass number, which is entirely inadmissible. The required accuracy can be achieved by first carefully calibrating all units separately from the telemetering system (with the registration of the spectrum and the measurement of SSV levels corresponding to certain given mass numbers) and then together with the telemetering system, but without spectrum registration. However, any intermediate calibrations would unavoidably increase the over-all error in determining the levels of the parameters to be telemetered, since the use of highly accurate devices for calibration purposes is difficult for many reasons. Therefore, it was very important to perform a so-called through calibration of the entire mass spectrometer — radiotelemetering channel with the recording of the mass spectrum of a mixture of gases of known composition. Initially, the performance of such calibrations was difficult, since a vacuum system for the analyzer evacuation was required for

*From the theory of the device, it follows that the mass number corresponding to a given peak of the collector ion current is related to the scanning saw-tooth voltage by the expression $M = V/k$, where V is the scanning saw-tooth voltage at the moment when the peak (in volts) appears, M is the mass number corresponding to the given peak (in atomic mass units), and k is a constant which depends on the analyzer parameters and the operating conditions. For the described device, $k = 7.2$ v/amu.

the device to operate and for producing the mass spectrum. The situation was improved when it was demonstrated by laboratory experiments that the analyzer can operate over a prolonged period of time when it is sealed off from the vacuum device. The degassing and filling operations on the mass spectrometer tube were performed for this purpose.

Work with a sealed-off tube which was filled with a mixture of known inert gases was convenient and offered great advantages. In the first place, it was possible to record the spectrum for telemetering at any stage of experiment preparation and thus obtain the recording to be used in plotting the calibration characteristic of the device with respect to masses. This made it possible to omit all the intermediate calibration stages, and, consequently, to increase considerably the accuracy in determining the mass numbers. The calibration characteristic was plotted with respect to five reference points on the Ar^{40}, Ar^{36}, Ne_2^{28}, Ne^{22}, and Ne^{20} mass scale. The presence of a stable filling in the mass spectrometer tube made it possible to adjust the device quickly by setting the desired operating conditions (resolution) with respect to the peaks of the Ne^{20} and Ne^{22} isotopes and also provided the possibility of thoroughly checking the operation of the device and making sure that the operating conditions and adjustment did not change.

In order to obtain reproducible results which could be compared with each other, all devices were filled with mixtures of a more or less equal composition under pressures which were determined as accurately as possible. Simple dosing devices, which ensured the required filling homogeneity in the first approximation, were used for composing the mixture and for filling the tubes. In filling, the mixture pressure in the tube was determined with the VI-3 ionization vacuummeter. It was possible to determine the pressure inside the tube also after it was sealed off. This determination was performed with respect to the ion current of the second analyzer grid for a fixed emission current under normal operating conditions for the first two grids, while the mass spectrometer electronic unit was disconnected. The calibration for determining the pressure was performed for one tube; the calibration graph is shown in Fig. 13.

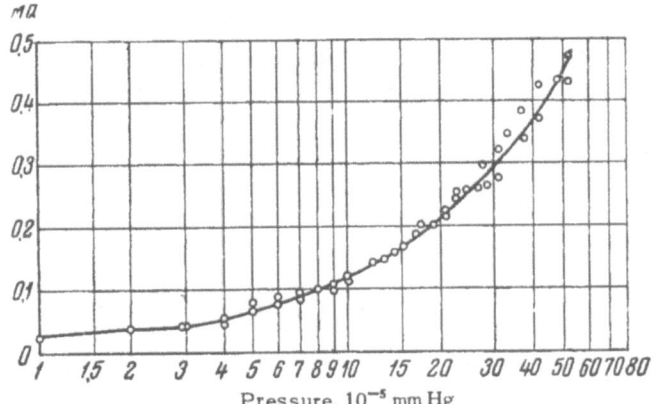

FIG. 13. Calibration graph for determining the pressure in the sealed-off mass spectrometer tube.

The tuning of all devices was performed in the same manner. The high-frequency level was set to the maximum value, which was equal to 12-14 V_{eff}. Then, the bias voltage was finely adjusted with respect to the Ne^{20} or Ne^{22} maximum peak. After this, the retarding potential was adjusted. In order to achieve results which could be compared with each other, all the devices were tuned in the same manner, namely, to a resolution of the order of 20 with respect to the peak base in the region of mass 20. The resolution was estimated visually with respect to the peaks of Ne^{20} and Ne^{22} isotopes, which were observed on the screen of a control oscillograph.

For laboratory work with the device and also for all subsequent stages in experiment preparation, special control panels (Fig. 14) have been developed and constructed. The control panel makes it possible to perform the following operations: a) switching on and off of the mass spectrometer; b) control of all feed unit voltages; c) control of the mass spectrometer operating conditions; d) control of voltages at the telemetering outputs; e) calibration of the mass spectrometer amplifier and control of the calibrator voltages (ten voltage steps in the range from 1.3 to 90 v); f) feed of the ion source of the mass spectrometer tube, control and regulation of the emission current, and control of the ionization voltage; g) calibration of the scanning saw-tooth voltage; h) mass spectrum observation on the cathode-ray oscillograph screen*; i) recording of the mass spectrum and of all other parameters to be telemetered by means of a loop oscillograph†.

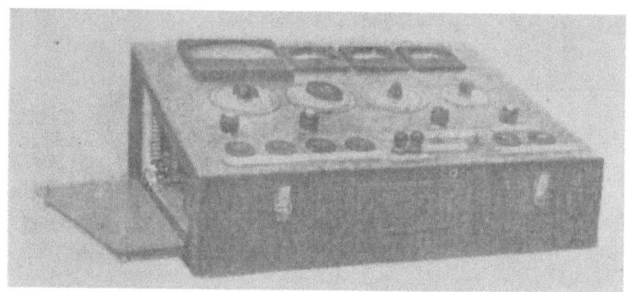

FIG. 14. Mass spectrometer control panel.

The characteristic mass spectrum of the gases filling the tube is shown in Fig. 15.

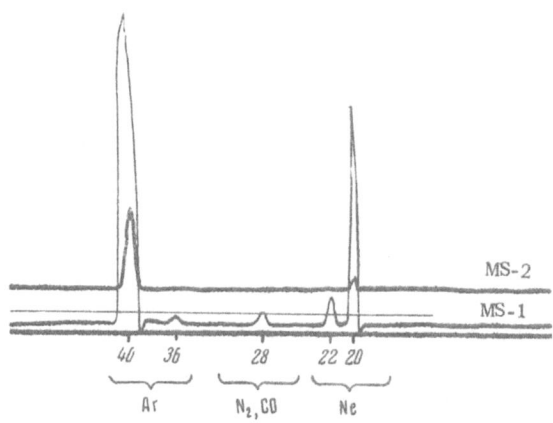

FIG. 15. Mass spectrum of the gas mixture filling the mass spectrometer tube.

*A standard ÉO-7 oscillograph was used as the control oscillograph, where additional dc current lead-ins for the vertical and horizontal axes were provided.
†A POB-12 loop oscillograph was used.

The normal operating conditions of the device are the following* (all voltages are given with respect to the tube frame):

Grid No. 1: 50 v ⎱ (pulling grids)
Grid No. 2: 130 v ⎰

Grids Nos. 3-11: (10-400) v (mass scanning saw-tooth voltage)

Grids Nos. 4, 7, and 10: 11.7 v_{eff} (the high frequency is 4.0 Mc)

Grids Nos. 12, 13, and 14: +63 v (retarding potential "stop")

Thus, the device operates under the $U_{stop}/U_{hf}^{eff} = 5.4$ regime.

Grid No. 15: (450-480) v.

Moreover, a bias potential of −23 v with respect to grids Nos. 8, 9, and 11 is supplied to grids Nos. 3, 5, and 6.

Under the above conditions, the mass spectrometer resolution is equal to 20 in the region of mass 20 (with respect to the peak base). The device mass range is from 6 to 48 amu. The scanning time for the entire mass range (mass scanning period) is approximately equal to 1.7 sec. The amplifier input resistance is 10^{10} ohm. The amplifier dynamic range is approximately equal to 500; it is secured by two outputs: the sensitive output of $0-5 \cdot 10^{-10}$ amp and the low-sensitivity output of $0-100 \cdot 10^{-10}$ amp.

ACKNOWLEDGMENT

In conclusion, the author extends his thanks to Z. G. Petrenko, O. F. Lantsman, A. I. Sedov, and R. P. Shirshov, who did a great amount of work in constructing, preparing, and testing the mass spectrometer tubes, to L. P. Chulkin, O. V. Rodionov, and A. A. Perno, who made and tested the mass spectrometer electronic units and the control panel, and to S. V. Vasyukov for his help in the work.

*These are the operating conditions of the mass spectrometer which was installed in the third Soviet satellite.

LITERATURE CITED

1. K. I. Gringauz and M. Kh. Zelikman, Uspekhi Fiz. Nauk, 63, 1, 239 (1957).
2. I. M. Imyanitov, Uspekhi Fiz. Nauk, 63, 1, 267 (1957).
3. P. A. Redhead, Canad. J. of Phys., 30, 2, 1 (1952).
4. W. H. Bennett, J. Appl. Phys., 21, 2, 143 (1950).
5. J. W. Townsend, Rev. Sci. Instr., 23, 10, 538 (1952).
6. J. W. Townsend, E. B. Meadows, and E. S. Pressley, Investigation of Upper Atmosphere by Means of Rockets (Russian translation), IL, 1957, p. 192.
7. C. Y. Johnson and E. B. Meadows, J. Geoph. Res., 60, 2, 193 (1955).
8. C. Y. Johnson and J. P. Heppner, J. Geoph. Res., 60, 4, 533 (1955).
9. C. Y. Johnson and J. P. Heppner, J. Geoph. Res., 61, 3, 575 (1956).
10. E. B. Meadows and J. W. Townsend, Diffusive Separation in the Winter Nighttime Arctic Upper Atmosphere, 112 to 150 km. Report to the Fifth Assembly of the IGY SK, Moscow, 1958.
11. J. W. Townsend, Jr., C. Y. Johnson, J. C. Holmes and E. B. Meadows, Atmospheric Composition at Arctic High Altitudes, Report to the Fifth Assembly of the IGY SK, Moscow, 1958.
12. V. G. Istomin, Investigation of the Earth's Atmosphere Ionic Composition by Means of Rockets and Earth Satellites, Report to the Fifth Assembly of the IGY SK, Moscow, 1958.
13. P. A. Redhead and C. R. Crowell, J. Appl. Phys., 24, 3, 331 (1953).
14. T. C. Wherry and F. W. Karasek, J. Appl. Phys., 26, 682 (1955).
15. M. Ya. Shcherbakova, Zhur. Tekh. Fiz., 27, 3, 599 (1957).
16. E. F. Doil'nitsyn, A. U. Trubetskoi, and M. Ya. Shcherbakova, Zhur. Tekh. Fiz., 27, 2, 404 (1957).
17. V. G. Istomin, Pribory i Tekhnika Éksperimenta, 2, 111 (1958).*
18. L. R. Tsvang, Pulse Measurement of Light Ion Spectra in Atmosphere. Dissertation, Institute of Applied Geophysics, Academy of Sciences, USSR (1956).
19. B. A. Mirtov and V. G. Istomin, Uspekhi Fiz. Nauk, 63, 1, 227 (1957).

*For translation see Instruments and Experimental Techniques, No. 2, p. 306 (1958).

THE MANOMETER ERROR CAUSED BY SMALL LEAKS IN THE CASING OF A SATELLITE

S. A. Kuchai

The apparatus on the third satellite includes manometers with a sensitivity capable of measuring static pressures of 10^{-6}-10^{-9} mm Hg under normal conditions on earth.

In measurements in the upper atmosphere the manometer may also encounter molecules which are brought along by the satellite itself and which are found in the upper atmosphere either because of gas emission from the surface of the casing or as a result of leaks. Desorption from the surface stops relatively quickly, but leakage from within becomes practically constant throughout the lifetime of the whole apparatus. Thus, the possibility of the occurrence of manometer errors imposes definite demands on the tightness of the casing seal.

We shall call the gas molecules within the casing "proper" molecules. The manometer error resulting from "proper" molecules getting into the manometer as evaluated below.

The concentration of gas within the manometer, for which the streams of gas passing through a cross section of the intake tubing in the manometer is the same inward as outward, is established. To rule out the influence of the walls, we assume that equality of the normal components of the streams is maintained. For low pressures and random motion of the molecules the component of the outward gas stream normal to the cross section of the tube is equal to

$$q_i = \frac{n_m v_M}{6} \pi b^2 \text{ molecules/sec,}$$

where n_m is the concentration of gas in the manometer, v_M

is the mean velocity of molecules at the temperature of the walls, and b is the radius of the cross section.

The inward stream has a normal component q_e, the magnitude of which depends on the distribution of velocities in the exterior gas stream. Considering that $q_i = q_e$, we can calculate how the "manometer pressure" depends on this stream.

$$P = \frac{6q_e}{\pi b^2 v_M} \frac{1}{N} \text{ mm Hg}.$$

Here $N = 3.6 \times 10^{16}$ is the concentration of molecules at a pressure of 1 mm Hg.

Thus, to calculate the absolute manometer error it is sufficient to know the normal component of "proper" molecules at a cross section of the intake tube.

Let us consider a sphere passing through rarefied gas, inside of which there is placed a manometer. If the pressure of the gas is sufficiently small, and a direct encounter of a molecule from the surface of the sphere with the manometer is prevented by locating the latter properly, then it must be considered that the "proper" molecules can strike the manometer only as a result of single scattering from molecules of the gaseous surroundings. Among these molecules we distinguish, in addition to the unperturbed molecules of the upper atmosphere, "proper" molecules and "acquired" molecules arising from the evaporation of molecules from the surface of the sphere. The manometer error is made up of errors resulting from scattering by each of these three components.

The molecules of the upper atmosphere, which the manometer is intended to measure, are in a system of coordinates moving together with the sphere at the same velocity V, which is significantly larger than the random velocity. We assume that the density of the upper atmosphere around the sphere is uniform.

The proper molecules, which are evaporated from the surface of the sphere and move away from it with a finite velocity v_p, make up a "proper" atmosphere. Its density decreases with the distance from the sphere.

The "acquired" atmosphere consists of molecules of the upper atmosphere, which first hit the sphere and then evaprate from its surface. The velocity with which molecules of this type are removed from the sphere is unknown within the limits mm $v_p < v_a < V$. Since the orientation of the satellite in flight does not remain constant, it is permissible to consider that the density of the "acquired" atmosphere is symmetric with respect to the center of the sphere. It is assumed that the evaporation of the molecules making up the "proper" and "acquired" atmosphere is describable by the cosine law.

In the problems considered below, the source of the scattered molecules is either a point or uniformly distributed over the surface of the sphere; and the velocity distribution of scattered molecules is describable at the surface by a cosine law or else the departure occurs along the normal to the surface. The manometer tube is directed along a radius of the sphere (Fig. 1).

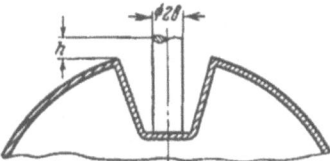

FIG. 1. Location of the manometer (h is the height of the manometer above the surface).

In the problems concerning a concentrated leak, we consider the worst possible position, as far as errors go, of a manometer "over a leak"; i.e., we consider a reverse stream through an area coinciding with the cross section of the intake tube and located over a leak. However, the screening effect with respect to the departure of proper molecules, which would occur if the housing of an actual satellite were located in such a position, is not taken into account.

We assume that the molecules are represented by rigid spheres with the same radius and that in the system of coordinates moving with the sphere the proper molecules are stationary, while molecules of the upper atmosphere move with a velocity V and are reflected along the line of centers at the moment of contact. The angular distribution of proper mole-

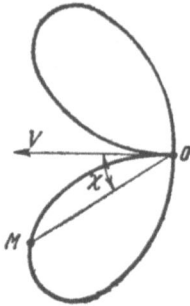

FIG. 2. Angular distribution of proper molecules scattered by molecules of the upper atmosphere. The radius vector OM is proportional to the probability of scattering through an angle κ with the direction of the velocity $\vec{V}$.

cules scattered by molecules of the upper atmosphere is shown in Fig. 2.

For scattering in "proper" and "acquired" atmospheres we assume that after a collision a molecule can be scattered in any direction with an equal probability. This simple model serves only to evaluate the maximum, since it clearly makes the reverse stream too large. Actually the velocities of the colliding molecules are directed in a sphere, so that the angular distribution should maintain the same direction even after scattering.

The numerical values used in the calculations are: a, the radius of the sphere, 100 cm; b, the radius of the intake tube, 1.5 cm; h, the height of the cross section above the surface, 10 cm; n_A, the concentration of molecules in the upper atmosphere, 3.6×10^{10} molecules/cm^3; V, the velocity of the satellite, 8×10^5 cm/sec; v_p, the mean velocity of the proper molecules, 3×10^4 cm/sec; v_m, mean velocity of molecules in the manometer, 3×10^4 cm/sec; μ_0, total magnitude of leakage, 3.6×10^{16} molecules/sec; σ, scattering cross section, 5×10^{-15} cm^2.

The value of n_A at room temperature corresponds to a pressure of 10^{-6} mm Hg, and the value of μ_0 to a leak of 1 liter-micron Hg/sec.

Taking into account the assumptions mentioned above about the source of the leak, the gaseous surroundings, and the laws of scattering, several particular cases were considered. The computations are not particularly difficult and are omitted to conserve space.

1. Dispersed Leak. Scattering by "Proper" Atmosphere

The density of a dispersed source of proper molecules in this case amounts to

$$\nu_p = \frac{\mu_0}{4\pi a^2}.$$

The normal component of the reverse stream on a unit surface of the sphere amounts to

$$\frac{Q}{4\pi a^2} = \frac{\mu_0^2 \sigma}{32\pi^2 a^3 v_p} K.$$

The factor K is equal to 0.27 for normally emitted molecules and 0.68 for tangentially emitted ones. In the worst case the error does not exceed

$$P_1 = 2.7 \cdot 10^{-15} \text{ mm Hg}.$$

2. Dispersed Leak. Scattering by "Acquired" Atmosphere

The problem reduces to the preceding one, since the "acquired" atmosphere can be considered to result from a distribution of leaks with a density $v_a = (n_A V/4)$. The error P_2 should be larger than P_1 in the ratio

$$\frac{\nu_a}{\nu_p} \frac{v_p}{v_a} = \frac{n_A V \pi a^2}{\mu_0} \frac{v_p}{v_a} = 2.5 \cdot 10^4 \frac{v_p}{v_a}.$$

In the worst case $v_a = v_p$,

$$P_2 = 6.8 \cdot 10^{-11} \text{ mm Hg}.$$

3. Point Leak. Scattering by "Proper" Atmosphere. Scattered Molecules Come off along a Normal. Manometer over the Leak

The normal component of the stream q_e through a cross section of the manometer amounts to

$$q_e = \frac{\mu_0 \sigma}{\pi v_p} - \frac{1}{b} f\left(\frac{h}{b}\right) \text{ molecules/sec,}$$

where h is the height of the cross section above the surface (see Fig. 1), and

$$f(x) = \int_0^\infty \frac{d\zeta}{(1+\zeta^2)(\zeta+x)^2}.$$

For h/b = 7 one obtains f(7) = 2.2 × 10^{-2} and correspondingly

$$P_3 = 7.9 \cdot 10^{-10} \text{ mm Hg.}$$

4. Point Leak. Scattering by "Acquired" Atmosphere. Cosine Angular Distribution of Molecules near the Leak

The magnitude

$$q_e = \frac{1}{2} \mu_0 \frac{n_A V a \sigma}{4 v_a} J\left(\frac{b}{a}\right),$$

where

$$J(x) = \int_{\xi=0}^\infty \int_{\eta=0}^1 \frac{1}{1+2\eta\xi+\xi^2}\left[1 - \frac{\xi^2 - x^2}{\sqrt{(\xi^2-x^2)^2 + 4x^2\xi^2\eta^2}}\right] \eta \, d\eta \, d\xi.$$

For the assumed values of b and a, the integral J = 1.8 × 10^{-2}.

In the worst case, $v_a = v_p$, one obtains

$$P_4 = 3.0 \cdot 10^{-8} \text{ mm Hg.}$$

Under the same conditions, but with the assumption that all the scattered molecules come off along a normal, one would obtain

$$'P_4 = 3.9 \cdot 10^{-8} \text{ mm Hg.}$$

5. Point Leak. Scattering by Molecules of the Upper Atmosphere. Scattered Molecules Come Off along Normal. Manometer Located over Point Leak at an Angle of 45°

The magnitude

$$q_e = n_A \delta \mu_0 bL,$$

where
$$L = \int_0^\infty d\xi \int_0^1 d\eta \frac{\eta \xi^2}{(\xi^2+\eta^2)^{5/2}} \int_0^{2\pi} \frac{(\xi-\eta\cos\alpha)^2(\xi+\eta\cos\alpha)}{(\xi-\eta\cos\alpha)^2+4\eta^2\sin^2\alpha} d\alpha = 4.6.$$

Correspondingly
$$P_5 = 3.5 \cdot 10^{-8} \text{ mm Hg}.$$

It is clear that leaks near the manometer are a more dangerous source of error, so that in the construction of the casing these parts should be sealed with particular care. In connection with this it is in order to evaluate the error under the following conditions.

In a region of radius D around the manometer the total leakage is no greater than μ_1. The leakage in the remainder of the surface is characterized by a total leak μ_2.

In an extremely unfavorable case the point leak μ_1 is located above the manometer, and the leak μ_2 is concentrated at a distance D from it. In the calculations we assume $\mu_1 \approx 10^{-2}$ liter-mm Hg/sec, $\mu_2 \approx 1$ liter-mm Hg/sec, D = 20 cm.

6. Scattering by an "Acquired" Atmosphere

The error P_6 is made up of $'P_6$ caused by the leak μ_1 and $''P_6$ caused by the leak μ_2.

The error $'P_6$ is calculated in the same way as P_2, but with a different value of the leak (μ_1 instead of μ_0)

$$'P_6 = P_2 \frac{\mu_1}{\mu_0} = 6.8 \cdot 10^{-12} \text{ mm Hg}.$$

The value of $''P_6$ is calculated for a normal emission of the scattered molecules.

$$q_e = \mu_2 \frac{n_A V}{64 v_a} \sigma a J,$$

where
$$J = ab^2 \int_0^\infty \frac{\zeta^2 d\zeta}{(a+h+\zeta)^2 (D^2+\zeta^2)^2}.$$

For the assumed values of a, b, h, D, and for the case $v_a = v_p$

$$J = 4.3 \cdot 10^{-5},$$
$$''P_6 = 2.9 \cdot 10^{-10} \text{ mm Hg}.$$

7. Scattering by Molecules of the Upper Atmosphere

The error P_7 is made up of $'P_7$ caused by the leak μ_1 and $''P_7$ caused by the leak μ_2.

In view of the similarity of this case with the problem of P_5

$$'P_7 = P_5 \frac{\mu_1}{\mu_0} = 3.5 \cdot 10^{-10} \text{ mm Hg}.$$

The value of $''P_7$ is calculated for normal emission of the scattered molecules

$$q_e = \frac{4b^2}{\pi} \mu_2 n_A \sigma \frac{1}{D} M,$$

where

$$M = \int_0^1 \frac{\zeta^2 d\zeta}{(1+\zeta^2)^{1/2}} = \frac{\sqrt{2}}{12}.$$

Correspondingly

$$''P_7 = 8.6 \cdot 10^{-11} \text{ mm Hg}.$$

The results obtained permit the formulation of qualitative requirements for sealing satellites so that the manometer errors will not exceed requirements.

ON THE PROBLEM OF THE INTERACTION BETWEEN A SATELLITE AND THE EARTH'S MAGNETIC FIELD

Yu. V. Zonov

The problem of the motion of the artificial earth satellite relative to its own center of mass is a complicated one and requires that a large number of factors be accounted for. In particular, it is of interest to consider the phenomena related to the interaction between the satellite and the earth's magnetic field. Of no less interest is an examination of the electric processes in the satellite hull, since they can influence, in one degree or another, the results of certain scientific experiments carried out on the satellite.

We report here the results of an investigation on the interaction between the satellite and the earth's magnetic field. We consider first the currents induced by the translational motion of the satellite relative to the magnetic field, then by the change in the speed of rotation of the satellite about its own axis due to eddy currents, and finally the disturbing forces exerted by the magnetic field on a satellite that has no rotation of its own.

The most substantial of the foregoing factors is the flow of eddy currents in the metallic hull of the satellite. These currents cause a noticeable reduction in the angular velocity of the satellite about its own axis. The charges and currents due to the translational motion of the satellite are without any significant effect on the nature of the satellite's motion, and need be taken into account only in some of the experiments conducted with the aid of satellites.

ORIGIN OF A POTENTIAL DIFFERENCE ON A MOVING SATELLITE

The earth's magnetic field is represented, in first approximation, as a dipole field, but in view of the relatively small size of the satellite the magnetic field in its vicinity can be considered as plane-parallel. The motion of the satellite relative to the magnetic field causes the free electrons of the metal hull of the satellite to move under the influence of the Lorentz forces

$$\mathbf{F} = \frac{e}{c}[\mathbf{VB}], \qquad (1)$$

where $\mathbf{V}$ is the satellite velocity relative to the magnetic field, $\mathbf{B}$ the magnetic induction, e the electron charge, and c the electromagnetic constant.

Thus, a potential difference is produced on the satellite by an increase in the number of free electrons on one side at the expense of the number of electrons on the other. The potential difference is

$$U = lE = \frac{1}{c} VB \sin \alpha, \qquad (2)$$

where α is the angle between $\mathbf{V}$ and $\mathbf{B}$, and l is the distance between two points on the body, measured parallel to the direction of the force $\mathbf{F}$.

The potential difference on a satellite may reach several tenths of a volt, and therefore in certain experiments, such as in the measurement of the earth's electrostatic field, must be accounted for (1).

If the increase in the number of electrons on one side of the satellite is considered as a certain electric surface charge σ, and if the rotation of the satellite about its own axis is taken into account, the result is a displacement of the electric charge, that is, an electric current over the surface of the satellite.

Let us assume that the satellite is spherical. Knowing the value of E, we can also determine the value of σ [2]

$$\sigma = \frac{3E}{4\pi} \cos \theta \qquad (3)$$

(The angle θ is shown in Fig. 1.)

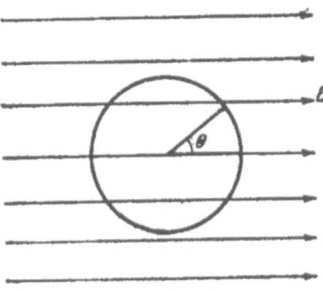

FIG. 1

Replacing angle θ with latitude φ and the longitude λ (Fig. 2), and considering a rotation with velocity ω about the axis perpendicular to the vector **E**, we obtain

$$\sigma = \frac{3E}{4\pi} \cos \varphi \cos (\lambda + \omega t). \tag{4}$$

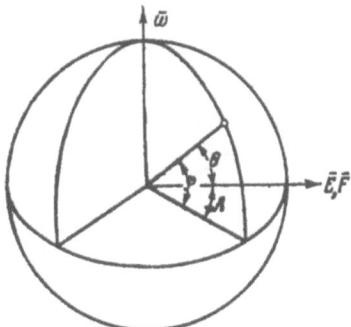

FIG. 2

The angle between $\bar{\omega}$ and **E** can be arbitrary, but it is convenient to use a value $\pi/2$ in the estimates. In view of the absence of electric sources and sinks, the continuity equation can be written for each surface element

$$\frac{3\omega E r}{4\pi} \cos^2 \varphi \sin (\lambda + \omega t) = \cos \varphi \frac{\partial i_\varphi}{\partial \varphi} + \frac{\partial i_\lambda}{\partial \lambda} - \sin \varphi \cdot i_\varphi, \tag{5}$$

where i_λ is the current flowing in the latitudinal direction through an infinitesimal section of surface of unit thickness,

i_φ is the similar current in the meridional direction, and r is the radius of the hull.

This equation has the following solution

$$\begin{cases} i_\lambda = -\dfrac{3\omega E r}{4\pi}\cos^2\varphi \cos(\lambda + \omega t), \\ i_\varphi = 0. \end{cases} \qquad (6)$$

Thus, the current through the surface element of the satellite hull, when viewed in this manner, reduces to a latitudinal current. The current will cause Joule heat to be liberated and dissipated in space. The heat energy is taken from the kinetic energy of the rotating body, the angular velocity of which must therefore decrease.

Knowing the current flowing through a surface element, we can obtain the power dissipated as heat from the entire surface of the sphere.

Differentiating the equation of kinetic energy of a rotating sphere

$$\mathscr{E} = \frac{J_M \omega^2}{2}, \qquad (7)$$

where J_M is the moment of inertia, and replacing $d\mathscr{E}$ by the heat-dissipation loss, W dt, where W is the power dissipated as heat, we obtain

$$\omega d\omega = -\frac{W}{J_M} dt, \qquad (8)$$

which when integrated yields

$$\omega = \omega_0 \exp\left(-\frac{9\rho h V^2 B^2 \sin^2\alpha}{8\rho_0 \pi^2 r^3 c^2} t\right), \qquad (9)$$

where ρ_0 is the density of the satellite body, ρ the resistivity of the hull, and h the thickness of the hull.

Let us apply this formula to an 83.6-kg satellite 0.58 m in diameter (corresponding to the size and weight of the first Soviet earth satellite). We assume that the satellite hull is aluminum. The hull will be assumed to be 1 mm thick. If the satellite moves along the geomagnetic equator ($\alpha = \pi/2$; B = 0.3 gauss) at 8 km/sec, we find that the angular velocity of the satellite about its own axis will decrease by a factor e in about 10^{14} seconds (one year is approximately 10^7 seconds), showing that this deceleration is practically negligible.

As the satellite moves in the ionosphere, the difference of potential on its opposite sides causes current to flow in the metallic hull. The calculations show that this current is weak and has hardly any effect on the motion of the satellite and on its thermal conditions, but to obtain unambiguous results in many experiments, for example magnetometric ones, it must be taken into account. Let a satellite one meter in radius move in an ionized region, where the electron density is of the order of $10^6/cm^3$. To estimate the effect, let us assume that one-half of the frontal section captures electrons and the other loses them by recombination with the ions encountered. If the satellite moves at 8 km/sec, a current on the order of 10^{-3} amp can flow in the hull of the satellite, and this can produce a magnetic field of several times 10 gammas (1 gamma = 10^{-5} oersted).

EDDY CURRENTS IN THE SATELLITE HULL AND REDUCTION OF THE ANGULAR VELOCITY OF ITS OWN ROTATION

In addition to motion of the satellite relative to the magnetic field, the satellite can also rotate at a certain angular velocity. In this case, eddy currents will be produced in the metallic hull of the satellite (Fig. 3), and they can lead to a noticeable reduction in the angular velocity of rotation of the satellite about its own axis.

The fact that the angular velocity of a conductor is reduced in a magnetic field is a well-known phenomenon. The magnitude of the retarding torque for a spherical shell was given by Smythe.[3] After transformation we get

$$M_T = \frac{6\pi\mu_0^2 H^2 R \omega_T a^4 h}{9R^2 + \mu_0^2 a^2 \omega_T^2 h^2}, \tag{10}$$

where a is the radius of the spherical shell, h the thickness of the shell, H the intensity of the magnetic field, R the

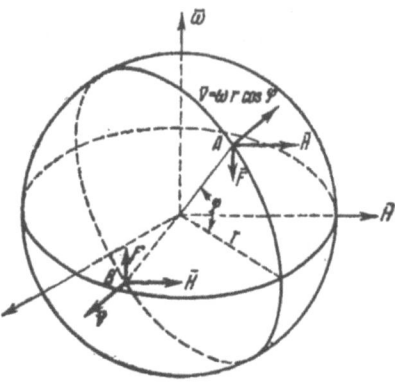

FIG. 3

volume resistivity of the shell material, μ_0 the permeability of vacuum, and ω_T the components of angular velocity of rotation perpendicular to the flux lines of the magnetic field.

Formula (10) is in mksa units.

The dependence of the angular velocity of rotation on the time is obtained from

$$J_M \frac{d\omega}{dt} = -M, \tag{11}$$

where J_M is the moment of inertia of the body, and M is the moment of the external forces.

Inserting for the external moment the expression for the retarding moment of the shell and integrating from ω_0T to ω_T, we get

$$9R^2 \ln \frac{\omega_T}{\omega_{0T}} + h^2\mu_0^2 a^2 \left(\frac{\omega_T^2}{2} - \frac{\omega_{0T}^2}{2} \right) = -\frac{6\pi\mu_0^2 H^2 a^4 Rh}{J_M} t. \tag{12}$$

Introducing a new variable $\Omega = \omega_T/\omega_{0T}$ and dividing both halves by the coefficient of the first term, we obtain after a slight transformation

$$\Omega e^{c_1 (\Omega^2 - 1)} = e^{-c_2 t}, \tag{13}$$

where

$$c_1 = \frac{\omega_{0T}^2 \mu_0^2 a^2 h^2}{18 R^2}, \tag{14}$$

$$c_2 = \frac{2\pi\mu_0^2 H^2 a^4 h}{3RJ_M}. \tag{15}$$

Inserting the numerical values of the quantities in (14), and using reasonable values for possible satellite designs, we find that $c_1 \ll 1$. Thus, for example, for the first Soviet satellite, taking $a = 0.29$ m, putting $h = 10^{-3}$ m, $R = 2.8 \times 10^{-8}$ ohm-m, and considering that $\omega_0 T$ does not exceed 10 degrees per second, we obtain $c_1 \leq 10^{-6}$.

In such a case, (13) can be rewritten

$$\Omega = e^{-c_2 t} \tag{16}$$

or, in other words, since $\Omega = \omega_T/\omega_0 T$,

$$\omega_T = \omega_{0T} e^{-c_2 t}. \tag{17}$$

To calculate the moment of inertia of the satellite we replace it by a sphere with uniformly distributed density. We obtain

$$J_M = \frac{8}{15} a^5 \rho_0 \pi,$$

where ρ_0 is the value assumed for the mean density.

Taking this expression into account, c_2 constant becomes

$$c_2 = \frac{5\mu_0^2 H^2 h}{4\rho_0 R a}. \tag{18}$$

For the first Soviet satellite we have

$$a = 2.9 \cdot 10^{-1} \text{m},$$
$$\rho_0 = 836 \text{ kg/m}^3.$$

We also assume $h = 10^{-3}$ m and $R = 2.8 \times 10^{-8}$ ohm-m (aluminum).

We take for $B = \mu_0 H$ a value 0.3 gauss or 0.3×10^{-4} weber/m^2.

After performing the calculations, we find that reduction in the angular velocity of the satellite about its own axis will be reduced by a factor e in 70 days. We note that if the moment of inertia of the antennas is taken into account, the deceleration is even slower.

For the American Vanguard satellite the deceleration was the more significant and can lead to a noticeable reduction in the angular velocity (which the satellite is intended to have). Considering that the weight of the Vanguard is 10 kg, and its radius 0.255 m, and assuming all the other quantities of (18) to be the same as for the Soviet satellite, we find that the Vanguard will experience a reduction in the rotation about its own axis by a factor e within approximately 11 days. A similar conclusion is found in Ref. 4.

Summarizing, we see that in a satellite that has a highly conducting hull and rotates with a certain angular velocity about its own axis, the eddy currents produced in the hull are quite significant, as is the associated reduction in the angular velocity.

We make the following comments:

(a) In view of the thinness of the satellite hull, in which the eddy currents are considered, the value of the magnetic induction $B = \mu H$ is taken equal to the value of the induction in the surrounding space.

(b) Equations (13), (16), and (17) were obtained under the assumption that the hull is of solid conducting material. This, however, is not always true, for individual parts of the hull may be made of dielectrics, and furthermore, for even an all-metal hull, it must necessarily be made of individual plates and the resistance at the points where these join is different than that of the material of the hull itself. This changes the accelerating torque and therefore the angular velocity about the satellite axis, compared with the values given by the formula.

(c) The variation of the satellite angular velocity about its own axis, given in (13), (16), and (17), was obtained for the component of angular velocity orthogonal to the magnetic intensity vector **H**. If the vector ω has a constant direction in absolute space, then the vector **H** will change its direction as the satellite moves in the earth's magnetic field (unless the satellite moves along the geomagnetic equator). It is seen therefore that the deceleration will affect not the one component ω_T, but the angular velocity ω as a whole.

PERTURBING ACTION OF THE MAGNETIC FIELD

In the preceding section we considered rapid rotation of the satellite about the flux lines of the field. Thus, we can consider that the direction of the field relative to the axis of rotation does not change after a few rotations of the satellite about its own axis. We might consider a different case, namely assume the satellite to be stationary relative to a system of coordinates moving with it in the orbit about the earth, with axes parallel to the axes of the absolute system. The direction of the flux lines of the field, determined by the vector H, is variable, i.e., again the hull of the satellite moves in relation to the magnetic field.

In this case the magnetic field will no longer slow down the rotation of the satellite, but exert a certain perturbing action on it. The strongest perturbation will be experienced by the satellite as it moves along an orbit passing through the geomagnetic poles. In this case the earth's magnetic vector, in view of its dipole character, will rotate through 4π degrees during one revolution of the satellite about the earth, in the above-mentioned system of coordinates. The mean angular velocity of rotation will be $4\pi/T$, where T is the period of rotation. When moving along the geomagnetic equator, as can readily be seen, no such perturbations will now appear.

As the satellite moves through a certain range of latitudes, the vector of magnetic intensity describes, in the above-mentioned system of coordinates, a complicated conical surface whose shape depends on the inclination of the orbit to the earth's equator, on the angle between the axis of the magnetic dipole and earth's axis, and on the time of the day. This conical surface can be determined if the projections of the vector of magnetic intensity on the axis of the system are known, in a certain system of coordinates, as a function of the satellite position. If one of the axes of the system, fixed in the satellite, is directed parallel to the earth's axis, and the other two are placed in the equatorial plane, then the shape of the conical surface described by the vector H during one-half revolution of the satellite about the earth within a certain range of latitudes, is as shown in Fig. 4.

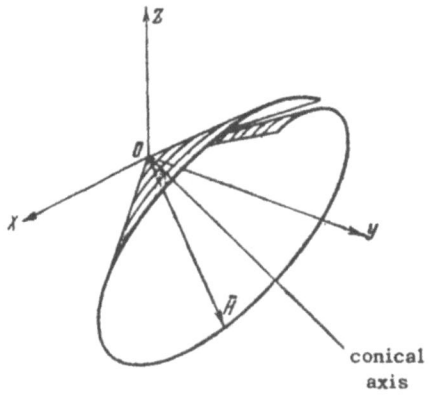

FIG. 4

With time, the magnetic field will drag the satellite and in final analysis it will start rotating about an axis close to the axis of the cone described by the vector **H**. The law of motion of the vector **H** depends on many conditions and is rather complicated; during each half revolution of the satellite about the earth the conical surface will differ from the one described in the preceding half revolution. The position of the axis of the conical surface will differ accordingly and since the flux lines rotate about the axis of the cone, the satellite will twist during each half revolution about a different axis. Since the conical figures of each half revolution differ slightly from each other, and accordingly the axes differ only slightly, the limiting motion of the satellite can be considered as a rotation about a certain axis and an oscillation of this axis within certain small limits. The magnitude of the speed of rotation of the satellite in this limiting motion will be of the order indicated above, $\sim 4\pi/T$.

LITERATURE

1. I. M. Imyanitov, Uspekhi fizicheskikh nauk (Progress in Physical Sciences), 63, 1b (1957), p. 267. (Transl. available from International Physical Index, New York.)
2. P. Frank and R. Von Mises, Die Differentialgleichungen und Integralgleichungen der Mechanik und Physik (Brunswick), 1930–1935.
3. W. R. Smythe, Static and Dynamic Electricity, McGraw-Hill Book Co., Inc., New York, 1950.
4. H. B. Rosenstock, Astronautica Acta, 3, 215 (1957).

Volume 4

THE MOTION OF AN ARTIFICIAL SATELLITE IN THE NORMAL GRAVITATIONAL FIELD OF THE EARTH

M. D. Kislik

The present paper undertakes a consideration of the problem of motion of an artificial satellite in the earth's normal gravitational field without taking into account the effect of air resistance or anomalies in the force of gravity. The results obtained can be used in calculating the orbits of high-altitude satellites and for qualitative analysis of the effect of the earth's oblateness on satellite motion.

The potential V of the normal field of the gravitational forces of the earth (potential of the gravitational force field of a general ellipsoid such as the earth[1, 2]) at a point with coordinates r and ψ can be written in series form:[3]

$$V = \frac{b_0}{r} - \frac{b_2}{r^3} P_2(\sin \psi) + \frac{b_4}{r^5} P_4(\sin \psi) - \dots, \tag{1}$$

where $b_0, b_2, b_4, \dots$, are constant coefficients, $P_2(\sin \psi)$, $P_4(\sin \psi), \dots$, are Legendre polynomials of order 2, 4, ..., r is the distance to the center of mass of the earth, ψ is the geocentric latitude.

In the expansion (1) terms containing Legendre polynomials of sixth and higher orders can be rejected. The forces corresponding to these terms amount to a fraction of a milligal[1] in the entire space external to the earth. The term $(b_4/r^5) P_4(\sin \psi)$ gives rise to an acceleration of the same order as the anomalies in the force of gravity. Therefore, it is permissible to neglect this term in the given problem, or, what amounts to the same thing, to slightly modify the coefficient b_4, since this leads to errors of the same order as the errors connected with neglecting the anomalies.

Having made note of this, we shall attempt to find a force

field, the potential $\overline{V}$ of which would permit the equations of motion of a satellite to be integrated and at the same time would be a sufficient approximation to the potential V.

Let us introduce an inertial system of dimensionless curvilinear coordinates q_1, q_2, q_3.

The coordinates q_1, q_2 are defined by the equations

$$q_1 = \frac{1}{2}\left[\left(\frac{r^2}{d^2}-1\right) + \sqrt{\left(\frac{r^2}{d^2}-1\right)^2 + 4\frac{r^2}{d^2}\sin^2\phi}\right],$$

$$q_2 = \frac{1}{2}\left[\left(\frac{r^2}{d^2}-1\right) - \sqrt{\left(\frac{r^2}{d^2}-1\right)^2 + 4\frac{r^2}{d^2}\sin^2\phi}\right], \quad (2)$$

where d is a specific, although arbitrary linear parameter.

The coordinate surfaces q_1 = const and q_2 = const are, as is not too difficult to see, ellipsoids of rotation and single-branch hyperboloids of rotation, respectively. The coordinates q_1, q_2 are able to vary within the limits

$$-1 \leqslant q_2 \leqslant 0 \leqslant q_1 \leqslant +\infty.$$

The coordinate q_3 represents the angle between the meridian plane containing the given point and some initial plane passing through the earth's axis of rotation (longitude).

The system q_1, q_2, q_3 is not a system of elliptical coordinates in space in the usual definition.[4] We shall, however, admitting a certain conditionality, call the coordinates q_1, q_2, q_3 dimensionless elliptical coordinates.

In C. Jacobi's classic work, Lectures on Dynamics,[5] it is remarked that the equations of motion of a point in a force field whose potential Π has the form

$$\Pi = \frac{F(q_1) + \Phi(q_2)}{q_1 - q_2}, \quad (3)$$

where $F(q_1)$, $\Phi(q_2)$ are arbitrary functions of the elliptical coordinates q_1, q_2, can always be integrated. We wish to show that a field with potential $\overline{V}$ equal to

$$\overline{V} = \frac{b_0}{d} \cdot \frac{\sqrt{q_1}}{q_1 - q_2} \quad (4)$$

and satisfying therefore the condition (3) by appropriate choice of the parameter d can in all outer space be made sufficiently near the gravitational force field of the generalized earth-ellipsoid.

Assuming that the ratio d/r is small for all considered values of r, we can write the potential $\overline{V}$ in the form of a series:

$$\overline{V} = \frac{b_0}{r}\left[1 - \frac{d^2}{r^2}P_2(\sin\psi) + \frac{d^4}{r^4}P_4(\sin\psi) - \ldots\right]. \tag{5}$$

The expansion (5) follows from (2) and (4). Letting the parameter d be equal to

$$d = \sqrt{\frac{b_2}{b_0}}, \tag{6}$$

the second term of the expansion (5) can be made equal to the second term of the expansion (1). (The first terms of these expansions coincide independently of the value of d.)

The value of the linear parameter d for such a choice amounts to about 211 km ($b_0 \approx 3.99 \times 10^{14}$ m^3/sec^2, $b_2 \approx 1.78 \times 10^{25}$ m^5/sec^2). Consequently, the ratio d/r in outer space will now satisfy the inequality d/r < 1/30 (r > 6370 km), which justifies the assumptions made above.

In the expansion (5) terms containing the ratio d/r in sixth and higher powers are associated with forces less than a fraction of a milligal, and they may be rejected. Therefore, the difference $\Delta V = \overline{V} - V$ can be described by the equality

$$\Delta V = \frac{b_0 d^4 - b_4}{r^5} P_4(\sin\psi) = \left(\frac{b_2^2}{b_0} - b_4\right) \cdot \frac{1}{r^5} P_4(\sin\psi).$$

For the values of the radial (Δg_r) and meridian (Δg_m) components of the acceleration created by the potential ΔV we obtain the estimates ($b_4 \cong 1.6 \times 10^{36}$ m^7/sec^2),

$$|\Delta g_r| \leqslant 6.1 \text{ milligal}, \qquad |\Delta g_m| \leqslant 3.3 \text{ milligal}.$$

Consequently, the potential $\overline{V}$ for the indicated choice of the parameter d differs only insignificantly from the potential V. The maximum error in determining the force of gravity of the generalized earth-ellipsoid with application of (4) is no more than 7 milligal at the surface of the earth. On moving farther away from the earth this error will decrease.

We emphasize that the quantities Δg_r and Δg_m should be considered errors only in the event that (4) is satisfied for a description of the field of the generalized earth-ellipsoid. If, on the other hand, we consider the actual gravitational force field of the earth and use as this quantity the new normal part of the field with potential $\bar{V}$, then the potential ΔV need only be included with appropriate sign into the potential of the anomalous field. Such a redistribution should not give rise to any additional errors.

We note in passing that still another field of rather simple structure might be indicated, one that provides a good approximation to the field of the generalized earth-ellipsoid. The potential of this field V' is described by the equation

$$V' = \frac{b_0}{d} \cdot \text{arctg} \, \frac{1}{\sqrt{q_1}}. \tag{7}$$

Here the parameter d should be chosen from the condition

$$d = \sqrt{\frac{3b_2}{b_0}}.$$

The equipotential surfaces of the field with potential V' are confocal ellipsoids of rotation. The expansion of the potential V' in Legendre polynomials has the form

$$V' = \frac{b_0}{r} \left[1 - \frac{1}{3} \frac{d^2}{r^2} P_2(\sin \phi) + \frac{1}{5} \frac{d^4}{r^4} P_4(\sin \phi) - \ldots \right].$$

The gravitational force field of the generalized earth-ellipsoid will be described by (7) with an error no greater than 1.5 milligal in outer space. This result could no doubt prove to be of interest in itself.

We describe the equations of motion of the center of mass of the satellite in canonical form. Let us consider the motion of the satellite relative to an inertial coordinate system. As the canonical variables we take the coordinates q_1, q_2, q_3 and the generalized momenta corresponding to these, $p_1 = (\partial T/\partial \dot{q}_1)$; $p_2 = (\partial T/\partial \dot{q}_2)$; $p_3 = (\partial T/\partial \dot{q}_3)$, where $T = T(q_1, q_2, q_3, \dot{q}_1, \dot{q}_2, \dot{q}_3)$ is the kinetic energy of the satellite (the dots indicating time differentiation).

The canonical equations of motion in Hamiltonian form become[4,6]

$$\begin{aligned}
\dot{q}_1 &= \frac{\partial}{\partial p_1}(\hat{T}-\overline{V}), & \dot{p}_1 &= -\frac{\partial}{\partial q_1}(\hat{T}-\overline{V}), \\
\dot{q}_2 &= \frac{\partial}{\partial p_2}(\hat{T}-\overline{V}), & \dot{p}_2 &= -\frac{\partial}{\partial q_2}(\hat{T}-\overline{V}), \\
\dot{q}_3 &= \frac{\partial}{\partial p_3}(\hat{T}-\overline{V}), & \dot{p}_3 &= -\frac{\partial}{\partial q_3}(\hat{T}-\overline{V}).
\end{aligned} \qquad (8)$$

The function $\hat{T} = \hat{T}(q_1, q_2, q_3, p_1, p_2, p_3)$ represents the kinetic energy of the satellite, expressed in terms of canonical variables.

Let us put the right hand sides of (8) into explicit form. In order to accomplish this we wish to find the dependence of the kinetic energy T on the coordinates q_1, q_2, q_3 and the generalized velocities $\dot{q}_1$, $\dot{q}_2$, $\dot{q}_3$. In the coordinates r, ψ, q_3 the quantity T is expressed by the equation

$$T = \frac{1}{2}[\dot{r}^2 + r^2\dot{\psi}^2 + r^2\cos^2\psi\,\dot{q}_3^2]$$

(we are assuming the mass of the satellite to be equal to unity).

Differentiating the equations

$$\left.\begin{aligned}
r^2 &= d^2(1+q_1+q_2), \\
\mathrm{tg}^2\psi &= -\frac{q_1 q_2}{(1+q_1)(1+q_2)},
\end{aligned}\right\} \qquad (9)$$

which follow from the relations (2), we obtain

$$\left.\begin{aligned}
\dot{r}^2 &= \frac{d^2(\dot{q}_1+\dot{q}_2)^2}{4(1+q_1+q_2)}, \\
r^2\dot{\psi}^2 &= -\frac{d^2[q_2(1+q_2)\dot{q}_1 + q_1(1+q_1)\dot{q}_2]^2}{4\,q_1 q_2(1+q_1)(1+q_2)(1+q_1+q_2)}, \\
r^2\cos^2\psi\,\dot{q}_3^2 &= d^2(1+q_1)(1+q_2)\dot{q}_3^2.
\end{aligned}\right\} \qquad (10)$$

After some elementary transformations we arrive at the equation

$$T = \frac{d^2}{8}\left[\frac{q_1-q_2}{q_1(1+q_1)}\dot{q}_1^2 + \frac{q_2-q_1}{q_2(1+q_2)}\dot{q}_2^2 + 4(1+q_1)(1+q_2)\dot{q}_3^2\right]. \qquad (11)$$

From this we get the following expressions for the generalized momenta:

$$\left.\begin{aligned}
p_1 &= \frac{d^2(q_1-q_2)}{4q_1(1+q_1)}\dot{q}_1, \\
p_2 &= \frac{d^2(q_2-q_1)}{4q_2(1+q_2)}\dot{q}_2, \\
p_3 &= d^2(1+q_1)(1+q_2)\dot{q}_3.
\end{aligned}\right\} \qquad (12)$$

Expressing the generalized velocities in terms of the momenta and substituting into (11), we obtain

$$\hat{T} = \frac{2}{d^2}\left[\frac{q_1(1+q_1)}{q_1-q_2}p_1^2 + \frac{q_2(1+q_2)}{q_2-q_1}p_2^2 + \frac{1}{4(1+q_1)(1+q_2)}p_3^2\right]. \quad (13)$$

As a result, the canonical equations of motion of the satellite in the inertial coordinate system q_1, q_2, q_3 assume the form

$$\dot{q}_1 = \frac{4q_1(1+q_1)}{d^2(q_1-q_2)}p_1,$$

$$\dot{q}_2 = \frac{4q_2(1+q_2)}{d^2(q_2-q_1)}p_2,$$

$$\dot{q}_3 = \frac{1}{d^2(1+q_1)(1+q_2)}p_3,$$

$$\dot{p}_1 = -\frac{2}{d^2}\left[p_1^2 - \frac{q_2(1+q_2)}{(q_1-q_2)^2}(p_1^2 - p_2^2) - \frac{1}{4(1+q_1)^2(1+q_2)}p_3^2\right] -$$

$$- \frac{b_0}{d} \cdot \frac{q_1+q_2}{2\sqrt{q_1}(q_1-q_2)^2},$$

$$\dot{p}_2 = -\frac{2}{d^2}\left[p_2^2 - \frac{q_1(1+q_1)}{(q_2-q_1)^2}(p_2^2 - p_1^2) - \frac{1}{4(1+q_1)(1+q_2)^2}p_3^2\right] +$$

$$+ \frac{b_0}{d} \cdot \frac{\sqrt{q_1}}{(q_1-q_2)^2},$$

$$\dot{p}_3 = 0.$$

$$\quad (14)$$

Integration of the system (14) can, by making use of Jacobi's method,[4,6] be replaced by a search for the total integral of the following partial differential equation:

$$\frac{2}{d^2}\left[\frac{q_1(1+q_1)}{q_1-q_2}\left(\frac{dW}{\partial q_1}\right)^2 + \frac{q_2(1+q_2)}{q_2-q_1}\left(\frac{dW}{\partial q_2}\right)^2 +$$

$$+ \frac{1}{4(1+q_1)(1+q_2)}\left(\frac{\partial W}{\partial q_3}\right)^2\right] - \frac{b_0}{d}\cdot\frac{\sqrt{q_1}}{q_1-q_2} = C_1. \quad (15)$$

Here $W = W(q_1, q_2, q_3)$ is the function sought, which we call the characteristic function, C_1 = const is the total energy of the satellite.

Equation (15) is obtained, as we know, by replacement in the left-hand side of the energy integral $\hat{T} - \overline{V} = C_1$ of the momenta p_1, p_2, p_3 by $(\partial W/\partial q_1)$, $(\partial W/\partial q_2)$, $(\partial W/\partial q_3)$.

The total integral of (15) is then the function $W(q_1, q_2, q_3, C_1, C_2, C_3)$, which satisfies this equation and contains, in addition to the additive constant, two arbitrary constants C_2, C_3.

Let us rewrite (15) in the following form:

$$q_1(1+q_1)\left(\frac{\partial W}{\partial q_1}\right)^2 - q_2(1+q_2)\left(\frac{\partial W}{\partial q_2}\right)^2 + \frac{q_1-q_2}{4(1+q_1)(1+q_2)}\left(\frac{\partial W}{\partial q_3}\right)^2 -$$
$$-\frac{b_0 d}{2}\sqrt{q_1} = \frac{C_1 d^2}{2}(q_1 - q_2). \tag{16}$$

From the form of (16) it follows that the function W can be written as follows:

$$W = W_1(q_1, q_2, C_1, C_3) + C_2 q_3. \tag{17}$$

Then we obtain the following equation, which must be satisfied by the function W_1:

$$q_1(1+q_1)\left(\frac{\partial W_1}{\partial q_1}\right)^2 - q_2(1+q_2)\left(\frac{\partial W_1}{\partial q_2}\right)^2 +$$
$$+ \frac{C_2^2(q_1-q_2)}{4(1+q_1)(1+q_2)} - \frac{b_0 d}{2}\sqrt{q_1} = \frac{C_1 d^2}{2}(q_1-q_2). \tag{18}$$

Bearing in mind the obvious equality

$$\frac{q_1-q_2}{(1+q_1)(1+q_2)} = \frac{1}{1+q_2} - \frac{1}{1+q_1},$$

we arrive at a modification of (18) in the form

$$q_1(1+q_1)\left(\frac{\partial W_1}{\partial q_1}\right)^2 - \frac{C_2^2}{4(1+q_1)} - \frac{b_0 d}{2}\sqrt{q_1} - \frac{C_1 d^2}{2}q_1 =$$
$$= q_2(1+q_2)\left(\frac{\partial W_1}{\partial q_2}\right)^2 - \frac{C_2^2}{4(1+q_2)} - \frac{C_1 d^2}{2}q_2. \tag{19}$$

The last equation breaks down into two equations:

$$q_1(1+q_1)\left(\frac{\partial W_1}{\partial q_1}\right)^2 - \frac{C_2^2}{4(1+q_1)} - \frac{b_0 d}{2}\sqrt{q_1} - \frac{C_1 d^2}{2}q_1 = C_3,$$
$$q_2(1+q_2)\left(\frac{\partial W_1}{\partial q_2}\right)^2 - \frac{C_2^2}{4(1+q_2)} - \frac{C_1 d^2}{2}q_2 = C_3. \tag{20}$$

It is easily seen that the function W_1, which satisfies the relations in (20), will then be the total integral of (19).

Solving (20) for $\partial W_1/\partial q_1$ and $\partial W_1/\partial q_2$:

$$\frac{\partial W_1}{\partial q_1} = \sqrt{\frac{\frac{C_2^2}{4(1+q_1)} + \frac{b_0 d}{2}\sqrt{q_1} + \frac{C_1 d^2}{2}q_1 + C_3}{q_1(1+q_1)}},$$

$$\frac{\partial W_1}{\partial q_2} = \sqrt{\frac{\frac{C_2^2}{4(1+q_2)} + \frac{C_1 d^2}{2}q_2 + C_3}{q_2(1+q_2)}}, \tag{21}$$

integrating, and recognizing the dependence (17), we obtain the following expression for the characteristic function:

$$W = \int dq_1 \sqrt{\frac{\frac{C_2^2}{4(1+q_1)} + \frac{b_0 d}{2}\sqrt{q_1} + \frac{C_1 d^2}{2} q_1 + C_3}{q_1(1+q_1)}} +$$

$$+ \int dq_2 \sqrt{\frac{\frac{C_2^2}{4(1+q_2)} + \frac{C_1 d^2}{2} q_2 + C_3}{q_2(1+q_2)}} + C_2 q_3. \quad (22)$$

After finding the function W the integrals of the canonical equations (14) can be specified by the equations

$$\frac{\partial W}{\partial C_2} = C_4, \quad \frac{\partial W}{\partial C_3} = C_5,$$

$$\frac{\partial W}{\partial q_1} = p_1, \quad \frac{\partial W}{\partial q_2} = p_2, \quad \frac{\partial W}{\partial q_3} = p_3,$$

$$\frac{\partial W}{\partial C_1} = t + C_6, \quad (23)$$

where C_4, C_5, C_6 are arbitrary constants, t is the time.

The first two integrals of the system (23) are geometric: they determine the form of the satellite's trajectory. The last integral is kinematic: it specifies the law of motion of the satellite along its trajectory. The remaining integrals serve to determine the momenta (velocities).

In explicit form the system of integrals (23) appears as follows:

$$\int \frac{C_2 dq_1}{4(1+q_1)\sqrt{q_1\left[\frac{C_2^2}{4} + (1+q_1)\left(\frac{b_0 d}{2}\sqrt{q_1} + \frac{C_1 d^2}{2} q_1 + C_3\right)\right]}} +$$

$$+ \int \frac{C_2 dq_2}{4(1+q_2)\sqrt{q_2\left[\frac{C_2^2}{4} + (1+q_2)\left(\frac{C_1 d^2}{2} q_2 + C_3\right)\right]}} + q_3 = C_4, \quad (24)$$

$$\int \frac{dq_1}{2\sqrt{q_1\left[\frac{C_2^2}{4} + (1+q_1)\left(\frac{b_0 d}{2}\sqrt{q_1} + \frac{C_1 d^2}{2} q_1 + C_3\right)\right]}} +$$

$$+ \int \frac{dq_2}{2\sqrt{q_2\left[\frac{C_2^2}{4} + (1+q_2)\left(\frac{C_1 d^2}{2} q_2 + C_3\right)\right]}} = C_5, \quad (25)$$

$$p_1 = \sqrt{\dfrac{\dfrac{C_2^2}{4(1+q_1)} + \dfrac{b_0 d}{2}\sqrt{q_1} + \dfrac{C_1 d^2}{2} q_1 + C_3}{q_1(1+q_1)}}, \qquad (26)$$

$$p_2 = \sqrt{\dfrac{\dfrac{C_2^2}{4(1+q_2)} + \dfrac{C_1 d^2}{2} q_2 + C_3}{q_2(1+q_2)}}, \qquad (27)$$

$$p_3 = C_2, \qquad (28)$$

$$\int \dfrac{q_1 \cdot d^2 \cdot dq_1}{4\sqrt{q_1\left[\dfrac{C_2^2}{4} + (1+q_1)\left(\dfrac{b_0 d}{2}\sqrt{q_1} + \dfrac{C_1 d^2}{2} q_1 + C_3\right)\right]}} +$$

$$+ \int \dfrac{q_2 \cdot d^2 \cdot dq_2}{4\sqrt{q_2\left[\dfrac{C_2^2}{4} + (1+q_2)\left(\dfrac{C_1 d^2}{2} q_2 + C_3\right)\right]}} = t + C_6. \qquad (29)$$

Equations (24-29) completely solve the problem of the motion of a satellite under the action of a gravitational force field with potential $\bar{V}$.

We now wish to bring the results obtained to a form more suitable for practical application. Let us assume the following at the initial instant of the satellite's motion:*

$$t = t_k = 0,$$

$$q_1 = q_{1k},\ q_2 = q_{2k},\ q_3 = q_{3k}, \qquad p_1 = p_{1k},\ p_2 = p_{2k},\ p_3 = p_{3k}.$$

Substituting the initial values of the variables into (24), (25), and (29), it is possible to find

$$q_3 = q_{3k} - \dfrac{1}{4} C_2 \left\{ \int_{q_{1k}}^{q_1} \dfrac{dq_1}{(1+q_1)\sqrt{q_1\left[\dfrac{C_2^2}{4} + (1+q_1)\left(\dfrac{b_0 d}{2}\sqrt{q_1} + \dfrac{C_1 d^2}{2} q_1 + C_3\right)\right]}} + \right.$$

$$\left. + \int_{q_{2k}}^{0} \dfrac{dq_2}{(1+q_2)\sqrt{q_2\left[\dfrac{C_2^2}{4} + (1+q_2)\left(\dfrac{C_1 d^2}{2} q_2 + C_3\right)\right]}} \right\}, \qquad (30)$$

$$\int_{q_{1k}}^{q_1} \dfrac{dq_1}{\sqrt{q_1\left[\dfrac{C_2^2}{4} + (1+q_1)\left(\dfrac{b_0 d}{2}\sqrt{q_1} + \dfrac{C_1 d^2}{2} q_1 + C_3\right)\right]}} +$$

$$+ \int_{q_{2k}}^{q_2} \dfrac{dq_2}{\sqrt{q_2\left[\dfrac{C_2^2}{4} + (1+q_2)\left(\dfrac{C_1 d^2}{2} q_2 + C_3\right)\right]}} = 0, \qquad (31)$$

*In all that follows the subscript k corresponds to the initial point.

$$t = \frac{d^2}{4} \left\{ \int_{q_{1k}}^{q_1} \frac{q_1 dq_1}{\sqrt{q_1 \left[\frac{C_2^2}{4} + (1+q_1)\left(\frac{b_0 d}{2}\sqrt{q_1} + \frac{C_1 d^2}{2} q_1 + C_3\right) \right]}} + \right.$$

$$\left. + \int_{q_{2k}}^{q_2} \frac{q_2 dq_2}{\sqrt{q_2 \left[\frac{C_2^2}{4} + (1+q_2)\left(\frac{C_1 d^2}{2} q_2 + C_3\right) \right]}} \right\}. \tag{32}$$

Let us determine the arbitrary constants C_1, C_2, C_3. The constant C_1 (total energy of the satellite) is expressed in terms of the initial conditions as follows:

$$C_1 = \hat{T}_k - \bar{V}_k = \frac{2}{d^2} \left[\frac{q_{1k}(1+q_{1k})}{q_{1k} - q_{2k}} p_{1k}^2 + \frac{q_{2k}(1+q_{2k})}{q_{2k} - q_{1k}} p_{2k}^2 + \right.$$

$$\left. + \frac{1}{4(1+q_{1k})(1+q_{2k})} p_{3k}^2 \right] - \frac{b_0}{d} \cdot \frac{\sqrt{q_{1k}}}{q_{1k} - q_{2k}}. \tag{33}$$

The constant C_2 is obviously equal to

$$C_2 = p_{3k}. \tag{34}$$

To determine the constant C_3 we make use of (27).

Considering this equation at the initial point and solving it for C_3, we arrive at the relation

$$C_3 = \frac{b_0 d}{2} \cdot \frac{q_{2k}\sqrt{q_{1k}}}{q_{1k} - q_{2k}} - \frac{p_{3k}^2 (1 + q_{1k} + q_{2k})}{4(1+q_{1k})(1+q_{2k})} -$$

$$- \frac{q_{1k} q_{2k}}{q_{1k} - q_{2k}} [p_{1k}^2(1+q_{1k}) - p_{2k}^2(1+q_{2k})]. \tag{35}$$

We now transform to the dimensionless arbitrary constants D_1, D_2, D_3 which are defined by the relations

$$D_1 = C_1 \cdot \frac{d}{b_0}; \quad D_2 = \frac{C_2}{\sqrt{b_0 d}}; \quad D_3 = \frac{C_3}{b_0 d}. \tag{36}$$

At the same time we replace the variables q_1, q_2 according to the equations

$$\xi = \sqrt{q_1}, \quad \eta = \sqrt{-q_2}. \tag{37}$$

Then the integrals of the equations of motion of the satellite can be written in the form

$$q_3 = q_{3k} + \frac{D_2}{\sqrt{2}}[I_2(\eta) - I_1(\xi)], \tag{38}$$

$$I_3(\xi) = I_4(\eta), \tag{39}$$

$$t = \sqrt{\frac{d^3}{2b_0}}[I_5(\xi) + I_6(\eta)], \tag{40}$$

$$p_1 = \sqrt{\frac{b_0 d}{2}} \cdot \frac{\sqrt{Q(\xi)}}{\xi(1+\xi^2)}, \tag{41}$$

$$p_2 = -\sqrt{\frac{b_0 d}{2}} \cdot \frac{\sqrt{P(\eta)}}{\eta(1-\eta^2)}, \tag{42}$$

$$p_3 = D_2\sqrt{b_0 d}, \tag{43}$$

with the following notation:

$$I_1(\xi) = \int_{\xi_k}^{\xi} \frac{d\xi}{(1+\xi^2)\sqrt{Q(\xi)}}, \quad I_2(\eta) = \int_{\eta_k}^{\eta} \frac{d\eta}{(1-\eta^2)\sqrt{P(\eta)}},$$

$$I_3(\xi) = \int_{\xi_k}^{\xi} \frac{d\xi}{\sqrt{Q(\xi)}}, \quad I_4(\eta) = \int_{\eta_k}^{\eta} \frac{d\eta}{\sqrt{P(\eta)}},$$

$$I_5(\xi) = \int_{\xi_k}^{\xi} \frac{\xi^2 d\xi}{\sqrt{Q(\xi)}}, \quad I_6(\eta) = \int_{\eta_k}^{\eta} \frac{\eta^2 d\eta}{\sqrt{P(\eta)}}.$$

The fourth-degree polynomials $Q(\xi)$ and $P(\eta)$ are defined by the equations

$$Q(\xi) = m\xi^4 + \xi^3 + n\xi^2 + \xi + s,$$

$$P(\eta) = -m\eta^4 + n\eta^2 - s.$$

The coefficients m, n, s of the polynomials are functions of the arbitrary constants D_1, D_2, D_3, namely:

$$m = D_1, \quad n = D_1 + 2D_3, \quad s = \frac{1}{2}D_2^2 + 2D_3.$$

The integrals I_1, I_2, I_3, I_4, I_5, I_6 are elliptic integrals. Without going into detail on their method of computation, we point out that in using an electronic computer it is most convenient to use an expansion of these integrals in power series with respect to the small parameters.

As shown by comparative calculations, a program utilizing the equations obtained allows the machine time necessary for carrying out the computations to be shortened by a factor of ten to a hundred (depending on the extent of the calculated part of the orbit and the number of points used for the determination) compared with a program based on numerical integration of the equations of motion.

It would be interesting to consider certain special cases of the motion investigated that are amenable to simple qualitative analysis.

We first wish to write out the equations connecting the generalized momenta p_1, p_2, p_3 with the projections v_1, v_2, v_3 on the positive directions of the coordinate lines of the spherical inertial coordinate system r, ψ, q_3. Making use of (10) and (12) and remembering that

$$v_1 = \dot{r}, \quad v_2 = r\dot{\psi}, \quad v_3 = r\cos\psi\,\dot{q}_3,$$

we obtain

$$v_1 = \frac{2}{r(q_1 - q_2)}\left[q_1(1+q_1)p_1 - q_2(1+q_2)p_2\right],$$

$$v_2 = \frac{r\sin 2\psi}{d^2(q_2 - q_1)}(p_2 - p_1), \qquad (44)$$

$$v_3 = \frac{p_3}{d\sqrt{(1+q_1)(1+q_2)}}.$$

Expressing the momenta p_1, p_2, p_3 explicitly in terms of the velocities v_1, v_2, v_3, we get

$$p_1 = \frac{r}{4q_1(1+q_1)}\left[2v_1\frac{q_1(1+q_1)}{1+q_1+q_2} + v_2\sin 2\psi\right],$$

$$p_2 = \frac{r}{4q_2(1+q_2)}\left[2v_1\frac{q_2(1+q_2)}{1+q_1+q_2} + v_2\sin 2\psi\right], \qquad (45)$$

$$p_3 = d\sqrt{(1+q_1)(1+q_2)} \cdot v_3.$$

Let us also express the arbitrary constants D_1, D_2, D_3 in terms of the quantities v_{1k}, v_{2k}, v_{3k}. Considering Eq. (45) at the initial point and substituting the expressions for p_{1k}, p_{2k}, p_{3k} into Eqs. (33), (34), (35), we obtain

$$D_1 = \frac{d \cdot v_k^2}{2b_0} - \frac{\sqrt{q_{1k}}}{q_{1k} - q_{2k}},$$

$$D_2 = \frac{v_{3k} r_k \cos\psi_k}{\sqrt{b_0 d}} = v_{3k}\sqrt{\frac{d}{b_0}} \cdot \sqrt{(1+q_{1k})(1+q_{2k})},$$

$$D_3 = \frac{d}{4b_0}\left[\frac{2b_0}{d} \cdot \frac{q_{2k}\sqrt{q_{1k}}}{q_{1k} - q_{2k}} - (1+q_{1k}+q_{2k})(v_{2k}^2 + v_{3k}^2) + \right.$$
$$\left. + (v_{1k}\sin\psi_k + v_{2k}\cos\psi_k)^2\right], \qquad (46)$$

where

$$v_k^2 = v_{1k}^2 + v_{2k}^2 + v_{3k}^2.$$

We first of all analyze the case $D_1 = D_2 = D_3 = 0$. The equality $D_1 = 0$ means that the total energy of the "satellite" has that minimum value which is necessary for the "satellite" to be transported from earth to infinity.* From the relation $D_2 = 0$ it follows that the motion proceeds in a meridian plane. The equality $D_3 = 0$, when $D_1 = 0$ and $v_{3k} = 0$, implies the condition

$$\frac{v_{2k}}{v_{1k}} = \frac{1 + q_{2k}}{q_{1k}} \operatorname{tg} \psi_k. \tag{47}$$

It is not difficult to see that in satisfying the condition (47) the initial velocity vector of the "satellite" will be directed along the normal to the ellipsoid of rotation defined by the equation $q_1 = q_{1k}$. The polynomial $Q(\xi)$ assumes the form

$$Q(\xi) = \xi(1 + \xi^2).$$

For the momentum p_1 the following expression will hold:

$$p_1 = \sqrt{\frac{b_0 d}{2}} \cdot \frac{1}{\sqrt{\xi(1 + \xi^2)}}. \tag{48}$$

The polynomial $P(\eta)$ is identically equal to zero. For $q_{2k} \neq 0$ and $q_{2k} \neq -1$ there follows directly from the integral (42)

$$p_2 = 0$$

and, consequently,

$$q_2 = q_{2k} = \text{const.}$$

For $q_{2k} = 0$ (initial point lying in the plane of the equator) or $q_{2k} = -1$ (initial point lying on the earth's axis of rotation), when in the right-hand side of (42) an indeterminacy makes its appearance, the same result is derived from consideration of the derivative

$$\frac{dq_2}{dq_1} = -\frac{q_2(1 + q_2)}{q_1(1 + q_1)} \cdot \frac{p_2}{p_1} = -\frac{\eta \sqrt{P(\eta)}}{\xi \sqrt{Q(\xi)}}.$$

*The term "satellite" in this case takes on a certain conditional nature.

Consequently, with removal of the "satellite" from earth to infinity with the least initial velocity required to do this, which is directed along the normal to the ellipsoid $q_1 = q_{1k}$, the motion will proceed in a hyperbola (with focal length 2d), for which the imaginary axis coincides with the axis of rotation of the earth, the real axis lying in the meridian plane of the initial point. This case is similar to the parabolic case in the theory of Kepler motion.

The velocity components v_1, v_2 and the flight time t turn out, for $D_1 = D_2 = D_3 = 0$ to be connected with the coordinate ξ by the relations

$$v_1 = \sqrt{\frac{2b_0}{d}} \cdot \frac{\xi}{\xi^2 + \eta_k^2} \sqrt{\frac{\xi(1+\xi^2)}{1+\xi^2-\eta_k^2}},$$

$$v_2 = \sqrt{\frac{2b_0}{d}} \cdot \frac{\eta_k}{\xi^2 + \eta_k^2} \sqrt{\frac{\xi(1-\eta_k^2)}{1+\xi^2-\eta_k^2}},$$

$$t = \sqrt{\frac{d^3}{2b_0}} \int_{\xi_k}^{\xi} \frac{(\xi^2 + \eta_k^2)\, d\xi}{\sqrt{\xi(1+\xi^2)}}.$$

These equations follow directly from (44), (48), and the expression for the generalized velocity $\dot{q}_1$ (see (14)):

$$\dot{q}_1 = 2\sqrt{\frac{2b_0}{d^3}} \cdot \frac{\xi\sqrt{\xi(1+\xi^2)}}{\xi^2 + \eta_k^2}.$$

Let us turn now to an analysis of the equations of motion in the case when the constants D_1, D_2, D_3 are connected by the relations

$$\left.\begin{array}{l} D_2^2 = 2(1+\xi_k^2)^2 \left(D_1 + \dfrac{1}{2\xi_k}\right), \\[6pt] D_3 = -\dfrac{1}{2} D_1 (1 + 2\xi_k^2) - \dfrac{1+3\xi_k^2}{4\xi_k}. \end{array}\right\} \quad (49)$$

One becomes easily convinced that with these limitations imposed on the initial conditions the motion of the satellite will proceed over the surface of the ellipsoid $q_1 = q_{1k}$. By analogy with a satellite in a circular orbit, or circular satellite, we shall call the present one an ellipsoidal satellite.

Thus, by substituting the value $\xi = \xi_k$ into the polynomial

$Q(\xi)$ and into its first and second derivatives $Q'(\xi)$ and $Q''(\xi)$ and expressing n, s according to (49) in terms of D_1 and ξ_k, we obtain

$$Q(\xi_k) = 0, \quad Q'(\xi_k) = 0,$$

$$Q''(\xi_k) = 8D_1\xi_k^2 + \frac{3\xi_k^2 - 1}{\xi_k}.$$

The polynomial $P(\eta)$ for $\eta = \eta_k$ and the same values of m, n, s is equal to

$$P(\eta_k) = -D_1(\xi_k^2 + \eta_k^2)^2 - \frac{(\xi_k^2 + \eta_k^2)^2 + (1 - \eta_k^2)(\eta_k^2 - \xi_k^2)}{2\xi_k}.$$

Since the inequalities

$$P(\eta_k) \geqslant 0, \quad D_2^2 \geqslant 0$$

must always hold true, the constant D_1 proves to be found within the limits

$$-\frac{1}{2\xi_k} \leqslant D_1 \leqslant -\frac{1}{2\xi_k}\left[1 - \frac{(1-\eta_k^2)(\xi_k^2 - \eta_k^2)}{(\xi_k^2 + \eta_k^2)^2}\right]. \tag{50}$$

The quantity

$$\Delta = \frac{(1-\eta_k^2)(\xi_k^2 - \eta_k^2)}{(\xi_k^2 + \eta_k^2)^2}$$

can never become less than zero, since $\xi_k^2 \cong (r_k^2/d^2) \gg 1 \geq \eta_k^2$. (We are assuming that the initial point lies above the earth's surface.) Consequently, the inequality (50) will always be non-contradictory. Making use of (50), we arrive at the relation

$$Q''(\xi_k) \leqslant -\frac{1+\xi_k^2}{\xi_k} + \frac{4\xi_k(1-\eta_k^2)(\xi_k^2 - \eta_k^2)}{(\xi_k^2 + \eta_k^2)^2} \leqslant$$

$$\leqslant -\frac{1+\xi_k^2}{\xi_k(\xi_k^2 + \eta_k^2)^2}[(\xi_k^2 + \eta_k^2)^2 - 4\xi_k^2(1-\eta_k^2)] < 0,$$

inasmuch as $\xi_k \gg 2$. This means that in the neighborhood of the point ξ_k the polynomial $Q(\xi)$ takes on negative values for $\xi \gtrless \xi_k$. Motion is therefore possible only under the condition that the coordinate ξ be identically equal to ξ_k, i.e., along the surface of the ellipsoid $q_1 = q_{1k}$.

For $D_1 = -(1/2\xi_k)$ the motion must proceed in a meridian plane, since on the basis of (49) $D_2 = 0$. The satellite's trajectory in this case is an ellipse (with focal length 2d), the minor axis of which coincides with the earth's axis of rotation, the major axis lying in the meridian plane passing through the initial point. The semimajor axis of this ellipse a and the eccentricity e are expressed in terms of the coordinate ξ_k and parameter d as follows:

$$a = d\sqrt{1+\xi_k^2},$$

$$e = \frac{1}{\sqrt{1+\xi_k^2}}.$$

The flight time of the satellite is connected with the coordinate η by the dependence

$$t = \sqrt{\frac{d^3\xi_k}{b_0}} \int_{\eta_k}^{\eta} \frac{(\xi_k^2 + \eta^2)\,d\eta}{\sqrt{(1-\eta^2)(\xi_k^2 - \eta^2)}}.$$

For $D_1 = -(1/2\xi_k)(1-\Delta)$ and $q_{2k} = 0$ we obtain a circular satellite with orbit lying in the equatorial plane.

The height of the ellipsoidal satellite over the surface of the generalized earth-ellipsoid will increase with displacement of the satellite from the equator to either pole. For $D_2 = 0$ the maximum difference in altitudes for an orbit located at a comparatively small distance from the earth is about 18 km.

In the general case, for each given position of the initial point a unique dependence can be established between the azimuth of the escape plane A_k and the magnitude of the velocity v_k necessary for creating an ellipsoidal satellite. The reverse dependence will not be unique, since to one and the same velocity there will obviously correspond directions with azimuths $\pm A_k$ and $(180° \pm A_k)$. These dependences follow from the relations (46) and (49). If, for example, we specify the quantity v_k and determine the value D_1, we can find D_2 and, consequently, v_{3k} from (49). The azimuth A_k is defined by the equation

$$\sin A_k = \frac{v_{3k}}{v_k}.$$

Here the azimuth A_k is understood to mean the angle between the direction of the meridian and initial velocity vector in the plane tangent to the ellipsoid $q_1 = q_{1k}$. (For an ellipsoidal satellite the initial velocity vector always lies in this plane.)

The longitude and flight time of the ellipsoidal satellite are in general equal to

$$q_3 = q_{3k} - \frac{D_2}{\sqrt{2}} \left[\frac{1}{1+\xi_k^2} I_4(\eta) - I_2(\eta) \right],$$

$$t = \sqrt{\frac{d^3}{2b_0}} [\xi_k^2 I_4(\eta) + I_6(\eta)].$$

The inequalities (50) and relations (49) are treated as conditions that define the minimum velocity $v_{k,\,min}$ necessary to create a satellite (if we understand the latter to mean a freely escaped body, whose coordinate q_1 throughout the entire time of motion satisfies the inequality $q_1 \geq q_{1k}$). This velocity turns out to depend not only on the distance to the center of the earth r_k, but also on the latitude ψ_k and azimuth of the escape plane A_k. For a specified value of q_{1k} the velocity $v_{k,\,min}$ has the greatest value for $q_{2k} = 0$ and $A_k = \pm 90°$, i.e., for the creation of a circular satellite moving in the plane of the equator.

Let us say a few words on the possibility of motion of the "satellite" in a field with potential $\overline{V}$ with the coordinate q_2 remaining constant. It was shown above that such a case will, in particular, exist when $D_1 = D_2 = D_3 = 0$. Let us pose the question in a more general form: by what conditions must the constants D_1, D_2, D_3 be connected in order that throughout the time of motion the following equality will hold:

$$q_2 = q_{2k} = \text{const.}$$

Omitting from the discussion for the present the cases $q_{2k} = 0$ and $q_{2k} = -1$, in the energy integral

$$\frac{2}{b_0 d} \left[\frac{\xi^2(1+\xi^2)}{\xi^2+\eta^2} p_1^2 + \frac{\eta^2(1-\eta^2)}{\xi^2+\eta^2} p_2^2 + \frac{1}{4(1+\xi^2)(1-\eta^2)} p_3^2 \right] - \frac{\xi}{\xi^2+\eta^2} = D_1$$

we assume $q_2 = q_{2k}$; $p_2 = 0$; $p_3 = D_2 \sqrt{b_0 d}$. Since for $q_2 = \text{const}$ we obtain $\dot{p}_2 = 0$, then, drawing on the fifth equation in the system (14), we obtain

$$p_1^2 \frac{\xi^2(1+\xi^2)}{\xi^2+\eta_k^2} = \frac{b_0 d}{2}\left[\frac{\xi}{\xi^2+\eta_k^2} + \frac{D_2^2(\xi^2+\eta_k^2)}{2(1+\xi^2)(1-\eta_k^2)^2}\right].$$

Substituting the expression for p_1 into the energy integral, we arrive at the following relation between D_1 and D_2, which does not contain the running coordinate ξ:

$$D_1 = \frac{D_2^2}{2(1-\eta_k^2)^2}. \tag{51}$$

Making use of (35), we express D_3 in terms of D_2:

$$D_3 = -\frac{D_2^2(1-2\eta_k^2)}{4(1-\eta_k^2)^2}. \tag{52}$$

Hence it follows that the condition $q_2 = q_{2k}$ = const is not fulfilled when $D_1 < 0$ (in this case the total energy of the satellite is insufficient to carry it from the earth to infinity). For $D_1 = 0$ we have $D_2 = 0$, $D_3 = 0$, i.e., the case already considered. For $D_1 > 0$, motion with a constant value of the coordinate q_2 will be feasible, since the initial conditions can be connected by the relations (51), (52). In this case the trajectory of the "satellite" will lie on the surface of the single-branch hyperboloid of rotation $q_2 = q_{2k}$.

If $q_{2k} = 0$ or $q_{2k} = -1$, it is obvious that for any value of D_1 we can obtain $q_2 = q_{2k}$ = const. In the former case for this we must assume $v_{2k} = 0$ (motion in the equatorial plane), in the latter case the initial velocity vector must be directed along the vertical (motion along the earth's axis of rotation). We note that the transformations by which (50) and (51) were derived become inapplicable for establishing connection between D_1, D_2, D_3.

The analogy with the case $q_1 = q_{1k}$ = const becomes fully and definitely evident from the analysis made. This is explained by the nature of the elliptical coordinates used in the solution.

The theory of satellite motion in a field with potential $\overline{V}$ as presented in its most basic features in the present paper may be regarded as a next approximation after the theory of Kepler motion* in describing the true picture of the flight of

―――――――
*The integrals of Kepler motion result as a special case of the integrals (38–43) when $d \to 0$.

an earth satellite in outer space. The potential $\bar{V}$ adopted in the basis of this theory is a considerably better approximation to the real gravitational field of the earth than the potential of a point mass, leaving unaccounted for essentially only the force-of-gravity anomaly. But the main difference in the earth's field from the field of a point mass is that difference elicited by the ellipsoidal character of the earth and is fully contained in the potential $\bar{V}$.

Besides the application of the relations obtained for purely analytic purposes, they can be used to solve the problem of the weights and periodic perturbation of the elements of elliptical orbits of artificial satellites under the influence of the earth's oblateness. In later publications the author hopes to quote the results of investigations in this direction.

LITERATURE

1. I. D. Zhongolovich, Transactions of the Institute of Theoretical Astronomy, No. III (in Russian), Academy of Sciences of the U.S.S.R. Press. 1952.
2. A. A. Mikhailov, Course in Gravimetry and Theory of the Earth's Figure (in Russian), GUGK Press, 1939.
3. N. I. Idelson, Potential Theory (Russian translation), ONTI, 1936.
4. G. K. Suslov, Theoretical Mechanics (in Russian), GTTI, 1946.
5. C. Jacobi, Lectures on Dynamics (Russian translation), GTTI, 1936.
6. N. N. Buchholtz, Fundamental Course in Theoretical Mechanics, Part II, (Russian translation), GTTI, 1939.

DETERMINATION OF UPPER-ATMOSPHERE DENSITY FROM THE RESULTS OF MEASUREMENTS OF THE FLIGHT OF THE THIRD SOVIET ARTIFICIAL EARTH SATELLITE

P. E. Él'yasberg and V. D. Yastrebov

As has been shown in a series of recent articles,[1,2] observation of artificial earth satellites appears to be one of the best means of determining upper-atmospheric density. However, observation of each individual artificial satellite gives very limited material concerning the state of the upper atmosphere, corresponding to a relatively narrow span of measurements of altitude and other conditions (illumination, solar activity, etc.) which determine the air density at high altitudes. In view of this, the systematic accumulation of data obtained through orbital element measurements of various artificial earth satellites assumes great importance in the task of acquiring raw material for the construction of a more complete dynamic model of the atmosphere (i.e., a model which takes into account its dependence not only on altitude, but also on the other atmospheric elements).

This paper presents results of upper-atmospheric density determination from the data of the most accurate measurements of orbital elements of the third Soviet artificial earth satellite, obtained during the first part of its life.

For determination of the satellite orbit, the following basic assumptions were made concerning the character of its motion:

a) The satellite was considered to be a material point moving under the influence of the earth's gravitational force and the retarding force of the air.

b) The potential of the gravitational accelerating force was determined by the equation

$$W = \alpha_{00}\frac{R}{r} + \alpha_{20}\left(\frac{R}{r}\right)^3 P_{20}, \tag{1}$$

where R is the average radius of the earth; r is the distance to its center; $P_{20} = \frac{1}{2}(3\sin^2\psi - 1)$ is the second-order Legendre polynomial; ψ is the geocentric latitude. In the expression for the gravitational-force potential of the ellipsoid, the component $5\alpha_{40}(R/r)^5 P_{40}$ is neglected since its value is less than that of the terms which are caused by gravitational force anomalies. The coefficients α_{00} and α_{20} appear as constants, whose values are determined by the geometric parameters and gravitational characteristics of the earth ellipsoid we have assumed for the earth's model.

c) The decelerating force of air resistance was calculated by the equation

$$a_{Rx} = S\rho v^2, \tag{2}$$

where v is the module of the satellite velocity vector with respect to a coordinate system rigidly fastened to the earth; ρ is the air density at the given altitude; S is a ballistic coefficient whose magnitude is determined by geometric, weight and aerodynamic characteristics of the satellite. The air density ρ as a function of satellite altitude is found by the following equation, which is in a form convenient for electronic computer calculations:

$$\rho = A_i e^{k_{1i}(h-h_i)^2 - k_{2i}(h-h_i)}, \tag{3}$$

where h is the satellite flight altitude, A_i, k_{1i}, and k_{2i} are constant coefficients used in the height interval

$$h_i \leqslant h \leqslant h_{i+1}. \tag{4}$$

The coefficients A_i, k_{1i}, and k_{2i} were computed in advance on the basis of the assumed atmospheric model; the number of intervals (4) was so chosen as to ensure acceptable accuracy in the approximate value of density found as a function of altitude by (3).

d) The flight altitude h was computed by the equation

$$h = r - \bar{a}(1 - \alpha \sin^2 \phi), \qquad (5)$$

where $\bar{a}$ and α are the semimajor and semiminor axes of the earth ellipsoid. The error in height produced by equation (5), related to the deviation of the true vertical from the radius-vector and to the noncoincidence of the surface of the spheroid with that of the ellipsoid, does not exceed 30 meters and can be neglected.

The satellite's motion was studied in a rectangular system of coordinates Oxyz rigidly connected to the earth. The origin of this system coincides with the earth's center; the axis Oz passes through the north pole; the axis Ox passes through the Greenwich meridian; the axis Oy points to the east.

We define the following quantities for use below:

$q_j(t_i)$, orbit elements in the coordinate system Oxyz ($j = 1, 2 \ldots 6$; t_i is the time of taking the ith measurement);

q_{j0}, initial conditions of motion relative to a certain time t_0;

r_i, measurable orbit parameters;

r_{im} and r_{ic}, their measured and computed values, respectively;

$\sigma(r_i)$, mean square deviation characterizing the accuracy of measurements of parameters r_i.

The satellite orbital elements are determined from the given measurements by solving boundary-value problems for the following system of differential equations of motion:

$$\begin{aligned}
\dot{v}_x &= -S\rho v v_x - g_r \frac{x}{r} + g_m \frac{z}{r} \frac{x}{r_1} + 2\Omega_3 v_y, & \dot{x} &= v, \\
\dot{v}_y &= -S\rho v v_y - g_r \frac{y}{r} + g_m \frac{z}{r} \frac{y}{r_1} - 2\Omega_3 v_x, & \dot{y} &= v_y, \\
\dot{v}_z &= -S\rho v v_z - g_r \frac{z}{r} - g_m \frac{r_1}{r}, & \dot{z} &= v_z,
\end{aligned} \qquad (6)$$

where

$$r = \sqrt{x^2 + y^2 + z^2}, \quad r_1 = \sqrt{x^2 + y^2}, \quad v = \sqrt{v_x^2 + v_y^2 + v_z^2},$$

g_r and g_m are the radial and meridianal components, respectively, of the gravitational acceleration (positive direction of g_r is toward the center of the earth, of g_m, toward the south), and Ω_3 is the angular velocity of the earth's rotation.

The values of g_r and g_m are determined by the relationships

$$g_r = \frac{1}{r}\left[\alpha_{00}\frac{R}{r} + 3\alpha_{20}\left(\frac{R}{r}\right)^3 P_{20}\right] - \Omega_3^2 r_1 \cos\psi,$$

$$g_m = -\frac{1}{r}\alpha_{20}\left(\frac{R}{r}\right)^3 P_{21} + \Omega_3^2 r_1 \sin\psi, \qquad (7)$$

where

$$P_{21} = 3\sin\psi\cos\psi, \quad \sin\psi = \frac{z}{r}, \quad \cos\psi = \frac{r_1}{r}.$$

The initial data used for solving the boundary-value problems come from measurements of t_i, r_i ($t_i \leq t_{i+1}$, where $i = 1, 2, \ldots N$, and N is the total number of individual measurements), and approximate values of the initial conditions of motion $\tilde{q}_{j0}$ relative to some given moment of time $t_0 \leq t_i$. Solution of the boundary-value problems should yield more precise values of the initial conditions of motions at the time t_0; this is done by introducing corrections to the approximate values through equations of the form

$$q_{j0}^{(l)} = \tilde{q}_{j0} + \sum_{i=1}^{l} \delta q_{j0}^{(i)}, \quad (j = 1, 2, \ldots, 6), \qquad (8)$$

where l is the required number of successive approximations for adequate convergence of the approximation process.

For the determination of the corrections δq_{j0} it was assumed that the variation of the measured parameters is related to the variation of the initial conditions of motion by linear dependence of the form

$$\Delta r_i = \sum_{j=1}^{6} \frac{\partial r_i}{\partial q_{j0}} \delta q_{j0}, \quad (i = 1, 2, \ldots, N), \qquad (9)$$

where $\partial r_i/\partial q_{j0}$ is the partial derivative of the measured parameters with respect to the initial conditions of motion. The validity of such an assumption is confirmed by the rapid convergence of the successive approximation process, even for very rough given values of $\tilde{q}_{j0}$ with errors of the order of several hundred meters per second in the velocity vector components, and of several hundred kilometers in the coordinates.

In the capacity of measurable parameters we used the slant distance D_i of the satellite from the initial point (IP), the azimuth A_i of the line of sight "IP-Satellite," and the location angle of this line γ_i.

The method of least squares was used to calculate the corrections to the initial conditions of motion; this permitted determination of the orbital elements with the highest probability of accuracy.

For measurements of the parameters D_i, A_i, and γ_i of unequal accuracy, we assumed the following conditional equations[3]:

$$\sum_{j=1}^{6} \frac{\partial r_i}{\partial q_{j0}} p_i \delta q_{j0} = \Delta r_i \cdot p_i, \quad (i = 1, 2, \ldots, N_1) \tag{9'}$$

where $p_i = \dfrac{1}{\sigma(r_i)}$ is the "weight" of the ith measurement; $\Delta r_i = r_{iN_1} - p_{ip}$; N_1 is the number of measurements used.

Using known rules[3] on the given system of conditional equations (9'), we obtain the normal equations

$$\sum_{n=1}^{6} a_{mn} \delta q_{n0} = b_m, \quad (m = 1, 2, \ldots, 6), \tag{10}$$

where

$$\left.\begin{array}{l} a_{mn} = \displaystyle\sum_{i=1}^{N_1} \dfrac{\partial r_i}{\partial q_{m0}} \cdot \dfrac{\partial r_i}{\partial q_{n0}} \cdot p_i^2, \\[1em] b_m = \displaystyle\sum_{i=1}^{N_1} \dfrac{\partial r_i}{\partial q_{m0}} \cdot \Delta r_i \cdot p_i^2. \end{array}\right\} (m, n = 1, 2, \ldots, 6) \tag{11}$$

In the system of six linear algebraic equations (10), the quantities entering into the expressions (11) are determined in each case by the initial conditions q_{j0} through the process of numerical integration. For more convenient computation with the electronic computer, seven sets of motion equations (6) were integrated in parallel; each of these corresponded to one of the seven aggregate conditions of motion:

$$\left.\begin{array}{l} v_{x0},\quad v_{y0},\,v_{z0},\,x_0,\,y_0,\quad z_0; \\ v_{x0}+\Delta v_{x0},v_{y0},\,v_{z0},\,x_0,\,y_0,\quad z_0; \\ \dots\dots\dots\dots\dots\dots\dots\dots\dots \\ v_{x0},\quad v_{y0},\,v_{z0},\,x_0,\,y_0+\Delta y_0, z_0; \\ v_{x0},\quad v_{y0},\,v_{z0},\,x_0,\,y_0,\quad z_0+\Delta z_0. \end{array}\right\} \quad (12)$$

Thus the normal elements plus six perturbations of the trajectory are obtained in each step of the numerical integration; this allows us to find the derivatives $(\partial q_m/\partial q_{j0})(t_i)$ of the varying orbital elements by the method of limiting finite differences. The quantities r_{ip} and $\partial r_i/\partial q_{j0}$ are determined by elementary finite-limit equations from the values of the derivatives $(\partial q_m/\partial q_{j0})(t_i)$ and the elements of the unperturbed orbit $q_m(t_i)$ (for known coordinates of a given IP).*

Through solving the system of equations (10) we obtain corrections $\delta q_{j0}^{(i)}$, and better accuracy in the ith approximation to the values of the conditions of motion through equation (8); these and the results of measurements are the necessary input data for carrying out the next approximation. Among all the data obtained on t_i and r_i (i = 1, 2, ... N_1), there would be, as a rule, certain measurements containing errors of a nonstatistical character. It is quite obvious that these must not be employed in the computation. This is so because the conditions of equation (9) hold for each approximation process only if the measurements used in the process satisfy the condition

$$|\Delta r_i| \leqslant \Delta r_{i,\,\text{per}} \quad (i=1, 2, \ldots, N_1),$$

where $\Delta r_{i\,\text{per}}$ is the permissible deviation for the corresponding measured parameter (D, A, or γ).

The following method of assigning values to the permissible deviation appears to be the most expedient. For the first approximation, they are called "rough" with respect to

---

*For accuracy, r_{ip} and $\partial r_i/\partial q_{j0}$ were computed by the interpolation of orbital element values to time t_i.

the accuracy assigned to the values of the conditions of motion. After each approximation, the normalized error $\bar{\sigma}(r)$ (see Ref. 3) and the permissible deviation $\Delta r_{i\,per}$ were found by the equations

$$\bar{\sigma}(r) = \sqrt{\frac{\sum_{i=1}^{N_1} \Delta r_i^2 p_i^2 - \sum_{j=1}^{6} b_j \delta q_{j0}}{k-6}}, \qquad (13)$$

and

$$\Delta r_{i,\,per} = \mu \cdot \sigma(r_i) \cdot \bar{\sigma}(r), \qquad (14)$$

where μ is a constant coefficient whose value for the solution of the problem is assigned on the basis of the specific nature of the measured systematic errors in a given case which are not susceptible to calculation (in the absence of systematic errors, $\mu = 3$). The value of the permissible deviation must obviously have a lower limit; this limit is naturally set by $\Delta r_{i\,min} = \mu \sigma(r_i)$. In the process of successive approximation, the normalized error constantly decreases because the initial conditions of motion become more accurate and because faulty measurements are excluded from the computation. For $\bar{\sigma}(r) \leq 1$, the subsequent approximations are made with a constant permissible deviation equal to $\Delta r_{i\,min}$.

As a criterion for proper convergence of the successive approximation process, we apply the inequality

$$\frac{|\delta q_{j0}|}{\sigma(q_{j0})} \leq 1 \quad (j = 1, 2, \ldots, 6). \qquad (15)$$

As is shown by actual trial on given measurements, the correction obtained by each successive approximation is 10-100 times smaller than that of the previous approximation. It therefore follows from equation (15) that with the use of such a criterion for proper convergence, the accuracy of the computation is more than an order of magnitude higher than the accuracy of determining the orbit.

During the flight of the third artificial earth satellite the elements of the actual orbit were determined by measurements taken for two successive circuits. For the method used to solve the boundary-value problems, such a method of processing the measured data appears to be most suitable. It is true that in determining the orbit on the basis of data measured during one circuit,* it is generally not possible to guarantee the required accuracy of several orbital elements, for example, the period of revolution. Furthermore, the presence of significant systematic measurement error will cause all orbital elements to show large errors. On the other hand, if we use measurements taken for several days, this makes the process of boundary-value-problem calculation take a prohibitively long time; furthermore, the accuracy of determining orbit elements decreases significantly through the influence of methodical errors (neglect of gravitational force anomalies, inaccurate calculation of air resistance, and inaccuracies in the numerical integration all lead to additional errors which can reach significant proportions; for example, with a slant distance, this error can reach several kilometers in the course of one day's flight).

The results of the boundary-value-problem solutions are final corrected values of the initial conditions of motion t_0, q_{j0} ($j = 1, 2, \ldots 6$). For easier analysis, these data were adduced for the initial circuit, i.e., for the time at which the satellite moving under these conditions crossed the earth's equatorial plane from south to north. The indicated adduction

TABLE 1

Circuit number	Orbital elements				
	a, m	i, deg	T, min	h_{min}, m	h_{max}, m
39	7 423 833	65.2062	105.9466	225 799	1 876 736
53	7 423 105	65.2024	105.9305	225 261	1 875 519
66	7 422 525	65.2053	105.9179	225 320	1 874 133
133	7 419 405	65.2027	105.8498	224 492	1 867 711
147	7 418 740	65.2025	105.8353	224 584	1 866 101
173	7 417 353	65.2044	105.8051	224 119	1 863 365
214	7 415 453	65.2035	105.7637	223 856	1 859 216
255	7 413 708	65.2036	105.7255	223 557	1 855 390

*The measurements were made from U.S.S.R. territory on a limited portion of the orbit.

was done by numerical integration with positive or negative steps, depending on the stage of the initial circuit.*

Table 1 shows the main orbital elements for a number of circuits (which we shall henceforth designate "supporting"): the semimajor axis of the precessing ellipse a; the inclination i; the period of revolution T; the minimum and maximum flight altitudes h_{min} and h_{max}. Note that the period of revolution is taken as the time difference between successive south to north passages of the satellite across the equatorial plane.

The question of evaluating the accuracy of determination of the orbital elements is most simply resolved by assuming that the measurements contain only accidental errors. The first thing done is evaluation of the accuracy attained in determining the initial conditions of motion. To do this, one obtains the elements of the inverse matrix $[Q_{ij}] = [a_{mn}]^{-1}$, where $[a_{mn}]$ is the matrix of the system of normal equations (10), and further, the mean square error $\sigma(q_{j0})$ is taken from the expression

$$\sigma(q_{j0}) = \bar{\sigma}(r)\sqrt{Q_{jj}}, \quad (j = 1, 2, \ldots, 6). \tag{16}$$

As was mentioned above, the value of $\sigma(q_{j0})$ is used to obtain the criteria for proper convergence by means of equation (15).

In order to obtain the mean square errors of the main orbital elements we shall investigate the latter as functions of the initial conditions of motion

$$f_k = f_k(q_{j0}) \quad (j = 1, 2, \ldots, 6). \tag{17}$$

By expanding $f_k(q_{j0})$ into a series in the neighborhood of the circuit origin, we shall obtain the following expression with accuracy up to terms linear in δq_{j0}:

$$\delta f_k \approx \sum_{j=1}^{6} \frac{\partial f_k}{\partial q_{j0}} \delta q_{j0}. \tag{18}$$

Then the values of $\sigma(f_k)$ can be found by means of an equation:[3]

---

*The boundary-value problems for orbital elements were solved for points near the beginning of the circuit.

$$\sigma(f_k) = \bar{\sigma}(r) \sqrt{\sum_{i,j=1}^{6} Q_{ij} \frac{\partial f_k}{\partial q_{i0}} \frac{\partial f_k}{\partial q_{j0}}}, \tag{19}$$

where $\partial f_k / \partial q_{j0}$ is the partial derivative of the corresponding orbital elements with respect to the initial conditions of motion, found by known equations.

Table 2 shows computed mean square errors for: the semimajor axis a; inclination i; precessing period of revolution* T_{prec}; perigee radius r_p and apogee radius r_a; also the normalized value of the mean square error.

TABLE 2

Circuit number	Accuracy characteristics					
	$\bar{\sigma}(r)$	$\sigma(a)$, m	$\sigma(i)$, deg	$\sigma(T_{prec})$, min	$\sigma(r_p)$, m	$\sigma(r_a)$, m
39	0.71	1	$1.0 \cdot 10^{-4}$	$0.2 \cdot 10^{-5}$	23	23
53	1.08	2	$0.7 \cdot 10^{-4}$	$0.4 \cdot 10^{-5}$	10	11
66	1.58	1	$1.7 \cdot 10^{-4}$	$0.3 \cdot 10^{-5}$	17	17
133	1.42	2	$2.1 \cdot 10^{-4}$	$0.4 \cdot 10^{-5}$	21	22
147	1.04	4	$1.8 \cdot 10^{-4}$	$0.9 \cdot 10^{-5}$	20	22
173	0.78	1	$0.5 \cdot 10^{-4}$	$0.2 \cdot 10^{-5}$	14	15
214	0.73	1	$0.7 \cdot 10^{-4}$	$0.1 \cdot 10^{-5}$	12	12
255	0.93	1	$0.8 \cdot 10^{-4}$	$0.3 \cdot 10^{-5}$	8	9

The data shown in Table 2 pertain to evaluation of the accuracy of orbital element determination by the so-called "internal convergence" method. This involves consideration of mainly accidental measurement errors. Systematic errors are not taken into account to any great degree. At the same time it is obvious that in the solution of problems by the method of least squares with a large number of measurements, it is the systematic errors which usually determine the accuracy. Therefore the values of mean square error for orbital elements as found in Table 2 must be regarded as values which have been greatly reduced.

To evaluate the actual accuracy of the problem solutions we use the fact that the orbital element data given in Table 1

---

* $T_{prec} = \dfrac{2\pi}{\sqrt{R\alpha_{00}}} a^{\frac{3}{2}}$.

are for various "supporting" circuits separated by appreciable time intervals. We can therefore assume that the errors of a systematic character occurring during a given group of measurements change in a random fashion from one group to the next. Accordingly, the individual orbital measurement errors appear to be random in nature. In order to evaluate the magnitude of these errors for the data of Table 1, we found the mean-rule changes in the orbital elements, and then the deviations from these changes.

We used linear expressions of the form shown below to express the normalizing rule for changes in the magnitudes of h_{min}, h_{max}, and T:

$$\xi = \xi_0 + A_\xi N,$$

where ξ represents a number of the orbital elements under investigation; ξ_0 represents its initial value; N is the circuit number, A_ξ is a constant coefficient determined by the dependence of ξ on T. As regards the inclination i, we can consider it constant for all portions of the orbit.

The magnitudes and likewise the average values of the coefficients ξ_0 and A_ξ were found from the data of Table 1 by the method of least squares. Table 3 shows the values obtained for these quantities.

TABLE 3

Orbital element	Dimensions	Coefficients	
		ξ_0	A_ξ
h_{min}	m	225 545	-9.80
h_{max}	m	1 876 657	-99.70
T	min	105.9438	$-1.029 \cdot 10^{-3}$
i	deg	65.2038	—

We also computed for all "supporting" orbits the deviation $\delta\xi_j$ of the measured orbital elements shown in Table 1 from the corresponding average values (ξ represents a number of orbital elements; j is the number of the "supporting" circuit). These deviations are shown in Table 4.

DETERMINATION OF UPPER-ATMOSPHERE DENSITY

TABLE 4

Circuit number	Deviation of measured orbital elements from average			
	δh_{min}, m	δh_{max}, m	δT, min	δi, deg
39	232	−113	$0.94 \cdot 10^{-3}$	0.0024
53	−167	59	$-0.77 \cdot 10^{-3}$	−0.0014
66	20	−31	$-0.01 \cdot 10^{-3}$	0.0015
133	−152	227	$0.77 \cdot 10^{-3}$	−0.0012
147	77	13	$0.77 \cdot 10^{-3}$	−0.0013
173	−123	−31	$-1.71 \cdot 10^{-3}$	0.0006
214	6	−192	$-1.98 \cdot 10^{-3}$	−0.0003
255	109	70	$2.02 \cdot 10^{-3}$	−0.0002
$\sigma(\xi)$	152	136	$1.5 \cdot 10^{-3}$	0.0014
$\sigma(\xi_0)$	88	80	$0.88 \cdot 10^{-3}$	0.00053
$\dfrac{\sigma(A)}{A}$	0.075	0.0066	0.0071	—
$\sigma_{min}(\xi)$	53	48	$0.53 \cdot 10^{-3}$	0.00053

Using these deviations $\delta\xi_j$, we computed their mean square value

$$\sigma(\xi) = \sqrt{\frac{\Sigma(\delta\xi_j)^2}{n-k}},$$

where n is the total number of "supporting" circuits; k is the number of independent parameters used in the normalizing rule (k = 1 for ξ = 1 and k = 2 for the remaining orbital elements).

Table 4 shows values of $\sigma(\xi)$ computed in this manner. From the table we can see that the deviations $\delta\xi_j$ are of random character. Furthermore, the calculation shows that no matter how high we make the order of the polynomials used to construct the normalizing rule for orbital element changes, the magnitudes $\sigma(\xi)$ do not diminish. This serves to verify our choice of normalizing rule.

In addition to the quantities already mentioned, Table 4 shows mean square errors in the coefficients found for the normalizing rule $\sigma(\xi_0)$ and $\sigma(A)/A$ and also the errors in the normalized values of orbital elements $\sigma_{min}(\xi)$ for the mean of the "supporting circuit" interval (i.e., for the 147 circuits in which these errors were minimal). The calculation was based on the same methods as were used to solve the least-squares problems.

The error evaluations in Table 4 for orbital elements indicate that the method we have used gives highly accurate solutions of the problems posed. However the error values thus obtained are significantly larger than those shown in Table 2. This indicates that systematic errors play a large role in these measurements, and so do possible inaccuracies in the values of the forces acting on the satellite in flight (of primary importance is the force of air resistance).

Let us now turn to determining the density of the upper atmosphere from the given data. For this we use the equation

$$\rho_f = \frac{\Delta T}{S \sqrt{H}} \Phi(a, e, H), \qquad (20)$$

where ρ_f is the air density at the minimum flight height of the satellite above the earth's surface; ΔT is the decrease in the period of the satellite in one circuit; S is the coefficient determining the influence of air resistance in the equations of motion for the satellite (6); $\Phi(a, e, H)$ is a function dependent on H, the semimajor axis a and the eccentricity e of the orbit, but one which does not change much with changes in these variables. It is furthermore assumed that the air density at altitude h can be found by the equation

$$\rho(h) = \rho_f \exp\left(-\frac{h - h_{min}}{H}\right). \qquad (21)$$

The expression (20) can easily be obtained from the known relation for finding ΔT (see Ref. 2). It follows from this expression that knowledge of the height of the uniform atmosphere H has an essential bearing in the determination of ρ_f by the measured change in period of revolution ΔT. However this quantity is not presently known with sufficient accuracy.

In order to decrease the influence of the error in H on the accuracy of our determination of the air density, we use the fact that this influence is a minimum at a certain altitude:

$$h = h_{min} + \Delta h. \qquad (22)$$

Indeed, it follows from expressions (20), (21) and (22) that

DETERMINATION OF UPPER-ATMOSPHERE DENSITY

$$\rho(h) = \frac{\Delta T}{S} \Phi(a, e, H) f(\Delta h, H), \qquad (23)$$

where

$$f(\Delta h, H) = \frac{1}{\sqrt{H}} \exp\left(-\frac{\Delta h}{H}\right). \qquad (24)$$

The magnitude of H does not appear to have any influence on the function $\Phi(a, e, H)$ (see Ref. 2). Therefore the dependence of the density f(h) on the altitude of the uniform atmosphere H is completely determined in the form of the function $f(\Delta h, H)$. Figure 1 shows graphically the dependence of this function on Δh for various values of H. These graphs show that for $\Delta h \approx 22$ km and for H held within the limits 25 km $\leq H \leq 100$ km, the function $f(\Delta h, H)$ may deviate from its average value by about 6%, while for $\Delta h = 0$, the corresponding relative deviation in $f(\Delta h, H)$ reaches 33%.

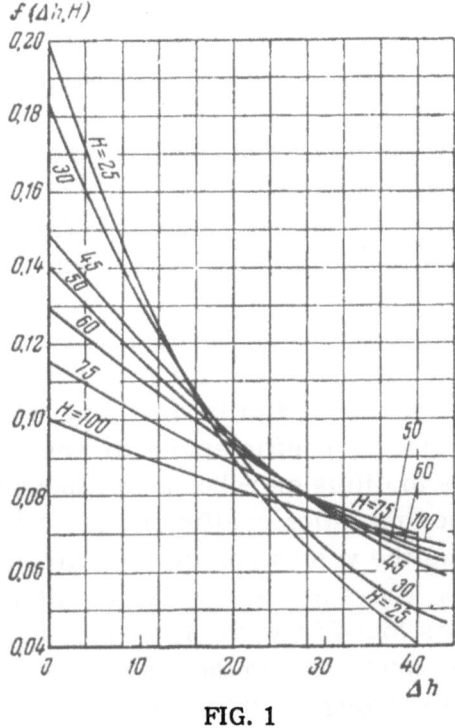

FIG. 1

If we take narrower limits for H, the proper selection of Δh will make the change in $f(\Delta h, H)$ even smaller. However in the course of the present work we shall continue to use the wider limits indicated above for the change in altitude of the uniform atmosphere, which it is known to have by present observation (for altitudes of the order of 200-250 km). Therefore the error in finding air density at altitude $h = h_{min} + 22$ km, caused by inaccuracy in the value of the altitude of the uniform atmosphere does not exceed about 6%.

For actual determination of air density it is expedient to use the method of comparison with the results of numerical integration of equation (6) for certain orbits close to the actual ones. Moreover, we exclude methodical errors having to do with approximations of the type of expression (20). If we assume that the function $\Phi(a, e, H)$ which enters into this dependence does not change when transferred to a real orbit, we obtain

$$\rho_f = \rho_{f_0} \frac{\Delta T}{\Delta T_0} \cdot \frac{S_0}{S}, \tag{25}$$

where the index 0 indicates quantities pertaining to the orbit being calculated.

In accordance with the results obtained above we shall determine the quantity ρ for the middle of the orbital section under study. Using the data of Table 3 we find for this circuit that $\Delta T = 1.029 \times 10^{-3}$ min, $h_{min} = 224.5$ km, and $T = 105.8346$ min.

In determining S, we shall assume that all orientations of the satellite are equally probable, and that the area of its midships section is equal to 1/4 of its total surface.[4] We shall further assume that the coefficient of air resistance $c_x = 2.1$; this corresponds to diffusion rebound of atmospheric particles with an accommodation coefficient $\alpha = 1$ (see Ref. 4).

Table 5 shows the values obtained for air density at altitude $h = 246$ km. There are also given in this table for comparison the values found from the first and second satellites.[2] All these values correspond to the mean of the dispersion interval for $\rho(h, H)$, using the interval for change of H indicated above (see Fig. 1).

Table 5 shows that the values of air density obtained from the second and third satellites practically coincide. The value obtained from the first satellite is somewhat lower (by approximately 20%). This may be explained by the fact that the perigees of the second and third satellites occurred during daytime, while that of the first satellite occurred at night.

TABLE 5

Designation of satellite	Air density at altitude $h = 246$ km, g/cm³
Satellite No. 1 and its rocket-carrier	$1.40 \cdot 10^{-13}$
Satellite No. 2	$1.73 \cdot 10^{-13}$
Satellite No. 3	$1.72 \cdot 10^{-13}$

In addition to the average values given in the table, there were computed densities for various "supporting" circuits. Moreover the values of ΔT and h_{min} were found by computations based on adjacent "supporting" circuits whose results were in general agreement. These values were taken for the center of the intervals between "supporting" circuits. The results obtained are given in Table 6.

TABLE 6

Circuit number	Local solar time		Air density at altitude $h = 246$ km, 10^{-13} g/cm³
	hours	minutes	
46	10	51	1.79
60	10	28	1.72
100	9	47	1.77
140	8	59	1.80
160	8	38	1.94
194	7	55	1.76
234	7	12	1.61

This table also gives the solar time for passage of the satellite through the point of minimum flight altitude.

It can be seen from Table 6 that the values of air density obtained for various intervals between "supporting" circuits are close to the average values given above (the maximum relative deviation does not exceed 13%). Furthermore, starting with the 160th circuit, there is a certain decrease in density which may be associated with the parallel decrease in the local solar time at the orbital point of minimum flight altitude. However, this decrease does not appear significant in the interval in question, and lies within the limits of accuracy of the calculation; there is therefore no basis for expecting a significant density change from this cause.

Relations (20) and (21) were used to evaluate the accuracy of the average density values given in Table 5. In this process, we neglected the influence of variations in the function $\Phi(a, e, H)$. As a result, we obtained for the relative errors in determination of density

$$\frac{\delta\rho}{\rho} < \frac{\delta_{max}(\Delta T)}{\Delta T} + \frac{\delta_{max}(h_{min})}{H} + \frac{\delta_{max}(S)}{S} + \frac{\delta_{max}(\rho_H)}{\rho}, \qquad (26)$$

where δ_{max} indicates the maximum absolute errors in determining the corresponding quantities, and $\delta_{max}(\rho_H)$ indicates the maximum error pertaining to the inaccuracy in knowledge of the value of H.

The data of Table 4 were used to determine the various components of the total error. Here we consider the maximum error to be equal to three times the mean-square value, while in determining the quantity $\delta_{max}(\Delta T)$ we use the evaluation of the calculation of the coefficient A_ξ in the linear expression given above for the period of revolution T ($\xi = T$). Furthermore, we assume H = 25 km for the minimum value of H. As a result we obtain

$$\frac{\delta_{max}(\Delta T)}{\Delta T} \simeq 2\%, \qquad \frac{\delta_{max}(h_{min})}{H} \approx 0.6\%.$$

Thus the errors pertaining to inaccuracy in measuring orbital elements are relatively small. A source of greater error may be due to certain assumptions made for the calculations, i.e., in the values of $\delta_{max}(S)/S$ and $\delta_{max}(\rho_H)/\rho$. As was shown above, $\delta_{max}(\rho_H)/\rho = 6\%$.

As regards error in δS, it is generally related to the assumptions made regarding the equal probability of all satellite orientations and also to the value of the accommodation coefficient $\alpha = 1$ assumed in finding the coefficient c_x. The first of these assumptions can be considered sufficiently accurate at present (in any case, for finding the average value of density for a large number of circuits). This view is supported by the good agreement between values of ρ obtained from various observed objects for the same conditions of passage through the region of perigee (first satellite and its rocket-carrier,[2] second and third satellites).

If we adopt for α the limiting value $\alpha_{lim} = 0$, then the quantity S increases by the ratio 4:3 (see Ref. 1). Then the air density decreases by 25%.

The various component errors $\delta\rho$ studied above appear to be independent and can therefore be combined by the rules for combination of independent random quantities. As a result we find that the maximum relative error in determining the density is

$$\frac{\delta_{max}\rho}{\rho} = 26\%.$$

This evaluation should be considered in all probability to be quite a high one, since the assumption that $\alpha = 0$ appears to be of low probability.

LITERATURE

1. M. L. Lidov, Artificial Earth Satellites, Volume 1, Plenum Press, New York, 1960.
2. P. N. Él'yasberg, Artificial Earth Satellites, Volume 1, Plenum Press, New York, 1960.
3. P. I. Shilov, Method of Least Squares (in Russian), Geodezizdat, 1941.
4. M. L. Lidov, Izv. Akad. Nauk SSSR, Geophysics Series, 13, 1957.

VARIATIONS IN DENSITY OF THE UPPER ATMOSPHERE FROM DATA OBTAINED FROM THE MEASURED PERIOD OF ROTATION OF ARTIFICIAL EARTH SATELLITES

G. A. Kolegov

Observations of artificial earth satellites have shown that the rate of decrease of their periods of rotation as a result of the braking action of the upper atmosphere does not grow monotonically with lowering of the orbit, but undergoes fluctuations instead. In view of the fact that this decrease depends on the density of the upper atmosphere, we can use the fluctuations in rate of decrease to assess the fluctuations in density. There are several factors which produce variations in atmospheric characteristics. The first and most obvious is the heating effect of the sun. A second no less important factor is the presence of atmospheric tides, whose amplitude in the upper layers can reach significant proportions.[1] Furthermore, changes in solar activity can influence the upper atmosphere. Thus for example, it has been shown[2] that there is significantly good correlation between the fluctuations in braking and the number of solar flares. However, these results require further corroboration.

All these and many other factors (rotation of the satellite about its center of mass, variation of density with latitude and flight altitude), interwoven produce the complex fluctuations of atmospheric braking effect which act on the satellites.[2,3]

This paper presents calculations based on data obtained by observations of Soviet satellites, and also some interpretations of the results.

In view of the fact that the orbits of the Soviet satellites were elliptical, the satellites underwent braking essentially only in the region of their perigees, at the lowest point of orbit. Using the relationships of the theory of ellipses,[4] it is possible to obtain the following expression for determining the density of the atmosphere in the region of perigee during a given circuit N:

$$\rho_p(N) = -\frac{2}{3b}\sqrt{\frac{e(N)}{2\pi r_p(N) H}} \cdot \frac{1-e(N)}{[1+e(N)]^{3/2}} \cdot \frac{\Delta T(N)}{T(N)},$$

where $b = (C_x S_{total})/4m$ is the ballistic coefficient; $C_x = 2.1$, in agreement with Lidov;[3] S_{total} is the surface of the satellite; m is its mass; $e(N)$ is the eccentricity of the orbit during the Nth circuit; $r_p(N)$ is the distance from the center of the earth at perigee during the Nth circuit; $T(N)$ is the orbital period during the Nth circuit; $\Delta T(N)$ is the rate of its deceleration during one circuit; H is the height of the uniform atmosphere.

The period of revolution and the rate of deceleration for a given circuit can be determined as follows:

$$T(N) = t(N+1) - t(N),$$

$$\Delta T(N) = T(N+1) - T(N).$$

Here $t(N)$ is the time of passage of the satellite past a certain latitude during the Nth circuit.

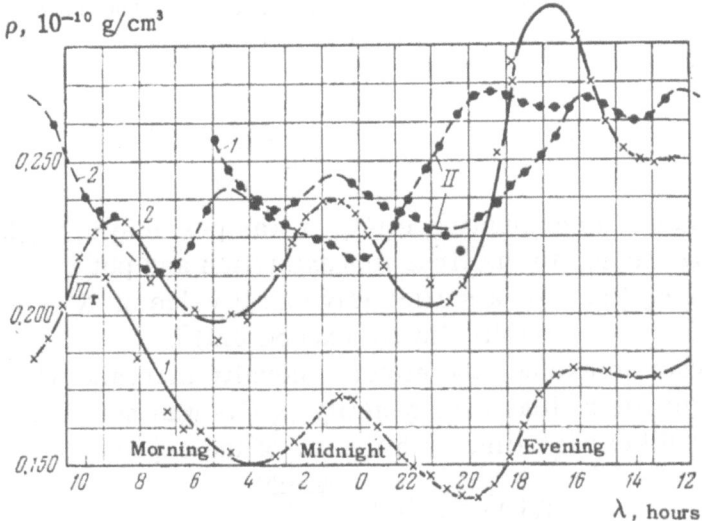

FIG. 1. Curves showing density during the course of a day from observations of the second satellite (dots) and of the rocket-carrier of the third satellite (crosses). 1) First circuit; 2) second circuit.

Experimental data on the time of passage of a satellite past a given latitude $t(N_1)$, $t(N_2)$ ... served to determine the value of $t(N)$. A third-power polynomial was derived on the basis of a number of circuits, using the method of least squares on measurements of $t(N_1)$, $t(N_2)$ .... The measurement interval was so chosen as to give a statistical character to deviations of the measured points from the polynomial values. Since the interpolation errors were minimal in the middle of an interval, the period of rotation T and the incremental decrease thereof ΔT were measured only during the middle rotation of the interval. For the determination of T and ΔT during the Nth circuit, the interpolation interval moved over by the corresponding amount $N' - N$ revolutions; an analogous computation was then made.

The eccentricity was determined by the relation

$$e(N) = 1 - \frac{(2\pi)^{2/3}}{\mu^{1/3}} \cdot r_p(N) T^{-2/3}(N),$$

where μ is the earth's gravitational constant.

The quantity $r_p(N)$ was measured accurately only at the beginning of the life span of the satellites; its subsequent changes were calculated with the aid of an approximation equation which is not difficult to derive from the relationship between ΔT and Δr_p given by Él'yasberg:[4]

$$r_p(N') = r_p(N) - \frac{H}{2} \ln \frac{1 + \frac{1}{e(N')}}{1 + \frac{1}{e(N)}}. \tag{1}$$

Furthermore, in computing $e(N')$, it was possible to neglect the change in r_p to the first approximation and put $r_p(N') \approx r_p(N)$. For a sufficiently small value of $N' - N$ the error in eccentricity did not exceed 0.01%.

In order to exclude the effect of density increase with lower satellite altitude, the quantity $\rho_p(N)$ was normalized to $h_0 = 226$ km each time with the aid of the expression

$$\rho_0(N) = \rho_p(N) e^{-\frac{h_0 - h(N)}{H}},$$

where $h(N)$ is the height of the perigee above the surface of the earth during the Nth circuit.

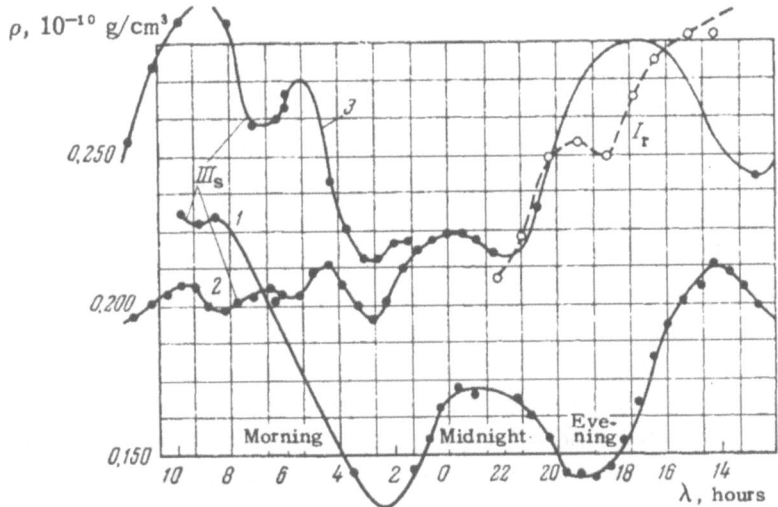

FIG. 2. Curves showing density change from the observed data for the rocket-carrier of the first satellite (circles) and for the third satellite (solid dots). 1) First circuit; 2) second circuit; 3) third circuit.

In view of the fact that the satellite orbits undergo precession (about 3 deg/day), the position of the orbit perigee does not remain fixed with respect to the sun. Thus, the orbital perigee can execute several circuits with respect to the sun during the lifetime of the satellite. Following Lidov,[3] we give the value of ρ_0 from its relation to λ, the difference between the right ascension of the perigee and the sun. It is not difficult to see that λ represents the local solar time of the satellite's passage through its perigee. In Figs. 1 and 2, curves I_r, II, III_s, and III_r show the densities calculated from data from observations of the rocket-carrier of the first satellite, the second and third satellites, and the rocket-carrier of the third satellite, respectively. Since the curves remained essentially the same for various values of H, we confined ourselves to one of the actual values, H = 50 km. The curves are drawn through points determined with errors not greater than 1%. For points with errors exceeding 1% the curves are drawn through averaged values. It can be seen from the figures that in addition to a noticeable density increase at noon and a decrease at midnight, which can be explained as the result of solar heating, there is also an evident, regular, 8-hour variation. All the curves except that for the second circuit of curve II (where the fluctuation phase changed by π) show maxima at midnight and minima at noon. These facts cannot be explained by either solar heating or by solar tidal action. The latter permits the existence of the 8-hour fluctuation, but with only a small amplitude in comparison with the 12-hour fluctuation.[5]

It should also be noted that although the courses of curves III_r and III_s are in general alike, the variations at the beginning of the second circuit for curve III_s cannot be explained from our point of view. It is possible that these variations are due to unsteady rotation of the third satellite about its center of mass. It must be pointed out that this fact does not support our conjectures concerning the correlation of braking fluctuations with solar activity, since curves III_r and III_s correspond to the same time period (within an accuracy of two days) in this respect.

LITERATURE

1. S. K. Mitra, Verkhnyaya Atmosfera (The Upper Atmosphere), IL, Moscow, 1955.
2. T. R. Nonweller, Nature, 182, 468, 1958.
3. M. L. Lidov, Artificial Earth Satellites, Volume 1, Plenum Press, New York, 1960.
4. P. E. Él'yasberg, Artificial Earth Satellites, Volume 1, Plenum Press, New York, 1960.
5. G. L. Pekeris, Proc. Roy. Soc. A158, 650, 1937.

DETERMINATION OF ILLUMINATION CONDITIONS AND PERIODS OF ILLUMINATION AND DARKNESS FOR THE ARTIFICIAL SATELLITE

I. M. Yatsunskii

In order to solve many of the problems associated with artificial satellites (operation of solar batteries, temperature cycle, orbital measurements, etc.), it is necessary to know how much of its time a satellite spends in sunlight and in darkness, and the possibility of observation by optical means. On the other hand, it is necessary to select the appropriate moment for launch into orbit in order to obtain the proper orbital plane with respect to the sun (for example, for the purpose of permitting solar observations).

Experience gained from the first artificial satellites shows that operational forecasting of visibility and illumination conditions during various periods of the satellite's lifetime, and the selection of a favorable launching time, is best done by graphical analysis methods which take into account the peculiarities of the satellite's motion with respect to the earth's shadow.

The present article describes one of the possible methods for determining the illumination conditions of a satellite, and gives a brief analysis of the motion of the first, second, and third Soviet satellites with respect to the earth's shadow.

GRAPHIC ANALYSIS METHOD OF EVALUATING THE ILLUMINATION CONDITIONS

Let us examine a unit sphere with its center coincident with that of the earth. For the spherical triangle with vertices corresponding to the sun, north pole, and satellite at a given moment it is easy to obtain the relations

$$\cos z = \sin \Delta = \sin \psi \sin \delta + \cos \psi \cos \delta \cos t_\odot. \tag{1}$$

Here z is the zenith angle to the sun from the point at which the satellite finds itself ($\Delta = 90° - z$ — sun's elevation), ψ is the geocentric latitude of the satellite; δ is the solar declination; $t_\odot$ is the horological angle of the sun with respect to the satellite. The angle $t_\odot$ can be related to the satellite longitude L_s and to the time t by the equation

$$t_\odot = t + 9^h + L_s - \eta, \tag{2}$$

where t is Moscow time, corresponding to the position of the satellite on the longitude meridian L_s; η is the "time relation," i.e., the difference between the average and true solar time.

Equations (1) and (2) solve the first part of the problem — the determination of the zenith angle to the sun from the given satellite coordinates and the time. Evidently, the solution is no different for any point of the earth's surface. Its inadequacy as regards satellite behavior consists of the absence of parameters dealing with satellite motion; this prevents advance long-term prediction of illumination conditions without previous accurate computation of the entire orbit. However, it can be shown that for determination of the illumination conditions it is only necessary to find the orbital plane and area; this makes the problem much simpler.

Let us introduce

$$t_\odot = \gamma + \lambda, \tag{3}$$

where

$$\gamma = \gamma_0 - (\eta - \eta_0) + \int_{t_0}^{t} \omega_1(\tau)\, d\tau - \omega_2(t - t_0), \tag{4}$$

$$\sin \lambda = \operatorname{ctg} i \operatorname{tg} \psi. \tag{5}$$

In equations (3)-(5), the angle γ is the difference between the latitudes of the ascending orbital node (L_Ω) and of the solar meridian ($L_\odot$) at time t, i.e., the horological solar angle with respect to the orbital node ($\gamma = L_\Omega - L_\odot$); λ is the difference between the latitudes of the satellite (L_s) and of the ascending node ($\lambda = L_S - L_\Omega$); ω_1 is the angular velocity of the orbital node; ω_2 is the angular velocity of the "average" equatorial sun; t is the running time of day; i is the inclination of the orbit; t_0, γ_0, η_0 are the initial values of t, γ, η.

According to (2) and (3) we have

$$\gamma_0 = t_0 + 9^h + L_s^{(0)} - \eta_0 - \lambda(\phi_0). \tag{6}$$

The rotational velocity of the node is obtained with adequate accuracy from the equation (if we take $t_0 = 0$)

$$\omega_1 = \omega_1^{(0)}(1 + kt), \tag{7}$$

whereupon we have for the initial precessional velocity of the node $\omega_1^{(0)} = -2\pi\epsilon(\cos i)/\mu T_0 p_0^2$, where ϵ and μ are terrestrial constants, related by $\epsilon\mu^{-1} = \bar{a}^2[\alpha - (m/2)]$; here $\bar{a} = 6378.1$ km is the radius of the earth's equator; α is the earth's flattening; m is the relation of the centrifugal force to the gravitational force at the equator; $\alpha - (m/2) = 0.001633$.

The coefficient k in equation (7) is determined by the relation

$$k = \frac{1}{T}\left(2\frac{p'}{p} + \frac{T'}{T}\right),$$

where the quantities p' and T' are reduced "gradients" corresponding to the orbital parameter p and to the rotational period T for a single circuit around the earth by the satellite.

With the passage of time, p' and T' increase monotonically on the average (approximately linearly), provided that the time t is sufficiently long. The observed fluctuations of these quantities, related by certain data with solar activity (sun spots) and with rotation of the sun about its own axis, have an insignificant amplitude (10-15%), and cannot exert any significant influence on the magnitude of the correction term kt in equation (7). Furthermore, the more significant periodic fluctuations related to latitude changes of the orbital perigee will average out provided that t is much longer than the fluctuation period ($T\psi$). If the two times are of the same order, or if $t < T\psi$, then although we might not observe a monotonic increase of p' and T', the absolute values of the changes in these quantities in a given case are so small that it is expedient to consider them linear functions of time as before in the computation of the correction term kt.

Therefore we adopt

$$p' = p'_0(1 + \alpha_1 t),$$

$$T' = T'_0(1 + \alpha_2 t),$$

where p'_0 and T'_0 are the initial values of the "gradient" orbital parameter and period of rotation; α_1 and α_2 are constant coefficients which can be approximated by $\alpha_1 \approx (1/2p'_0)(dp'/dt)$, $\alpha_2 \approx (1/2T'_0)(dT'/dt)$.

In actual practice, for known values of the parameter p and the period T during a given initial time interval t_1 ($t_1 < t$), it is expedient to use average values of p', T', α_1 and α_2 during this interval.

After substituting ω_1 (equation 7) into equation (4) and integrating (for $t_0 = 0$, $p = p_0$, $T = T_0$), we obtain for the angle γ

$$\gamma = \gamma_0 + \omega_1^{(0)} t \left(1 + \frac{k_0}{2} t + bt^2\right) - \omega_2 t - (\eta - \eta_0). \qquad (4')$$

where

$$\frac{k_0}{2} = \frac{1}{T_0}\left(\frac{p'_0}{p_0} + \frac{T'_0}{2T_0}\right),$$

$$b = \frac{2}{3T_0}\left(\frac{p'_0}{p_0}\alpha_1 + \frac{T'_0}{2T_0}\alpha_2\right).$$

Thus, knowing at a given moment the satellite coordinates $L_s^{(0)}$ and $\psi^{(0)}$, the initial values of the orbital elements T_0 and p_0, the average "gradients" p' and T' and their rates of change dp'/dt and dT'/dt, it is possible to use equations (1)-(5) to calculate the zenith angle of the sun for a given orbital inclination and for any given time with fixed limits of the terms $(k_0/2)t$ and bt^2 (it is required that $(k_0/2)t < 0.2$, $bt^2 < 0.2$). For satellites close to the earth, it is practical to predict the orbital plane position for one-third the life of the satellite, reckoning from an arbitrary beginning.

The illumination conditions depend not only on the position of the orbital plane with respect to the sun, but also on the elevation of the satellite flight path as it enters and leaves the earth's shadow. The flight elevation determines the depression of the horizon for the satellite (ϵ), which is computed by the equation

$$\cos \epsilon = \frac{R}{r} = \frac{R}{p}[1 + e \cos(u - \omega)], \qquad (8)$$

where r is the radius vector of the satellite; R is the average radius of the earth; p, e, ω are the corresponding parameter, eccentricity and perigee argument of the elliptical orbit,

$$\sin u = \frac{\sin \psi}{\sin i}. \qquad (9)$$

The zenith angle to the sun at the moment when the satellite crosses the edge of the earth's shadow will be equal to

$$z_\odot^{max} = 90° + \epsilon. \qquad (10)$$

Comparing $z_\odot^{max}$ to the sun's zenith angle as found by equation (1), we obtain an equation for determining the latitude at which the satellite enters and leaves the earth's shadow.

$$\cos z_\odot^{max} = -\sin \epsilon = -\sqrt{1 - \left(\frac{R}{r}\right)^2} =$$
$$= \sin \psi \sin \delta + \cos \psi \cos \delta \cos [\gamma(t) + \lambda(\psi)]. \qquad (11)$$

Equation (11) likewise permits us to find the time (horological angle for the point at which the satellite crosses the shadow boundary at a given latitude) and to solve the reverse problem—determination of the initial position of the orbital plane (angle γ_0) for which the satellite, at a given moment, enters (or leaves) the earth's shadow at latitude ψ. With the aim of finding simple methods of solving various problems concerning illumination conditions arising during the launching of the first Soviet satellites, special graphs were constructed on the basis of equation (1); on these, the abscissa gave the solar horological angle $t_\odot$, the ordinate gave the geocentric latitude (ψ), and the parameter was the solar elevation (Δ). The graph was constructed with a constant declination δ (it was sufficient to choose a spacing of δ equal to 5°). Besides lines of equal solar elevation, the graph shows (dotted lines) discrete positions of the satellite orbital plane for an inclination i = 65°. Each of the dotted curves corresponds to a fixed value of the horological angle for the ascending orbital node (γ), measured along the equator from the sun's meridian (Δ = +90°) to the meridian of the ascending node; each point on the curve with latitude ψ has its own value of horological angle. Figure 1 is an example of such a graph for δ = −5°.

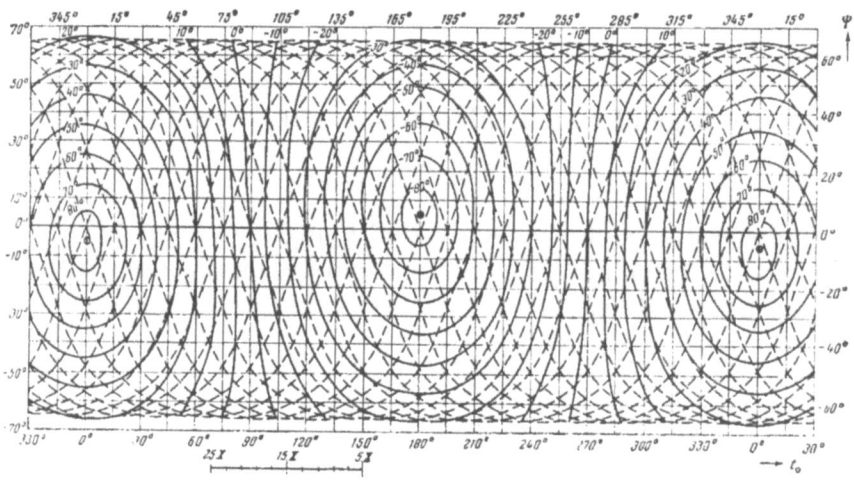

FIG. 1. Values of solar elevation (declination δ = −5°) and their dependence on the geographic latitude (ψ) and horological angle ($t_\odot$); the dotted curves show discrete (each 10°) positions of the orbital plane with inclination 65°.

In order to solve the basic problem—determining the solar elevation for a given satellite, and in particular, the place where it enters and leaves the earth's shadow—it is necessary to start with certain satellite coordinates ($\psi_s^{(0)}$, $L^{(0)}$) at a certain time and then calculate the angle γ_0 by means of equation (6). Having selected the graph with the appropriate solar inclination δ, we introduce the angle γ_0 in the form of a vertical line which will signify the position of the ascending node of the orbit with respect to the sun's meridian. We use equation (4') to compute subsequent positions of the node (angle γ) after a certain amount of time has passed (1-2 days), up to a given moment t_k (not longer than one-third the satellite's lifetime), and introduce corresponding marks onto graphs of the type of Fig. 1 having the proper values values of δ.

Each mark corresponds to a certain determined position of the orbital plane, found by interpolating between the bracketing dotted curves. The positions of the orbital plane thus found permit direct determination from the graph of solar elevations for any latitude, for both the ascending and descending branches of the orbital loop. If the value of δ on a graph does not correspond exactly to the observed datum, then an interpolation is made for δ.

Knowing the change in elevation of the satellite flight path (or its radius vector r) by latitude, we find on the basic graph for the given orbital plane position two latitudes ψ_1 and ψ_2 (entrance into and departure from shadow), for which the solar elevation Δ is equal to the lowering of the horizon ϵ [equation (8)]. In practice, it is possible to construct a graph of $\epsilon(\psi)$ for each satellite, which makes it easy to carry out the required comparison of Δ and ϵ.

The time of flight on the dark side of the orbit is determined by the equation

$$\Delta t_{\text{shadow}} = \frac{T}{2\pi}[E_2 - E_1 - e(\sin E_2 - \sin E_1)], \tag{12}$$

where E_2 and E_1 are the eccentric anomalies of the shadow exit and entrance points respectively (determined from the known values of ψ_1 and ψ_2).

The time of emergence into sunlight will be

$$\Delta t_{\text{sun}} = T - \Delta t_{\text{shadow}}.$$

In order to carry out optical observations of the satellite it is important to determine the "twilight" zones, i.e., zones for which the satellite is illuminated but the observation point is in shadow. The "twilight" zone can be found directly from graphs of the type of Fig. 1 for every position of the orbital plane once the latitudes of entering and leaving the shadow (ψ_1 and ψ_2) have been found. We shall isolate the portions of the orbit from points having latitudes ψ_1 and ψ_2 to points with $\Delta \approx -3°$. The region bounded on one side approximately by latitude ψ_1 (or ψ_2), on the other by the curve $\Delta \approx -3°$, and along the orbit by two equidistant curves (at such a distance from the orbital plane that the satellite location angle β is not less than 15-20°) will represent a zone (tied to the earth's surface by the position of the given orbital loop) within which optical observations are possible.

In most cases (low flight elevation) the "twilight" zone is found to be simply a latitude band bounded by ψ_1 (or ψ_2) on one side and by latitude $\psi_{\Delta = -3°}$ on the other, corresponding on the graph to the point of intersection of the curve $\Delta = -3°$ with the appropriate projection of the orbital plane under investigation. Naturally, in order to determine the possibility of observing the satellite at a given point on a given loop it is likewise necessary to verify that the satellite location angle β satisfies the condition given above ($\beta > 15-20°$).

The accuracy of the method described above is determined by the errors introduced through neglect of the earth's oblate shape, of the refraction of final values, and of the sun's angular diameter. For angles of orbital plane intersection with the shadow boundary which exceed 20°, the corresponding error in determination of the geographic latitude of entering (or leaving) the shadow does not exceed 1°.

A second source of error is the inaccuracy in the value of orbital parameters—the orientation of its plane (γ, i) in space, the elements of the ellipse (p, e), and the position of the perigee (ω) with respect to the equator; this latter determines the flight altitude at the moment of crossing the shadow boundary. These errors may become significant for predictions made ahead for one-third of the remaining lifetime of the satellite. However, in most cases, for crossings of the shadow boundary at angles above 20°, errors in flight altitude of 20-30 km and in orbital position of 1-2° give latitude errors for the intersection point again not over 1°.

ANALYSIS OF THE ILLUMINATION CONDITIONS FOR SOVIET SATELLITES

The illumination conditions for the first, second, and third artificial satellites differ significantly because of their launchings at different times of the year and day, and with various apogee elevations. Figure 1 shows the position of the orbital plane for the first satellite in its initial period of flight. Figure 2 shows the corresponding periods of time spent in shadow

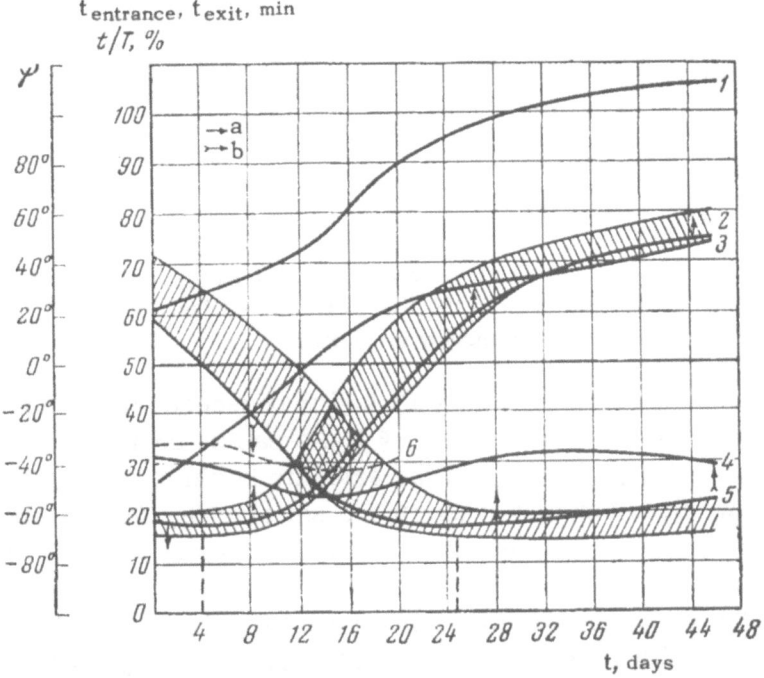

FIG. 2. Conditions for visibility of the third satellite at various geographic latitudes and the time spent in shadow and in sunlight: 1) change of t_{exit}; 3) change of $t_{entrance}$; 2, 5) corresponding latitudes; 4, 6) dependence of t/T for the third and first satellites. The arrows a indicate time of entrance into shadow, arrows b the time of exit. The shaded area is "twilight." The vertical dotted lines show the limits of regions with opposite directions of entrance and exit.

and in sunlight by this satellite. It is seen that these times did not change significantly; this is explained by the large angle of intersection between the orbital plane and the shadow boundary during the period in question.

The second satellite was launched in such a way that its orbital plane orientation was close to that of the shadow boundary. During the course of several days the satellite generally did not enter the earth's shadow. Figure 3 shows

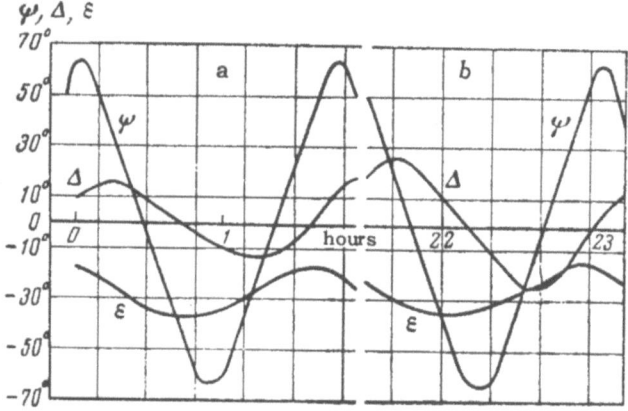

FIG. 3. Change of solar elevation (Δ) and lowering of the horizon (ϵ) for the second satellite in the first days of its flight. a) Nov. 3, 1957; b) Nov. 7, 1957. Region of shadow is omitted.

the actual solar elevation (Δ) and that necessary for contact with the earth's shadow (ϵ). As can be seen, only on the fourth day after launch did the satellite begin to enter the shadow. Such conditions are more satisfactory for the operation of equipment earmarked for study of solar ultraviolet radiation.

Figure 2 shows the motion characteristics of the third satellite with respect to the earth's shadow during its first month and a half of life (the curves of latitude change and of times of entering and leaving shadow, and the corresponding time spent in shadow as a percent of its circuit time around the earth). The twilight zones during which observation was possible are also indicated. Analysis of the curves shows that just as in the case of the first satellite, the time spent in shadow and in sunlight remained relatively constant; this is related to the large angles at which the orbital plane intersects the shadow zone during the period studied. Movement of the shadow zone itself (and of the twilight zone) in latitude was quite significant; this is explained by the relatively rapid rotation (precession) of the orbital plane.

In conclusion we wish to note that the selection of orbits inclined to the equator (at 65°) for the Soviet satellites established two periods during the year when the satellites do enter shadow: a summer period from the end of April to the middle of August, and a winter period from the end of October to the middle of February. For high flight altitudes in the shadow region (over 800 km), short periods of not entering the shadow can occur during any time of the year. In accord with the period of rotation of the orbital node with respect to the sun (approximately three months for the Soviet satellites), the period of staying out of shadow will repeat itself with the same or doubled (in the case of high flight altitudes) frequency.

DETERMINATION OF THE PARAMETERS OF THE ORBIT OF AN ARTIFICIAL SATELLITE FROM THE RESULTS OF GROUND MEASUREMENT

T. M. Éneev, A. K. Platonov, and R. K. Kazakova

In the present paper we shall describe briefly a method for determining the elements of the orbit of a satellite and for making predictions about its motion on the basis of the data obtained from optical and radiotechnical observations.

The method is based on the assumption that it will be applied with fast electronic computers.

DESCRIPTION OF THE PROBLEM

Let us consider a Cartesian system of coordinates x, y, z, immobile with respect to stars, and whose origin is located in the center of the earth: the x axis is directed towards the point of the spring equinox, the z axis along the axis of rotation of the earth, and the y axis so as to form with x and z a right-hand system of coordinates. Thus the motion of the satellite will be completely determined if the dependence of x, y, and z upon the time t are completely known. In the general case, x, y, and z will depend both upon the time t and upon some parameters $\alpha_1, \alpha_2, \ldots \alpha_n$, characterizing the orbit of the given satellite, i.e., we have

$$x = x(t; \alpha_1, \alpha_2, \ldots, \alpha_n),$$
$$y = y(t; \alpha_1, \alpha_2, \ldots, \alpha_n), \qquad (1)$$
$$z = z(t; \alpha_1, \alpha_2, \ldots, \alpha_n).$$

The problem of predicting the motion of the satellite consists in determining the values of the parameters in question from the data of direct observations and via their subsequent elaboration and to calculate the ephemerides for more or less considerable intervals of time.

The observation data are the satellite coordinates in the coordinate system of the observation point for given instances of time.

In astronomical observations equatorial coordinates, right ascension α and declination δ are commonly used. In radiotechnical observations spherical coordinates are usually used: the distance D, the azimuth A, the altitude γ.

The problem of elaborating the data consists in making a choice of the orbital parameters α_1, α_2, ... α_n such that the calculated coordinates of the satellite agree as well as possible with the observed ones. The well-known method of least squares is used for this purpose.[1]

Let us consider the case where the distance D, the azimuth A, and the altitude γ of the satellite are measured directly. Let us suppose that we know more or less accurately the orbital parameters α_1, α_2, ... α_n. Then for a given observation point and for an arbitrary instant t we can calculate the distance D, the azimuth A, and the altitude γ according to the formulas

$$\left. \begin{array}{l} D = \sqrt{(x')^2 + (y')^2 + (z')^2}, \\ A = \text{arc tg} \dfrac{y'}{x'}, \\ \gamma = \text{arc tg} \dfrac{z'}{\sqrt{(x')^2 + (y')^2}}, \end{array} \right\} \quad (2)$$

where x', y', z' are the Cartesian coordinates of the satellite in the coordinate system of the observation point. We have assumed that the origin of the coordinates of this local system coincides with the observation point, that the axis x' is directed towards the south along the tangent of the meridian, the axis y' is directed towards the west along the tangent to the parallel, and the axis z' is perpendicular to the plane x', y' (i.e., is perpendicular to the surface of the earth at the given point). The relative coordinates of the satellite x', y', and z' are related to the absolute coordinates x, y, and z by the formulas

$$\left.\begin{array}{l} x' = a_{10} + a_{11}x + a_{12}y + a_{13}z, \\ y' = a_{20} + a_{21}x + a_{22}y + a_{23}z, \\ z' = a_{30} + a_{31}x + a_{32}y + a_{33}z, \end{array}\right\} \quad (3)$$

where the coefficients a_{10}, a_{11}, a_{12}, ... a_{33} are in turn determined by the formulas

$$\begin{array}{lll}
a_{10} = -R_k \sin \varepsilon_k & a_{20} = 0, & a_{30} = -R_k \cos \varepsilon_k, \\
a_{11} = \cos(\lambda_k + \psi) \sin \varphi_k, & a_{21} = -\sin(\lambda_k + \psi), & a_{31} = \cos(\lambda_k + \psi)\cos \varphi_k, \\
a_{12} = \sin(\lambda_k + \psi) \sin \varphi_k, & a_{22} = \cos(\lambda_k + \psi), & a_{32} = \sin(\lambda_k + \psi) \cos \varphi_k, \\
a_{13} = -\cos \varphi_k, & a_{23} = 0, & a_{33} = \sin \varphi_k. \quad (4)
\end{array}$$

In the latter formulas φ_k and λ_k are the geographical coordinates of the observation point (latitude and longitude, respectively); R_k is the distance of the observation point from the center of the earth; ε_k is the angle between the local vertical at the observation point and the radius vector drawn from the center of the earth; ψ is the angle between the plane of the Greenwich meridian in the instant of observation and the direction of the spring equinox point. The angle ψ is a known function of time. For the purpose of predicting the motion of a satellite it is sufficient to assume that the function $\psi(t)$ is linear:

$$\psi = \psi_0 + \Omega(t - t_0),$$

where Ω is the angular velocity of rotation of the earth.

Let us examine the differences between the calculated coordinates D_ν, A_ν, γ_ν and the observed ones D_ν^{obs}, A_ν^{obs}, γ_ν^{obs} for different instants of time t_ν. Let us multiply these differences by some weight coefficients W_ξ, W_η and W_ζ:*

$$\begin{array}{l} \xi_\nu = W_\xi (D_\nu - D_\nu^{obs}), \\ \eta_\nu = W_\eta (A_\nu - A_\nu^{obs}), \quad \nu = 1, 2, \ldots, N \\ \zeta_\nu = W_\zeta (\gamma_\nu - \gamma_\nu^{obs}), \end{array} \quad (5)$$

where ν denotes the number of each observation, and N denotes the total number of observations.

*The weight coefficients W_ξ, W_η, and W_ζ are defined by the following formulas:

$$\dot{W}_\xi = \frac{1}{\sigma_D}, \quad W_\eta = \frac{\cos \gamma}{\sigma_A}, \quad W_\zeta = \frac{1}{\sigma_\gamma},$$

where σ_D, σ_A, and σ_γ are the corresponding mean square errors of measurement.

It is evident that ξ_ν, η_ν, and ζ_ν are functions of the parameters α_1, α_2, ... α_n. According to the method of least squares, these parameters must be chosen so that the sum of the squares of ξ_ν, η_ν, and ζ_ν should be minimum. With this purpose, following the method of least squares, we can select a first approximation for the parameters in question, $\alpha_1^{(0)}$, $\alpha_2^{(0)}$, ... $\alpha_n^{(0)}$, and try to find the corrections $\Delta\alpha_1$, $\Delta\alpha_2$, ... $\Delta\alpha_n$ in a first approximation, solving the system of normal equations

$$\left.\begin{aligned} A_{11}\Delta\alpha_1 + A_{12}\Delta\alpha_2 + \cdots + A_{1n}\Delta\alpha_n &= B_1, \\ A_{21}\Delta\alpha_1 + A_{22}\Delta\alpha_2 + \cdots + A_{2n}\Delta\alpha_n &= B_2, \\ \cdots\cdots\cdots\cdots\cdots\cdots\cdots\cdots\cdots\cdots \\ A_{n1}\Delta\alpha_1 + A_{n2}\Delta\alpha_2 + \cdots + A_{nn}\Delta\alpha_n &= B_n, \end{aligned}\right\} \quad (6)$$

where the coefficients A_{ij} and B_i can be determined by the formulas

$$\left.\begin{aligned} A_{ij} &= \sum_{\nu=1}^{N} \cdot \left(\frac{\partial \xi_\nu}{\partial \alpha_i} \cdot \frac{\partial \xi_\nu}{\partial \alpha_j} + \frac{\partial \eta_\nu}{\partial \alpha_i} \cdot \frac{\partial \eta_\nu}{\partial \alpha_j} + \frac{\partial \zeta_\nu}{\partial \alpha_i} \cdot \frac{\partial \zeta_\nu}{\partial \alpha_j} \right), \\ B_i &= -\sum_{\nu=1}^{N} \cdot \left(\xi_\nu \frac{\partial \xi_\nu}{\partial \alpha_i} + \eta_\nu \frac{\partial \eta_\nu}{\partial \alpha_i} + \zeta_\nu \frac{\partial \zeta_\nu}{\partial \alpha_i} \right). \end{aligned}\right\} \quad (7)$$

The same formulas will allow us to find the third approximation after having found the second approximation, and so on, until the iteration procedure for obtaining the precise values of the parameters α_1, α_2, ... α_n reaches the required accuracy.

A vital point in the method for elaboration of observations is the choice of the parameters α_1, α_2, ... α_n as is also the choice of the method of calculation of the absolute Cartesian coordinates x, y, z.

As orbital parameters one can choose, for instance, the three Cartesian coordinates x, y, z and the three components of the velocity $\dot{x}$, $\dot{y}$, $\dot{z}$ at some instant t_0, or the six osculating elements of the orbit at some instant; for instance, the major semiaxis a, the eccentricity e, the perigee longitude ω, the inclination i, the node longitude Ω, and the mean anomaly M_0.

These elements can serve as initial data for the system of differential equations describing the satellite motion. By integrating this system numerically one can predict the motion of the satellite.

It must be noted, however, that for satellites with a height of the perigee of 200-300 km the predictions obtained in this way can be very rough, especially if they are given for considerable intervals of time (e.g., many days in advance). The reason for this is essentially the inaccurate knowledge of the parameters of the upper layers of the atmosphere.

In order to get a reliable and long-range forecasting of the satellite motion it would be necessary, in general, to include in the number of unknown parameters $\alpha_1, \alpha_2, \ldots \alpha_n$ also the upper atmosphere parameters, and, possibly, those relating to the gravitational field of the earth. One must remember, however, that the immediate determination of the parameters of the atmosphere and of the gravitational field from an elaboration of the observations is technically difficult. It is apparently most convenient to include the inaccuracy of the knowledge of the geophysical constants into the orbit elements themselves, and then remove this inaccuracy during the elaboration.

The calculation of the absolute coordinates x, y, z of a satellite can be carried out in different ways. One of these ways is the direct integration of the differential equations for x, y, z. Another method consists in calculating x, y, z through the osculating elements $a, e, \omega, i, \Omega, M$ by the formulas

$$\begin{aligned} x &= r(\cos u \cos \Omega - \sin u \sin \Omega \cos i), \\ y &= r(\cos u \sin \Omega + \sin u \cos \Omega \cos i), \\ z &= r \sin i \cdot \sin u, \\ u &= \omega + \vartheta, \end{aligned} \qquad (8)$$

$$\vartheta = M + 2e \sin M + \frac{5}{4} e^2 \sin 2M + \frac{1}{12} e^3 (13 \sin 3M - 3 \sin M) + \ldots,$$

where ϑ is the true anomaly, M is the mean anomaly. The latter method has its advantages, because it permits applying the most labor-saving method for calculating the coordinates x, y, z.

CHOICE OF THE PARAMETERS DETERMINING THE ORBIT

In studying the method under consideration for elaborating the observations we took into account the following properties of the orbit of a satellite:

a) During one satellite revolution the orbit can be described sufficiently well in first approximation by means of the elliptical theory of unperturbed motion;

b) The perturbations of the satellite orbit due to the earth's atmosphere and to the unknown components of the gravitational field are very small and, with the majority of the observation instruments, they become noticeable only after many revolutions;

c) At present the accuracy with which we know the main geophysical constants is such that, if the initial data are known in an absolutely precise way, the orbit of a satellite can be calculated from one revolution with an error not exceeding a few tens of meters;

d) The values of the osculating elements of the earth a, e, ω, i, Ω, M_0 for a given geographical latitude φ change very slowly from revolution to revolution, so that a discrete sequence of values for different times of each of the elements in question can be connected by a smooth line.

The latter property of a satellite orbit determines the choice of the parameters $\alpha_1, \alpha_2, \ldots \alpha_n$.

As parameters $\alpha_1, \alpha_2, \ldots \alpha_n$ we chose the parameters $\bar{a}_q, \bar{e}_q, \bar{\omega}_q, \bar{i}_q, \bar{\Omega}_q, \overline{M}_{0q}$ (q = 0, 1, ...) which determine the values of the osculating elements a, e, i, ω, Ω, M_0 for a given geographical latitude φ_0. The discrete sequences of values of the osculating elements for the given geographical latitude φ_0 were approximated by formulas of the type:

$$\left. \begin{aligned} \bar{a} &= \sum_{q=0}^{k_a} \bar{a}_q (t-t_0)^q, \\ \bar{e} &= \sum_{q=0}^{k_e} \bar{e}_q (t-t_0)^q, \\ &\cdots \cdots \cdots \cdots \\ \overline{M}_0 &= \sum_{q=0}^{k_M} \overline{M}_{0q} (t-t_0)^q, \end{aligned} \right\} \quad (9)$$

where $k_a, k_e, \ldots, k_M$ are integral numbers, $\bar{a}_q, \bar{e}_q, \ldots, \overline{M}_{0q}$ are parameters determined on the basis of the elaboration of the observation data.

The properties of the orbit of a satellite indicated under (b) and (c) made it possible to work out a most economical method for calculating the coordinates x, y, z for those cases when the observations relating to a long time interval are elaborated simultaneously. Since the perturbations of the orbit by the earth's atmosphere and by the unknown components of the gravitation field are small, revolutions close to each other will be similar to each other both geometrically and dynamically; therefore, in order to calculate the coordinates x, y, z for different turns of the orbit it is not necessary to consider accurately all revolutions of the time interval under study; it will be enough to consider only some selected revolutions. The coordinates of the satellite for the remaining revolutions could be later determined by interpolation. As to the number of basic revolutions and to the order of the interpolation formulas, they would depend upon the length of the time interval under consideration. Experiment shows that, for orbits similar to the orbits of the Soviet satellites, it is sufficient to select three basic revolutions for the simultaneous elaboration of observations relating to two to three weeks.

The calculations of the method under consideration were made in the following order. The orbit elements a, e, ω, i, Ω, and also the mean anomaly M were calculated by numerical integration for the basic turns of the orbit. The initial data for this calculation were chosen with a latitude φ_0, and formula (9) was used for determining these initial data. As a result of the integration a table giving the values with time of the osculating elements a, e, ω, i, Ω, and of the mean anomaly M was prepared for each basic revolution. Using these tables it was easy to determine the osculating elements for any instant t by double interpolation, once within the individual tables for the basic revolutions, once between the tables themselves. After the elements a, e, ω, i, Ω, and M had been determined, the coordinates x, y, and z were easily found from formulas (8).

The method applied permitted us to shorten very much the time required to compute the coordinates x, y, z to the necessary degree of accuracy. Since the method for calculating the osculating elements a, e, ..., M has a fundamental importance in the procedure under consideration, the following section will be devoted to a more detailed description of this method.

DETERMINATION OF THE VALUES OF THE OSCULATING ELEMENTS FOR A GIVEN INSTANT OF TIME

As has been mentioned above, the osculating elements of the orbit of a satellite for a reference geographical latitude φ_0 can be determined with the help of formulas (9). It has also been noted that the values of the osculating elements calculated according to formulas (9) have a meaning only for special instants t_{0N} corresponding to the moments of passage of the satellite through the given latitude φ_0, since only at these moments do the elements $\bar{a}, \bar{e}, ..., \bar{M}_0$ actually define the position of the satellite.

On the other hand, knowing t_{0N} and calculating from formula (9) the values of $\bar{a}, \bar{e}, ..., \bar{M}_0$ corresponding to these instants, we obtain the initial data for calculating the motion on special orbit turns. It can be easily seen that the calculation of the satellite motion in an arbitrary orbit turn can be carried out completely independent of the other turns. This circumstance suggests a very economical method for calculating the elements a, e, ..., M for any instant.

Let us suppose that the observations of the satellite motion are carried out during the interval of time (t_0, t_k). Let us consider for this time interval the set of the instants $t_0^{(N)}$ corresponding to the passages of the satellite at the given latitude φ_0, and let us choose from this set three instants $t_0^{(1)}$, $t_0^{(m)}$, $t_0^{(k)}$, where $t_0^{(1)}$ is the earliest instant of the set $t_0^{(N)}$, $t_0^{(m)}$ is the latest moment of the set, and $t_0^{(k)}$ is some intermediate instant. For each of the moments in question, $t_0^{(1)}$, $t_0^{(m)}$, $t_0^{(k)}$, let us calculate the orbital elements $\bar{a}, \bar{e}, \bar{\omega}, \bar{i}, \bar{\Omega}, \bar{M}$.

Thus we obtain

$$\left. \begin{aligned} &\text{for} \quad t = t_0^{(1)} \quad a = \bar{a}_0^{(1)}, \; e = \bar{e}_0^{(1)}, \ldots, M = \bar{M}_0^{(1)}, \\ &\text{for} \quad t = t_0^{(m)} \quad a = \bar{a}_0^{(m)}, \; e = \bar{e}_0^{(m)}, \ldots, M = \bar{M}_0^{(m)}, \\ &\text{for} \quad t = t_0^{(k)} \quad a = \bar{a}_0^{(k)}, \; e = \bar{e}_0^{(k)}, \ldots, M = \bar{M}_0^{(k)}. \end{aligned} \right\} \quad (10)$$

Here $\bar{a}, \bar{e}, \bar{\omega}, \bar{i}, \bar{\Omega},$ are calculated according to formula (9) and $\overline{M}_0^{(1)}, \overline{M}_0^{(m)}, \overline{M}_0^{(k)}$ are given by the formulas

$$\overline{M}_0^{(1)} = \overline{M}_{00}^{(1)} + \frac{\sqrt{\mu}}{(a_0^{(1)})^{3/2}} (t_0^{(1)} - t_0),$$

$$\overline{M}_0^{(m)} = \overline{M}_{00}^{(m)} + \frac{\sqrt{\mu}}{(a_0^{(m)})^{3/2}} (t_0^{(m)} - t_0),$$

$$\overline{M}_0^{(k)} = \overline{M}_{00}^{(k)} + \frac{\sqrt{\mu}}{(a_0^{(k)})^{3/2}} (t_0^{(k)} - t_0),$$

and (see also (9))

$$\overline{M}_{00}^{(1)} = \sum_{q=0}^{k_M} \overline{M}_{0q} (t_0^{(1)} - t_0)^q,$$

$$\overline{M}_{00}^{(m)} = \sum_{q=0}^{k_M} \overline{M}_{0q} (t_0^{(m)} - t_0)^q,$$

$$\overline{M}_{00}^{(k)} = \sum_{q=0}^{k_M} \overline{M}_{0q} (t_0^{(k)} - t_0)^q.$$

Using the data of (10) as initial data, let us integrate the equations of motion in the osculating elements for the first, the intermediate, and the last revolution, and let us represent the results of this integration as a table with a constant time increment Δt:

$$\left.\begin{array}{c}\Delta a_1^{(1)}, \Delta a_2^{(1)}, \ldots, \Delta a_i^{(1)}, \ldots \\ \Delta e_1^{(1)}, \Delta e_2^{(1)}, \ldots, \Delta e_i^{(1)}, \ldots \\ \cdots\cdots\cdots\cdots\cdots\cdots\cdots \\ \Delta M_1^{(1)}, \Delta M_2^{(1)}, \ldots, \Delta M_i^{(1)}, \ldots \end{array}\right\}$$

$$\left.\begin{array}{c}\Delta a_1^{(m)}, \Delta a_2^{(m)}, \ldots, \Delta a_i^{(m)}, \ldots \\ \Delta e_1^{(m)}, \Delta e_2^{(m)}, \ldots, \Delta e_i^{(m)}, \ldots \\ \cdots\cdots\cdots\cdots\cdots\cdots\cdots \\ \Delta M_1^{(m)}, \Delta M_2^{(m)}, \ldots, \Delta M_i^{(m)}, \ldots \end{array}\right\} \quad (11)$$

$$\left.\begin{array}{c}\Delta a_1^{(k)}, \Delta a_2^{(k)}, \ldots, \Delta a_i^{(k)}, \ldots \\ \Delta e_1^{(k)}, \Delta e_2^{(k)}, \ldots, \Delta e_i^{(k)}, \ldots \\ \cdots\cdots\cdots\cdots\cdots\cdots\cdots \\ \Delta M_1^{(k)}, \Delta M_2^{(k)}, \ldots, \Delta M_i^{(k)}, \ldots \end{array}\right\}$$

Here $\Delta a_i^{(1)} = a_i^{(1)} - \bar{a}_0^{(1)}$, $\Delta a_i^{(m)} = a_i^{(m)} - \bar{a}_0^{(m)}$ etc., and $a_i^{(1)}$, $a_i^{(m)}$, ... are the values of the osculating elements corresponding to the ith point of the table. Using the tables obtained it is possible to calculate now the osculating elements for any instant t_ν intermediate between the extremes of the measurement interval (t_0, t_k), and also for instants outside the interval (t_0, t_k). The latter calculation is necessary in order to predict the satellite motion.

Let us consider first the case when the observation moment t_ν lies in one of the time intervals $(t_0^{(1)}, t_0^{(2)})$, $(t_0^{(m)}, t_0^{(m+1)})$ or $(t_0^{(k)}, t_0^{(k+1)})$. The values of the osculating elements can be obtained in this case with the help of (11), using a single interpolation. Thus, for instance, let

$$t_0^{(1)} < t_\nu < t_0^{(2)}.$$

Then, using a quadratic interpolation, we obtain for the element q (where q is one of the elements a, e, ..., M)

$$q\vert_{t=t_\nu} = q_0^{(1)} + \Delta q^{(1)},$$

$$\Delta q^{(1)} = \Delta q_{i-1}^{(1)} 0{,}5\lambda(\lambda - 1) + \Delta q_i^{(1)}(1 - \lambda^2) + \Delta q_{i+1}^{(1)} 0{,}5\lambda(\lambda + 1),$$

$$\lambda = \frac{\Delta t_i}{\Delta t} - 1;$$

Δt_i is defined by the formula

$$\Delta t_i = (t_\nu - t_0^{(1)}) - (i - 1)\Delta t),$$

where i is an integer chosen in such a way that the condition $0 < \Delta t_i \leq \Delta t$ be fulfilled, and Δt is, as before, the interval of time for which the tables of the osculating elements are calculated.

Let us consider now the case when t_ν does not lie in any of the intervals $(t_0^{(1)}, t_0^{(2)})$, $(t_0^{(m)}, t_0^{(m+1)})$, $(t_0^{(k)}, t_0^{(k+1)})$. Let us suppose that

$$t_0^{(N)} < t_\nu < t_0^{(N+1)}.$$

In this case we can determine the element q by quadruple interpolation. Let us first calculate $\Delta q^{(1)}$, $\Delta q^{(m)}$, $\Delta q^{(k)}$ according to the formulas

$$\Delta q^{(1)} = \Delta q_{i-1}^{(1)} 0.5\lambda(\lambda-1) + \Delta q_i^{(1)}(1-\lambda^2) + \Delta q_{i+1}^{(1)} 0.5\lambda(\lambda+1),$$
$$\Delta q^{(m)} = \Delta q_{i-1}^{(m)} 0.5\lambda(\lambda-1) + \Delta q_i^{(m)}(1-\lambda^2) + \Delta q_{i+1}^{(m)} 0.5\lambda(\lambda+1),$$
$$\Delta q^{(k)} = \Delta q_{i-1}^{(k)} 0.5\lambda(\lambda-1) + \Delta q_i^{(k)}(1-\lambda^2) + \Delta q_{i+1}^{(k)} 0.5\lambda(\lambda+1),$$
$$\lambda = \frac{\Delta t_i}{\Delta t} - 1,$$

where Δt_i is given by the formula

$$\Delta t_i = (t_v - t_0^{(N)}) - (i-1)\Delta t,$$

and i, as before, is chosen so that the condition $0 < \Delta t_i \le \Delta t$ be fulfilled. Finally, using a fourth (intertabular) interpolation, we determine Δq

$$\Delta q = \Delta q^{(1)} \frac{(t_0^{(k)} - t_0^{(N)})(t_0^{(m)} - t_0^{(N)})}{(t_0^{(k)} - t_0^{(1)})(t_0^{(m)} - t_0^{(1)})} + \Delta q^{(m)} \frac{(t_0^{(k)} - t_0^{(N)})(t_0^{(N)} - t_0^{(1)})}{(t_0^{(k)} - t_0^{(m)})(t_0^{(m)} - t_0^{(1)})} + \Delta q^{(k)} \frac{(t_0^{(N)} - t_0^{(m)})(t_0^{(N)} - t_0^{(1)})}{(t_0^{(k)} - t_0^{(m)})(t_0^{(k)} - t_0^{(1)})},$$

and

$$q\big|_{t=t_v} = q\big|_{t=t_0^{(N)}} + \Delta q.$$

Let us revert now to the problem of determining $t_0^{(N)}$, that is, to the problem of determining the instants of passage of the satellite at a given latitude φ_0. According to the formulas of spherical trigonometry at the instants in question the following relation should be fulfilled:

$$\sin \varphi_0 = \sin u \sin i. \qquad (12)$$

On the other hand, as has been noted before, the orbit elements for the instants $t_0^{(N)}$ can be calculated according to formulas (9) and, consequently, relation (12) can also be written in the form

$$\sin \varphi_0 = \sin \bar{u} \sin \bar{i}, \qquad (13)$$

where

$$\bar{u} = \bar{\omega} + \bar{M} + 2\bar{e}\sin\bar{M} + \frac{5}{4}\bar{e}^2\sin 2\bar{M} + \frac{1}{12}\bar{e}^3(13\sin 2\bar{M} - 3\sin\bar{M}),$$
$$\bar{M} = \bar{M}_0 + \frac{\sqrt{\mu}}{\bar{a}^{3/2}}(t_0^{(N)} - t_0),$$

and the elements $\bar{i}$, $\bar{\omega}$, $\bar{e}$, $\overline{M}_0$, $\bar{a}$ are determined from the formula (9).

The relation (13) is a transcendental equation with respect to $t_0^{(N)}$. In the general case it will have an infinite number of roots, which, first, lie in the interval (t_0, t_k), and, second, correspond to the motion of the satellite in a certain direction, e.g., from south to north. The practical determination of the roots of equation (13) can be best carried out by some iteration method (e.g., by Newton's method). Without discussing this question in more detail, we shall remark that in general the determination of the roots of equation (13) is not very difficult.

In concluding this section we shall consider the problem of the integration of the equations in the osculating elements for the basic revolutions. In general, the equations for the osculating elements must involve perturbing forces of various natures, namely perturbing gravitational forces, which appear as a result of the difference between the gravitational field of the earth and a central field, the perturbing aerodynamic forces, or perturbing forces of other celestial bodies. For the practical purpose of making predictions these equations can in most cases be simplified. If the accuracy of the calculations is limited to a few tens of meters, the calculation of one revolution of orbits having a maximum height not exceeding a few thousand kilometers and a minimum height not lower than 250 km can be carried out without taking into account the aerodynamic forces* and without taking into account the influence of other celestial bodies. In addition, it is also possible in this case to neglect the influence of the anomalies of the attraction field of the earth.

The equations in the osculating elements, in which only the terms corresponding to the fact that the earth's field is not central are retained, take therefore the form [2,3]

$$\frac{da}{dt} = \left(\frac{v}{r}\right)^4 \frac{2a\lambda_7}{1-e^2} [e\lambda_5\lambda_8 - (1+e\lambda_6)\lambda_9],$$

$$\frac{de}{dt} = \left(\frac{v}{r}\right)^4 \lambda_7 \left\{\lambda_5\lambda_8 - \left[\left(1+\frac{r}{p}\right)\lambda_6 + e\frac{r}{p}\right]\lambda_9\right\}.$$

*For satellites for which the values of the transverse load are not less than for the first Soviet artificial satellites.

$$\frac{d\omega}{dt} = \left(\frac{\nu}{r}\right)^4 \frac{\lambda_7}{9}\left[-\lambda_6\lambda_8 - \left(1+\frac{r}{p}\right)\lambda_5\lambda_9 + 2e\frac{r}{p}\lambda_2^2\lambda_3^2\right],$$

$$\frac{di}{dt} = -\left(\frac{\nu}{r}\right)^4 2\lambda_7 \frac{r}{p}\lambda_1\lambda_2\lambda_3\lambda_4,$$

$$\frac{d\Omega}{dt} = -\left(\frac{\nu}{r}\right)^4 2\lambda_7 \frac{r}{p}\lambda_2\lambda_3^2,$$

$$\frac{dM}{dt} = \frac{\sqrt{\mu}}{a^{3/2}} + \left(\frac{\nu}{r}\right)^4 \lambda_7 \frac{\sqrt{1-e^2}}{e}\left[\left(\lambda_6 - 2e\frac{r}{p}\right)\lambda_8 + \left(1+\frac{r}{p}\right)\lambda_5\lambda_9\right],$$

where $\lambda_1 = \sin i$, $\lambda_2 = \cos i$, $\lambda_3 = \sin u$, $\lambda_4 = \cos u$, $\lambda_5 = \sin \vartheta$, $\lambda_6 = \cos \vartheta$,

$$\lambda_7 = \sqrt{\frac{a(1-e^2)}{\mu}}, \quad \lambda_8 = 3\lambda_1^2\lambda_3^2 - 1, \quad \lambda_9 = 2\lambda_1^2\lambda_3\lambda_4.$$

The constant ν characterizes the extent to which the earth's attraction field is noncentral and may be expressed through the flattenings α in the following way

$$\nu = \sqrt[4]{\mu R_e^2\left(\alpha - \frac{\Omega^2 R_e}{2g_e}\right)},$$

where R_e is the equatorial radius of the earth, Ω is the angular velocity of the daily rotation of the earth, g_e is the acceleration of the attraction force of the earth at the equator.

CALCULATION OF THE MATRIX ELEMENTS OF THE SYSTEM OF NORMAL EQUATIONS

It has been shown that the matrix elements of the system of normal equations have the form (7), and, consequently, they are related to the parameters α_1 through the sequence of functions ξ, η, ζ, and their derivatives $(\partial\xi/\partial\alpha_i)$, $(\partial\eta/\partial\alpha_i)$, $(\partial\zeta/\partial\alpha_i)$. Thus, in order to be able to calculate the matrix elements A_{ij} and B_i it is first of all necessary to elaborate a method for computing the derivatives $(\partial\xi/\partial\alpha_i)$, $(\partial\eta/\partial\alpha_i)$, $(\partial\zeta/\partial\alpha_i)$.

A sufficiently simple method for calculating the derivatives can be formulated on the basis of the remark that a satellite moves along an almost elliptical orbit, and that the influence of the main perturbing forces has in general a regular character.

We describe now the method for calculating the derivatives. From formulas (5), (2), (3), and (8) it is clear that

$$\xi = \xi[a(\alpha_1, \alpha_2, \ldots, \alpha_n), e(\alpha_1, \alpha_2, \ldots, \alpha_n), \ldots, M(\alpha_1, \alpha_2, \ldots, \alpha_n)], \tag{14}$$

and, consequently,

$$\frac{\partial \xi}{\partial \alpha_i} = \frac{\partial \xi}{\partial a}\frac{\partial a}{\partial \alpha_i} + \frac{\partial \xi}{\partial e}\frac{\partial e}{\partial \alpha_i} + \cdots + \frac{\partial \xi}{\partial M}\frac{\partial M}{\partial \alpha_i}. \tag{15}$$

The relationship between $a, e, \ldots, M$ and $\alpha_1, \alpha_2, \ldots, \alpha_n$ can be represented in the form

$$\left.\begin{array}{l} a = \bar{a} + \Delta a(\alpha_1, \alpha_2, \ldots, \alpha_n), \\ e = \bar{e} + \Delta e(\alpha_1, \alpha_2, \ldots, \alpha_n), \\ \omega = \bar{\omega} + \Delta \omega(\alpha_1, \alpha_2, \ldots, \alpha_n), \\ i = \bar{i} + \Delta i(\alpha_1, \alpha_2, \ldots, \alpha_n), \\ \Omega = \bar{\Omega} + \Delta \Omega(\alpha_1, \alpha_2, \ldots, \alpha_n), \\ M = \overline{M}_0 + \dfrac{\sqrt{\mu}}{\bar{a}^{3/2}} \cdot (t - t_0) + \Delta M(\alpha_1, \alpha_2, \ldots, \alpha_n), \end{array}\right\} \tag{16}$$

where $\bar{a}, \bar{e}, \ldots, \overline{M}_0$ are calculated according to formulas (9), and $\Delta a, \Delta e, \ldots, \Delta M$ are corrections to the main terms $\bar{a}, \bar{e}, \ldots, \bar{M}$ of formulas (16). Therefore, differentiating formulas (16) with respect to α_1, it is possible to neglect in first approximation the derivatives of the corrections $\Delta a, \Delta e, \ldots, \Delta M$, i.e., it is possible to assume that $(\partial a/\partial \alpha_i) \approx (\partial \bar{a}/\partial \alpha_i)$, $(\partial e/\partial \alpha_i) \approx (\partial \bar{e}/\partial \alpha_i)$, etc. Since the parameters of the trajectory α_i are the parameters $a_q, e_q, \ldots, M_{0q}$ ($q = 1, 2, \ldots$) mentioned on p. 241, the derivatives $(\partial \xi/\partial \alpha_i)$ can be represented in the form

$$\left.\begin{array}{l} \dfrac{\partial \xi}{\partial a_q} = \left[\dfrac{\partial \xi}{\partial a} - \dfrac{3}{2} \cdot \dfrac{\sqrt{\mu}}{\bar{a}^{5/2}} \cdot (t-t_0)\dfrac{\partial \xi}{\partial M}\right](t-t_0)^q, \\[6pt] \dfrac{\partial \xi}{\partial e_q} = \dfrac{\partial \xi}{\partial e}(t-t_0)^q, \qquad \dfrac{\partial \xi}{\partial i_q} = \dfrac{\partial \xi}{\partial i}(t-t_0)^q, \\[6pt] \dfrac{\partial \xi}{\partial \omega_q} = \dfrac{\partial \xi}{\partial \omega}(t-t_0)^q, \qquad \dfrac{\partial \xi}{\partial \Omega_q} = \dfrac{\partial \xi}{\partial \Omega}(t-t_0)^q, \\[6pt] \dfrac{\partial \xi}{\partial M_{0q}} = \dfrac{\partial \xi}{\partial M}(t-t_0)^q. \end{array}\right\} \tag{17}$$

Similar relations can be obtained for $(\partial\eta/\partial\alpha_i)$ and $(\partial\zeta/\partial\alpha_i)$. The derivatives $(\partial\xi/\partial a)$, $(\partial\eta/\partial a)$, $(\partial\zeta/\partial a)$, $(\partial\xi/\partial e)$, ..., $(\partial\zeta/\partial M)$ are calculated according to the formulas

$$\frac{\partial \xi}{\partial q_i} = \xi_0 \left(a'_{11} \frac{\partial x}{\partial q_i} + a'_{12} \frac{\partial y}{\partial q_i} + a'_{13} \frac{\partial z}{\partial q_i} \right),$$

$$\frac{\partial \eta}{\partial q_i} = \eta_0 \left(a'_{21} \frac{\partial x}{\partial q_i} + a'_{22} \frac{\partial y}{\partial q_i} + a'_{23} \frac{\partial z}{\partial q_i} \right), \quad (i = 1, 2, \ldots, 6) \quad (18)$$

$$\frac{\partial \zeta}{\partial q_i} = \zeta_0 \left(a'_{31} \frac{\partial x}{\partial q_i} + a'_{32} \frac{\partial y}{\partial q_i} + a'_{33} \frac{\partial z}{\partial q_i} \right),$$

$$q_1 = a, \quad q_2 = e, \quad q_3 = \omega, \quad q_4 = i, \quad q_5 = \Omega, \quad q_6 = M,$$

where

$$\xi_0 = \frac{W\xi}{\sqrt{(x')^2 + (y')^2 + (z')^2}}, \quad \eta_0 = \frac{W\eta}{(x')^2 + (y')^2},$$

$$\zeta_0 = \frac{W\zeta}{[(x')^2 + (y')^2 + (z')^2] \cdot \sqrt{(x')^2 + (y')^2}},$$

and

$$a'_{11} = a_{11}x' + a_{21}y' + a_{31}z', \quad a'_{21} = a_{21}x' - a_{11}y',$$
$$a'_{12} = a_{12}x' + a_{22}y' + a_{32}z', \quad a'_{22} = a_{22}x' - a_{12}y',$$
$$a'_{13} = a_{13}x' + a_{33}z', \quad a'_{23} = -a_{13}y',$$

$$a'_{31} = a_{31}(x'^2 + y'^2) - a_{21}y'z' - a_{11}z'x',$$
$$a'_{32} = a_{32}(x'^2 + y'^2) - a_{22}y'z' - a_{12}z'x',$$
$$a'_{33} = a_{33}(x'^2 + y'^2) - a_{13}z'x'.$$

In these formulas x', y', z' and a_{11}, a_{22}, ... a_{33} are calculated according to formulas (3) and (4). Finally, the derivatives $(\partial x/\partial q_i)$, $(\partial y/\partial q_i)$, $(\partial z/\partial q_i)$ can be calculated by the formulas

$$\frac{\partial x}{\partial a} = \frac{x}{a}, \quad \frac{\partial x}{\partial e} = C_5 x - C_1 C_6, \quad \frac{\partial x}{\partial \omega} = -C_1,$$

$$\frac{\partial y}{\partial a} = \frac{y}{a}, \quad \frac{\partial y}{\partial e} = C_5 y - C_2 C_6, \quad \frac{\partial y}{\partial \omega} = -C_2,$$

$$\frac{\partial z}{\partial a} = \frac{z}{a}, \quad \frac{\partial z}{\partial e} = C_5 z - C_3 C_6, \quad \frac{\partial z}{\partial \omega} = -C_3,$$

$$\frac{\partial x}{\partial i} = r \sin u \sin \Omega \sin i, \quad \frac{\partial x}{\partial \Omega} = -y, \quad \frac{\partial x}{\partial M} = C_7(C_4 x - C_1),$$

$$\frac{\partial y}{\partial i} = -r \sin u \cos \Omega \sin i, \quad \frac{\partial y}{\partial \Omega} = x, \quad \frac{\partial y}{\partial M} = C_7(C_4 y - C_2),$$

$$\frac{\partial z}{\partial i} = r \sin u \cos i, \quad \frac{\partial z}{\partial \Omega} = 0, \quad \frac{\partial z}{\partial M} = C_7(C_4 z - C_3),$$

where

$$C_1 = r(\sin u \cos \Omega + \cos u \sin \Omega \cos i),$$
$$C_2 = r(\sin u \sin \Omega - \cos u \cos \Omega \cos i),$$
$$C_3 = -r \cos u \sin i,$$
$$C_4 = \frac{e \sin \vartheta}{1 + e \cos \vartheta},$$
$$C_5 = C_4 C_6 - \frac{\cos \vartheta}{1 + e \cos \vartheta} - \frac{2e}{1 - e^2},$$
$$C_6 = 2\sin M + \frac{5}{2} e \sin 2M + \frac{1}{4} e^2 (13 \sin 3M - 3 \sin M),$$
$$C_7 = 1 + 2e \cos M + \frac{5}{2} e^2 \cos 2M + \frac{1}{4} e^3 (13 \cos 3M - \cos M).$$

It must be remembered that the formulas given for the derivatives are applicable to the case of an orbit with small eccentricity (not more than 0.2 ~ 0.3). For higher eccentricities the formulas for C_6 and C_7 must be changed.

APPLICATION OF THE METHOD OF STEEPEST DESCENT TO THE ELABORATION OF THE OBSERVATIONS

As has been remarked before, the method for determining the orbit of an artificial earth satellite consists in an iterative refinement of the parameters $\alpha_1, \alpha_2, ..., \alpha_n$. It is therefore essential that this process be always convergent. However, it is well known that the process of successive approximations, when the classic method of least squares is used, can diverge in a number of cases, especially if the initial approximation is quite rough. Since, in order to predict the satellite motion, the ephemerides must be given promptly, and the first approximation may be rather rough, the problem of ensuring reliably the convergence of the successive-approximation process is very important in our problem.

In order to increase the reliability of the procedure under consideration, the method of least squares was completed by another method, which would ensure the convergence of the refinement process for very rough initial approximations. This method is a development of the well-known steepest-descent method, which is applicable to the problem of statistical analysis.[4] Without stopping to consider the details of the derivation and the basis for this method, we shall give its algorithm.

Let a first approximation $\alpha_1^{(0)}$, $\alpha_2^{(0)}$, ..., $\alpha_n^{(0)}$ of the parameters α_1, α_2, ..., α_n be given. Then, the final values for the parameters in question can be obtained from the solution of the following system of differential equations:

$$\begin{aligned} A_{11}\frac{d\alpha_1}{dS} + A_{12}\frac{d\alpha_2}{dS} + \ldots + A_{1n}\frac{d\alpha_n}{dS} &= \frac{B_1}{\sqrt{\Phi}}, \\ A_{21}\frac{d\alpha_1}{dS} + A_{22}\frac{d\alpha_2}{dS} + \ldots + A_{2n}\frac{d\alpha_n}{dS} &= \frac{B_2}{\sqrt{\Phi}}, \\ \cdots \cdots \cdots \cdots \cdots \cdots \cdots \cdots \cdots \cdots \\ A_{n1}\frac{d\alpha_1}{dS} + A_{n2}\frac{d\alpha_2}{dS} + \ldots + A_{nn}\frac{d\alpha_n}{dS} &= \frac{B_n}{\sqrt{\Phi}}, \end{aligned} \qquad (19)$$

where the coefficients A_{ij} and B_i are determined according to formulas (7), and Φ is given by the formula

$$\Phi = \sum_{\nu=1}^{N} (\xi_\nu^2 + \eta_\nu^2 + \zeta_\nu^2).$$

The system (19) is integrated over S with initial conditions $s = 0$ and $\alpha_1 = \alpha_1^{(0)}$, $\alpha_2 = \alpha_2^{(0)}$, ..., $\alpha_n = \alpha_n^{(0)}$, up to the value $S = S_k$ for which the quantity $F = 0$, F being given by the formula

$$F = \sum_{i=1}^{n} B_i^2.$$

The values of the parameters α_1, α_2, ..., α_n obtained for $S = S_k$ will be the values looked for, namely those best fitting the observations in the sense of the least squares method.

As the practical application has shown, the method described ensures a reliable convergence of the method of successive approximations of the orbit for very rough initial values of the parameters α_1, α_2, ..., α_n. In order to accelerate the elaboration process for determining the orbit, it is convenient to combine the method just described with the classical least squares method.

CONCLUSION

A procedure for predicting the motion of a satellite follows immediately from the above sections. In fact, as soon as the parameters a_q, e_q, ..., M_q ($q = 0, 1, 2, ...$), which give the law of variation of the osculating elements for a given latitude φ_0, have been determined, the osculating elements $a, e, ..., M$ for any instant of time can be determined according to the procedure. Then, on the basis of the elements $a, e, ..., M$ thus found, it is possible to calculate the ephemerides for a longer or shorter period of time with the help of formulas (2), (3), and (8). The length of the period covered by sufficiently accurate ephemerides calculated in this way will depend both upon the quality and quantity of the observations subjected to elaboration, and upon the length of the interval of time (t_0, t_k) in which these observations are distributed.

LITERATURE

1. A. N. Krylov, Lectures on Approximated Calculations, Gostekhizdat, Moscow, 1938.
2. G. N. Duboshin, Introduction to Celestial Mechanics, Gostekhizdat, Moscow, 1938.
3. M. F. Subbotin, Textbook of Celestial Mechanics, Volume 2, Gostekhizdat, 1937.
4. A. S. Householder, Principles of Numerical Analysis, McGraw-Hill, New York, 1953.

METHODS FOR THE NUMERICAL SOLUTION OF FINITE DIFFERENCE EQUATIONS AND THEIR APPLICATION TO COMPUTATIONS OF SATELLITE ORBITS

G. P. Taratynova

This paper presents methods for the numerical solution of finite difference equations and their application to the computation of orbits of satellites. The necessity for creating such methods arose in the solution of certain problems in celestial mechanics in which it was required to obtain the solution of a system of nonlinear differential equations describing the motion for long intervals of time. This is the same sort of problem as those of determining the length of "life" of a satellite, determining the evolution of a satellite's orbit during its lifetime, of the long-range forecasting of a satellite's orbit, etc.

The methods developed are of a general character, and may be used for computing a wide class of nonlinear oscillatory systems.

In the case of a satellite we use, as the equations of motion, the equations in the orbit's osculating elements ordinarily used in celestial mechanics.

As the disturbing forces acting on the satellite, it is necessary to take into account the resistance of the atmosphere, deviations of the earth's gravitational field from a central one due to the flattening of the earth and, in certain cases perhaps, anomalies in the gravitational forces. The disturbing forces due to the gravitational attraction of the sun and the moon are small, and may ordinarily be neglected.

The equations of motion of a satellite are nonlinear differential equations whose right members contain small parameters. When one is investigating the motion of a satellite in the presence of the basic disturbing forces (atmospheric resistance and deviations of the earth's gravitational field from a central one) such parameters are the density of the atmosphere at some fixed altitude, and the flattening of the earth. For zero values of the small parameters, the equations of motion can be integrated in finite form: the orbit is an ellipse. For nonzero values of the parameters, the satellite's orbit is the envelope of a family of osculating ellipses, wherein the system of equations does not have an analytical solution.

For the solution of a system of equations in osculating elements in the general case for comparatively short intervals of time, one ordinarily uses various methods of numerical integration. However, in the case of long intervals of time (for example, for hundreds and thousands of revolutions of a satellite around the earth), use of these methods requires a very long time when the computations are done on electronic computers. Indeed, the change of the osculating elements during one period, equal to 2π, has an oscillatory character and the integration step cannot be large.

In connection with this,[1,2] it was first suggested that the transition be made from the systems of equations, ordinarily used in celestial mechanics, in the true anomaly ϑ or the argument of latitude u, to the system of equations obtained as the result of integrating the first system over a period equal to 2π.

The left members of the equations thus obtained are the values of the disturbances of the osculating orbit elements during one revolution. By virtue of the smallness of these disturbances during one revolution, they were, in the works cited, taken approximately to be equal to the derivatives of the osculating elements with respect to the number of revolutions N, where $N = (\vartheta - \vartheta_0)2\pi$ or $N = (u - u_0)/2\pi$. Instead of the original system of equations in osculating elements with respect to the true anomalies or to the latitude argument, there was integrated a system of differential equations in N— the number of revolutions. The right members of such equations are some set of a system of integrals. This system of integrals can be replaced by a corresponding system of differential equations. In such a case, solution of the transformed system of equations reduces to a two-stage algorithm: an external integration with respect to the argument N—the number of revolutions—and an internal integration with respect to the true anomaly ϑ or the latitude argument u over a period of 2π, i.e., during one revolution, where this latter step must be executed each time in computing the right members of the external system of equations.

The suggested method of a two-stage integration turned out to be very effective in computing the orbits of satellites. In computations of standard satellite orbits, the initial step for integrating the external system of equations with respect to the number of revolutions was on the order of 100 revolutions. This made it possible, without reducing the accuracy, to obtain the values of the orbit's osculating elements over a long period of time and on the basis of a comparatively small volume of computational work.

However, the method of a two-stage integration just cited was not as completely correct as, for example, the well-known methods of numerical integration of ordinary differential equations. Indeed, the replacement of the increments of the osculating elements over one period by the derivatives of these elements with respect to the number of revolutions entailed a methodical error. Obviously, such a methodical error is larger the smaller the integration step for the external system of equations. For this reason, for comparatively low altitudes, when the step of integration ΔN of the external system was decreased, the aforementioned method of two-stage integration was inapplicable. Strictly speaking, instead of the aforementioned external system of differential equations, one should seek the solution of a system of equations in finite differences with respect to the orbit's osculating elements, the right members of which are some set of a system of integrals.

By the development and application of methods of numerical solution, directly applicable to finite difference equations, the aforementioned methodical error was completely eliminated.

Until recently, the question was not raised in the literature

FINITE DIFFERENCE EQUATIONS FOR SATELLITE ORBITS

as to the numerical solution of finite difference equations in general form. Exact solutions of certain classes of finite difference equations of special forms are known (for example, equations with constant coefficients, with sufficiently simple right members[3]).

By analogy with the methods for the numerical integration of ordinary differential equations, there were developed two types of methods for the numerical solution of finite difference equations: the Runge type methods and the Adams type interpolation methods.

In the limiting case when the small differences on which the difference equations are defined tend to zero, and the difference equations themselves are transformed to ordinary differential equations, the methods obtained correspond to the methods for the numerical integration of ordinary differential equations of the cited types. In this sense, the methods developed for the numerical solution of difference equations can be considered as generalizations of the methods for the numerical integration of differential equations.

GENERALIZED RUNGE METHODS

We have a first-order finite difference equation:

$$y_{n+1} - y_n = f(x_n, y_n) \quad (n = 1, 2, \ldots), \tag{1.1}$$

where $x_{n+1} = x_n + h$, $f(x, y)$ are differentiable functions. We assume that (1.1) satisfies the initial condition: for $x = x_0$, $y = y_0$. We formulate the problem of the numerical solution of (1.1) in the following way: to find the value of the function y being sought which satisfies (1.1) for $x = x_n + H$ (where $H = Nh$, and N is a positive integer), if the value of the function y for $x = x_n$ is known.

Using Newton's interpolation polynomial, we present the function y being sought in the form of an expansion:

$$y(x_n + H) = y_n + Nf_n + \frac{N(N-1)}{2!}\delta f_n +$$
$$+ \frac{N(N-1)(N-2)}{3!}\delta^2 f_n + \frac{N(N-1)(N-2)(N-3)}{4!}\delta^3 f_n + \ldots \tag{1.2}$$

$$f_n = f(x_n; y_n),$$
$$\delta f_n = f(x_n + h; y_{n+1}) - f(x_n; y_n),$$
$$\delta^2 f_n = f(x_n + 2h; y_{n+2}) - 2f(x_n + h; y_{n+1}) + f(x_n; y_n)$$
. . . .

By analogy with the Runge methods of fourth-order numerical integration of ordinary differential equations, we present the increments of the function y being sought in the form of combinations of the four expressions

$$k_1 = Nf(x_n; y_n),$$
$$k_2 = Nf(x_n + \alpha_1 H; y_n + \beta_1 k_1),$$
$$k_3 = Nf(x_n + \alpha_2 H; y_n + \beta_2 k_1 + \beta_3 k_2),$$
$$k_4 = Nf(x_n + \alpha_3 H; y_n + \beta_4 k_1 + \beta_5 k_2 + \beta_6 k_3). \qquad (1.3)$$

We now show that one can choose the parameters α_1 to α_3, β_1 to β_6 and the weight functions R_1 to R_4, depending on N, such that, with an accuracy to within fourth-order terms, the following equation will hold:

$$\Delta y = y(x_n + h) - y(x_n) = R_1 k_1 + R_2 k_2 + R_3 k_3 + R_4 k_4. \qquad (1.4)$$

(Fourth-order terms are those containing expressions of the type $f(\partial^3 f/\partial x^3)$, $f(\partial^3 f/\partial x^2 \partial y)$, ..., etc.)

It is known that, in the case of the numerical integration of ordinary differential equations, the aforementioned parameters and weight functions are some constant numbers.

The form of the functions k_i (i = 1,2,3,4) in (1.3) is more general than that ordinarily adopted in the Runge method for integrating ordinary differential equations. When it is attempted to present the functions k_i, as in the case of differential equations, in the simpler forms

$$k_1 = Nf(x_n; y_n),$$
$$k_2 = Nf\left(x_n + \frac{1}{2} H; y_n + \frac{1}{2} k_1\right),$$
$$k_3 = Nf\left(x_n + \frac{1}{2} H; y_n + \frac{1}{2} k_2\right),$$
$$k_4 = Nf(x_n + H; y_n + k_3), \qquad (1.5)$$

it turns out that it is impossible to so choose solely the weight functions R_i (i = 1,2,3,4) that (1.4) hold with an accuracy up to the fourth order. This is given mathematical expression in the fact that the system of algebraic equations used to define the weight functions is not completely satisfied. Therefore, the auxiliary parameters α_i (i = 1,2,3) and β_j (j = 1,2, ..., 6) were introduced, defining the points at which the functions k_i are computed, and the latter were presented in the most general form of (1.3).

The finite differences δf_n, $\delta^2 f_n$, ... can be expressed, by well-known formulas, in terms of the functions $f(x_n, y_n)$, $f(x_n + h, y_n + f_n)$, ... By expanding the latter, as well as the functions k_i (i = 1,2,3,4), in Taylor series in two variables about the point (x_n, y_n), limiting ourselves to terms no higher than fourth order, and by comparing coefficients of homologous derivatives with equation (1.4), we obtain a system of fourth-order algebraic equations for determining the parameters α_1, α_2, ... β_6 and the weight functions R_i (i = 1,2,3,4). In view of its complexity, this system will not be given here.

In investigating the system of equations thus obtained, it turned out that the parameters α_i and β_i were necessarily connected by the relationships

$$\alpha_1 = \beta_1, \quad \alpha_2 = \beta_2 + \beta_3, \quad \alpha_3 = \beta_4 + \beta_5 + \beta_6.$$

It further turned out that two of the parameters being sought could be given arbitrarily. To the thus found two-parameter family of solutions of the system of algebraic equations, depending on two arbitrary parameters, correspond the two-parameter family methods of numerical solution of finite difference equations.

It is convenient to choose α_1 and α_2 as the free parameters. In this case, the solution has the form

$$\alpha_3 = \frac{N-1}{N}, \quad \beta_1 = \alpha_1, \quad \beta_2 = \frac{\alpha_2[\alpha_1(4\alpha_1 N - 3N + 5) + \alpha_2(N-3)]}{2\alpha_1(2\alpha_1 N - N + 1)},$$

$$\beta_3 = \frac{\alpha_2(\alpha_1 - \alpha_2)(N-3)}{2\alpha_1(2\alpha_1 N - N + 1)}, \quad R_3 = \frac{(N-1)(N-2)(2\alpha_1 N - N + 1)}{12\alpha_2 N^2(\alpha_2 - \alpha_1)(\alpha_2 N - N + 1)},$$

$$R_4 = \frac{3N(N-1) + \alpha_1 N(3\alpha_2 N - 4N + 2) + \alpha_2 N(3\alpha_1 N - 4N + 2)}{12(\alpha_1 N - N + 1)(\alpha_2 N - N + 1)},$$

$$R_2 = \frac{(N-1) - 2(\alpha_2 R_3 + \alpha_3 R_4)}{2\alpha_1 N}, \quad R_1 = 1 - R_2 - R_3 - R_4.$$

$$\beta_6 = \frac{(N-1)(N-2)(N-3)}{24\alpha_1\beta_3 N^3 R_4}, \quad \beta_5 = \frac{(N-1)(N-2) - 6N^2(\alpha_2\beta_6 R_4 + \alpha_1\beta_3 R_3)}{6\alpha_1 N^2 R_4},$$

$$\beta_4 = \frac{N-1}{N} - \beta_5 - \beta_6. \tag{1.6}$$

To obtain a general solution of (1.6), we eliminated the cases which we call singular:

1) $\alpha_1 = 0$; 2) $\alpha_2 = 0$; 3) $\alpha_1 = \alpha_2$;
4) $\alpha_2 = \alpha_3$; 5) $\alpha_1 = \frac{N-1}{N}$; 6) $\alpha_1 = \frac{N-1}{2N}$;
7) $\alpha_2 = \frac{N-1}{N}$; 8) $R_2 = 0$; 9) $R_3 = 0$; 10) $R_4 = 0$.

It can be shown that if, in the cases cited, the system of equations for determining the parameters $\alpha_1, \alpha_2, \ldots \beta_6$ and $R_1, \ldots, R_4$ is solvable, then always, except for the case $\alpha_2 = (N-1)/N$, there is a family of solutions depending on one parameter and, consequently, to it corresponds the one-parameter family of methods of numerical solution of finite difference equations.

We now give the solutions for the singular cases.

1) $\alpha_1 = 0$. The system of equations for determining $\alpha_1, \alpha_2, \ldots R_4$ is not satisfied for any finite values of the parameters sought.

2) $\alpha_2 = 0$. We have the one-parameter family of solutions

$$\begin{aligned}
&\alpha_1 = \frac{N-1}{2N}, \quad \alpha_3 = \frac{N-1}{N}, \quad \beta_1 = \alpha_1, \\
&\beta_2 = -\beta_3, \quad \beta_4 = \alpha_3 - \beta_5 - \beta_6, \\
&\beta_5 = \frac{(N-2)(3N-1)}{2N(N+1)}, \quad \beta_6 = \frac{(N-1)(N-2)(N-3)}{2N^2\beta_3(N+1)}, \\
&R_1 = 1 - R_2 - R_3 - R_4, \quad R_2 = \frac{2(N-2)}{3(N-1)}, \\
&R_3 = \frac{(N-2)(N-3)}{12\beta_3 N(N-1)}, \quad R_4 = \frac{N+1}{6(N-1)},
\end{aligned} \quad (1.7)$$

β_3 is the free parameter.

3) $\alpha_1 = \alpha_2$. The one-parameter family of solutions has the form

$$\begin{aligned}
&\alpha_1 = \alpha_2 = \frac{N-1}{2N}, \quad \alpha_3 = \frac{N-1}{N}, \quad \beta_1 = \alpha_1, \\
&\beta_2 = \alpha_2 - \beta_3, \quad \beta_4 = \frac{1}{N}, \\
&\beta_5 = \frac{N-2}{N} - \beta_6, \quad \beta_6 = \frac{(N-1)(N-2)(N-3)}{2\beta_3 N^2(N+1)},
\end{aligned} \quad (1.8)$$

FINITE DIFFERENCE EQUATIONS FOR SATELLITE ORBITS

$$R_1 = \frac{N+1}{6(N-1)}, \quad R_2 = \frac{2(N-2)}{3(N-1)} - R_3,$$
$$R_3 = \frac{(N-2)(N-3)}{6\beta_3 N(N-1)}, \quad R_4 = \frac{N+1}{6(N-1)}. \quad (1.8)$$

β_3 is the free parameter.

Among the family of methods of numerical solution of difference equations corresponding to this one-parameter family of solutions, there is a method which is an immediate generalization of the most widely used method of numerical integration of ordinary differential equations (cf. (1.5)). The formulas for this method are obtained for $\beta_3 = (N-3)/2N$ and have the following form:

$$\Delta y_n = \frac{N+1}{6(N-1)} k_1 + \frac{N-2}{3(N-1)} k_2 + \frac{N-2}{3(N-1)} k_3 + \frac{N+1}{6(N-1)} k_4, \quad (1.9)$$

where

$k_1 = Nf(x_n; y_n),$

$k_2 = Nf\left(x_n + \frac{N-1}{2N} H; \; y_n + \frac{N-1}{2N} k_1\right),$

$k_3 = Nf\left(x_n + \frac{N-1}{2N} H; \; y_n + \frac{1}{N} k_1 + \frac{N-3}{2N} k_2\right),$

$k_4 = Nf\left(x_n + \frac{N-1}{N} H; \; y_n + \frac{1}{N} k_1 + \frac{2(N-2)}{N(N+1)} k_2 + \frac{(N-1)(N-2)}{N(N+1)} k_3\right).$

4) $\alpha_2 = \alpha_3$. The corresponding family of solutions has the form

$$\alpha_1 = \frac{1}{N}, \quad \alpha_2 = \frac{3N-1}{4N}, \quad \alpha_3 = \frac{3N-1}{4N}, \quad \beta_1 = \alpha_1,$$

$$\beta_2 = \frac{(3N-1)(-3N+13)}{32N}, \quad \beta_3 = -\frac{(3N-1)(3N-5)}{32N},$$

$$\beta_4 = \frac{3N-1}{4N} - \beta_5 - \beta_6, \quad \beta_5 = \frac{(-8\beta_6 N + 3N - 5)(3N-1)}{32N},$$

$$R_1 = 1 - R_2 - R_3 - R_4, \quad R_2 = \frac{N^2-1}{6(3N-5)}, \quad (1.10)$$

$$R_3 = \frac{4(N-1)(N-2)(4\beta_6 N - N + 3)}{3\beta_6 N(3N-1)(3N-5)},$$

$$R_4 = \frac{4(N-1)(N-2)(N-3)}{3\beta_6 N(3N-1)(3N-5)}.$$

β_6 is the free parameter.

5) $\alpha_1 = (N-1)/N$. The one-parameter family of solutions has the form

$$\alpha_2 = \frac{N-1}{2N}, \quad \alpha_3 = \frac{N-1}{N}, \quad \beta_1 = \frac{N-1}{N},$$

$$\beta_2 = \frac{3N-1}{8N}, \quad \beta_3 = \frac{N-3}{8N},$$

$$\beta_4 = \frac{N-1}{N} - \beta_5 - \beta_6, \quad \beta_5 = -\frac{\beta_6(N-3)}{4(N-1)},$$

$$R_1 = \frac{N+1}{6(N-1)}, \quad R_2 = \frac{N+1}{6(N-1)} - R_4,$$

$$R_3 = \frac{2(N-2)}{3(N-1)}, \quad R_4 = \frac{N-2}{3\beta_6 N}.$$

(1.11)

β_6 is the free parameter.

6) $\alpha_1 = (N-1)/2N$. It is easily shown that this case reduces to case 3.

7) $\alpha_2 = (N-1)/N$. For this value of parameter α_2, we have, not a one-parameter family of solutions, but a unique solution, corresponding to the case when $R_3 = 0$. This solution has the form

$$\alpha_1 = \frac{1}{N}, \quad \alpha_3 = \frac{3N-1}{4N}, \quad \beta_1 = \frac{1}{N}, \quad \beta_2 = -\frac{(N-1)(N-4)}{2N},$$

$$\beta_3 = \frac{(N-1)(N-2)}{2N}, \quad \beta_4 = -\frac{(3N-1)(3N^2 - 18N + 11)}{64N(N-1)},$$

$$\beta_5 = \frac{(N-1)(3N-1)(3N-5)}{64N(N-2)}, \quad \beta_6 = \frac{(N-3)(3N-1)(3N-5)}{64N(N-1)(N-2)},$$

$$R_1 = -\frac{(N-7)(N+1)}{6(3N-1)}, \quad R_2 = \frac{N^2-1}{6(3N-5)},$$

$$R_4 = \frac{16(N-1)(N-2)}{3(3N-1)(3N-5)}, \quad R_3 = 0.$$

(1.12)

8) $R_2 = 0$. We have the one-parameter family of solutions

$$\alpha_2 = \frac{2\alpha_3(2N-1) - 3N + 3}{6\alpha_3 N - 4N + 2}, \quad R_3 = \frac{(N-1)(3\alpha_3 N - 2N + 1)}{6\alpha_2 N^2 (\alpha_3 - \alpha_2)},$$

$$R_4 = \frac{-2\alpha_2 N R_3 + N - 1}{2\alpha_3 N}, \quad R_1 = 1 - R_3 - R_4,$$

$$\beta_6 = \frac{R_3(\alpha_3 - \alpha_2)(N-3)}{R_4(4\alpha_3 N - 3N + 1)}, \quad \beta_3 = \frac{(N-1)(N-2)(N-3)}{24\alpha_1 \beta_6 R_4 N^3},$$

$$\alpha_1 = \frac{12\alpha_2^2 \beta_6 R_4 N^2 - (N-1)^2(N-2)}{12\alpha_2 \beta_6 R_4 N^3 - 2N(N-1)(N-3)}, \quad \beta_1 = \alpha_1, \quad \beta_2 = \alpha_2 - \beta_3,$$

$$\beta_4 = \alpha_3 - \beta_5 - \beta_6, \quad \beta_5 = \frac{(N-1)(N-2)}{6\alpha_1 R_4 N^2} - \frac{\alpha_2}{\alpha_1} \beta_6 - \frac{R_3}{R_4} \beta_3.$$

(1.13)

α_3 is the free parameter.

9) $R_3 = 0$. We do not give the corresponding solution, which is expressed in terms of radicals and, in practical cases, is inconvenient.

10) $R_4 = 0$. The system of equations for determining the paparameters α_i, β_j and the weights R_k is not satisfied for any finite values of the parameters being sought.

The set of two-parameter family methods, depending on the parameters α_1 and α_2, plus the one-parameter family methods given, contain in themselves all possible methods of fourth-order numerical solution of finite difference equations of the Runge type.

We can analogously obtain the methods for third-order numerical solution of finite difference equations.

We present the increments of the function being sought in terms of combinations of the three expressions

$$\left.\begin{aligned} k_1 &= Nf(x_n; y_n), \\ k_2 &= Nf(x_n + \alpha_1 H; \ y_n + \beta_1 k_1), \\ k_3 &= Nf(x_n + \alpha_2 H; \ y_n + \beta_2 k_1 + \beta_3 k_2). \end{aligned}\right\} \quad (1.14)$$

We so choose the parameters α_1, α_2, β_1, β_2, β_3 and the weight functions R_1, R_2, R_3 that, with an accuracy to within third-order terms, the following equation holds:

$$\Delta y_n = y_{n+1} - y_n = R_1 k_1 + R_2 k_2 + R_3 k_3. \quad (1.15)$$

Proceeding the same as in the case of the fourth-order methods, we obtain a system of third-order algebraic equations for determining the parameters α_i, β_j and the weight functions R_k. From this system of equations, which we do not give here, it necessarily follows that

$$\alpha_1 = \beta_1, \quad \alpha_2 = \beta_2 + \beta_3.$$

Further, in this case two of the parameters sought can also be given arbitrarily.

As these free parameters, it is convenient to choose the parameters α_1 and α_2. In such a case, we have the following two-parameter family of solutions:

$$\left.\begin{aligned} \beta_3 &= \frac{\alpha_2(\alpha_1 - \alpha_2)(N-2)}{\alpha_1(3\alpha_1 N - 2N + 1)}, \quad R_1 = 1 - R_2 - R_3, \\ R_2 &= \frac{(N-1)(3\alpha_2 N - 2N + 1)}{6\alpha_1 N^2 (\alpha_2 - \alpha_1)}, \quad R_3 = \frac{(N-1)(3\alpha_1 N - 2N + 1)}{6\alpha_2 N^2 (\alpha_1 - \alpha_2)}. \end{aligned}\right\} \quad (1.16)$$

To this two-parameter family of solutions corresponds the family of third-order methods depending on two parameters.

Family (1.16) of solutions does not contain in itself those solutions for which some of the parameters are not finite. To these appertain the following:

1) $\alpha_1 = 0$, 2) $\alpha_2 = 0$, 3) $\alpha_1 = \alpha_2$, 4) $\alpha_1 = \dfrac{2N-1}{3N}$.

We consider these cases:

1) $\alpha_1' = 0$. The system of equations for determining the parameters $\alpha_1, \alpha_2, \ldots, R_3$ is not satisfied.

2) $\alpha_2 = 0$. We have the one-parameter family of solutions

$$\alpha_1 = \frac{2N-1}{3N}, \quad R_1 = 1 - R_2 - R_3, \quad R_2 = \frac{3(N-1)}{2(2N-1)},$$

$$R_3 = \frac{(N-1)(N-2)}{2\beta_3 N(2N-1)}, \qquad (1.17)$$

β_3 is the free parameter.

3) $\alpha_1 = \alpha_2$. The one-parameter family of solutions has the form

$$\alpha_1 = \frac{2N-1}{3N}, \quad \alpha_2 = \frac{2N-1}{3N}, \quad R_1 = \frac{N+1}{2(2N-1)}, \qquad (1.18)$$

$$R_2 = \frac{(N-1)(3\beta_3 N - N + 2)}{2\beta_3 N(2N-1)}, \quad R_3 = \frac{(N-1)(N-2)}{2\beta_3 N(2N-1)},$$

β_3 is the free parameter.

4) $\alpha_1 = (2N-1)/3N$. This case reduces either to case 2 or to case 3.

In using the Runge type methods of numerical solution of finite difference equations, it is necessary to know how to choose correctly the step $H = Nh$, and for this it is necessary to estimate the magnitude of the error of the function being sought. We do not pose here the problem of an exact estimate of the error in the solution for the function being sought. (We note that an accurate estimate of the error in the general case is also not given for ordinary differential equations, cf., for example, Ref. 4.) We give only a sufficiently simple method for determining certain terms of possibly higher order in Expansion (1.2).

By varying the choice of just the weight functions, one can arrive at a position where the expression

$$\Delta^{(k)}y_n = R_1^* k_1 + R_2^* k_2 + R_3^* k_3 + R_4^* k_4 \tag{1.19}$$

will contain only terms of third-order smallness and certain other higher-order terms.

In this case the formulas for determining such weight functions have the form

$$R_1^* = \frac{D_1}{D}, \quad R_2^* = \frac{D_2}{D}, \quad R_3^* = \frac{D_3}{D}, \quad R_4^* = \frac{D_4}{D}, \tag{1.20}$$

where

$$D_1 = \frac{N-1}{6N^2}[(2N-1)(\alpha_2-\alpha_1)(\alpha_1\beta_5+\alpha_2\beta_6) -$$
$$- \alpha_1\beta_3(2N-1)(\alpha_3-\alpha_1) - (N-2)(\alpha_3-\alpha_2)(\alpha_3-\alpha_1)],$$

$$D_2 = \frac{N-1}{6N^2}[\alpha_2\alpha_3(N-2)(\alpha_3-\alpha_2) + \alpha_1\alpha_3\beta_3(2N-1) -$$
$$- \alpha_2(2N-1)(\alpha_1\beta_5+\alpha_2\beta_6)],$$

$$D_3 = \frac{\alpha_1(N-1)}{6N^2}[(\alpha_1\beta_5+\alpha_2\beta_6)(2N-1) + \alpha_3(N-2)(\alpha_1-\alpha_3)],$$

$$D_4 = \frac{\alpha_1(N-1)}{6N^2}(\alpha_2(\alpha_2-\alpha_1)(N-2) - \alpha_1\beta_3(2N-1)],$$

$$D = \alpha_1\alpha_2(\alpha_2-\alpha_1)(\alpha_1\beta_5+\alpha_2\beta_6) + \alpha_1^2\alpha_3\beta_3(\alpha_1-\alpha_3).$$

We assume that we have some concrete method of solving finite difference equations. There correspond to it some parameters $\alpha_1, \alpha_2, \ldots, \beta_6$ and weight functions R_k ($k = 1,2,3,4$) depending on N. By defining new weight functions R_k^* ($k = 1,2,3,4$) by Formulas (1.20) with the previous values of the functions k_i ($i = 1,2,3,4$), we can determine from Formula (1.19) the magnitude of the third-order term in Expansions (1.2). From the magnitude of this latter, one can judge the magnitude of the error of solution.

Analogous formulas can be obtained in the case of third-order methods.

We have just given the two-parameter family of Runge type methods of fourth and third order for the numerical solution of finite difference equations. The choice of method from these families of methods is determined by conditions of simplicity, symmetry, the presence of some constants or other in the machine's memory, in the case of programming for a computer, etc. However, the attempt can be made to so order the two free parameters that the remainder term is minimized, namely, that the difference between the principal and the neglected terms of Expansion (1.2) and the part of these terms taken into account in the formula for the increment of the function being sought, Formula (1.4), would equal zero.

We consider the family of third-order methods; in this case the corresponding formulas are simpler. The increment Δy of the function being sought, defined by Formula (1.15), contains part of the fourth-order terms.

The expression for this part has the form

$$\Phi = (R_2 \alpha_1^3 + R_3 \alpha_2^3) A_1 + R_3 \alpha_1^2 \beta_3 A_2 + R_3 \alpha_1 \alpha_2 \beta_3 A_3,$$

where

$$A_1 = \frac{1}{6} NH \left(\frac{\partial}{\partial x} + \frac{f}{h} \frac{\partial}{\partial y} \right)^3 f,$$

$$A_2 = \frac{1}{2} N^2 H^2 \frac{\partial f}{\partial y} \left(\frac{\partial}{\partial x} + \frac{f}{h} \frac{\partial}{\partial y} \right)^2 f,$$

$$A_3 = N^2 H^2 \left(\frac{\partial}{\partial x} + \frac{f}{h} \frac{\partial}{\partial y} \right) f \left(\frac{\partial}{\partial x} + \frac{f}{h} \frac{\partial}{\partial y} \right) \frac{\partial f}{\partial y}.$$

In these formulas, we use a differential operator of the form

$$\left(\frac{\partial}{\partial x} + \frac{f}{h} \frac{\partial}{\partial y} \right) f = \frac{\partial f}{\partial x} + \frac{f}{h} \frac{\partial f}{\partial y}.$$

On the other hand, the fourth-order term in exact Formula (1.2) equals

$$\Psi = \frac{3}{2} \frac{(N-1)^2}{N^2} A_1 + \frac{1}{12} \frac{(N-1)^2(N-2)}{N^3} A_2 +$$

$$+ \frac{(N-1)(N-2)(3N-1)}{24N^3} A_3 + \frac{(N-1)(N-2)(N-3)}{24N^3} A_4,$$

where

$$A_4 = N^3 H \left(\frac{\partial f}{\partial y} \right)^2 \left(\frac{\partial}{\partial x} + \frac{f}{h} \frac{\partial}{\partial y} \right) f.$$

We now express R_2, R_3 and β_3 in terms of the free parameters α_1 and α_2 by Formulas (1.16). We have

$$R_2 \alpha_1^3 + R_3 \alpha_2^3 = \frac{(N-1)[-3N\alpha_1\alpha_2 + (2N-1)(\alpha_1+\alpha_2)]}{6N^2},$$

$$R_3 \beta_3 \alpha_1^2 = \frac{(N-1)(N-2)\alpha_1}{6N^2}, \quad R_3 \alpha_1 \alpha_2 \beta_3 = \frac{(N-1)(N-2)\alpha_2}{6N^2}.$$

We so choose the free parameters α_1 and α_2 that the following condition is met:

$$\Psi - \Phi = 0. \tag{1.21}$$

This will correspond to the fact that the accuracy of the method of numerical solution of finite differences is increased from the third to the fourth order.

Condition (1.21) can be brought to the form:

$$\alpha_1\alpha_2 T_1 + \alpha_1 T_2 + \alpha_2 T_2 + T_3 = 0, \tag{1.22}$$

where

$$T_1 = \frac{N(N-1)}{2N^2} A_1,$$

$$T_2 = -\frac{(N-1)(2N-1)}{6N^2} A_1 - \frac{(N-1)(N-2)}{6N^2} A_2, \; T_3 = \frac{3(N-1)^2}{2N^2} A_1 + \frac{(N-1)^2(N-2)}{12N^3} A_2 + \frac{(N-1)(N-2)(3N-1)}{24N^3} A_3 + \frac{(N-1)(N-2)(N-3)}{24N^3} A_4.$$

From Eq. (1.22) we obtain

$$\alpha_2 = -\frac{\alpha_1 T_2 + T_3}{\alpha_1 T_3 + T_1}. \tag{1.23}$$

We thus obtain that, for any value of free parameter α_1, there exists an optimal, in the aforementioned sense, value of parameter α_2, defined by Formula (1.23). With this, the optimal value of parameter α_2 is changed along the solution of the original finite difference equation, since it depends on first and higher-order derivatives of the right member of equation (1.1). It is possible in principle, in the numerical solution of a finite difference equation, to choose the optimal value of parameter α_2 at each step.

In the case of fourth-order methods of numerical solution of finite difference equations, the problem of the optimal determination of the free parameters α_1 and α_2 is also solved, but in this case the solution is so much more complicated that it will not be given here.

GENERALIZED ADAMS METHODS

In the present section we present the interpolation methods of numerical solution of finite difference equations which are the generalizations of Adams' method. Two cases are considered: the case of equally spaced points and the case of unevenly spaced points.

We first consider the case of equally spaced points. We assume that we know the values of the increment δy of the function being sought, which satisfies finite difference Equation (1.1), for four values of the argument:

$$x_{n-3N} = x_n - 3H, \quad x_{n-2N} = x_n - 2H, \quad x_{n-N} = x_n - H \text{ and } x_n,$$

where $H = Nh$. It is required to determine the value y_{n+N} of the function being sought for the argument value $x_{n+N} = x_n + H$. The same limitations are placed on the function $f(x, y)$ as in the section on generalized Runge methods.

By definition, the known increments of the function being sought for the aforementioned argument values equal

$$f_{n-3N} = y(x_n - 3H + h) - y(x_n - 3H),$$
$$f_{n-2N} = y(x_n - 2H + h) - y(x_n - 2H),$$
$$f_{n-N} = y(x_n - H + h) - y(x_n - H),$$
$$f_n = y(x_n + h) - y(x_n).$$

We so choose the coefficients α, β, and γ that, with an accuracy to within fourth-order terms, the following equality holds:

$$\Delta y_n = y_{n+N} - y_n = N f_n + \alpha N \Delta f_{n-N} + \beta N \Delta^2 f_{n-2N} + \gamma N \Delta^3 f_{n-3N}, \quad (2.1)$$

where Δy_n is the increment of the function y being sought, defined by Formula (1.2).

By expanding the corresponding functions in Taylor series about the point x_n and by comparing coefficients of homologous derivatives y'', y''', ..., we obtain the equations for determining α, β, and γ. By solving the equations thus obtained for α, β, γ, we finally get

$$\alpha = \frac{N-1}{2N}, \quad \beta = \frac{(N-1)(5N-1)}{12N^2}, \quad \gamma = \frac{(N-1)(3N-1)}{8N^2}.$$

By expressing the increment Δy_n of the function being sought directly in terms of the functions f_{n-3N}, f_{n-2N}, f_{n-N} and f_n, we will have

$$\Delta y_n = R_1 N f_{n-3N} + R_2 N f_{n-2N} + R_3 N f_{n-N} + R_n N f_n, \quad (2.2)$$

where
$$R_1 = -\frac{(N-1)(3N-1)}{8N^2}, \quad R_2 = \frac{(N-1)(37N-11)}{24N^2},$$

$$R_3 = -\frac{(N-1)(59N-13)}{24N^2}, \quad R_4 = \frac{55N^2 - 36N + 5}{24N^2}.$$

The given Formulas (2.1) and (2.2) are the generalizations of Adams' interpolation method for sloping lines. Below, we present the corresponding formulas for horizontal and broken lines.

In the case of horizontal lines we have

$$\Delta y_n = N f_{n-3N} + \alpha' N \Delta f_{n-3N} + \beta' N \Delta^2 f_{n-3N} + \gamma' N \Delta^3 f_{n-3N}, \quad (2.3)$$

where
$$\alpha' = \frac{N-1}{2N}, \quad \beta' = -\frac{N^2-1}{12N^2}, \quad \gamma' = \frac{N^2-1}{24N^2},$$

or, in another form,

$$\Delta y_n = R_1' N f_{n-3N} + R_2' N f_{n-2N} + R_3' N f_{n-N} + R_4' N f_n, \quad (2.4)$$

where
$$R_1' = -\frac{(N+1)(3N+1)}{8N^2}, \quad R_2' = \frac{(N-1)(19N+7)}{24N^2},$$

$$R_3' = -\frac{5(N^2-1)}{24N^2}, \quad R_4' = \frac{N^2-1}{24N^2}.$$

In the case of broken lines we have

$$\Delta y_n = N f_{n-N} + \alpha'' N \Delta f_{n-N} + \beta'' N \Delta^2 f_{n-2N} + \gamma'' N \Delta^3 f_{n-3N}, \quad (2.5)$$

where
$$\alpha'' = \frac{N-1}{2N}, \quad \beta'' = -\frac{N^2-1}{12N^2}, \quad \gamma'' = -\frac{N^2-1}{24N^2};$$

and, correspondingly, another form

$$\Delta y_n = R_1'' N f_{n-3N} + R_2'' N f_{n-2N} + R_3'' N f_{n-N} + R_4'' N f_n, \quad (2.6)$$

where
$$R_1'' = \frac{N^2-1}{24N^2}, \quad R_2'' = -\frac{5(N^2-1)}{24N^2}, \quad R_3'' = \frac{(N+1)(19N-7)}{24N^2},$$

$$R_4'' = \frac{(N-1)(9N-1)}{24N^2}.$$

To use the given methods of numerical solution of difference equations, one must have four initial values of the function being sought for certain values of the argument. One must have such initial values both at the very beginning of the process of numerical solution, and in changing (increasing or decreasing) the step H. With this, one can proceed analogously to the way this is done in the theory of ordinary differential equations.[4] It is possible to find the initial values of the function being sought by one of the Runge type methods; one can use the so-called "initial approaches." In the case of finite difference equations, the "initial approaches" have the following form.

Approach I. With the assumption that the value of the function sought y_{n-3N} for the argument value x_{n-3N} is known, the increment Δy_{n-3N} is defined by the formula

$$\Delta y_{n-3N} = N f_{n-3N}, \text{ where } f_{n-3N} = f(x_{n-3N}, y_{n-3N}),$$

and, consequently,

$$y_{n-2N} = y_{n-3N} + N f_{n-3N}.$$

There are then determined

$$f_{n-2N} = f(x_{n-2N}, y_{n-2N}) \text{ and } \Delta f_{n-3N} = f_{n-2N} - f_{n-3N}.$$

Approach II. The increments Δy_{n-3N} and Δy_{n-2N} are determined from the formulas

$$\Delta y_{n-3N} = N f_{n-3N} + \frac{N-1}{2N} N \Delta f_{n-3N},$$

$$\Delta y_{n-2N} = N f_{n-2N} + \frac{N-1}{2N} N \Delta f_{n-3N},$$

and, consequently,

$$y_{n-2N} = y_{n-3N} + \Delta y_{n-3N},$$

$$y_{n-N} = y_{n-2N} + \Delta y_{n-2N}.$$

Then there are determined

$$f_{n-2N} = f(x_{n-2N}, y_{n-2N}), \quad f_{n-N} = f(x_{n-N}, y_{n-N}),$$

$$\Delta f_{n-3N} = f_{n-2N} - f_{n-3N}, \quad \Delta f_{n-2N} = f_{n-N} - f_{n-2N},$$

$$\Delta^2 f_{n-3N} = f_{n-N} - 2 f_{n-2N} + f_{n-3N}.$$

FINITE DIFFERENCE EQUATIONS FOR SATELLITE ORBITS

Approach III. The increments Δy_{n-3N}, Δy_{n-2N} and Δy_{n-N} are determined:
  a) by Formula (2.3) for horizontal lines:
$$\Delta y_{n-3N} = N f_{n-3N} + \frac{N-1}{2N} N \Delta f_{n-3N} - \frac{N^2-1}{12N^2} N \Delta^2 f_{n-3N};$$
  b) by Formula (2.5) for broken lines:
$$\Delta y_{n-2N} = N f_{n-2N} + \frac{N-1}{2N} N \Delta f_{n-2N} - \frac{N^2-1}{12N^2} N \Delta^2 f_{n-3N}.$$
  c) by Formula (2.1) for sloping lines:
$$\Delta y_{n-N} = N f_{n-N} + \frac{N-1}{2N} N \Delta f_{n-2N} + \frac{(N-1)(5N-1)}{12N^2} N \Delta^2 f_{n-3N}$$

and, consequently,
$$y_{n-2N} = y_{n-3N} + \Delta y_{n-3N},$$
$$y_{n-N} = y_{n-2N} + \Delta y_{n-2N},$$
$$y_n = y_{n-N} + \Delta y_{n-N}.$$

Thereafter, the functions $f_{n-2N} = f(x_{n-2N}, y_{n-2N})$, $f_{n-N} = f(x_{n-N}, y_{n-N})$ and $f_n = f(x_n, y_n)$ are determined. With this, the approaches are terminated.

We now consider the case of unevenly spaced points. We assume that we know the values of the increment δy of the function being sought for four values of the argument:

$$x_{n-3N} = x_n - \alpha_1 H, \quad x_{n-2N} = x_n - \alpha_2 H, \quad x_{n-N} = x_n - \alpha_3 H \text{ and } x_n,$$

where $H = Nh$; α_1, α_2, α_3 are arbitrary numbers, where $\alpha_1 \neq \alpha_2 \neq \alpha_3 \neq 0$.

It is required to determine the value y_{n+N} of the function being sought for the argument value $x_n + H$. By proceeding in a way analogous to that for equally spaced points, one can obtain the following formula for determining Δy_n (for the case of sloping lines):

$$\Delta y_n = R_1 N f_{n-3N} + R_2 N f_{n-2N} + R_3 N f_{n-N} + R_4 N f_n, \qquad (2.7)$$

where
$$R_1 = \frac{(N-1)\{[4(\alpha_2+\alpha_3)+3(1+2\alpha_2\alpha_3)]N - [2(\alpha_2+\alpha_3)+3]\}}{12N^2\alpha_1(\alpha_1-\alpha_2)(\alpha_3-\alpha_1)},$$

$$R_2 = \frac{(N-1)\{[4(\alpha_3+\alpha_1)+3(1+2\alpha_3\alpha_1)]N - [2(\alpha_3+\alpha_1)+3]\}}{12N^2\alpha_2(\alpha_2-\alpha_3)(\alpha_1-\alpha_2)},$$

$$R_3 = \frac{(N-1)\{[4(\alpha_1+\alpha_2) + 3(1+2\alpha_1\alpha_2)]N - [2(\alpha_1+\alpha_2)+3]\}}{12N^2\alpha_3(\alpha_3-\alpha_1)(\alpha_2-\alpha_3)},$$

$$R_4 = 1 - R_1 - R_2 - R_3.$$

It is easily remarked that the formulas for determining the weight functions R_1, R_2 and R_3 can be obtained one from another by cyclically permuting the elements α_1, α_2 and α_3. This circumstance can be used, for example, in programming this method for an electronic computer.

* * *

Previously, we have given methods of the Runge type and the Adams type for the numerical solution of finite difference equations. The question arises as to which cases are most efficiently handled by one or another of the concrete methods. The question is answered in the same way as in the case of the numerical integration of ordinary differential equations.

On the one hand, the Adams type interpolation methods have an advantage over the Runge type methods in that they require fewer computations in determining the increment at each step of the function being sought (in the case of methods of fourth-order accuracy, the right members must be computed once with the Adams methods, and four times with the Runge methods). On the other hand, when programming these algorithms for an electronic computer, Adams type methods for solving finite difference equations require a significantly greater volume of memory than do the Runge type methods.

These two circumstances must be taken into account in each concrete case.

LIMITING CASES

It was shown earlier that, in the case of methods for the numerical solution of finite difference equations, in contradistinction to methods for the numerical integration of differential equations, the parameters α_i, β_j and the weight functions R_k turn out to be dependent on the ratio N of the step H of the numerical solution to the elementary step h in terms of which the original difference equation, (1.1) was defined.

We now consider the limiting cases: $h \to 0$ or $N \to \infty$. Finite difference Equation (1.1), after division by h and passage to the limit, is transformed to an ordinary differential equation of the form

$$\frac{dy}{dx} = F(x, y),$$

where

$$F(x, y) = \lim_{h \to 0} \frac{f(x, y)}{h}$$

(it is assumed that the limit in the right member exists).

In this limiting case, the family of Runge type methods of fourth- and third-order numerical solution of finite difference equations corresponds to the family of methods of numerical integration of ordinary differential equations. Individual representatives of these families are well known in the literature.

The Runge type methods for the numerical solution of finite difference equations are, by their structures, identical. Among the limiting methods for the case of ordinary differential equations are such which differ in form from those ordinarily used.

We first consider a fourth-order Runge type method. From Formulas (1.6) we have, as $N \to \infty$:

$$\alpha_3 = 1, \quad \beta_1 = \alpha_1, \quad \beta_2 = \frac{\alpha_2[\alpha_1(3 - 4\alpha_1) - \alpha_2]}{2\alpha_1(1 - 2\alpha_1)},$$

$$\beta_3 = \frac{\alpha_2(\alpha_2 - \alpha_1)}{2\alpha_1(1 - 2\alpha_1)}, \quad \beta_4 = 1 - \beta_5 - \beta_6,$$

$$\beta_5 = \frac{(1 - \alpha_1)[2(1 - \alpha_2)(2\alpha_2 - 1) - (\alpha_2 - \alpha_1)]}{2\alpha_1(\alpha_2 - \alpha_1)(6\alpha_1\alpha_2 - 4\alpha_1 - 4\alpha_2 + 3)},$$

$$\beta_6 = \frac{(1 - \alpha_1)(1 - \alpha_2)(1 - 2\alpha_2)}{\alpha_2(\alpha_2 - \alpha_1)(6\alpha_1\alpha_2 - 4\alpha_1 - 4\alpha_2 + 3)},$$

$$R_1 = 1 - R_2 - R_3 - R_4,$$

$$R_2 = \frac{2\alpha_2 - 1}{12\alpha_1(1 - \alpha_1)(\alpha_2 - \alpha_1)}, \quad R_3 = \frac{1 - 2\alpha_1}{12\alpha_2(1 - \alpha_2)(\alpha_2 - \alpha_1)},$$

$$R_4 = \frac{6\alpha_1\alpha_2 - 4\alpha_1 - 4\alpha_2 + 3}{12(1 - \alpha_1)(1 - \alpha_2)}.$$

The parameters α_1 and α_2 can be given arbitrarily (with account taken of the limitations cited in section 1). By setting, for example, $\alpha_1 = 1/4$ and $\alpha_2 = 1/2$, we get one of the fourth-order algorithms:

$$\Delta y_n = \tfrac{1}{6} k_1 + \tfrac{2}{3} k_3 + \tfrac{1}{6} k_4,$$

where

$$k_1 = HF(x_n, y_n),$$
$$k_2 = HF\left(x_n + \tfrac{1}{4} H;\ y_n + \tfrac{1}{4} k_1\right),$$
$$k_3 = HF\left(x_n + \tfrac{1}{2} H;\ y_n + \tfrac{1}{2} k_2\right),$$
$$k_4 = HF(x_n + H;\ y_n + k_1 - 2k_2 + 2k_3).$$

We also give the limiting values, as $N \to \infty$, for the weights R_i^*, used for estimating the error in the function being sought. Going to the limit in Expression (1.19), we have, for the case of differential equations:

$$\Delta^{(k)} y = R_1^* k_1 + R_2^* k_2 + R_3^* k_3 + R_4^* k_4,$$

where

$$R_1^* = \frac{D_1}{D},\ R_2^* = \frac{D_2}{D},\ R_3^* = \frac{D_3}{D},\ R_4^* = \frac{D_4}{D},$$

$$D_1 = \tfrac{1}{6}[2(\alpha_2 - \alpha_1)(\alpha_1\beta_5 + \alpha_2\beta_6) - 2\alpha_1\beta_3(\alpha_3 - \alpha_1) - (\alpha_2 - \alpha_1)(\alpha_3 - \alpha_2)(\alpha_3 - \alpha_1)],$$

$$D_2 = \tfrac{1}{6}[\alpha_2\alpha_3(\alpha_3 - \alpha_2) + 2\alpha_1\alpha_3\beta_3 - 2\alpha_2(\alpha_1\beta_5 + \alpha_2\beta_6)],$$

$$D_3 = \tfrac{\alpha_1}{6}[2(\alpha_1\beta_5 + \alpha_2\beta_6) + \alpha_3(\alpha_1 - \alpha_3)],$$

$$D_4 = \tfrac{\alpha_1}{6}[\alpha_2(\alpha_2 - \alpha_1) - 2\alpha_1\beta_3],$$

$$D = \alpha_1\alpha_2(\alpha_2 - \alpha_1)(\alpha_1\beta_5 + \alpha_2\beta_6) + \alpha_3\alpha_1^2\beta_3(\alpha_1 - \alpha_3).$$

The values of the remaining parameters $\alpha_1, \alpha_2, \ldots \beta_6$ are advantageously taken to be the same as in the determination of the increment of the function y being sought. In such a case, the check term $\Delta^{(k)}y$ will be expressed in terms of the same functions k_i ($i = 1,2,3,4$) as the increment Δy_n.

The limiting methods for integrating ordinary differential equations which correspond to singular solutions in the case of methods for difference equations will not be given here. They can be obtained without difficulty by a limiting passage in Formulas (1.7)-(1.13). We give only the limiting methods for the case $R_2 = 0$, since to this case there correspond limiting methods which, by their structure, differ from those ordinarily used. By going directly to the limit in (1.4), where α_1, α_2, ..., R_4 are expressed by (1.13), we have

$$\Delta y_n = R_1 k_1 + R_2 k_2 + R_3 k_3 + R_4 k_4,$$

where

$$k_1 = HF(x_n; y_n),$$
$$k_2 = HF(x_n; y_n),$$
$$k_3 = HF\left(x_n + a_1 H; y_n + a_1 k_1 + a_2 H^2 \left(\frac{dF}{dx}\right)_n\right),$$
$$k_4 = HF\left(x_n + a_3 H; y_n + a_4 k_1 + a_5 k_3 + a_6 H^2 \left(\frac{dF}{dx}\right)_n\right),$$

$$a_1 = \lim_{N \to \infty} \alpha_2, \quad a_2 = \lim_{N \to \infty} \alpha_1 \beta_3, \quad a_3 = \lim_{N \to \infty} \alpha_3$$

$$a_4 = \lim_{N \to \infty} (\beta_4 + \beta_5), \quad a_5 = \lim_{N \to \infty} \beta_6, \quad a_6 = \lim_{N \to \infty} \alpha_1 \beta_5,$$

and the weight functions R_k ($k = 1, 2, 3, 4$) assume the following limiting values:

$$R_1 = \frac{6a_3^2 - 6a_3 + 1}{6a_3(4a_3 - 3)}, \quad R_2 = 0, \quad R_3 = \frac{2(3a_3 - 2)^3}{3(4a_3 - 3)(6a_3^2 - 8a_3 + 3)},$$

$$R_4 = \frac{1}{6a_3(6a_3^2 - 8a_3 + 3)}.$$

Here, α_3 is the free parameter.

The one-parameter family of fourth-order methods for integrating differential equations which we have given is characterized by the fact that, in it, the increment of the function being sought depends, not on four, but on three functions k_i and, in addition, on the total derivative with respect to x of the right member of the original equation. Such methods are profitably used in those cases when dF/dx is a comparatively simple function (for example, when F is a polynomial, etc.).

For example, a concrete method of this family, corresponding to the value $\alpha_3 = 1/2$, has the form

$$\Delta y_n = \frac{1}{6} k_1 + \frac{1}{6} k_3 + \frac{2}{3} k_4,$$

where

$$k_1 = HF(x_n; y_n),$$
$$k_3 = HF\left(x_n + H; y_n + k_1 + \frac{1}{2}H^2\left(\frac{dF}{dx}\right)_n\right),$$
$$k_4 = HF\left(x_n + \frac{1}{2}H; y_y + \frac{3}{8}k_1 + \frac{1}{8}k_3\right).$$

The limiting third-order methods are not given here.

To the interpolation methods for the numerical solution of finite difference equations in the limiting case $N \to \infty$, correspond the well-known methods of numerical integration of ordinary differential equations.

In the case of equally spaced points, in the limit as $N \to \infty$, we obtain, from Formulas (2.2), (2.4) and (2.6):

a) for sloping lines:
$$\Delta y_n = H\left(-\frac{3}{8}F_{n-3N} + \frac{37}{24}F_{n-2n} - \frac{59}{24}F_{n-N} + \frac{55}{24}F_n\right),$$

b) for horizontal lines:
$$\Delta y_n = H\left(\frac{3}{8}F_{n-3N} + \frac{19}{24}F_{n-2N} - \frac{5}{24}F_{n-N} + \frac{1}{24}F_n\right),$$

c) for broken lines:
$$\Delta y_n = H\left(\frac{1}{24}F_{n-3N} - \frac{5}{24}F_{n-2N} + \frac{19}{24}F_{n-N} + \frac{9}{24}F_n\right).$$

The "initial approach" formulas for determining the four initial values of the function being sought go over, in the limit, to the well-known "initial approach" formulas for the case of ordinary differential equations.

In the case of unevenly spaced points, for the increment of the function being sought in the limit as $N \to \infty$, we get from Formula (2.7)

$$\Delta y_n = H(R_1 F_{n-3N} + R_2 F_{n-2N} + R_3 F_{n-N} + R_4 F_n),$$

where

$$R_1 = \frac{4(\alpha_2 + \alpha_3) + 3(1 + 2\alpha_2\alpha_3)}{12\alpha_1(\alpha_1 - \alpha_2)(\alpha_3 - \alpha_1)},$$

$$R_2 = \frac{4(\alpha_3 + \alpha_1) + 3(1 + 2\alpha_3\alpha_1)}{12\alpha_2(\alpha_2 - \alpha_3)(\alpha_1 - \alpha_2)},$$

$$R_3 = \frac{4(\alpha_1 + \alpha_2) + 3(1 + 2\alpha_1\alpha_2)}{12\alpha_3(\alpha_3 - \alpha_1)(\alpha_2 - \alpha_3)},$$

$$R_4 = 1 - R_1 - R_2 - R_3.$$

APPLICATION OF TWO-STAGE INTEGRATION METHODS TO THE CALCULATION OF SATELLITE ORBITS

For describing the disturbed motion of a satellite, we used differential equations in the osculating elements with respect to the independent variable u —the latitude argument:[1,2]

$$\frac{dp}{du} = F2r\tilde{T},$$

$$\frac{de}{du} = F\left[\tilde{S}\sin\vartheta + \left(1+\frac{r}{p}\right)\tilde{T}\cos\vartheta + e\frac{r}{p}\tilde{T}\right],$$

$$\frac{d\omega}{du} = F\left[-\frac{1}{e}\tilde{S}\cos\vartheta + \frac{1}{e}\left(1+\frac{r}{p}\right)\tilde{T}\sin\vartheta - \frac{r}{p}\cot i\,\tilde{W}\sin u\right], \quad (4.1)$$

$$\frac{d\Omega}{du} = F\frac{r}{p}\frac{\sin u}{\sin i}\tilde{W},$$

$$\frac{di}{du} = F\frac{r}{p}\tilde{W}\cos u,$$

$$\frac{dt}{du} = F,$$

where

$$F = \frac{1}{\frac{\sqrt{pfM}}{r^2} - \frac{r}{p}\cot i\,\tilde{W}\sin u}, \quad \vartheta = u - \omega, \quad r = \frac{1}{1+e\cos\vartheta},$$

$$\tilde{S} = \frac{\sqrt{p}}{\sqrt{fM}}S, \quad \tilde{T} = \frac{\sqrt{p}}{\sqrt{fM}}T, \quad \tilde{W} = \frac{\sqrt{p}}{\sqrt{fM}}W,$$

f is the gravitational constant, M is the mass of the earth, ϑ is the true anomaly, p is the orbit parameter, e is the eccentricity, ω is the angular distance of the perigee from the ascending node, Ω is the longitude of the ascending node, i is the orbit's inclination, S, T, and W are the projections of the disturbing acceleration on the radius vectors perpendicular to it, in the plane of the orbit, and on the normal to the orbital plane, respectively.

The projections of the disturbing acceleration S, T, and W can be presented in the form

$$S = S_1 + S_2 + S_3,$$
$$T = T_1 + T_2 + T_3,$$
$$W = W_1 + W_2 + W_3,$$

where S_1, T_1, and W_1 are the projections of the disturbing acceleration due to the fact that the earth's gravitational field is not a central one because of the flattening of the earth; S_2, T_2, W_2 are the projections of the disturbing acceleration created by the resistance of the atmosphere; S_3, T_3, W_3 are the projections of the disturbing acceleration created by anomalous gravitational forces.

The earth's gravitational potential, within an accuracy of the order of the square of the oblateness, has the form[5]

$$\begin{aligned}V_1 = \frac{fM}{r}\Big\{ &1 + \left(\frac{r_0}{r}\right)^2 [c_{20} P_{20}(\sin\psi) + (c_{22}\cos 2\lambda + d_{22}\sin 2\lambda) P_{22}(\sin\psi)] + \\&+ \left(\frac{r_0}{r}\right)^3 [c_{30} P_{30}(\sin\psi) + (c_{31}\cos\lambda + d_{31}\sin\lambda) P_{31}(\sin\psi) + \\&\quad + (c_{32}\cos 2\lambda + d_{32}\sin 2\lambda) P_{32}(\sin\psi) + \\&\quad + (c_{33}\cos 3\lambda + d_{33}\sin 3\lambda) P_{33}(\sin\psi)] + \\&+ \left(\frac{r_0}{r}\right)^4 [c_{40} P_{40}(\sin\psi) + (c_{41}\cos\lambda + d_{41}\sin\lambda) P_{41}(\sin\psi) + \\&\quad + (c_{42}\cos 2\lambda + d_{42}\sin 2\lambda) P_{42}(\sin\psi) + \\&\quad + (c_{43}\cos 3\lambda + d_{43}\sin 3\lambda) P_{43}(\sin\psi) + \\&\quad + (c_{44}\cos 4\lambda + d_{44}\sin 4\lambda) P_{44}(\sin\psi)]\Big\},\end{aligned}$$

(4.2)

where ψ and λ are the satellite's geocentric latitude and longitude, $P_{ik}(\sin\psi)$ are spherical functions of the corresponding order; the constants r_0, f, M, c_{ik}, d_{ik} are given in Ref. 5.

In Expansion (4.2) for the earth's gravitational potential, if we limit ourselves to the principal terms with c_{20} and c_{40}, we get

$$V_1 = \frac{fM}{r}\left[1 + \left(\frac{r_0}{r}\right)^2 c_{20} P_{20}(\sin\psi) + \left(\frac{r_0}{r}\right)^4 c_{40} P_{40}(\sin\psi)\right].$$

For the principal spherical functions or Legendre polynomials, we have the expressions [5]

$$P_{20}(\sin\psi) = \frac{1}{2}(3\sin^2\psi - 1),$$

$$P_{40}(\sin\psi) = \frac{1}{8}(35\sin^4\psi - 30\sin^2\psi + 3).$$

Differentiating the expression for V_1 with respect to r and ψ and projecting the projections of acceleration of the S, T, W axes, we obtain the formulas for determining the projections S_1, T_1, W_1 of the disturbing acceleration:

$$S_1 = -\frac{fM}{r^2}\frac{r_0^2}{r^2}\left[3c_{20}P_{20}(\sin\psi)+5\frac{r_0^2}{r^2}c_{40}P_{40}(\sin\psi)\right],$$

$$T_1 = \frac{fM}{r^2}\frac{r_0^2}{r^2}\left[c_{20}P'_{20}(\sin\psi)+\frac{r_0^2}{r^2}c_{40}P'_{40}(\sin\psi)\right]\frac{\cos u}{\cos\psi}\sin i,$$

$$W_1 = \frac{fM}{r^2}\frac{r_0^2}{r^2}\left[c_{20}P'_{20}(\sin\psi)+\frac{r_0^2}{r^2}c_{40}P'_{40}(\sin\psi)\right]\frac{\cos i}{\cos\psi},$$

where by $P'_{20}(\sin\psi)$ and $P'_{40}(\sin\psi)$ we denote the derivatives with respect to ψ of the Legendre polynomials $P_{20}(\sin\psi)$ and $P_{40}(\sin\psi)$:

$$P'_{20}(\sin\psi) = 3\sin\psi\cos\psi,$$

$$P'_{40}(\sin\psi) = \frac{5}{2}\sin\psi\cos\psi(7\sin^2\psi - 3),$$

$$\cos\psi = \sqrt{1-\sin^2\psi}.$$

For the projections of the disturbing acceleration of the resistance force, we use the formulas[1,2]

$$S_2 = -\frac{c_x s}{2m}\rho v_{rel}v_r,$$

$$T_2 = -\frac{c_x s}{2m}\rho v_{rel}(v_n - \Omega_E r\cos i),$$

$$W_2 = \frac{c_x s}{2m}\rho v_{rel}\Omega_E r\sin i\cos u,$$

where m is the mass of the satellite, v_{rel} is the satellite's velocity relative to the atmosphere, Ω_E is the angular velocity of the earth's rotation, c_x is the coefficient of aerodynamic resistance, s is the area to which the aerodynamic coefficient is referred, ρ is the density of the atmosphere; v_r and v_n are the radial and transversal components of the absolute velocity:

$$v_r = \frac{\sqrt{fM}}{\sqrt{p}}e\sin\vartheta, \quad v_n = \frac{\sqrt{fM}}{\sqrt{p}}(1+e\cos\vartheta).$$

As the curve of the distribution of atmospheric density by altitude, we adopted the laws obtained on the basis of the data on physical parameters of the atmosphere which were given in

Ref. 6. In connection with the fact that, on the basis of the trajectory measurements of the first Soviet satellite, the atmospheric density at the perigee of the orbit, i.e., at an altitude of h = 225 km, turned out to be greater than was indicated by the data of Ref. 6 by a factor of about 1.3,[7, 8] the atmospheric density for all altitudes was correspondingly increased by a factor of 1.3, while the law of density variation with altitude remained as before. For approximating the law of density change with altitude, we used functions of the same form as in Refs. 1 and 2.

The disturbing potential for the anomalous gravitational forces has the form [9]

$$V_2 = \gamma_m \sum_{n=2}^{8} \left(\frac{R_m}{r}\right)^{n+1} \sum_{m=0}^{n} (\alpha_{nm} \cos m\lambda + \beta_{nm} \sin m\lambda) P_{nm}(\sin \varphi), \quad (4.3)$$

where R_m and γ_m are the mean values of the radius vector and the surface gravitational force of the terrestrial ellipsoid, φ and λ are the geographic latitude and longitude, $P_{nm}(\sin\varphi)$ are spherical functions, α_{nm} and β_{nm} are constant coefficients. The values of these coefficients are given in Ref. 9.

On the basis of Formula (4.3) for V_2, it is easy to obtain the expressions for S_3, T_3, W_3—the projections of the acceleration due to gravitational anomalies:

$$S_3 = \frac{\partial V_2}{\partial r},$$

$$T_3 = \frac{1}{r} \frac{\partial V_2}{\partial \varphi} \frac{\cos u}{\cos \psi} \sin i + \frac{1}{r \cos \varphi} \frac{\partial V_2}{\partial \lambda} \frac{\cos i}{\cos \varphi},$$

$$W_3 = \frac{1}{r} \frac{\partial V_2}{\partial \varphi} \frac{\cos i}{\cos \varphi} - \frac{1}{r \cos \varphi} \frac{\partial V_2}{\partial \lambda} \frac{\cos u}{\cos \varphi} \sin i,$$

where

$$\frac{\partial V_2}{\partial r} = -\frac{\gamma_m}{R_m} \sum_{n=2}^{8} (n+1) \left(\frac{R_m}{r}\right)^{n+2} \sum_{m=0}^{n} (\alpha_{nm} \cos m\lambda + \beta_{nm} \sin m\lambda) P_{nm}(\sin \varphi),$$

$$\frac{1}{r} \frac{\partial V_2}{\partial \varphi} = \frac{\gamma_m}{R_m} \sum_{n=2}^{8} \left(\frac{R_m}{r}\right)^{n+2} \sum_{m=0}^{n} (\alpha_{nm} \cos m\lambda + \beta_{nm} \sin m\lambda) P'_{nm}(\sin \varphi),$$

$$\frac{1}{r \cos \varphi} \frac{\partial V_2}{\partial \lambda} = \frac{\gamma_m}{R_m \cos \varphi} \sum_{n=2}^{8} \left(\frac{R_m}{r}\right)^{n+2} \sum_{m=0}^{n} (-\alpha_{nm} \sin m\lambda + \beta_{nm} \cos m\lambda) m P_{nm}(\sin \varphi).$$

To determine the spherical functions, one can use the recursion formulas

$$P_{n+1,m} = \frac{(2n+1)\sin\varphi\, P_{nm} - (n+m)P_{n-1,m}}{n-m+1} \quad \text{for } n+1 \neq m,$$

$$P_{n,n} = (2n-1)\cos\varphi\, P_{n-1,n-1} \quad \text{for } m = n+1,$$

where

$$P_{00} = 1, \quad P_{10} = \sin\varphi, \quad P_{11} = \cos\varphi.$$

By differentiating these expressions for P_{nm} with respect to φ, we obtain the corresponding formulas for the polynomials P'_{nm}:

$$P'_{n+1,m} = \frac{(2n+1)\cos\varphi\, P_{nm} + (2n+1)\sin\varphi\, P'_{nm} - (n+m)P'_{n-1,m}}{n-m+1}$$

$$\text{for } n+1 \neq m,$$

$$P'_{nn} = (2n-1)\cos\varphi\, P'_{n-1,n-1} - (2n-1)\sin\varphi\, P_{n-1,n-1} \quad \text{for } m = n+1,$$

where

$$P'_{00} = 0, \quad P'_{10} = \cos\varphi, \quad P'_{11} = -\sin\varphi.$$

We now derive the formula for the energy integral, which can be used as a check in carrying out the computations.

In our case, the system of forces under consideration is not conservative. Therefore, the ordinary law as to the conservation of mechanical energy, equal to the sum of the potential and kinetic energies, will not hold. However, by knowing at each moment of time the amount of energy expended due to the atmospheric braking of the satellite, we will have

$$E = T + V + E_\rho = \text{const}, \tag{4.4}$$

where T is the kinetic energy, V is the potential energy of the satellite and E_ρ is the energy dissipated.

We have

$$E_\rho = \int_{u_0}^{u} ds \cdot \mathbf{R},$$

where the integrand is the expression for the element of work performed by the resistance force.

Since d**s** = **v** dt, then

$$E_\rho = \int_{u_3}^{u} (v_r S_2 + v_n T_2) F \, du, \qquad (4.5)$$

where

$$F = \frac{dt}{du}.$$

By substituting the expressions for the kinetic and the potential energy in the right member of Formula (4.4), and also by using Formula (4.5), we obtain the definitive expression for the energy integral:

$$E = \frac{v^2}{2} + V_1 - V_2 + \int_{u_0}^{u} (v_r S_2 + v_n T_2) F \, du.$$

For solving System (4.1) of differential equations, describing the motion of the satellite for comparatively short intervals of time, one can use the methods of numerical integration. In the case of long intervals of time, such methods are of little value.

In connection with this, we go from the original System (4.1) of equations to the system of equations which is obtained as the result of integrating System (4.1) with respect to u over one revolution (i.e., over the period 2π). Such a system of equations can also serve for the description of the changes of the osculating elements of the satellite's orbit in the course of time. As a result, we obtain a system of finite difference equations with respect to the osculating orbital elements and time t of the form

$$\begin{aligned}
\delta p &= \chi_1(u, p, e, \ldots), \\
\delta e &= \chi_2(u, p, e, \ldots), \\
\delta \omega &= \chi_3(u, p, e, \ldots), \\
\delta \Omega &= \chi_4(u, p, e, \ldots), \\
\delta i &= \chi_5(u, p, e, \ldots), \\
\delta t &= \chi_6(u, p, e, \ldots),
\end{aligned} \qquad (4.6)$$

where, by δp, δe, ... δt we have denoted the changes in the osculating orbital elements p, e, ... and the time t during one revolution, i.e., for a change of 2π in u and, by the functions χ_j (j = 1, 2, ... 6), we have denoted the integrals

$$X_1 = \int_u^{u+2\pi} \frac{dp}{du} du, \quad X_2 = \int_u^{u+2\pi} \frac{de}{du} du, \quad X_3 = \int_u^{u+2\pi} \frac{d\omega}{du} du, \quad X_4 = \int_u^{u+2\pi} \frac{d\Omega}{du} du,$$

$$X_5 = \int_u^{u+2\pi} \frac{di}{du} du, \quad X_3 = \int_u^{u+2\pi} \frac{dt}{du} du,$$

where the derivatives in the integrands are defined by Equations (4.1).

We assume that the given system of equations, (4.6), must satisfy the initial conditions $u = u_0$, $p = p_0$, $e = e_0$, $\omega = \omega_0$, $\Omega = \Omega_0$, $t = t_0$.

System (4.6) of finite difference equations, the right members of which are expressed in terms of the values of a set of systems of integrals, describes the changes in the elements of the satellite's orbit with the passage of time. The solution of this system is a discrete sequence of the orbit's osculating elements for integral values of N or for values of the latitude argument $u = u_0 + 2\pi k$ ($k = 1, 2, \ldots$).

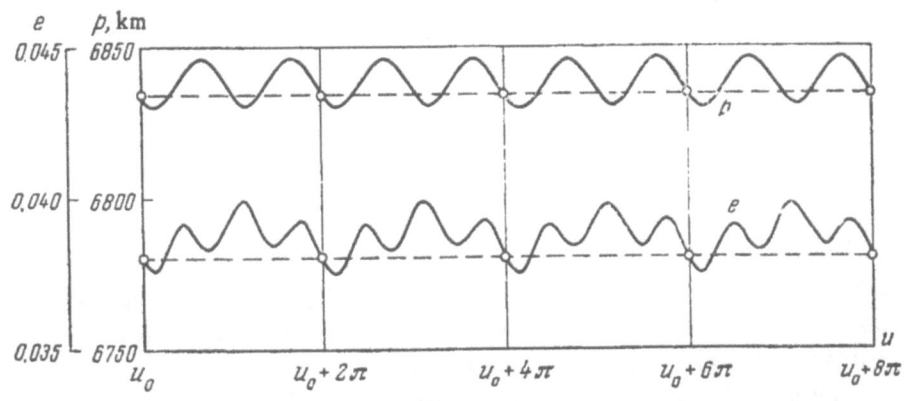

FIG. 1

The transition from the solution of System (4.1) of differential equations to the solution of System (4.6) of finite difference equations is shown geometrically in Fig. 1.

The solid lines present the curves of the changes in osculating elements p and e for some orbit, described by System (4.1) of equations, during the first four revolutions. The dotted lines show the curves for the changes in these same osculating elements from the corresponding system of difference equations.

The integrals in the right members of the equations should be replaced by the corresponding differential equations. In this case, the solution of System (4.6) of finite difference equations leads to the two-stage algorithm: external—numerical solution in terms of the argument N (in the given case, h must equal unity) and internal—integration with respect to the argument u, where this latter must be executed at each point of the external algorithm in the computation of the equations' right members.

The step $H = Nh$ in the numerical solution of System (4.6) of finite difference equations must be so chosen that the error in determining the functions sought does not exceed a given magnitude (with the condition that the integration step of the internal system Δu be chosen correspondingly) and, with this, the step H can be variable.

The next section will give the results of computing certain satellite trajectories. With this, the step in the numerical solution of the system of difference equations was significant for high altitudes and attained a magnitude on the order of 100 revolutions, after which, with the passage of time, the step was decreased.

Further, it should be kept in mind that the step H must be such that the functions k_j ($j = 1,2,3,4$), used for the definition of the increment of the function being sought in the case of finite difference Equations (4.6), are computed for synphased points with respect to the argument u, for example, for the points $u + 2\pi k$, where k is an integer.

Obviously, if the internal system of differential equations is integrated from some value of the independent variable u to the value $u + 2\pi$, h should be set equal to unity. In this case, N should be understood to be the number of integral revolutions with respect to u. The internal system of equations can be integrated from the value u to the value $u + 4\pi$, etc. Then, h must equal 2, etc.

CERTAIN RESULTS OF ORBIT COMPUTATIONS

The methods of two-stage solution were used for calculating satellite orbits.

In this section we present certain results of such computations with physical design parameter values close to those of the first Soviet satellite.

The initial values of the osculating elements of the orbit were taken equal to

$$p_0 = 6846.7 \text{ km}, \quad e_0 = 0.03881,$$
$$\Omega = 0.5940, \quad i_0 = 1.132.$$

To these correspond the following values of apogee altitude, perigee altitude and velocity at perigee:

$$h_{a_0} = 752.0 \text{ km}, \quad h_{\pi_0} = 219.7 \text{ km},$$
$$v_{\pi_0} = 7.922 \text{ km/sec}.$$

The external system of finite difference equations was solved numerically by means of a fourth-order Runge type algorithm. The integration step of the external system was

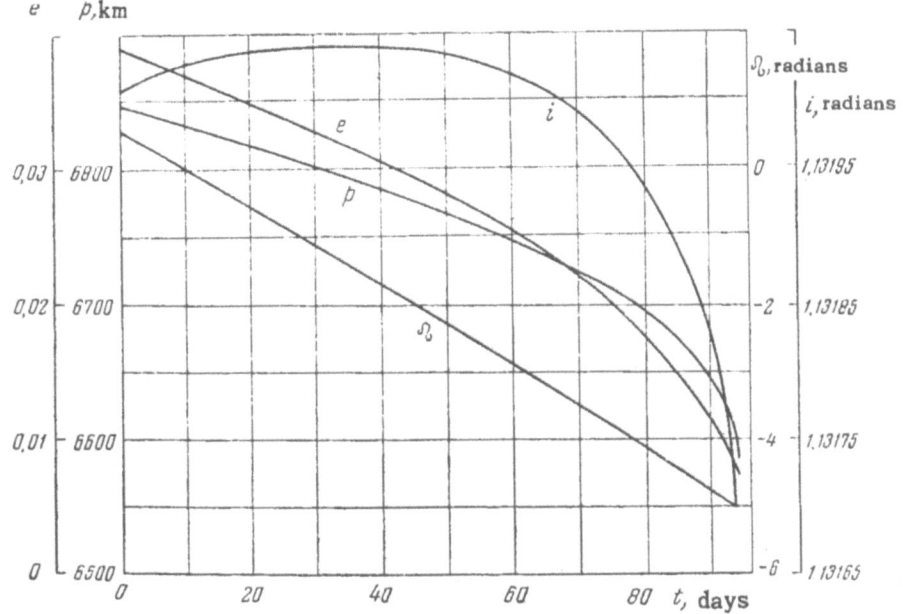

FIG. 2

originally equal to 128 revolutions. As the altitude decreased, the step was decreased.

The calculations were carried out for the cases: 1) all the aforementioned disturbing forces were operative; 2) atmospheric resistance was lacking; 3) the earth's flatness was lacking.

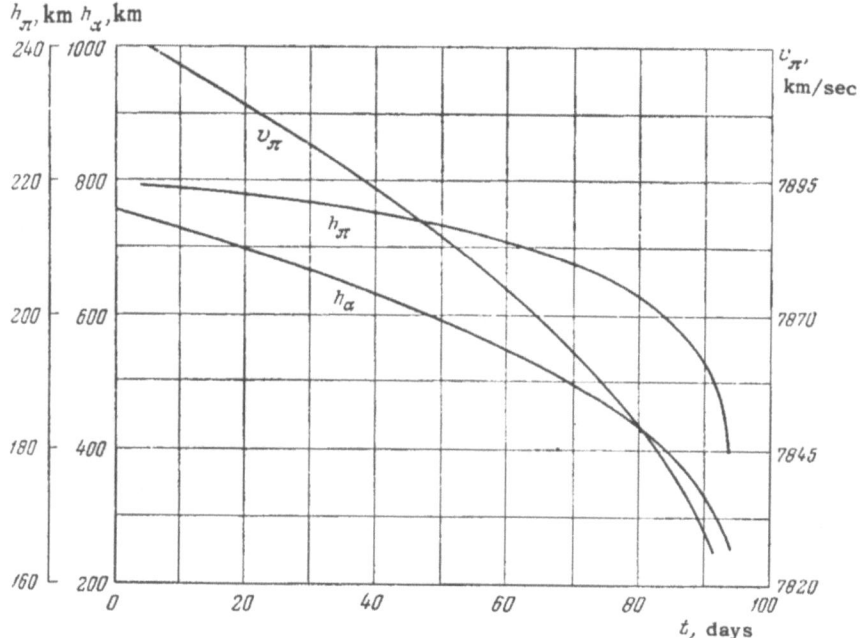

FIG. 3

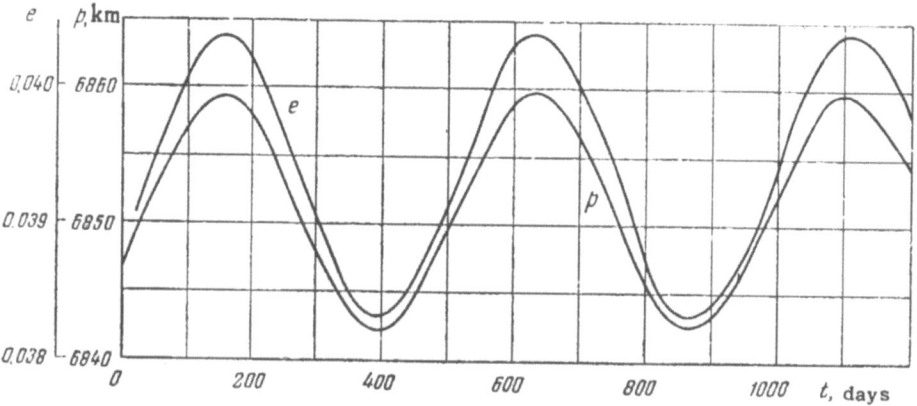

FIG. 4

The computational results are given in Figs. 2-4.

Figure 2 shows the curves for the variation of parameter p, eccentricity e, orbital inclination i and longitude of the ascending node Ω during a period of time equal to 94 days.

Figure 3 gives the curves for the variations in the altitude of the apogee h_α, altitude of the perigee h_π and velocity at perigee v_π for the same case.

Figure 4 gives the curves for the variations of parameter p and eccentricity e during a period of time equal to 1200 days, for case 2.

The curves for the variations in the orbit's osculating elements corresponding to case 3 are analogous to those given in Figs. 2 and 3, and will therefore not be given here.

ACKNOWLEDGMENT

The author wishes to express her gratitude to D. E. Okhotsimskii for his valuable comments and interest in this work, and also to V. I. Shelukhina for having performed the mathematical computations.

LITERATURE

1. D. E. Okhotsimskii, T. M. Eneev, and G. P. Taratynova, UFN, 63, 33 (1957).
2. G. P. Taratynova, UFN, 63, 51 (1957).
3. A. O. Gel'fand, The Calculus of Finite Differences (in Russian), 1952.
4. L. Collatz, Numerical Methods of Solving Differential Equations (Russian translation from the German), ILI, 1953.
5. I. D. Zhongolovich, Byulleten' ITA 6, 8, 81 (1957).
6. S. K. Mitra, The Upper Atmosphere (Russian translation), ILI, 1955.
7. M. L. Lidov, Artificial Earth Satellites, vol. 1, p. 10, Plenum Press, N.Y., 1960.
8. P. E. Él'yasberg, Artificial Earth Satellites, vol. 1, p. 25, Plenum Press, N.Y., 1960.
9. I. D. Zhongolovich, Tr. ITA AN SSSR 3 (1952).

EQUATIONS OF DISTURBED MOTION IN THE KEPLER PROBLEM

A. I. Lur'e

The equations of the disturbed motion of a planet were partially known to Newton; the history of this question and the derivation of these equations are presented in Tisserand's course in celestial mechanics[1] and in the work of A. N. Krylov.[2]

Tisserand, following the general methods of the theory of perturbed motion, computes the Lagrange brackets for the orbit's elliptical elements; the derivation of A. N. Krylov is based on geometric constructions. The derivation of these equations is given in Ref. 3.

The derivation suggested in this communication is based on the direct use of the method of variation of arbitrary constants. The equation of an elliptical orbit is written in vector form

$$\mathbf{r} = \frac{a(1-e^2)}{1 + e \cos \varphi} \mathbf{e}_r = r\mathbf{e}_r, \qquad (1)$$

where $\mathbf{e}_r$ is the unit vector from the center of gravity to the moving point; a and e are, respectively, the semimajor axis and the eccentricity of the orbit; $\cos \varphi = \mathbf{e}_r \cdot \mathbf{e}_1$, where $\mathbf{e}_1$ is the unit vector directed towards the perigee (along the orbit's semimajor axis).

We introduce now the orthogonal trihedral angle of the unit vectors $\mathbf{e}_r, \mathbf{e}_\varphi, \mathbf{e}_3 = [\mathbf{e}_r \mathbf{e}_\varphi]$; $\mathbf{e}_\varphi$ is the unit vector directed on the side of an increasing angle φ in the plane of the orbit, perpendicular to $\mathbf{e}_r$; $\mathbf{e}_3$ defines the plane of the orbit in undisturbed motion. In undisturbed motion, this trihedral angle has the angular velocity $\dot{\varphi} \mathbf{e}_3$, such that

$$\dot{\mathbf{e}}_r = \dot{\varphi} \mathbf{e}_\varphi, \quad \dot{\mathbf{e}}_\varphi = -\dot{\varphi} \mathbf{e}_r, \quad \dot{\mathbf{e}}_3 = 0, \qquad (2)$$

and, by the law of areas,

$$\dot{\varphi} = \frac{\sqrt{\mu a (1 - e^2)}}{r^2}, \tag{3}$$

where μ is the proportionality factor in the law of attraction.

The position of the orbital plane is defined by the longitude of the ascending node Ω on the ecliptic, giving the direction of the unit vector $\mathbf{n}$ of the line of nodes, and by the angle of inclination i of the orbital plane to the plane $O\xi\eta'$ in the system of fixed axes $O\xi\eta\zeta$: the position of the perigee in the orbital plane is given by the angular distance of the perigee from the node ω, so that $\cos \omega = \mathbf{n} \cdot \mathbf{e}_1$.

The velocity vector in undisturbed motion, as follows from (1), (2), and (3), equals

$$\mathbf{v} = \dot{\mathbf{r}} = \sqrt{\frac{\mu}{a}} \cdot \frac{1}{\sqrt{1-e^2}} [e \sin \varphi \mathbf{e}_r + (1 + e \cos \varphi) \mathbf{e}_\varphi], \tag{4}$$

and the acceleration vector is

$$\mathbf{w} = \dot{\mathbf{v}} = -\frac{\mu}{r^2} \mathbf{e}_r. \tag{5}$$

Following the method of variation of arbitrary constants, we retain the same expressions, (1) and (4), for the vectors $\mathbf{r}$ and $\mathbf{v}$ in the disturbed motion as in the undisturbed motion; but the elliptical orbit's elements a, e, Ω, i, ω will not be constant quantities but rather unknown functions of time. In disturbed motion, the angular velocity $\boldsymbol{\omega}$ of the trihedron $\mathbf{e}_r, \mathbf{e}_\varphi, \mathbf{e}_3$ equals

$$\boldsymbol{\omega} = \mathbf{k}\dot{\Omega} + \mathbf{n}\frac{di}{dt} + \mathbf{e}_3(\dot{\omega} + \dot{\varphi}), \tag{6}$$

where $\mathbf{k}$ is the unit vector along the $O\zeta$ axis. The projections on the trihedral axes $\mathbf{e}_r, \mathbf{e}_\varphi$ and $\mathbf{e}_3$ are defined by the well-known formulas

$$\omega_r = \dot{\Omega} \sin i \sin u + \frac{di}{dt} \cos u,$$

$$\omega_\varphi = \dot{\Omega} \sin i \cos u - \frac{di}{dt} \sin u,$$

$$\omega_3 = \dot{\Omega} \cos i + \dot{\omega} + \dot{\varphi} = \omega_3' + \dot{\varphi}, \tag{7}$$

where $u = \omega + \varphi$. We note that $\dot{\varphi}$ in these formulas for the disturbed motion differs from the quantity defined by (3); we shall denote the latter by $\dot\varphi^0$ (in general, a superscript "0" will refer to undisturbed motion).

By the formulas for the differentiation of unit vectors, we have

$$\dot e_r = [\omega e_r] = -\omega_\varphi e_3 + (\omega_3' + \dot\varphi) e_\varphi,$$
$$\dot e_\varphi = [\omega e_\varphi] = \omega_r e_3 - (\omega_3' + \dot\varphi) e_r, \qquad (8)$$
$$\dot e_3 = [\omega e_3] = -\omega_r e_\varphi + \omega_\varphi e_r.$$

By now requiring that the following equalities hold

$$\dot{\mathbf r} = \mathbf v = \mathbf v^0, \quad \dot{\mathbf v} = \mathbf w^0 + \mathbf F, \qquad (9)$$

where $\mathbf F$ is the additional force acting on the point in disturbed motion, we obtain the following equality, after differentiating with account taken of Relationships (8):

$$\mathbf v = \dot{\mathbf r} = e_r \left(\frac{\partial r}{\partial \varphi}\dot\varphi + \frac{\partial r}{\partial a}\dot a + \frac{\partial r}{\partial e}\dot e\right) + r[(\omega_3' + \dot\varphi)e_\varphi - \omega_\varphi e_3] =$$
$$= \left(e_r \frac{\partial r}{\partial \varphi} + e_\varphi r\right)\dot\varphi^0 = \sqrt{\frac{\mu}{a}} \frac{1}{\sqrt{1-e^2}} [e\sin\varphi\, e_r + (1 + e\cos\varphi) e_\varphi] = \qquad (10)$$
$$= v_r e_r + v_\varphi e_\varphi.$$

$$\dot{\mathbf v} = \left(\frac{\partial v_r}{\partial a}\dot a + \frac{\partial v_r}{\partial e}\dot e + \frac{\partial v_r}{\partial \varphi}\dot\varphi\right)e_r + \left(\frac{\partial v_\varphi}{\partial a}\dot a + \frac{\partial v_\varphi}{\partial e}\dot e + \frac{\partial v_\varphi}{\partial \varphi}\dot\varphi\right)e_\varphi +$$
$$+ v_r[-\omega_\varphi e_3 + (\omega_3' + \dot\varphi)e_\varphi] + v_\varphi[\omega_r e_3 - (\omega_3' + \dot\varphi)e_r] = -\frac{\mu}{r^2}e_r + \mathbf F. \quad (11)$$

From (10) we obtain the three equations

$$\omega_\varphi = 0; \quad \omega_3' + \dot\varphi = \dot\varphi^0; \quad -\frac{\partial r}{\partial\varphi}\omega_3' + \frac{\partial r}{\partial a}\dot a + \frac{\partial r}{\partial e}\dot e = 0. \qquad (12)$$

In expanded form, the last of these equations will be

$$\omega_3' e\sin\varphi - \frac{\dot a}{a}(1 + e\cos\varphi) + \frac{2e + e^2\cos\varphi + \cos\varphi}{1-e^2}\dot e = 0. \qquad (13)$$

By using Relationships (12) we can write the equations obtained from vector Equation (11) in the form

$$-\frac{\dot a}{2a}e\sin\varphi + \frac{\dot e}{1-e^2}\sin\varphi - \omega_3' e\cos\varphi = \sqrt{\frac{a}{\mu}}\sqrt{1-e^2}\,F_r,$$

$$-\frac{\dot a}{2a}(1+e\cos\varphi) + \frac{\dot e}{1-e^2}(\cos\varphi + e) + \omega_3' e\sin\varphi = \sqrt{\frac{a}{\mu}}\sqrt{1-e^2}\,F_\varphi,$$

$$\omega_r = \sqrt{\frac{a}{\mu}}\frac{\sqrt{1-e^2}}{1+e\cos\varphi}F_3. \qquad (14)$$

From the first equation of (12) and the last of (14) we find, by substituting the values of ω_r and ω_φ from (7), the equations of disturbed motion for the elements Ω and i

$$\frac{di}{dt} = \sqrt{\frac{a}{\mu}} \frac{\sqrt{1-e^2}}{1+e\cos\varphi} F_3 \cos u; \quad \dot{\Omega} \sin i = \sqrt{\frac{a}{\mu}} \frac{\sqrt{1-e^2}}{1+e\cos\varphi} F_3 \sin u. \quad (15)$$

From (13) and (14) we get

$$\dot{e} = \sqrt{\frac{a}{\mu}} \sqrt{1-e^2} \left(F_r \sin\varphi + \frac{e + 2\cos\varphi + e\cos^2\varphi}{1+e\cos\varphi} F_\varphi \right)$$

$$\frac{\dot{a}}{2a} = \sqrt{\frac{a}{\mu}} \frac{1}{\sqrt{1-e^2}} [F_r e \sin\varphi + (1 + e\cos\varphi) F_\varphi] \quad (16)$$

$$\omega_3' = \sqrt{\frac{a}{\mu}} \cdot \frac{\sqrt{1-e^2}}{e} \left(-F_r \cos\varphi + \frac{2+e\cos\varphi}{1+e\cos\varphi} F_\varphi \sin\varphi \right) = \dot{\Omega} \cos i + \dot{\omega}.$$

Finally, the second equation of (12) gives the relationship defining the disturbance for the time t_0 of passage through the perigee. If we denote

$$\zeta = n(t - t_0) = \frac{\sqrt{\mu}}{a\sqrt{a}} (t - t_0),$$

then this relationship is written in the form

$$\frac{\dot{\zeta} - n}{\sqrt{1-e^2}} + \omega_3' = -2\sqrt{\frac{a}{\mu}} \frac{\sqrt{1-e^2}}{1+e\cos\varphi} F_r. \quad (17)$$

Thus, the derivations are based on the application of the method of variation of arbitrary constants to the expressions for the planet's radius vector and its velocity vector and the use of the very well-known formulas for differentiation of the unit vectors of a moving trihedral system of axes. This derivation is the most simple and economical of all extant derivations.

LITERATURE

1. F. Tisserand. Traité de mécanique céleste, Vol. I, ch. X, 1889.
2. A. N. Krylov, Sur la variation des éléments des orbites élliptiques de planètes, 1915. Collected Works, Volume VI.
3. G. N. Duboshin. Introduction to Celestial Mechanics (in Russian), Gostekhizdat, Moscow, 1938.

ELEMENTS OF THE THEORY OF THE IMPACT OF SOLID BODIES WITH HIGH (COSMIC) VELOCITIES

K. P. Stanyukovich

In the present work we consider the question of the impact of meteorites with cosmic velocities on the surfaces of planets. This question is also relevant to the investigation of the impact of micrometeorites on the surface of artificial earth satellites and cosmic rockets.

The second question is somewhat simpler than the first. In fact, the impacts of micrometeorites with any surface causes such microscopic explosions that the force of gravity may be neglected in the investigation of the process of crater formation on the surface, and also in the investigation of the scattering of the products of explosion.

In the first case, when we consider the impact of a relatively large meteorite with the surface of a planet, the gravitational force must be taken into account in the investigation of the scattering of the products of explosion. In the case of different planets with all other conditions the same (the initial mass and velocity of the meteorite, its composition and form, and also the medium into which it penetrates), the crater dimensions and the character of the scattering of the explosion products will depend strongly on the gravitational force at the surface of the planet in question.

The problems to be considered can be simplified to a certain extent, if we assume that instead of an exploding meteorite penetrating some medium, we have an explosive charge with an equivalent amount of energy located at a certain distance under the surface (the initial hole produced by the meteorite's penetration into the medium can always be neglected).

We thus have before us the following problems:

To consider the process of the explosion of charges with varying power and calorific content, with the aim of investigating the dependence of dimensions and shapes of the craters formed and the character of the scattering of the explosion products on the properties of the medium and the gravitational force of the given planet; as a limiting case, to consider the same questions for micrometeorites, when the gravitational force can be neglected; to investigate the depth of penetration of meteorites in the given medium.

To reduce the problem of the explosion of a meteorite to that of the explosion of an equivalent explosive charge.

To analyze the results obtained in order to know how to calculate the dimensions and shape of the crater, and also the dimensions and shape of the so-called filled crater, if we understand by the latter the formation obtained after all or part of the material displaced by the explosion has fallen back on the planet; to consider how the pressures, forces, and temperatures act in the process of explosion and scattering, and also to draw conclusions on the basis of this analysis concerning cosmic phenomena on the one hand, and concerning the technical problem of protecting satellites and rockets from the impact of meteorites on the other. We will consider the impact of meteorites with velocities such that an explosive process will be known to take place accompanied by the destruction of all or part of the meteorite and of a definite area of the surface subjected to the explosion.

An explosive process will occur when the normal component of the impact velocity exceeds a certain critical value.

We will consider in the present work the general "over-all" effects of an ordinary explosion and of meteorite explosions, and will omit any consideration of the effect produced by the waves following the penetration and explosion. These effects are very interesting, but their careful consideration can not greatly alter the results that are obtained here.

THE EXPLOSION AND THE SCATTERING OF DISPLACED MATERIAL

Explosions in an Infinite Medium

When the detonation of any explosive occurs, then at the end of the detonation process, when the detonation wave reaches the boundary of the explosive medium, the surrounding medium begins to move under the action of the expanding explosion products. A shock wave is thus initiated in this latter medium.

The pressure at the shock front in strong metallic media is greater than the pressure at the detonation-wave front. In rock that is not so strong, and also in water, the pressure at the shock front caused by an explosion is less than at the detonation front.[1] The shock-front pressure falls as the distance increases.

Here we will not consider the laws related to the propagation of the shock wave or take into account the influence of the pressure wave in the medium, but will occupy ourselves with the investigation of the general character of the explosion.

On the explosion of an explosive charge in an unbounded inert medium, the volume occupied by the products of the explosion (V_∞) is proportional to the mass of explosive and depends on the properties of the medium.

The pressure of the expanding products of detonation varies from its initial value

$$\bar{p}_i = (\bar{k} - 1)\rho_0 Q, \tag{1}$$

to a final value $\bar{p}_a$, that is dependent on the properties of the medium; (here $\bar{k}$ is the polytropic coefficient for the explosion products, Q the calorific content of the explosive, and ρ_0 the density of the explosive). During the initial stage of the expansion when $\bar{p}_k \leq p \leq \bar{p}_i$,[1] the expansion proceeds according to the law $p \sim v^{-\bar{k}} \sim r^{-3\bar{k}}$, and then for $\bar{p}_a \leq p \leq \bar{p}_k$ according to the law $p \sim v^{-k} \sim r^{-3k}$, where k c_p/c_v is the isentropic exponent.

For a typical explosive $\bar{k} = 3$, $k = 1.4$-1.25, $\bar{p}_i = 10^5 \text{kg/cm}^2$, $\bar{p}_k = 10^3 \text{kg/cm}^2$. For expansion in air and water, when practically no viscous forces must be overcome and there is therefore no energy loss in the deformation and break-up of the medium, we can write $\bar{p}_a = p_a$, where p_a is the back-pressure of the medium. If

$$p_a = 1 \text{ kg/cm}^2$$

then

$$V_\infty = \bar{A} V_0, \qquad (2)$$

where V_0 is the initial volume of the explosive and $\bar{A} \approx 1000$.

In the case of expansion of the detonation products in a solid body $\bar{p}_a > p_a$. For solids of varying tensile strength the value of $\bar{p}_a$ can vary from 1 kg/cm² to 10,000 kg/cm². For $p_a = 10,000$ the value of V_∞ decreases to a value that is only $10,000^{1/k^*}$ times its value for a tensile strength of 1 kg/cm². In the general case we can write

$$V_\infty = \bar{A} V_0 \left(\frac{p_a}{\bar{p}_a}\right)^{\frac{1}{k^*}}, \qquad (3)$$

where k^* is the effective value of k depending on the properties of the medium. For typical soils, for example,

$$V_\infty \approx 50 V_0. \qquad (4)$$

The volume V_R of the expansion zone of the soil in any medium significantly exceeds the volume that can be occupied by

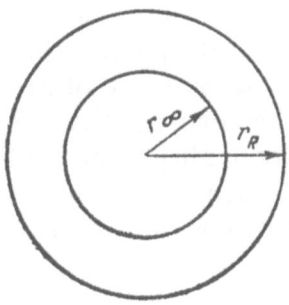

FIG. 1

the detonation products, but it is always proportional to V_∞ (Fig. 1), i.e.,

$$\frac{V_R}{V_\infty} = \bar{\alpha} = \text{const.} \qquad (5)$$

As experiment shows, $1 < \bar{\alpha} < 10$, for various media, soils, and rocks, where the smallest value is obtained for the strongest metals, and the greatest for the weakest soils.

Thus it can be confirmed that

$$V_R = \bar{\alpha} V_\infty = \bar{\alpha} \bar{A} V_0 \left(\frac{p_a}{\bar{p}_a}\right)^{\frac{1}{k^*}}. \qquad (6)$$

Finally, we can express the mass of deformed soil $M = \rho V_R$, where ρ is the soil density, in terms of the mass of explosive m_B as

$$M = \frac{\rho}{\rho_0}\left(\frac{p_a}{\bar{p}_a}\right)^{\frac{1}{k^*}} \cdot \bar{\alpha}\bar{A} m_B = A m_B. \qquad (7)$$

Here for a typical explosive $\bar{A} \approx 1000$, and the values of $\bar{\alpha}$ and $\bar{p}_a$ are obtained by experiment. For "pliable" media the value of k^* is nearer to the value of c_p/c_v than for stronger media, when the value of k is near to $\bar{k}$ (for soils and weak metals $p_a < \bar{p}_a$ and so $k^* = k$, while for strong metals $\bar{k} > k^* > k$, and for the strongest metals $k^* \approx \bar{k}$).

Explosions in a Bounded Medium

The most interesting case is that of the explosion of a charge located at a certain depth h_0 below the earth's surface, which we will assume to be horizontal.

In the investigation of deep explosions, it is usually necessary to take into account the force of gravity.

We will calculate the energy that is expended in overcoming the force of gravity during an explosion of a charge at depth h_0, in the case when the hollow formed after the explosion is in the form of a cone with radius of the base R_0 (see Fig. 2).

We will find the gravitational energy for the case when the soil is transported to the surface of the earth AA'. Since the element of soil mass bounded by the cones with $R + dR$ and R as the radii of their bases is

$$dM = \frac{2}{3}\pi\rho h_0 R\, dR = \frac{2}{3}\pi\rho h_0^3 \frac{\sin\varphi\, d\varphi}{\cos^3\varphi}, \qquad (1)$$

then the total mass is

$$M = \frac{\pi}{3} \rho h_0^3 \tan^2 \varphi_0, \qquad (2)$$

and the mass element bounded in addition by the sections at depths h + dh and h is

$$dM_h = 2\pi \rho h^2 \, dh \, \frac{\sin \varphi \, d\varphi}{\cos^3 \varphi}. \qquad (3)$$

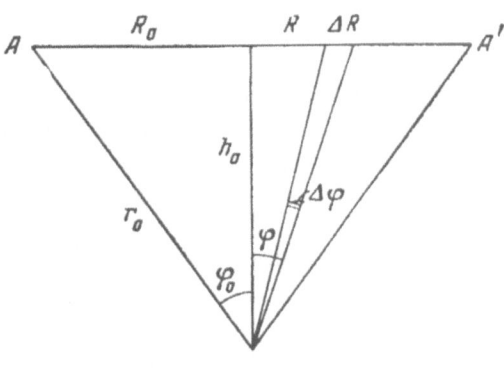

FIG. 2

It is obvious that the gravitational energy necessary to transport the element of mass dM to the surface AA′ is given by the expression

$$dE_g = g(h_0 - h) \, dM_h = 2\pi \rho g \, \frac{\sin \varphi \, d\varphi}{\cos^3 \varphi} (h_0 - h) h^2 \, dh, \qquad (4)$$

where g is the acceleration of gravity.

Thus when we integrate from h_0 to 0 and from 0 to φ_0 we find that

$$E_g = \frac{\pi \rho g h_0^4 \tan^2 \varphi_0}{12} = \frac{M g h_0}{4}. \qquad (5)$$

In the region bounded by the cones with base radii R + dR and R we have

$$dE_g = \frac{\pi \rho g h_0^4 \sin \varphi d\varphi}{6 \cos^3 \varphi} = \frac{g h_0}{4} dM. \qquad (6)$$

If we assume that the explosive energy is propagated isotropically (which is evidently not completely true), then the energy propagated in this region will be

$$dE_\text{B} = Q dm_\text{B} = m_\text{B} Q \frac{\sin\varphi\, d\varphi}{2} = \frac{E_\text{B}}{2}\sin\varphi\, d\varphi, \qquad (7)$$

where E_B is the total energy of the explosion.

The explosive energy propagated inside the cone will be

$$E_{\varphi_0} = E_\text{B}\frac{1-\cos\varphi_0}{2}. \qquad (8)$$

Thus

$$\frac{dE_g}{dE_\text{B}} = \frac{\pi\rho g h_0^4}{3E_\text{B}\cos^3\varphi}. \qquad (9)$$

It can be seen that if we do not consider the energy losses in the break-up of the soil, then the ejection of the soil will be accomplished under the condition $dE_\text{B} \geq dE_g$, and so

$$\frac{\pi\rho g h_0^4}{3\cos^3\varphi} \leq E_\text{B}.$$

Therefore for the given depth h_0

$$\cos\varphi \geq \cos\varphi_0 = \left(\frac{\pi\rho g h_0^4}{3E_\text{B}}\right)^{\frac{1}{3}}. \qquad (10)$$

The greatest possible depth is determined from this relation with $\varphi_0 = 0$

$$h_{0m} = \left(\frac{3E_\text{B}}{\pi\rho g}\right)^{\frac{1}{4}}. \qquad (11)$$

If (11) is used, then (10) can be written as

$$\cos\varphi_0 \geq \left(\frac{h_0}{h_{0m}}\right)^{\frac{4}{3}}. \qquad (12)$$

The relations (10) and (11) yield the optimal value for h_0 and the semi-vertical angle of the cone φ_0. If irreversible losses are taken into account, then the values of h_0 and φ_0 will be smaller.

We will now calculate the residual velocity a_0 with which a part of the soil will be ejected from the hollow in a direction making an angle φ with the normal. We will consider the maximum rate of ejection in more detail below, but at the present we will only assume that the soil particles gain velocity fairly rapidly and that a part of this velocity is lost in overcoming the force of gravity.

It is evident that by using the law of conservation of energy, we can write

$$\frac{a_0}{2} dM + dE_g = dE_B,$$

and thus

$$a_0 = \sqrt{2\left(\frac{dE_B}{dM} - \frac{dE_g}{dM}\right)} = \sqrt{\frac{3}{2} \frac{E_B \cos^3 \varphi}{\pi \rho h_0^3} - \frac{g h_0}{2}} \qquad (13)$$

or

$$a_0 = \sqrt{\frac{g h_0}{2}\left[\left(\frac{h_{0m}}{h_0}\right)^4 \cos^3 \varphi - 1\right]}. \qquad (14)$$

For $\cos \varphi = (h_0/h_{0m})^{4/3}$ we obtain $a_0 = 0$, while for $\varphi = 0$

$$a_0 = a_{0m} = \sqrt{\frac{g h_0}{2}\left(\frac{h_{0m}^4}{h_0^4} - 1\right)}. \qquad (15)$$

If $h_0 = h_{0m}$, then for $\varphi = 0$ we have $a_{0m} = 0$. The ratio

$$\frac{E_g}{E_B} = \frac{\pi \rho g h_0^4 \tan^2 \varphi}{12 E_B} = \left(\frac{h_0}{h_{0m}}\right)^4 \left(\frac{\tan \varphi}{2}\right)^2 \qquad (16)$$

for $\varphi = \varphi_0$ becomes

$$\frac{E_g}{E_B} = \left(\frac{h_0}{h_{0m}}\right)^4 \left(\frac{\tan \varphi_0}{2}\right)^2 = \frac{1}{4}\left[\left(\frac{h_0}{h_{0m}}\right)^{\frac{4}{3}} - \left(\frac{h_0}{h_{0m}}\right)^4\right], \qquad (17)$$

while the ratio

$$\frac{E_g}{E_{\varphi_0}} = \frac{\pi \rho g h_0^4 \tan^2 \varphi_0}{6 E_B (1 - \cos \varphi_0)} = \frac{\pi \rho g h_0^4 (1 + \cos \varphi_0)}{6 E_B \cos^2 \varphi_0} \qquad (18)$$

for $\cos \varphi_0 = (\pi \rho g h_0^4/3 E_B)^{1/3}$ becomes

$$\frac{E_g}{E_{\varphi_0}} = \frac{1}{2}\left(\frac{\pi \rho g h_0^4}{3 E_B}\right)^{\frac{1}{3}}\left[1 + \left(\frac{\pi \rho g h_0^4}{3 E_B}\right)^{\frac{1}{3}}\right] = \frac{1}{2} \cos \varphi_0 (1 + \cos \varphi_0). \qquad (19)$$

For a certain definite value $(h_0/h_{0m}) = \bar{\lambda}$, the value of (E_g/E_B) reaches its maximum. If we start with (17), it is easy to show that

$$\bar{\lambda} = \left(\frac{1}{3}\right)^{\frac{3}{8}} \approx \frac{2}{3}.$$

Thus

$$\frac{E_g}{E_B} = \frac{\sqrt{3}}{2\cdot 9} \approx 0.1; \quad \cos\varphi_0 = \left(\frac{1}{3}\right)^{\frac{1}{2}} = \frac{\sqrt{3}}{3} \approx 0.58; \quad \frac{R_0}{h_0} = \tan\varphi_0 = \sqrt{2} \approx 1.4.$$

On the basis of the relations derived, we can calculate the parameters of the crater formed by the explosion and the subsequent ejection of the "shattered" mass (the ejection crater). We will nevertheless occupy ourselves at first with improving the accuracy of the picture considered above of the explosion and ejection.

We will first improve our description of the ejection. The original form taken for the crater was cone-shaped, which was completely natural as the ejection of material occurs along a radius r. At the surface of the crater, however, an expansion wave (a discharging wave) arises, and this wave can cause supplementary ejection of a certain mass of the medium from the surface layer (Fig. 3). Because of this, the radius of

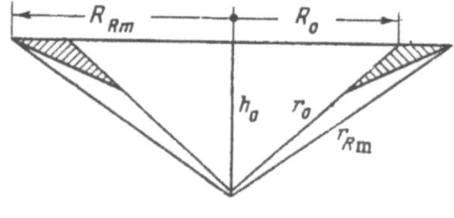

FIG. 3

the crater is somewhat increased (from R_0 to R_{Rm}), and the shape of the crater is changed. This effect cannot be taken into account very easily numerically, and we will not consider it here.

All the relations derived above have a meaning if $h \leq h_{0m} \leq r_{Rm}$, where r_{Rm} is the maximum radius of the disintegration zone, taking into consideration the influence of the free surface (the zone of non-linear deformation).

It follows from this condition and (11) that

$$\frac{3E_B}{\pi\rho g} \leq r_{Rm}^4 = \left(\frac{3V_{Rm}}{4\pi}\right)^{\frac{4}{3}} = \left(\frac{3M_{Rm}}{4\pi\rho}\right)^{\frac{4}{3}}, \tag{20}$$

where V_{Rm} and M_{Rm} are the volume and mass of the maximal zone of disintegration, respectively

Since

$$M_{Rm} = A \frac{E'_B}{Q} = A m_B \tag{21}$$

[$A = (\rho/\rho_0) \overline{\overline{\alpha}} \overline{A}(p_a/\overline{p}_a)^{1/k*}$, and $\overline{\overline{\alpha}}$ is a coefficient defining the zone of non-linear deformation], then (20) can be written in the form

$$\frac{g r_{ex}}{4Q} \geqslant \left(\frac{\rho}{\rho_0}\right)^{\frac{1}{3}} A^{-\frac{4}{3}}, \tag{22}$$

where $r_{ex} = [(3/4\pi)(m_B/\rho_0)]^{1/3}$ is the radius of the original explosive.

It therefore follows that if the calorific content Q of the explosive is increased, then the inequality (22) can be satisfied for a greater gravitational force. In the absence of any gravitational force, (22) does not hold in general.

We now consider various possibilities in the formation of craters by explosions.

1. If $h_0 > r_{Rm}$, then for $h_0 \lessgtr h_{0m}$ no crater will be formed, and we will have a so-called camouflet.

2. If $h_0 < r_{Rm}$ and $h_0 > h_{0m}$, then a partial heaving of the surface will be observed in the neighborhood of the epicenter (over the location of the explosion).

3. If $h_0 < r_{Rm}$ and $h_0 < h_{0m}$, then we will have the case when a crater is formed that was considered above. As h_0 decreases, the profile of the crater will decrease in area (the angle φ_0 will increase).

The first two cases are of no interest, and we will consider in more detail the analysis of the third case.

Let $h_0 \leq h_{0m} < r_{Rm}$. The semi-vertical angle of the crater cone is obtained from relation (9), and is given by

$$\cos \varphi_0 = \left(\frac{\pi \rho g h_0^4}{3 E_B}\right)^{\frac{1}{3}} = \left(\frac{h_0}{h_{0m}}\right)^{\frac{4}{3}} \tag{23}$$

where $(\pi \rho g h_0^4 / 3 E_B) \leq 1$.

The crater radius is

$$R_0 = h_0 \tan \varphi_0 = h_0 \sqrt{\left(\frac{3 E_B}{\pi \rho g h_0^4}\right)^{\frac{2}{3}} - 1}. \tag{24}$$

The relations indicated will be valid only if the condition

$$r_0 = \frac{h_0}{\cos \varphi_0} \leqslant r_{Rm} \tag{25}$$

is satisfied.

The semi-vertical angle of the cone is thus obtained from the relation

$$\cos \varphi_0 = \cos \varphi_{0l} = \frac{h_0}{r_{Rm}}. \tag{26}$$

If $\varphi_{0l} \leq \varphi_0$, then $\cos \varphi_0 \geq h_0/r_{Rm}$ and the crater radius will be

$$R_{0l} = h_0 \tan \varphi_{0l} = \sqrt{r_{Rm}^2 - h_0^2}. \tag{27}$$

The limiting condition (26) leads to the relation

$$\cos \varphi_{0l} = \frac{h_0}{\left(\frac{3}{4\pi} \cdot \frac{AE_B}{\rho Q}\right)^{\frac{1}{3}}} = \frac{h_0}{r_{ex}} \sqrt[3]{\frac{p}{p_0 A}}. \tag{28}$$

Thus in the integration the angle φ varies from 0 to φ_0 if $r_0 \leq r_{Rm}$, and from 0 to φ_{0l} if $r_0 > r_{Rm}$.

The relations derived above completely characterize the crater that is formed by a charge of explosive of mass m_B at a depth h_0, if $h_0 < r_{Rm}$.

If $r_0 \leq r_{Rm}$, then $\cos \varphi_0$ is calculated from the relation (23), while if $r_0 > r_{Rm}$, then $\cos \varphi_0 = \cos \varphi_{0l}$ is calculated from (28).

We can also calculate $R_0 = h_0 \tan \varphi_0$ and $M = (\pi/3)\rho h_0^3 \times \tan^2 \varphi_0$.

The quantity that determines the case we will have in the formation of the crater is the dimensionless number

$$\eta_\varphi = \frac{\cos^3 \varphi_0}{\cos^3 \varphi_{l0}} = \frac{Agh_0}{4Q}. \tag{29}$$

If $\eta_\varphi > 1$, then $\varphi_{0l} > \varphi_0$, and we have the case when $r_0 < r_{Rm}$ (here the gravitational force is predominant); if $\eta_\varphi < 1$, then $\varphi_{0l} < \varphi_0$, and we have the case when $r_0 > r_{Rm}$ (with the radius of the disintegrated medium playing the most important part).

If the gravitational field is absent, then we will always have the second case.

The same situation will occur when a gravitational field is present, if a high-calorific explosive is used. Only in the case of a very "yielding" medium, when the value of A is large even for large values of Q, will we have the first case.

In the presence of a very strong gravitational field, either case can occur, depending on the relation between g, Q, and A.

We will now carry out a final determination of the relation between the mass of the medium "thrown out" from the crater by the explosive and the mass of explosive, its calorific content (or the energy of the explosion), the initial depth of the explosion center, and we will also investigate the optimal depth, for which the ejected mass is the largest possible, in both the cases $\eta_\varphi > 1$ and $\eta_\varphi < 1$.

For $\eta_\varphi > 1$ we will have

$$M = \frac{\pi}{3} \rho h_0^3 \tan^2 \varphi_0, \qquad (30)$$

where $1 \geq \cos \varphi_0 \geq (\pi \rho g h_0^4 / 3 E_B)^{1/3}$.

If $\cos^3 \varphi_0 = (\pi \rho g h_0^4 / 3 E_B)$, then

$$M = \frac{\pi}{3} \rho h_0^3 \left[\left(\frac{3 E_B}{\pi \rho g h_0^4} \right)^{2/3} - 1 \right]. \qquad (31)$$

It is evident that the mass of ejected material will be a maximum for a depth $\bar{h}_0$ if

$$\bar{h}_0 = \left(\frac{E_B}{9 \pi \rho g} \right)^{\frac{1}{4}} = \frac{h_{0m}}{27^{\frac{1}{4}}} = \frac{3^{\frac{1}{4}} h_{0m}}{3}$$

(the inequality $(\pi \rho g h_0^4 / 3 E_B) < 1$ is obviously satisfied in this case).

If the value $\bar{h}_0$ is substituted in (23), (26), and (31), we obtain the results

$$\cos \bar{\varphi}_0 = \frac{1}{3}; \quad \bar{\varphi}_0 = 70°; \quad \bar{R}_0 = 2\sqrt{2} \bar{h}_0 = 2\sqrt{2} \left(\frac{E_B}{9 \pi \rho g} \right)^{\frac{1}{4}}; \qquad (32)$$

$$M = \frac{8\pi}{3} \rho \left(\frac{E_B}{9 \pi \rho g} \right)^{\frac{3}{4}}.$$

For $\eta_\varphi < 1$ we will have

$$M = \frac{\pi}{3}\rho r_0^3 \sin^2\varphi_0 \cos\varphi_0 = \frac{\pi}{3}\rho h_0^3 \tan^2\varphi_0, \qquad (33)$$

where $1 \geq \cos\varphi_0 \geq (h_0/r_{Rm})$. If $\cos\varphi_0 = (h_0/r_{Rm})$, then

$$M = \frac{\pi}{3}\rho r_{Rm}^2 h_0 \left(1 - \frac{h_0^2}{r_{Rm}^2}\right). \qquad (34)$$

It is obvious that as in the previous case, the mass of ejected material for a depth $\bar{h}_0$ will have a maximum value. This depth will be $\bar{h}_0 = (\sqrt{3}/3)r_{Rm}$. Here $\tan\bar{\varphi}_0 = \sqrt{2}$,

$$\cos\bar{\varphi}_0 = \sqrt{\frac{1}{3}};\ \sin\bar{\varphi}_0 = \sqrt{\frac{2}{3}};\ \bar{\varphi}_0 = 55°;$$

$$M = \frac{2\sqrt{3}\,\pi}{27}\rho r_{Rm}^3 = \frac{2\pi}{9}\rho \bar{h}_0^3. \qquad (35)$$

It can be seen that for different values of the parameter η_φ we will obtain different optimum cone angles ($\bar{\varphi}_0 \approx 70°$ for $\eta_\varphi > 1$ and $\bar{\varphi}_0 \approx 55°$ for $\eta_\varphi < 1$). We now relate the results obtained by pure theory and those obtained from experimental data.

It is known that for some depths (which are not always optimal in the sense of yielding the greatest mass of ejected material), the radius of the crater is

$$R_e = \lambda m_B^{\frac{1}{3}}, \qquad (36)$$

where λ is a coefficient obtained from experiment.

If we set $R_e = R_{0m} = R_0$, we can write (36) in the form

$$R_{0m}^3 = \lambda^3 m_B. \qquad (37)$$

It is also evident that

$$M_{Rm} = \frac{4}{3}\pi\rho R_{0m}^3 = \frac{4}{3}\pi\rho\lambda^3 m_B. \qquad (38)$$

If we compare (21) and (38), we find that

$$A = \frac{4}{3}\pi\rho\lambda^3. \qquad (39)$$

On the other hand,

$$A = \frac{p}{p_0}\bar{\bar{\alpha}}\bar{A}\left(\frac{p_a}{p_a}\right)^{\frac{1}{k^*}},$$

and so

$$\frac{4}{3}\pi\rho_0 \lambda^3 = \overline{\overline{\alpha}}\overline{A}\left(\frac{p_a}{\overline{p}_a}\right)^{\frac{1}{k^*}}.\tag{40}$$

These relations establish the relation between the empirical parameters $\overline{\overline{\alpha}}$, $\overline{A}$, A, and λ.

Since the value of $\overline{A}$ is fairly well known and λ can be determined by experiment, then we can obtain A from (39), and from (40) we can obtain the very important coefficient $\overline{\overline{\alpha}}$. We mention that $\overline{\overline{\alpha}} = f(p_a/\overline{p}_a)$. We will now find the total (scalar) momentum J_t of the material ejected by the explosion.

It is obvious that if the term $(gh_0/2)$ in (13) is neglected, then

$$dJ_t = a\,dM = a_0\,dM = \sqrt{\frac{2\pi\rho h_0^3 E_B}{3\cos^3\varphi}}\,d\varphi\cdot\sin\varphi,\tag{41}$$

and so if we integrate with respect to φ from 0 to φ_0 (setting $\varphi_0 = \varphi_{0l}$), we find that

$$J_t = 2\sqrt{\frac{2}{3}\pi\rho h_0^3 E_B}\left(\frac{1-\sqrt{\cos\varphi_0}}{\sqrt{\cos\varphi_0}}\right) = \overline{\theta}\sqrt{2ME_B},\tag{42}$$

where

$$\overline{\theta} = \frac{2}{\sqrt{\cos\varphi_0}}\frac{1-\sqrt{\cos\varphi_0}}{\tan\varphi_0}.\tag{43}$$

It is reasonable to introduce, instead of the total energy of the explosion E_B, that part of the energy E_{φ_0} which is dispersed from inside the cone with vertical angle $2\varphi_0$.

Starting from (7), we find that

$$E_{\varphi_0} = E_B\frac{1-\cos\varphi_0}{2},\tag{44}$$

and (42) takes the form

$$J_t = \theta\sqrt{2ME_{\varphi_0}},\tag{45}$$

where

$$\theta = \frac{2\sqrt{2\cos\varphi_0}}{\sqrt{1+\cos\varphi_0}\,[1+\sqrt{\cos\varphi_0}]}.\tag{46}$$

For $\varphi_0 = 0$ we have $\theta = 1$ and $(E_{\varphi_0}/E_B) = 0$, while for $\varphi_0 = 60°$ we have $\theta \approx 0.95$, $E_{\varphi_0}/E_B = 1/4$.

We now find the projection J_z of the momentum on the z axis, perpendicular to the earth's surface AA'. It is clear that

$$dJ_z = a_0 \cos\varphi\, dM = \sqrt{\frac{2\pi\rho h_0^3 E_B}{3\cos\varphi}} \sin\varphi \cdot d\varphi. \qquad (47)$$

Hence

$$J_z = 2\sqrt{\frac{2}{3}\pi\rho h_0^3 E_B}\,[1 - \sqrt{\cos\varphi_0}] = \bar{\theta}_1\sqrt{2ME_B} \qquad (48)$$

where

$$\bar{\theta}_1 = \frac{2(1-\sqrt{\cos\varphi_0})}{\tan\varphi_0}. \qquad (49)$$

Further, if E_B is replaced by E_{φ_0}, we find that

$$J_z = \theta_1\sqrt{2ME_{\varphi_0}}, \qquad (50)$$

where

$$\theta_1 = \frac{2\sqrt{2}\cos\varphi_0}{(1+\sqrt{\cos\varphi_0})\sqrt{1+\cos\varphi_0}}. \qquad (51)$$

Therefore for $\varphi_0 = 0$ we have $\theta_1 = 1$, and for $\varphi_0 = 60°$ we have $\theta_1 \approx 2/3$.

We can write $\theta_1 = \theta\sqrt{\cos\varphi_0}$, where the coefficient θ is a measure of the distribution of velocity for various angles, depending on the amount of ejected mass. If there is no variation ($\theta = 1$), then

$$J_{t_0} = \sqrt{2ME_{\varphi_0}}. \qquad (52)$$

Finally we can now write

$$J_t = \theta J_{t_0};\quad J_z = J_t\sqrt{\cos\varphi_0} = \theta\sqrt{2ME_{\varphi_0}\cos\varphi_0} = \theta J_{t_0}\sqrt{\cos\varphi_0}. \qquad (53)$$

In practical calculations, however, it is more convenient to use the relations (42) and (48).

As an example, it follows from these relations that for $\varphi_0 = 60°$ we have

$$J_t \approx 0.48\sqrt{2ME_B};\quad J_z \approx \tfrac{1}{3}\sqrt{2ME_B} = \tfrac{2}{3}\sqrt{2ME_{\varphi_0}}.$$

In general, without any previously assumed law for the

distribution of velocity according to mass, we can write

$$J_t = \int_0^{\varphi_0} a(\varphi) dM(\varphi). \tag{54}$$

$$J_z = \int_0^{\varphi_0} \cos\varphi\, a(\varphi) dM(\varphi), \tag{55}$$

where $a = a[M(\varphi)] = a(\varphi)$ gives the relation between velocity and mass.

IMPACT OF METEORITES WITH HARD SURFACES

The General Theory of Impact with High Velocities and Deep Penetration

It is known that the phenomena occuring during the impact of hard bodies with velocities exceeding a few kilometers per second are similar to those accompanying an explosion. For velocities greater than 3-5 km/sec, the crystalline structure of the meteorite and of some of the medium struck by the meteorite is destroyed, and in the true sense of the word an explosion has occurred.[2,3]

In the present work we will first of all refine certain results that have been obtained before, and then develop a more accurate theory of the relevant phenomena.

We will consider which processes may be observed during impacts at velocities more than a few kilometers per second.

A strong shock wave is propagated from the position of the impact, and the initial parameters for this shock can be determined from the relations[1]

$$p_{sh} = \frac{\rho_a u_0^2}{(1-\alpha_2)\left(1+\sqrt{\frac{\rho_a}{\rho_H}\frac{1-\alpha_1}{1-\alpha_2}}\right)^2}, \tag{1}$$

$$\frac{u_{sh}}{u_0} = \frac{1}{1+\sqrt{\frac{\rho_a}{\rho_H}\frac{1-\alpha_1}{1-\alpha_2}}}, \tag{2}$$

where $\alpha_1 = (\rho_H/\rho_{1y})$, $\alpha_2 = (\rho_a/\rho_{2y})$, ρ_a and ρ_H are the initial densities of the first and second medium, ρ_{1y} and ρ_{2y} are the densities of these media at the front of the shock wave, p_{sh} is the pressure at the shock front, u_{sh} the velocity of motion of the boundary between the media, and u_0 the velocity of impact.

The pressure and density are related by the equation of state. The propagation velocities of the shock waves in the two media are given by the relations

$$\frac{D_{1y}}{u_0} = 1 + \frac{u_{sh}}{(1-\alpha_1)u_0} \sqrt{\frac{\rho(1-\alpha_1)}{\rho_H(1-\alpha_2)}}; \qquad \frac{D_{2y}}{u_0} = \frac{u_{sh}}{(1-\alpha_2)u_0}. \tag{3}$$

The relation (1) can be used to find the minimum pressure necessary to vaporize or melt, or simply to disintegrate the medium "explosively," if we know the corresponding values of $u_0 = u_k^*$.

We will assume that for an impact velocity such that

$$u_0 \geqslant \sqrt{2\varepsilon_k}, \tag{4}$$

where ε_k is the energy density of the crystal lattice (the energy necessary to convert 1 g of the medium into a liquid, including the latent heat of fusion), the solid lattice is destroyed and the medium is converted into a type of liquid. In this case the kinetic (ordered) energy of the impact is transformed into the random motion of the particles of the liquid, and this will create an internal pressure; the liquid will expand and explosively disintegrate the surrounding medium. The value of ε_k for metallic bodies is, on the average, less than for rocks (silicates).

However, even for

$$u_0 = \bar{u}_k = \sqrt{2\bar{\varepsilon}_k} \tag{5}$$

($\bar{\varepsilon}_k$ is the energy necessary to pulverize the medium), when there is insufficient energy to convert the solid phase into a liquid, the medium will be converted into very small, hard, elastic particles which, on the whole, have properties similar to a quasi-gas, and which also produce explosive phenomena (mechanical explosions, when the ordered energy of the colliding bodies is converted into the random energy of their particles). For metallic bodies the value of $\bar{u}_k$ is, on the average, greater than for rocks.

For impact velocities

$$u_0 = u_{ki} \geqslant \sqrt{2\varepsilon_i}, \qquad (6)$$

where ε_i is the energy necessary for the vaporization of 1 g of the medium, including the heat of vaporization, the material of the medium will be vaporized. For metallic bodies ε_i is usually less than for rocks.

Values of $\bar{\varepsilon}_k$, ε_k, and ε_i (erg/g) are given in Table 1.

TABLE 1

Medium	$\bar{\varepsilon}_k$	ε_k	ε_i
Sand	$5 \cdot 10^7$	$5 \cdot 10^9$	10^{11}
Clay	10^8	$5 \cdot 10^9$	10^{11}
Granite	10^9	$7 \cdot 10^9$	$2 \cdot 10^{11}$
Aluminum	10^9	$4 \cdot 10^9$	10^{11}
Iron	10^9	$3 \cdot 10^9$	$7 \cdot 10^{10}$

It can be assumed that the mean dimensions of the "particles" of the quasi-gas are the same as those of the so-called nuclei of the solid phase that are present in the transition to this phase from the liquid phase. The dimensions of these nuclei are a few orders greater than the dimensions of the molecules.

During the expansion of the gas, its temperature will fall according to the relation

$$\left(\frac{V_H}{V}\right)^k = \frac{p}{p_H} = \left(\frac{T}{T_H}\right)^{\frac{k}{\kappa-1}}, \qquad (7)$$

where V_H, p_H, T_H are the initial specific volume, pressure, and temperature at the front of the shock wave.

The condensation process is described by the Clausius-Clapeyron equation

$$\frac{dp}{dT} = \frac{Q_i^*}{T(V_s - V.)}, \qquad (8)$$

where Q_i^* is the heat of vaporization, V_G is the specific volume of the gas, and V_S is the specific volume of the solid phase.

Since $V \ll V_G$ and $V_G = (RT/p)$, then $(dp/dT) = (Q_i^* p/RT^2)$, and so

$$\frac{p}{p_a} = e^{\frac{Q_i^*}{RT_0}\left(1-\frac{T_0}{T}\right)}, \qquad (9)$$

where T_0 is the condensation temperature for atmospheric pressure p_a.

Since $Q_i^* = c_V T^*$, where T^* is the effective temperature of vaporization, then

$$\frac{p}{p_a} = e^{\frac{T^*}{(k-1)T_0}\left(1-\frac{T_0}{T}\right)}. \qquad (10)$$

Using (7) and (10), we arrive at a relation that gives the conditions for condensation

$$\frac{T}{T_0} = \left(\frac{p_a}{p_H}\right)^{\frac{k-1}{k}} \frac{T_H}{T_0} e^{\frac{T^*}{kT_0}\left(1-\frac{T_0}{T}\right)}. \qquad (11)$$

It is evident that this equation has a solution for $\overline{T}_0 > T_0$, i.e., for $(\overline{T}_0/T_0) = 1 + \Delta$ where $\Delta < 1$, and $\overline{p} \ll p_H$.

The process of condensation thus begins at relatively low pressures (in comparison with the initial pressures), when the process of expansion has practically finished, and it can no longer influence the expansion and, in particular, it does not appreciably alter the reactive back-impulse (the pressure of the emerging explosion products).

Therefore at collision velocities of the order of 10 km/sec and above, the whole of the impacting body and some material from the larger body are converted into gas. The energy balance here is the following: part of the kinetic energy goes into the conversion of the body into gas, including the latent heat of formation of the gas (vapor formation). Then, after the expansion, part of this heat returns again to the medium.

In connection with the dimensions of the condensing droplets, the part of the latent energy of the gas returning to the medium that becomes kinetic energy is greater for larger rocks, and smaller for smaller rocks. The total surface energy of the particles approximates the volume energy only when scattering of the particles is molecular.

Even for particles having dimensions one or two orders larger than molecular, their surface energy is much smaller than the volume energy. During the expansion of the medium, the dimensions of the condensing particles significantly exceed molecular dimensions.[4] Therefore practically all the energy of vaporization returns to the medium. The energy used in the evaporation thus satisfies the relation

$$E_i = E_s + E_{ret},$$

where E_i is the latent heat of evaporation (and fusion), E_s the total surface energy of the particles, and E_{ret} is the energy returned to the medium. For $E_s \ll E_i$ we have $E_{ret} \approx E_i$.

We will now consider to what depth h the body (a meteorite) will penetrate on impact at a high velocity.

Since the resistance is[1,5]

$$F = M_0 \frac{du}{dt} = -\frac{c_x}{2} S\rho u^2 = M_0 u \frac{du}{dx},$$

then

$$u = u_0 \exp\left(-\frac{c_x}{2} \frac{\rho S x}{M_0}\right) = u_0 \exp\left(-\frac{c_x}{2} \frac{\rho x}{\delta l}\right), \quad (12)$$

where u_0 and u are the initial and current velocities respectively, M_0 is the mass of the meteorite, x is the distance travelled in the course of the impact, δ and ρ are the densities of the meteorite and the medium respectively, l is the mean dimension of the meteorite, S the area of its middle section, and c_x a dimensionless "streamline" coefficient.

In the case being considered, we can always assume that $c_x \approx 2$. Therefore

$$\frac{u}{u_0} = e^{-\frac{\rho x}{\delta l}}. \quad (13)$$

It can be assumed that for velocities of 3-5 km/sec, the penetration practically ends when the explosion phase begins. The "explosion" starts at the instant of impact, but we will assume that the explosion phase begins after the end of the penetration. If we set

$$\frac{u}{u_0} = e^{-2} \tag{14}$$

(which for $u_0 = 30$ km/sec yields $u \approx 4$ km/sec), we obtain $(\rho x / \delta l) = 2$, and so

$$x = \frac{2\delta l}{\rho}. \tag{15}$$

For $\delta \approx \rho$ we have $x = 2l$, i.e., the depth of penetration is of the order of the radius of the meteorite (r^*) [for a sphere $l = (r^*/3)$ and $x = (2/3)r^*$].

After penetration to the indicated depth, i.e., for a velocity of less than 3-5 km/sec, the impacting body (the meteorite) is already practically vaporized. The gases formed begin to expand, this increases the middle area S (the quantity l decreases), and this leads to a more rapid deceleration of the impacting body. We must therefore assume that the depth of penetration will rarely (depending on the impact velocity) exceed 5-10 $\bar{r}^*$, where the effective radius of the body is

$$\bar{r}^* = \sqrt[3]{\frac{M_0}{\delta}} \approx \frac{M_0}{S\delta}. \tag{16}$$

We must investigate in more detail how the depth of penetration depends on the impact velocity.

For low impact velocities of hard bodies (rocks or metals), up to 1 km/sec, the radius of the hollow $R \sim \bar{r}^*$.

The distance to which the body penetrates is

$$x \approx \frac{M_0 u_0^2}{2\rho (\bar{r}^*)^2 Q_{co}} \approx \frac{\delta \cdot \bar{r}^* u_0^2}{\rho Q_{co}}, \tag{17}$$

that is

$$\frac{x}{\bar{r}^*} \approx \frac{\delta \cdot \bar{u}_0^2}{p_{co}} = \frac{\bar{p}_{im}}{p_{co}},$$

where $\bar{p}_{im}$ is the pressure generated at the impact, Q_{co} is the density of the cohesive energy, and $\bar{p}_{co}$ is the pressure characterizing the cohesive properties of the medium.

The depth of penetration is

$$h = x \cos z, \tag{18}$$

where z is the complement of the declination of the meteorite (the angle between its trajectory and the vertical).

If the impact of the meteorite occurs at an angle $90° - z$ to a plane surface, then for large values of the angle z, the normal projection of the velocity of the meteorite $u_p = u_0 \cos z$ can be rather small, and then the explosive phenomena may not be observed.

Since an explosion is produced for $(u_0^2/2) \geq \varepsilon_k = (u_k^*)^2/2$, it is essential that $u_0 \cos z \geq u_k^*$. Since u_k^* is approximately equal to the velocity of sound c_k^* in the medium, we may say that explosive phenomena will be observed for $u_0 \cos z > c_k^*$.

For larger velocities, when the pulverization of the material begins, the middle section, and consequently the specific resistance (proportional to S/M_0) increases; here the depth increases more slowly than u_0^2, and part of the kinetic energy of the impacting body is dissipated in motion of the medium in the lateral directions, while the quantity of material "thrown out" increases as before, and is proportional to the kinetic energy of the impact.

For still higher velocities, when explosive phenomena commence, the lateral ejection becomes more important in comparison with the motion in the direction of the impact trajectory, since the gas expands in all directions, and the relation (12) remains valid as long as $u_0 > u_k^*$, where u_k^* is the velocity at which the explosive phenomena cease.

Thus

$$h = \frac{2M_0}{c \cdot S\rho} \ln \frac{u_0}{u_k^*}. \tag{19}$$

Meteorite Explosions in Infinite Media

We consider first of all "meteorite explosions" in an unbounded medium. It is clear that we may pose the problem in the following way.

Let an amount of energy $E_0 = (M_0 u_0^2/2)$ be instantaneously released in a volume equal to that of the meteorite. The mass $m_{\overline{K}}$ inside which will occur the "evaporation" of the material can be obtained from the relation

$$M_0 \varepsilon_k + m_{\bar{k}} \varepsilon_{\bar{k}} = \eta \frac{M_0 u_0^2}{2} = \eta E_0, \qquad (1)$$

where η is a coefficient giving the utilizable available energy ($\eta < 1$),

$$\varepsilon_k = \frac{u_k^2}{2}, \quad \varepsilon_{\bar{k}} = \frac{u_{\bar{k}}^2}{2} \qquad (2)$$

are the quantities of energy necessary for the evaporation of the material of the meteorite and the medium respectively, and u_k and $u_{\bar{k}}$ are the limiting velocities necessary for the "evaporation" of a unit mass of the material of the meteorite and the medium.

If $u_0 > u_k \gtrless u_{\bar{k}}$, then there is no appreciable difference between the quantities ε_k and $\varepsilon_{\bar{k}}$ for the impacting body and the body receiving the impact.

Then (1) takes the form

$$M_0 + m_k = \frac{\eta M_0 u_0^2}{u_k^2}. \qquad (3)$$

Since the quantity ε_k^* is close to the energy density given out in the explosion of a condensed (solid or liquid) explosive ($\varepsilon_k^* \approx Q$), and since, for the effect of an explosion on the surface or inside various bodies, experimental relations are known that relate the mass of explosive (m_B) and the energy density (Q) with the radius, depth, and shape of the crater in various media, then it is reasonable in relations (1) and (3) to replace ε_k and $\varepsilon_{\bar{k}}$ by Q. (For typical explosions $Q \approx 1\text{-}1.5 \text{ Cal/g}$, while ε_i for iron, aluminum, and granite has the values 2, 2.5, and 5 Cal/g).

We can therefore write

$$m_B Q = \eta E_0 = \frac{\eta M_0 u_0^2}{2} = E_B. \qquad (4)$$

It follows that

$$m_B = \frac{\eta M_0 u_0^2}{2Q}. \qquad (5)$$

After the gas has expanded to a stage where the energy density is approximately equal to Q, the further stage of expansion can be compared to the expansion of the products of the explosion of a charge of explosive, but with a density different from that of the usual explosive.

The density of rock is actually around $4\,\mathrm{g/cm^3}$, that of iron $8\,\mathrm{g/cm^3}$, while that of the standard explosive is $1.6\,\mathrm{g/cm^3}$.

Thus the volume energy density in the case considered will be a few times greater than in the case of the commoner explosives.

We will now make certain aspects of the problem more precise.

During the impact and explosion a shock wave is formed, which will be propagated through the medium. Since at the shock front part of the energy will be lost to the "disintegration" of the medium, the law of conservation of energy for the front of this shock wave must contain a term that takes this loss into account.

For a strong shock these conditions have the form[1] (E_H is is the energy at the wave front)

$$E_H = \frac{p_H}{2}(V_0 - V_H) - \varepsilon_k^*, \qquad (6)$$

$$u_H^2 = p_H(V_0 - V_H), \qquad (7)$$

$$D_y^2 = \frac{V_0^2 p_H}{V_0 - V_H} = \left(\frac{V_0 u_H}{V_0 - V_H}\right)^2. \qquad (8)$$

Here the relation $E_H = f(p_H, V_H)$ depends on the equation of state of the medium and in a certain sense on the initial velocity of impact, since for different pressures the equation of state can be different.

From (6) and (7) we have

$$E_H = \frac{u_H^2}{2} - \varepsilon_k^*. \qquad (9)$$

In these relations ε_k^* must be understood to be the energy consumed in any breaking-up or disintegration of the crystal lattice. It is clear that this process of disintegration of the lattice will continue until at the shock front

$$E_H + \frac{u_H^2}{2} \geqslant \varepsilon_k^*. \qquad (10)$$

It follows from (9) and (10) that the process of disintegration will take place for

$$\frac{u_H^2}{2} \geqslant \varepsilon_k^*. \tag{11}$$

We will introduce the notation $u_k^* = \sqrt{2\varepsilon_k^*}$ and investigate more thoroughly the condition $u_0 \geq u_k^*$ formulated above.

Since it follows from (11) that

$$u_H \geqslant u_k^*, \tag{12}$$

and $u_H \approx (u_0/2)$, then $u_0 \geq 2u_k^*$ or

$$\frac{u_0^2}{2} \geqslant 2u_k^{*2} = 4\varepsilon_k^*. \tag{13}$$

After the impact, in the first short interval of time when the regime of motion and "streamline flow" has not yet been established, "vaporization" will begin for $u_0 \geq u_k^*$ and then proceed rather rapidly; during the time interval of length

$$\tau = \frac{2l}{u_0} \tag{14}$$

a regime will be established such that the "vaporization" will occur for the condition

$$u_0 \geqslant 2u_k^*.$$

We will again increase the precision in the definition of ε_k^*. For vaporization $\varepsilon_k^* = \varepsilon_i$, for fusion $\varepsilon_k^* = \varepsilon_k$, and for simple "pulverization" (dispersion) of the medium $\varepsilon_k^* = \overline{\varepsilon}_k$.

In the first two cases in the process of expansion, the heat consumed in the latent heat of vaporization and fusion returns to the medium, except for the small part of the medium that expanded during its ejection from the crater into the atmosphere or into outer space. These losses can in general, however, be neglected, especially as the initial energy in this case will be relatively greater than in the case of simple dispersion of the medium.

Finally, therefore, the irreversibly lost energy will be $\varepsilon_k^* = \overline{\varepsilon}_k$. It can be assumed, as was done before, that $\varepsilon_k^* \approx Q$ for various media. We can therefore introduce, as in the case of the investigation of atomic explosions, the TNT equivalent, starting from the law of conservation of energy, i.e., the kinetic energy (E_0) of the impacting body less any losses (the efficiency can be used) can be compared to the energy of a charge of explosive (TNT) with a mass m_B and a calorific content Q, if we set $m_B Q = \eta E_0$.

It can be assumed for simplicity that the detonation process in a dense medium is instantaneous. The "disintegration" process will customarily be called the process of "vaporization."

For a detonation the energy density at the detonation front is

$$\varepsilon_{Hd} = E_{Hd} + \frac{u_{Hd}^2}{2}, \qquad (15)$$

where

$$E_{Hd} = \frac{p_{Hd}(V_{0d} - V_{Hd})}{2} = \frac{u_{Hd}^2}{2} \qquad (16)$$

is the potential energy; thus

$$\varepsilon_{Hd} = u_{Hd}^2 = 2Q. \qquad (17)$$

For instantaneous detonation, the mean energy density will be

$$\bar{\varepsilon}_d = \frac{\overline{u_{Hd}^2}}{2} = Q. \qquad (18)$$

In the second case, when $\bar{\varepsilon}_d = Q$, the problem is reduced to that of comparing the action of the impulse to that of the explosion of a charge of mass $2m_B$ and a calorific content $Q/2$, assuming instantaneous detonation.

In the first case the problem reduces approximately to that of a real detonation of a charge with mass m_B and calorific content Q, if by a real detonation is understood the case when the pressure at the front of the detonation wave is twice the mean pressure, and the energy density at the front is $u_H^2 = 2Q$.

The two cases are quite similar, as follows from the theory of explosions.

The first case yields to a simpler analysis, and we shall examine it in detail later on.

In what follows, the equivalent mass m_B of explosive with calorific content Q will be determined by means of equation (5) of this section.

The Explosion of a Meteorite in a Bounded Medium

We will first examine how to determine the maximum depth of the crater formed by the impact of a meteorite with a surface of any material.

It is evident that this depth depends on the depth of penetration and also on the radius of the mass of medium vaporized. The fact must also be taken into account, that after the con-

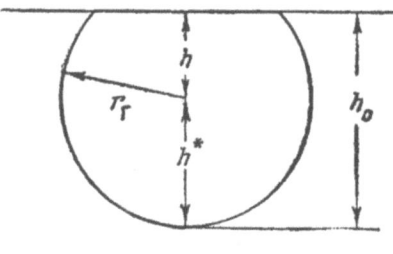

FIG. 4

clusion of the vaporization process, a further simple disintegration will take place at the front of the attenuating shock wave, and that a part of the resulting broken-up material will be ejected, thus further deepening the crater. For metals, however, this addition to the depth will be less than that of the evaporated zone; for soil and rock it will be greater.

The total depth of the crater will thus be (see Fig. 4)

$$h_0 = h + h^*, \qquad (1)$$

where

$$h = \left(\frac{2M_0}{c_x \cdot p \cdot S} \ln \frac{u_0}{u_x^*}\right) \cos z, \quad h^* = r_g = \sqrt[3]{A_0 \frac{3m_B}{4\pi p}}, \quad A_0 > 1, \qquad (2)$$

(r_g is the radius of the "vaporization"–"disintegration" zone).

Since $m_B = (\eta M_0 u_0^2 / 2Q)$, then

$$h_0 = \frac{2M_0}{c_x \cdot p \cdot S} \cos z \ln \frac{u_0}{u_x^*} + \sqrt[3]{A_0 \frac{3\eta M_0 u_0^2}{8\pi p\, Q}} \qquad (3)$$

($u_x^* \approx u_k^*$). (Here and in what follows we take $Q = (u_k^{*2}/2) = \varepsilon_k^*$ to be the density of energy necessary for breaking the bonds of the crystal lattice or for the fine pulverization of the rock.)

Since $M_0 = \delta \cdot l \cdot S$, we can write

$$\frac{M_0}{\rho \cdot S} = l \cdot \frac{\delta}{\rho} = \overline{\alpha}^* \frac{\delta}{\rho} \sqrt[3]{\frac{3M_0}{4\pi\delta}}, \qquad (4)$$

where $\overline{\alpha}^*$ is a form coefficient for the meteorite; for a sphere, for example, $\overline{\alpha}^* = 1$.

It is now convenient to write (3) in the form

$$h_0 = \sqrt[3]{\frac{3\overline{\eta}\eta M_0 u_0^2}{8\pi\rho Q}}, \qquad (5)$$

where

$$\frac{1}{\eta^{\frac{1}{3}}} = A_0^{\frac{1}{3}} + \Delta = A_0^{\frac{1}{3}} + \cos z \sqrt[3]{\frac{2Q\delta^2}{\eta\rho^2 u_0^2}} \cdot \frac{2}{c_x} \cdot \overline{\alpha}^* \ln \frac{u_0}{u_x^*}.$$

The mass of material ejected from the crater (we will assume that the crater is cone-shaped) is given by the formula

$$M = \frac{\pi}{3} \rho h_0^3 \tan^2 \varphi_0 = \frac{\overline{\eta}\eta M_0 u_0^2}{8Q} \tan^2 \varphi_0. \qquad (6)$$

We now determine the normal projection of the explosion impulse (see (48), page 306).

$$J_z = \overline{\theta}_1 \sqrt{2ME_\text{B}}. \qquad (7)$$

Since

$$E_\text{B} = \eta E_0, \quad M = \frac{\overline{\eta}\eta E_0}{4Q} \tan^2 \varphi_0,$$

we have

$$J_z = \sqrt{\frac{2\overline{\eta}}{Q}} \cdot \eta E_0 (1 - \sqrt{\cos \varphi_0}). \qquad (8)$$

If $\eta_\varphi = (Agh_0/4Q) \leq 1$ then

$$\cos \varphi_0 = \frac{h_0}{r_{Rm}} = \left(\frac{\overline{\eta}}{A}\right)^{\frac{1}{3}}, \qquad (9)$$

while if $\eta_\varphi \geq 1$ we have

$$\cos \varphi_0 = \frac{h_0}{r_0} = \left(\frac{\pi\rho g h_0^4}{3E_\text{B}}\right)^{\frac{1}{3}} = \left(\frac{\overline{\eta}gh_0}{4Q}\right)^{\frac{1}{3}} = \left(\frac{\overline{\eta}g}{4Q}\right)^{\frac{1}{3}} \cdot \left(\frac{3\overline{\eta}\eta M_0 u_0^2}{8\pi\rho Q}\right)^{\frac{1}{9}}. \qquad (10)$$

We now estimate the size of the dimensionless quantities A, $\overline{\eta}$, and η by using these relations.

As we pointed out above, the value of η is a little less than

$$\eta^{-\frac{1}{3}} = A_0^{\frac{1}{3}} + \frac{2\bar{\alpha}^*}{c_x}\cos z \sqrt[3]{\frac{2\delta^2 Q}{\eta p^2 u_0^2}} \cdot \ln\frac{u_0}{u_x^*} ;$$

for $u_0 = u_X^*$ we obtain $\bar{\eta} = A_0$. If $z = 0$, $c_x = 2$, $\bar{\alpha}^* = 1$, $\rho = \delta$ we will have

$$\bar{\eta}^{-\frac{1}{3}} = A_0^{\frac{1}{3}} + \left(\ln\frac{u_0}{u_x^*}\right)\sqrt[3]{\frac{2Q}{\eta u_0^2}} = A_0^{\frac{1}{3}} + \left(\frac{u_x^*}{\sqrt{\eta u_0}}\right)^{\frac{2}{3}} \ln\frac{u_0}{u_x^*} .$$

If we let $u_0 = e^3 u_X^* \approx 20 u_X^*$, $\eta \approx 1$, then $\bar{\eta}^{1/3} \approx 1 + (3/7.5)$ ≈ 1.4 ($A_0 \approx 1$). The quantity $\bar{\eta}^{1/3} = \bar{\eta}^{1/3}_{\max} \approx 1 + \bar{e}^{2/3} \approx 1.5$ for $u_0 = e u_X^*$. The value of A_0 depends on the properties of the medium; for metals it is closer to unity than for more yielding materials. The range of variation of $\bar{\eta}$ is not very large.

We now calculate A. Since

$$A = \bar{\alpha}\bar{A}\frac{p}{p_0}\left(\frac{p_a}{\bar{p}_a}\right)^{\frac{1}{k^*}} = \frac{4}{3}\pi\rho\lambda^3,$$

then by using the experimentally determined values of λ, it is easy to calculate A and the quantity $\bar{\alpha}(p_a/\bar{p}_a)^{1/k^*}$.

In Table 2 we give the values of ρ, λ, A and other parameters for various media, where we have taken $p_a = 1$ kg/cm^2, $\bar{A} = 1000$, $k^* = 7/5$, and $\rho_0 = 1.6$.

In the columns for λ and A, the first sub-column is for

TABLE 2

Medium	ρ	λ		A		$A_0^{\frac{1}{3}}\left(\frac{p_a}{\bar{p}_a}\right)^{\frac{1}{k^*}}$	$\bar{\alpha}\left(\frac{p_a}{\bar{p}_a}\right)^{\frac{1}{k^*}}$	$\left(\frac{=p_a}{\alpha\frac{}{p_a}}\right)^{\frac{1}{k^*}}$
Sand	2—3	10	100	10^4	10^7	8—5	0.64	6.4	$6.4 \cdot 10^2$
Clay	3.5	8	50	$7 \cdot 10^3$	$1.7 \cdot 10^6$	5—4	0.4	3.3	800
Granite	4	5	10	$2 \cdot 10^3$	$1.6 \cdot 10^4$	3—2	0.05	0.8	6.4
Aluminum	2.6	2	10	80	10^4	1.5	0.033	$5 \cdot 10^{-2}$	6.25
Iron	8	1	5	30	$4 \cdot 10^3$	1	$6.4 \cdot 10^{-3}$	$6.4 \cdot 10^{-3}$	0.8

the ejection crater, the second for the zone of non-linear deformation.

The rate of ejection of the crater material is given by the relation

$$a_0 = \left(\frac{3E_B \cos^3\varphi}{2\pi\rho \cdot h_0^3} - \frac{gh_0}{2}\right)^{\frac{1}{2}} = \left[\frac{gh_0}{2}\left(\frac{3E_B \cos^3\varphi}{\pi\rho \cdot g \cdot h_0^4} - 1\right)\right]^{\frac{1}{2}}. \quad (11)$$

If we substitute the values of h_0 and E_B in this formula, we arrive at the expression

$$a_0 = \left[\frac{2Q}{\bar{\eta}} \cos^3 \varphi - \frac{g}{2}\left(\frac{3\bar{\eta}\bar{\eta}_i M_0 u_0^2}{8\pi p Q}\right)^{\frac{1}{3}}\right]^{\frac{1}{2}}. \tag{12}$$

For $\varphi = 0$ we have

$$a_0 = a_{0\max} = \sqrt{2Q}\left[\frac{1}{\bar{\eta}} - \frac{g}{4Q}\left(\frac{3\bar{\eta}\bar{\eta}_i M_0 u_0^2}{8\pi p Q}\right)^{\frac{1}{3}}\right]^{\frac{1}{2}}. \tag{13}$$

The mass originating in the region bounded by the cones with base radii $R + dR$ and R, i.e., inside a given solid angle, is given by the relation (see (1), page 296).

$$dM = \frac{2}{3}\pi\rho h_0^3 \frac{\sin\varphi\, d\varphi}{\cos^3\varphi} = \frac{\bar{\eta}\bar{\eta}_i M_0 u_0^2}{2Q} \cdot \frac{\sin\varphi\, d\varphi}{2\cos^3\varphi}. \tag{14}$$

This mass will have the velocity a_0 given by (12).
It is easy to see that since

$$\eta_\varphi = \frac{Agh_0}{4Q} = \frac{Ag}{4Q}\left(\frac{3\bar{\eta}\bar{\eta}_i E_0}{4\pi p Q}\right)^{\frac{1}{3}}, \tag{15}$$

then for relatively low energies of fall we will have $\eta_\varphi < 1$, and the semi-vertical angle of the crater formed by the impact and explosion will not depend on the force of gravity.

On the other hand, for a high initial energy the limiting angle depends on the force of gravity on the planet in question.

For $\eta \leq 1$ we will have

$$J_z = \frac{\eta E_0}{\sqrt{Q}}\sqrt{2\bar{\eta}}\left[1 - \left(\frac{\bar{\eta}}{A}\right)^{\frac{1}{6}}\right]. \tag{16}$$

The minimum velocity will be for $\cos^3 \varphi_0 = \bar{\eta}/A$, when

$$a_0 = a_{0\min} = \left[\frac{2Q}{A} - \frac{gh_0}{2}\right]^{\frac{1}{2}} \geqslant 0, \tag{17}$$

and the total ejected mass will be

$$M = \frac{\bar{\eta}\eta M_0 u_0^2}{8Q}\left[\left(\frac{A}{\bar{\eta}}\right)^{\frac{2}{3}} - 1\right]. \tag{18}$$

For $\eta_\varphi \geq 1$

$$J_z = \frac{\eta E_0}{V_Q}\sqrt{2\bar{\eta}}\left[1-\left(\frac{\bar{\eta}g}{4Q}\right)^{\frac{1}{6}}\left(\frac{3\bar{\eta}\eta M_0 u_0^2}{8\pi p Q}\right)^{\frac{1}{18}}\right] = \eta E_0 \sqrt{\frac{2\bar{\eta}}{Q}}\left[1-\left(\frac{\bar{\eta}g h_0}{4Q}\right)^{\frac{1}{6}}\right]. \quad (19)$$

The minimum velocity will be equal to zero for

$$\cos^3 \varphi_0 = \frac{\bar{\eta}g h_0}{4Q}, \quad (20)$$

and the total ejected mass

$$M = \frac{\bar{\eta}\eta M_0 u_0^2}{8Q}\left[\left(\frac{4Q}{\bar{\eta}g}\right)^{\frac{2}{3}}\left(\frac{8\pi p Q}{3\bar{\eta}\eta M_0 u_0^2}\right)^{\frac{2}{9}}-1\right] = \frac{\bar{\eta}\eta M_0 u_0^2}{8Q}\left[\left(\frac{4Q}{\bar{\eta}g h_0}\right)^{\frac{2}{3}}-1\right]. \quad (21)$$

We now compare the projection on the normal of the momentum of the falling body

$$J_{0z} = M_0 u_0 \cos z \quad (22)$$

with the ejection reaction J_z:

$$\frac{J_z}{J_{0z}} = \sqrt{\frac{2\bar{\eta}}{Q}}\frac{\eta M_0 u_0^2 (1-\sqrt{\cos\varphi_0})}{2M_0 u_0 \cos z} = \sqrt{\frac{2\bar{\eta}}{Q}}\frac{\eta u_0 (1-\sqrt{\cos\varphi_0})}{2\cos z}. \quad (23)$$

Since $Q = (u_k^{*2}/2)$, we finally obtain

$$\frac{J_z}{J_{0z}} = \sqrt{\bar{\eta}}\frac{\eta(1-\sqrt{\cos\varphi_0})}{\cos z}\frac{u_0}{u_k^*}. \quad (24)$$

Since on the average $\sqrt{\bar{\eta}} \cdot \eta \approx 2$, $\varphi_0 \approx 60°$, then for $z = 0$

$$\frac{J_z}{J_{0z}} \approx 0.6 \frac{u_0}{u_k^*}. \quad (25)$$

For example for a rock meteorite striking an aluminum medium at a velocity of $u_0 = 40$ km/sec, for which $u_k^{*2} \approx 2$ km per sec, we have

$$\frac{J_z}{J_{0z}} \approx 12.$$

For high impact velocities, the reactive impulse always exceeds the momentum of the falling body, and therefore the total momentum gained by the medium during the impact is practically independent of the angle.

Now let us examine the scattering of the explosion products, and in particular the ejection of material into outer space.

THE MOTION OF THE MATERIAL EJECTED FROM THE CRATER

The Scattering of the Ejected Material

If the meteorite explosion takes place on a planet having an atmosphere, then we must take into account its resistance, which will slow down the ejected particles. All the particles, as a result of this (except in the case when the explosion is extremely large, as for example when the meteorite and the planet have similar dimensions), will be decelerated by the atmosphere. It follows from this that the great majority of these particles fall back on the surface of the planet in the neighborhood of the impact, and partially fill in the ejection crater, changing its shape and dimensions.

The total mass of the planet, following the fall and impact, is therefore increased by almost all the mass of the falling meteorite.

We will not consider here the laws of dispersion in a resistant medium, either in the case of an ordinary explosion or in the case of a meteorite.

We will consider the problem of the dispersion of particles from an exploding medium into a vacuum, which is more interesting from the astrophysical and cosmological point of view.

In its most general form, this problem involves the classical investigation of celestial mechanics on the motion of a material point under the action of a central gravitational force (the mass ejected during the explosion is extremely small compared with that subjected to the impact).

We will not consider the case when the two bodies are of comparable size.

The orbital equation will first of all be written in polar coordinates (Fig. 5) as

$$r = \frac{p}{1 - e \cos \theta}, \tag{1}$$

where $p = (b^2/a)$ is a parameter, $e = \sqrt{1 - (b^2/a^2)}$ the eccentricity, θ the polar angle, r the radius vector, and a and b the major and minor axes of the ellipse.

In rectangular coordinates Equation (1) takes the form

$$y^2 = p^2 + 2pex + x^2(e^2 - 1). \quad (2)$$

The parameter p and the eccentricity e are obtained from the obvious conditions that at the point $M(x_0, y_0)$, where $r = R_0$ and $\theta = \theta_0$, the velocity of ejection v_0 and the angle of inclination of the orbit to the horizontal plane α_0 are given. Since

$$v_0^2 = GM\left(\frac{2}{R_0} - \frac{1}{a}\right), \quad (3)$$

(M is the mass of the body, G the gravitational constant), and

$$\text{tg } \beta_0 = \frac{dy}{dx} = \tan(90° + \theta_0 + \alpha_0) = \frac{pe + (e^2 - 1)x_0}{y_0}, \quad (4)$$

where

$$\cos \theta_0 = \frac{R_0 - p}{eR_0}, \quad (5)$$

then we find that

$$\frac{p}{R_0} = \frac{v_0^2 R_0}{GM} \cos^2 \alpha_0; \quad e^2 = 1 - \frac{2v_0^2 R_0}{GM} \cos^2 \alpha_0 + \frac{v_0^4 R_0^2}{(GM)^2} \cos^2 \alpha_0. \quad (6)$$

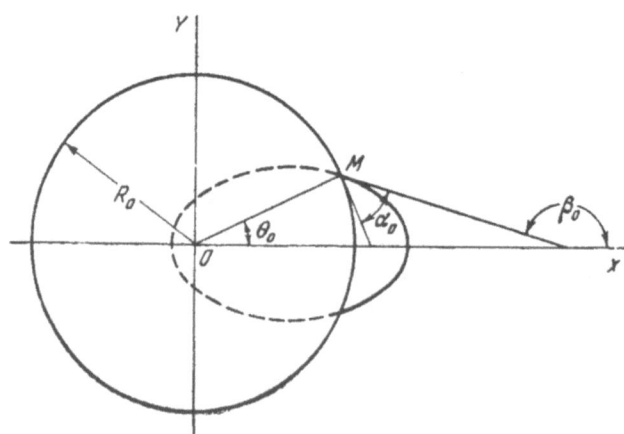

FIG. 5

Since the acceleration at the surface of the body receiving the impact is

$$g = \frac{GM}{R_0^2}, \quad (7)$$

and the circumferential velocity satisfies

$$v_c^2 = \frac{GM}{R_0},$$

then

$$\frac{p}{R_0} = \left(\frac{v_0}{v_c}\right)^2 \cos^2 \alpha_0, \tag{8}$$

$$e^2 = 1 + \left(\frac{v_0}{v_c}\right)^2 \left[\left(\frac{v_0}{v_c}\right)^2 - 2\right] \cos^2 \alpha_0. \tag{9}$$

The coordinates of the point $M(x_0, y_0)$ are obtained from

$$x_0 = \frac{R_0 - p}{e}; \quad y_0 = R_0 \sqrt{1 - \left(\frac{R_0 - p}{eR_0}\right)^2}, \tag{10}$$

while the coordinates of the point of impact of the particles for $1 > e \geq 0$ will be

$$x = x_0 = \frac{R_0 - p}{e}, \quad y = -y_0 = -R_0 \sqrt{1 - \left(\frac{R_0 - p}{eR_0}\right)^2}. \tag{11}$$

For $e \geq 1$ a body does not fall back to the surface of the planet. The relations written above completely solve the problem posed of determining the orbit of the particles ejected from the crater of the "exploded" material.

If $(v_0/v_c) \ll 1$, then the curvature of the planet's surface can be neglected, and we can also neglect the variation of the gravitational force with distance. In this case the main equations can be transformed to a simpler form, but it is simpler to obtain the equation of the orbit in Cartesian coordinates directly.

This equation is obtained from the obvious relations

$$y = v_0 t \sin \alpha_0 - \frac{gt^2}{2}; \quad x = v_0 t \cos \alpha_0, \tag{12}$$

from which we obtain

$$y = x \tan \alpha_0 - \frac{gx^2}{2v_0^2 \cos^2 \alpha_0}. \tag{13}$$

The material falls back to the surface of the planet at the point

$$\bar{x} = \frac{v_0^2}{g} \sin 2\alpha_0. \tag{14}$$

It is not difficult in either case to determine the distance at which the material ejected from the crater will collect or the height of the heap of this material on the surface of the planet, i.e., to determine the profile of the crater and its surroundings.

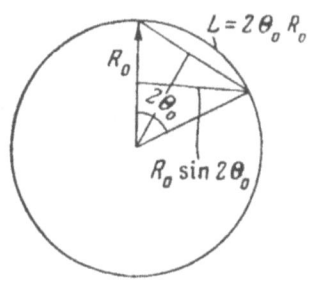

FIG. 6

The element of mass

$$dM = -\frac{2\pi}{3}\rho h_0^3 \frac{d(\sin\alpha_0)}{\sin^3\alpha_0} = -\frac{2\pi}{3}\rho h_0^3 \frac{\cos\alpha_0 d\alpha_0}{\sin^3\alpha_0},\quad (15)$$

ejected from the crater at an angle of $\alpha_0 = 90° - \varphi$ collects on a surface element dS at a distance L from the point of the explosion. The height of the pile will be

$$H = \frac{dM}{\rho dS}.\quad (16)$$

In the first case (Fig. 6)

$$L = 2\theta_0 R_0,\quad dL = 2R_0 d\theta_0,\quad dS = 4\pi R_0^2 \sin 2\theta_0 d\theta_0.$$

In what follows it is convenient to write

$$dS = -4\pi R_0^2 d(\cos^2\theta_0).\quad (17)$$

From Equation (1) we have

$$\cos\theta_0 = \frac{1}{e}\left(1 - \frac{p}{R_0}\right) = \frac{1 - \left(\frac{v_0}{v_c}\right)^2 \cos^2\alpha_0}{\sqrt{1 + \frac{v_0^2}{v_c^2}\cos^2\alpha_0\left(\frac{v_0^2}{v_c^2} - 2\right)}}.\quad (18)$$

We will consider two variants, the first when v_0 = const, and the second when $v_0 = v_0(\varphi) = v_0(\alpha_0)$ with the concrete form of this relation [see (13), page 299]

since
$$a_0^2 = v_0^2 = \frac{3}{2} \frac{E_B \cos^3 \varphi}{\pi \rho h_0^3} - \frac{gh_0}{2};$$

$$\frac{4}{3} \pi \rho h_0^3 = \frac{E_B}{Q\eta} = \frac{E_B}{Q^*},$$

then
$$v_0^2 = 2Q^* \sin^3 \alpha_0 - \frac{gh_0}{2}. \tag{19}$$

In the first variant

$$dS = \pi R_0^2 d\alpha_0 \frac{\sin 2\alpha_0 \left\{ \sin^2 2\alpha_0 \left(\frac{v_0}{v_c}\right)^2 \left[\left(\frac{v_0}{v_c}\right)^2 - 2\right] + 4\cos 2\alpha \left[1 + \cos^2 \alpha \left(\frac{v_0}{v_c}\right)^2 \left[\left(\frac{v_0}{v_c}\right)^2 - 2\right]\right] \right\}}{\left\{ 1 + \left(\frac{v_0}{v_c}\right)^2 \cos^2 \alpha_0 \left[\left(\frac{v_0}{v_c}\right)^2 - 2\right] \right\}^2}; \tag{20}$$

while in the second

$$\cos^2 \theta_0 = 1 - \frac{\left(\frac{v_0}{v_c}\right)^4}{4} \frac{\sin^2 2\alpha_0}{1 + \cos^2 \alpha_0 \left(\frac{v_0}{v_c}\right)^2 \left[\left(\frac{v_0}{v_c}\right)^2 - 2\right]} = \tag{21}$$

$$= 1 - \frac{\sin^2 2\alpha_0}{(2 \cdot g \cdot R_0)^2} \frac{\left(2Q^* \sin^3 \alpha_0 - \frac{gh_0}{2}\right)^2}{1 + \cos^2 \alpha_0 \left(\frac{2Q^* \sin^3 \alpha_0 - \frac{gh_0}{2}}{gR_0}\right) \left(\frac{2Q^* \sin^3 \alpha_0 - \frac{gh_0}{2}}{gR_0} - 2\right)};$$

we have used the notation
$$-d\cos^2 \theta_0 = f(\alpha_0) d\alpha_0, \tag{22}$$

where for $f(\alpha_0)$ we obtain a very complicated expression which we will not reproduce.

We thus have
$$dS = 4\pi R_0^2 f(\alpha_0) d\alpha_0. \tag{23}$$

In the second case we will also consider two variants. In the first variant

$$L = \bar{x} = \frac{v_0^2}{g} \sin 2\alpha_0, \quad dL = \frac{2v_0^2}{g} \cos 2\alpha_0 d\alpha_0, \tag{24}$$

$$dS = 2\pi L dL = 2\pi \left(\frac{v_0^2}{g}\right)^2 \sin 4\alpha_0 d\alpha_0,$$

while in the second

$$L = \bar{x} = \frac{2Q^*}{g} \sin^3 \alpha_0 \sin 2\alpha_0 - \frac{h_0}{2} \sin 2\alpha_0,$$

$$dL = d\alpha_0 \left[\frac{4Q^*}{g} \sin^3 \alpha_0 (4\cos^2 \alpha_0 - \sin^2 \alpha_0) - h_0 \cos 2\alpha_0 \right], \quad (25)$$

$$dS = 2\pi d\alpha_0 \left[\frac{4Q^*}{g} \sin^4 \alpha_0 \cos \alpha_0 - h_0 \sin \alpha_0 \cos \alpha_0 \right] \times$$

$$\times \left[\frac{4Q^*}{g} \sin^3 \alpha_0 (4\cos^2 \alpha_0 - \sin^2 \alpha_0) - h_0 \cos 2\alpha_0 \right].$$

It is of value in all cases to calculate

$$\alpha_0 = \alpha_0(\theta_0) = \alpha_0(L). \quad (26)$$

In the first case it is necessary in both variants to determine $\alpha_0(\theta_0)$, starting from the relation (18) for the first variant or from the first relation of (21) for the second variant.

To obtain a similar result for the second case, in the first variant we must use the first of the relations in (24) and in the second the first relation of (25).

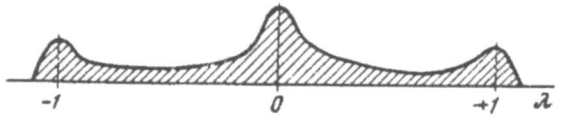

FIG. 7

Up to the present, we have been giving the law of distribution of the mass ejected from the crater in terms of the angle by using (15). However, when the explosion occurs at the surface, we obtain another approximation, i.e.,

$$dM = \frac{M_B}{2} \cos \alpha_0 d\alpha_0. \quad (27)$$

Thus for the first variant of the second case we obtain the relation

$$H = \frac{M_B \cos \alpha_0}{8\pi \rho L_0^2 \sin 2\alpha_0 \cos 2\alpha_0}, \quad (28)$$

where $L_0 = (v_0^2/g)$ is the maximum ejection distance.

Since $2\alpha_0 = (L/L_0) = \lambda$, then $\cos 2\alpha_0 = \pm\sqrt{1-\lambda^2}$ and $\cos^2 \alpha_0 = (1/2)(1\pm\sqrt{1-\lambda^2})$, and from (28) we have

$$\frac{H}{L_0} = \frac{M_B}{8\pi \rho L_0^3 \lambda} \sqrt{\frac{1 \pm \sqrt{1-\lambda^2}}{2(1-\lambda^2)}}. \quad (29)$$

For $0 \leq \alpha_0 \leq (\pi/4)$ we must use the positive sign, and for $(\pi/4) \leq \alpha_0 \leq (\pi/2)$ the negative sign. We finally arrive at the relation

$$\frac{H}{L_0} = \frac{M_{\text{B}}}{8\pi p h_0^3 \lambda \sqrt{2(1-\lambda^2)}} \left(\sqrt{1+\sqrt{1-\lambda^2}} + \sqrt{1-\sqrt{1-\lambda^2}} \right). \quad (30)$$

This relation was obtained by us before.[4] It is easy to see that for $\lambda = 0$ and $\lambda = \pm 1$ we have $H \to \infty$, while for $\lambda = 0.9$ we have $H = H_{\min}$.

In reality, of course, the height of the heap of ejected products is finite, and it will not tend to infinity for $\lambda = 0$ and $\lambda = \pm 1$.

The masses of ejection products accumulated in small neighborhoods of $\lambda = 0$ and $\lambda = \pm 1$ are small, and therefore the value of H obtained for $\lambda = 0$ and $\lambda = \pm 1$ as a consequence of the gravitational force is finite. The shape of the curve $H = H(\lambda)$ is shown in Fig. 7.

We have thus obtained a profile that is similar to that of a lunar circle and crater. We will call this shape an "embankment" crater. Almost identical results are also obtained in the other variants considered above.

The Possibility of the Ejection of Material into Outer Space

It is fairly evident that when a meteorite strikes a celestial body that is not too large (the moon, a small planet, but for the moment excluding the asteroids), part and sometimes all the ejected material can attain a velocity greater than the limiting parabolic velocity for the body in question, and travel on a hyperbolic trajectory into cosmic space.

Since the limiting parabolic velocity is

$$a_\infty = \sqrt{\frac{2GM}{R_0}} = \sqrt{2gR_0}, \quad (1)$$

then the condition for ejection into free space is

$$a_0^2 = \frac{3E_{\text{B}} \cos^3 \varphi}{2\pi p h_0^3} - \frac{gh_0}{2} \geqslant 2gR_0. \quad (2)$$

It follows from this that

$$\cos^3 \varphi \geqslant \cos^3 \varphi_\infty = 4\pi g R_0 \left(1 + \frac{h_0}{4R_0}\right) \frac{\rho h_0^3}{3E_B}. \tag{3}$$

In the case when it is a relatively large body suffering the impact, $h_0 \ll R_0$ (for small bodies the gravitational force is insignificant, and the whole mass escapes), and since $h_0^3 = (3\bar{\eta} E_B / 4\pi\rho Q)$ we will have

$$\cos \varphi_\infty = \sin \alpha_{0\infty} \geqslant \left(\frac{\bar{\eta} g R_0}{Q}\right)^{\frac{1}{3}}. \tag{4}$$

It follows that a mass

$$M_\infty = \frac{\pi}{3} \rho h_0^3 \tan^2 \varphi_\infty = \frac{\bar{\eta} E_B}{4Q}\left[\left(\frac{Q}{\bar{\eta} g R_0}\right)^{\frac{2}{3}} - 1\right], \tag{5}$$

where $(Q/\bar{\eta} g R_0) \geq 1$ escapes "to infinity."

For the earth, this inequality does not hold (in fact $(Q/\bar{\eta} g R_0) \approx 1/20$), and all the mass, even if there were no atmosphere, would remain on the earth, except for a small quantity of material which escapes in the region of the front.

For the moon, $(Q/\bar{\eta} g R_0) \approx 1.4$, and part of the ejected material escapes "to infinity." For small planets the relative quantity of material escaping "to infinity" is still greater.

It is evident that, even commencing with the moon, the determination of the semi-vertical angle of the crater cannot in practice take into account the influence of the force of gravity, and therefore the total ejected mass will be

$$M_t = \frac{\bar{\eta} E_B}{4Q}\left[\left(\frac{A}{\bar{\eta}}\right)^{\frac{2}{3}} - 1\right]. \tag{6}$$

For the condition $(Q/\bar{\eta} g R_0) \leq A$ we obtain the relations

$$\frac{M_\infty}{M_t} = \frac{\left(\frac{Q}{\bar{\eta} g R_0}\right)^{\frac{2}{3}} - 1}{\left(\frac{A}{\bar{\eta}}\right)^{\frac{2}{3}} - 1}$$

$$\frac{M_\infty}{M_0} = \frac{\bar{\eta}\eta u_0^2}{8Q}\left[\left(\frac{Q}{\bar{\eta} g R_0}\right)^{\frac{2}{3}} - 1\right], \tag{7}$$

where M_0 is the mass of the meteorite.

For $(Q/\bar{\eta}gR_0) \geq A$ we obtain

$$\frac{M_\infty}{M_0} = 1. \tag{8}$$

For the moon in the case of the impact of a stone meteorite $(M_\infty/M_0) \approx 1/10$; and $(M_\infty/M_t) = 10$ (for $u_0 \approx 40$ km/sec). For asteroids and smaller bodies $(M_\infty/M_0) = 1$.

We will now consider the question of the possible time for the complete destruction of relatively small cosmic bodies under the action of meteoric impacts. We considered this problem in a more approximate fashion previously.[4]

We will assume that the area of the "mean" section of the body moving in cosmic space is S, that the density of the material making up the meteorite is ρ_M, and that the mean velocity of impact, determined in general relative to the velocity of the body, is u_0; then the element of meteoric mass striking the body during an interval of time dt will be

$$dm_M = S \cdot \rho_M u_0 dt. \tag{9}$$

The change in mass of the body in time dt that is caused as a result of the collision, the explosion, and the subsequent ejection of material will be

$$dM_a = -\frac{\bar{\eta}S \cdot \rho_M u_0^3}{8Q}\left[\left(\frac{A}{\bar{\eta}}\right)^{\frac{2}{3}} - 1\right]dt. \tag{10}$$

Since $S = \bar{\alpha}*(M_a/\rho)^{2/3}$, where M_a is the mass and $\bar{\alpha}*$ the form factor of the body—for a sphere the form factor is $\bar{\alpha}* = (3\sqrt{\pi}/4)^{2/3}$—then

$$\frac{dM_a}{M_a^{\frac{2}{3}}} = -\frac{\bar{\alpha}\,\bar{\eta}\rho_M u_0^3}{8Q\rho^{\frac{2}{3}}}\left[\left(\frac{A}{\bar{\eta}}\right)^{\frac{2}{3}} - 1\right]dt, \tag{11}$$

and on integration this yields

$$M_a^{\frac{1}{3}} = M_{0a}^{\frac{1}{3}} - \frac{\bar{\alpha}\,\bar{\eta}\rho_M u_0^3 t}{24Q\rho^{\frac{2}{3}}}\left[\left(\frac{A}{\bar{\eta}}\right)^{\frac{2}{3}} - 1\right]; \tag{12}$$

here M_{0a} is the initial mass of the body.

The body is completely destroyed after a time

$$T = \frac{24 M_{0a}^{\frac{1}{3}} p^{\frac{2}{3}} Q}{\bar{a}\bar{\eta} u_0^3 p_M \left[\left(\frac{A}{\bar{\eta}}\right)^{\frac{2}{3}} - 1\right]} = \frac{12 M_{0a}^{\frac{1}{3}} p^{\frac{2}{3}} u_k^{*2}}{\bar{a}\bar{\eta} p_M u_0^3 \left[\left(\frac{A}{\bar{\eta}}\right)^{\frac{2}{3}} - 1\right]}. \quad (13)$$

For a sphere

$$T = \frac{16 R_0 \rho u_k^{*2}}{\bar{\eta} p_M u_0^3 \left[\left(\frac{A}{\bar{\eta}}\right)^{\frac{2}{3}} - 1\right]}. \quad (14)$$

In particular, if we use the value $\bar{\eta} = 2$, then

$$T = \frac{8 R_0 \rho u_k^{*2}}{p_M u_0^3 \left[\left(\frac{A}{2}\right)^{\frac{2}{3}} - 1\right]}. \quad (15)$$

If we take as a mean for rock bodies the values $A = 16$ and $u_k^* = 2 \cdot 10^5$ cm/sec, then

$$T = \frac{32 R_0 \rho \cdot 10^{10}}{3 u_0^3 p_M} \approx 10^{11} \frac{R_0}{u_0^3} \cdot \frac{\rho}{p_M}. \quad (16)$$

For bodies having a radius $R_0 = 1$ cm and a mean velocity of impact of $u_0 = 4 \cdot 10^6$ cm/sec,

$$T = \frac{10^{11}}{64 \cdot 10^{18}} \cdot \frac{\rho}{p_M} \approx 1.5 \cdot 10^{-9} \frac{\rho}{p_M}. \quad (17)$$

If we use the values

$$\rho = 4 \text{ g/cm}^3 \quad , \quad p_M = 10^{-22} \text{ g/cm}^3$$

we find that

$$T = \frac{6 \cdot 10^{-9}}{10^{-22}} = 6 \cdot 10^{13} \text{ sec} = 2 \cdot 10^6 \text{ years},$$

i.e., a rock body with a radius of 1 cm and a mass of 16 g will be destroyed in a time of the order of a million years.

The important cosmological conclusion can be drawn that small bodies in the solar system, as well as small bodies in space in general, are being continuously destroyed—disintegrated as a result of mutual collisions.

LITERATURE

1. K. P. Stanyukovich, Non-steady Motion of Continuous Media (in Russian), Gostekhizdat, 1955.
2. K. P. Stanyukovich and V. V. Fedynskii, Doklady Akad. Nauk SSSR, 56, 2 (1947).
3. K. P. Stanyukovich, Meteoritika, 7 (1950).
4. Ya. B. Zel'dovich and Yu. P. Raizer, ZhÉTF, 35, 1402, (1958), Soviet Physics—JETP, 8, 980 (1958).

METEORIC MATTER AND SOME GEOPHYSICAL PROBLEMS OF THE UPPER ATMOSPHERE

B. A. Mirtov

The present paper attempts to relate certain phenomena which arise in the upper atmosphere with the presence there of fast-moving particles of meteoric origin. It should be noted that the role of meteoric matter in the life of the upper layers of the atmosphere has been quite inadequately studied. The reason for this lies in the great difficulty of studying the finely dispersed meteoric material which enters the relatively dense layers of the earth's atmosphere (100 km and higher). Therefore, in addition to the usual meteor observations (photography and radar), investigations carried out over a long period of time on artificial earth satellites*[1-3] should yield very important data for the understanding of phenomena due to the interaction of meteoric material with the earth's atmosphere.

METEORIC MATTER AND SCATTERING OF LIGHT IN THE UPPER ATMOSPHERE

The dustiness of the earth's atmosphere is determined by dust entering it from the surface of the earth, and by meteoric particles which enter it from space.

---

*The study of "primary" meteoric particles at altitudes in excess of 100 km, with a receiver sensitivity no greater than 10^{-10} g, is not feasible for ordinary research rockets having a flight duration limited to 5–10 min.[1,2] The frequency of collisions derived from satellite measurements is 2×10^{-3} to 2×10^{-2} collisions per second per m² of surface (with a sensitivity of 10^{-9} g). Under these conditions, a receiver having an area of even 3000 cm² would receive one impact in 30 or 8 min respectively.

The distribution of dust in the lower layers of the atmosphere is extremely variable and depends on a number of factors which are very difficult to take into account. A considerably more definite picture emerges with regard to the dustiness of the upper layers of the atmosphere, caused exclusively by particles of meteoric matter. We shall return to consideration of this picture, but first we must consider that layer of the atmosphere above which the dustiness is actually determined by meteoric matter.

Dust of Terrestrial Origin

This dust consists of particles of the most varied sizes, and it is conceivable that, because of continual mixing, dust particles of terrestrial origin would be encountered even at considerable heights in the atmosphere. The altitude limit for the spreading of these dust particles in the atmosphere is set by the density of the latter, as well as by the presence of rather strong vertical air currents, which are capable of raising dust particles to great heights. It is natural for the higher layers of the earth's atmosphere to be enriched by particles of smaller size.*

If the dust for some reason finds itself above the layer in which it can float, the particles begin to "settle" until they reach denser layers, where their rate of fall becomes comparable with the velocity of possible vertical mixing. This "settling" velocity may be estimated by making use of Stokes' formula and introducing the correction made by Millikan:[4]

$$v = \frac{2g(\delta - \rho)}{9\eta} \left[1 + \frac{\lambda}{r}(A + Be^{-\frac{lr}{\lambda}}) \right] r^2; \qquad (1)$$

here g is the gravitational acceleration,† $\eta = 2 \cdot 10^{-4}$ is the coefficient of viscosity for air, independent of pressure,

*This phenomenon, except for certain peculiarities, is quite analogous to the phenomenon of gravitational separation of gases in the earth's atmosphere.
†The change in gravitational acceleration up to heights of 200–500 km may be neglected.

δ is the density of the material composing the dust particle ($\delta \approx 4$), ρ is the density of the air, r is the radius of the dust sphere, λ is the mean free path of an air molecule at a particular altitude, and A, B, and C are dimensionless constants having the approximate values: $A = 1.0110$, $B = 6.5 \times 10^{-3}$, $C = 3$.

TABLE 1

H, km	Radius of dust particle r				
	$1 \cdot 10^{-6}$ cm	$2 \cdot 10^{-5}$ cm	$1 \cdot 10^{-5}$ cm	$5 \cdot 10^{-5}$ cm	$5 \cdot 10^{-4}$ cm
	$0.01\ \mu$	$0.02\ \mu$	$0.1\ \mu$	$0.5\ \mu$	$5\ \mu$
130	$1.6 \cdot 10^3$	$3 \cdot 10^3$	$1.6 \cdot 10^4$	$1 \cdot 10^5$	$8 \cdot 10^5$
100	40	75	$4 \cdot 10^2$	$2.4 \cdot 10^3$	$2 \cdot 10^4$
90	8	15	80	$4.8 \cdot 10^2$	$4 \cdot 10^3$
80	1.2	2.2	12	72	$6 \cdot 10^2$
70	$2.6 \cdot 10^{-1}$	$5 \cdot 10^{-1}$	2.6	15	$1.3 \cdot 10^2$

Substituting in (1) varying values of r and λ, we obtain values for the rates of fall attained by various dust particles (in cm $\times$ sec^{-1}) in media of different densities (in the absence of mixing). The results of the calculations are shown in Table 1.

In the atmosphere, and especially in its upper layers, dust particles do not of course achieve such high velocities,* so that under real conditions the rate of fall will be smaller. However, even on the basis of the higher estimates given in Table 1, it can be seen that, starting at a height of 80-90 km, the rate of fall of particles rapidly decreases, and slight vertical movements of the air may retard or even halt the descent of particles, creating favorable conditions for the accumulation of dust.

At the present time, on the basis of work on the determination of the degree of diffuse separation of gases in the atmosphere,[5, 6] it can be quite confidently stated that even such small vertical movements of the air as might mix and smooth

*A dust particle borne aloft begins its fall with zero velocity, and in the gravitational field of the earth it must traverse (even in a vacuum) a very long path (H), in order to gather the speed indicated in the table in accordance with the condition $v = \sqrt{2gH}$.

the gaseous composition of the atmosphere terminate at altitudes of 90-100 km. The upward transport of the relatively heavy dust particles, as a consequence of the mixing factor, becomes unlikely at still lower altitudes. Therefore, it would hardly be expected that an appreciable amount of dust of terrestrial origin would systematically penetrate to altitudes in excess of 80-90 km.

Meteoric Particles

Above 80-90 km in the earth's atmosphere there are solid particles of cosmic matter. The sizes of particles of cosmic origin, like those of terrestrial dust particles, are extremely variegated; however, in contrast to the latter, there are, on the one hand, no particles among them with $r < 0.25\mu$* and on the other, even at the highest altitudes there may be among them particles of very appreciable size. Because of their high velocities these large particles are subjected to the influence of high braking temperatures and pressure heads in the earth's atmosphere, hence they may melt and break up to form particles of other sizes, among them very fine ones, for which $r < 0.25\mu$. In spite of this, however, it may be considered, as before, that above 90-100 km particles with $r < 0.25\mu$ are absent, since the processes which lead to their formation are in general sufficiently well-developed only at altitudes less than 100 km.[7] Another distinction between particles of cosmic origin and terrestrial dust particles is the fact that the former fly into the earth's atmosphere with enormous velocities of the order of 11-70 km/sec and reach altitudes of 160-140 km practically unretarded. But these particles have only to penetrate into denser layers of the atmosphere and braking soon begins, and their velocities are sharply reduced. Particles with $r < 1\mu$ lose their cosmic velocities in the altitude interval 130-95 km.

At heights greater than 140 km, the velocities with which the particles fall will be determined by their cosmic velocities.

*Dust particles with a radius smaller than 0.25 μ are driven out of the solar system by radiation pressure.

Modern data[7] permit us to make a rough estimate of the amount of meteoric material falling to the earth. Results of investigations by different authors differ widely, but nevertheless make it possible to establish some upper limit to the amount of material received every day by the earth's surface.

TABLE 2

Author, date	ρ, g/cm^2	M, $\frac{\text{tons}}{\text{day}}$	Author, date	ρ, g/cm^2	M, $\frac{\text{tons}}{\text{day}}$
Photometric methods			Estimates from sediments		
Fesenkov (1946)	$6 \cdot 10^{-23}$	20	De Jager (1955)	$5 \cdot 10^{-21}$	1500
Allen (1947)	$4 \cdot 10^{-23}$	12	Thomsen (1952)	$2 \cdot 10^{-20}$	6000
Van de Hulst (1947)	$3 \cdot 10^{-21}$	1000	Pettersson and Rotschi (1950)	$1 \cdot 10^{-20}$	4000
Siedentopf (1953)	$2 \cdot 10^{-22}$	100	Meteor observations		
Elsässer (1954)	$2 \cdot 10^{-23}$	8	Levin (1956)	$5 \cdot 10^{-23}$	15
Minnaert (1955)	$6 \cdot 10^{-22}$	200	Watson (1942)	$2 \cdot 10^{-23}$	5
			Whipple (1952)	$3 \cdot 10^{-21}$	1000
				$2 \cdot 10^{-20}$	5000

Table 2 presents data concerning the space density of meteoric material (ρ) derived by various authors. In addition, this table gives the mass of meteoric matter (M) falling on the entire surface of the earth each day. This mass is calculated on the assumption that the mean geocentric velocity of the meteors is 40 km/sec and without taking into account the slight increase in effective diameter in the process of capture of meteors by the earth.

Recently, with the aid of artificial earth satellites, new data[1-3] has been obtained on the amount of meteoric material that falls to the earth's surface. This amount lies somewhere in the range 24-4000 tons/day.

From the data of Table 2, a figure of 5000 tons/day may safely be considered the maximum possible estimate of the amount of cosmic material falling to the surface of the earth.*

*After a geological period (2×10^9 years) of such heavy fallout, cosmic dust would cover the earth with a layer less than two meters thick.

It is hardly likely that further refinement of this figure will raise it markedly (say, by an order of magnitude); some reduction in the adopted estimate might rather be expected.

At present, the mass distribution of particles of meteoric material has been studied very little. One of the most probable hypotheses describing the spectrum of sporadic meteor particles is that due to Watson,[8] according to which the amount of material brought in by meteors of a given size (a given stellar magnitude) remains constant for all meteors which enter the earth's atmosphere. This means that, in the mass distribution function for meteoric bodies,

$$f(m) = \frac{b}{m^s}$$

(m is the mass of the meteoroid, b is a coefficient of proportionality, and the exponent s = 2*). Furthermore, Watson assumes that meteoric particles entering the atmosphere are distributed, according to their size, over an interval of forty stellar magnitudes. According to these ideas and the maximum estimates of the amount of meteoric matter falling (5000 tons/day), we find that meteors of each stellar magnitude bring to the earth 125 tons of matter per day. As a direct consequence of the chosen theory, it appears that the overwhelming majority of meteoric particles entering the earth's atmosphere are of the smallest possible sizes: thus, dust particles from 0.30 to 1μ in diameter constitute 99% of the total number of particles arriving (among these, 60% are particles of radius 0.3μ and only 1% are particles of radius 1μ). The number of larger particles is small.

Specific Dust Content of the Upper Atmosphere

In considering cosmic dust as a geophysical factor which in some measure influences certain properties of the upper atmosphere, special attention should be paid to the specific dust content (K) at a given altitude

$$K = \frac{n}{N}, \tag{2}$$

*As is well known, this distribution is not necessarily the correct one, However, it is employed in many similar computations, especially since for sporadic meteors there are no serious objections to the condition that s = 2.

where n is the number of molecules per unit volume and N is the number of dust particles of a given size in the same volume. It is precisely this specific dust content in relation to altitude which, together with the energies of the incoming particles, determines the various geophysical effects which are due to the presence of micrometeors. These may be the scattering of light in the upper atmosphere, ionization, temperature effects, and some other phenomena related to the self-luminescence of the atmosphere.

Expressing the concentration of particles N in terms of the flux D and velocity v' of these particles, we obtain

$$K = \frac{nv'}{D}. \quad (3)$$

The flux D of cosmic particles penetrating the earth's atmosphere is a quantity which is constant for all layers of the atmosphere where mixing is absent. Indeed, with a steady fallout of particles (meteor streams are not being considered), the meteoric material does not accumulate in the atmosphere. As a result, the influx and efflux of cosmic dust are the same in equal intervals of time. This circumstance makes for the constancy of D from the upper boundary of the atmosphere down to a height of about 100 km.

The velocity of fall v' which enters into (3) is determined in different ways in the various layers of the upper atmosphere:

a) from the upper limit of the atmosphere down to a height of 200-150 km, where the incoming meteor particle does not yet undergo appreciable retardation, its speed will be determined by its initial space velocity;

b) from a height of 150 km down to 100-95 km, the rate of fall will decrease slowly at first, then rapidly; the decrease in velocity here obeys the drag equation;*[7]

*$v' = v_0 \cdot \exp\left(-\frac{\Gamma H^* S}{\mu} \rho\right)$, where v' is the velocity of the particle at a given altitude; v_0 is its initial velocity ($3 \cdot 10^6$ cm/sec); Γ is the coefficient of resistance ($\Gamma \approx 1$); $H^* = f(H)$ is the height in a uniform atmosphere; M is the mass of the particle ($6.3 \cdot 10^{-13}$ g); S is its cross section ($3.8 \cdot 10^{-9}$ cm^2); $\rho = \varphi(H)$ is the atmospheric density. Substituting the values quoted, we obtain

$$v' = 3 \cdot 10^6 \cdot \exp(-6 \cdot 10^3 \rho H^*).$$

c) from a height of 100-95 km downwards, after the dust particle has entirely lost its cosmic velocity, its motion will be determined by the law of free fall (Stokes-Millikan formula, Table 1).

The rates of fall of particles 0.3-0.4 μ in radius are listed in Table 3 (fifth column). In the calculation of Table 3, averaged values have been taken for the density (ρ) and number of particles per cm^3 (n) which were given in Ref. 9 for the height in a uniform atmosphere (H*): up to 200 km altitude, on the basis of data in Ref. 10; above 200 km, on the basis of

TABLE 3

H, km	ρ, g×cm^{-3}	H*, 10^5 cm	n, cm^{-3}	v', cm/sec	nv'
300	3.4·10^{-14}	42	1.1·10^9	3.00·10^6	3.3·10^{15}
275	7.0·10^{-14}	37	2.1·10^9	3.00·10^6	6.3·10^{15}
225	3.5·10^{-13}	29	1.0·10^{10}	2.98·10^6	3.0·10^{16}
200	6.1·10^{-13}	25	1.7·10^{10}	2.97·10^6	5.0·10^{16}
150	4.0·10^{-13}	16	1.1·10^{11}	2.88·10^6	3.2·10^{17}
140	7.0·10^{-12}	14	2.0·10^{11}	2.82·10^6	5.6·10^{17}
130	1.4·10^{-11}	12	4.0·10^{11}	2.70·10^6	1.1·10^{18}
120	3.7·10^{-11}	10	1.1·10^{12}	2.40·10^6	2.6·10^{18}
110	1.2·10^{-10}	8	3.3·10^{12}	1.68·10^6	5.6·10^{18}
100	5.5·10^{-10}	7	1.5·10^{13}	3.00·10^5	4.5·10^{18}
95	1.0·10^{-9}	7	2.7·10^{13}	3.00·10^4	8.0·10^{17}
90	4.0·10^{-9}	6	8.6·10^{13}	3.84·10^2	3.3·10^{16}
80	2.5·10^{-8}	6	5.7·10^{14}	5.80·10^1	3.3·10^{16}

data in Ref. 9. As the figures in the sixth column of Table 3 indicate, the product nv', and therefore K, exhibits a maximum at altitudes near 105 km. For higher altitudes the product nv' rapidly decreases, and for lower ones it declines steeply toward a constant value (in the region of free fall, where the rate of fall of the particles is inversely proportional to the density of the medium, i.e., below 90 km). Thus, the specific content of cosmic particles in the upper layers of the atmosphere is different at different altitudes. This circumstance permits us to suppose that in the region 95-150 km the earth's atmosphere may possess some specific properties which distinguish it from the other layers of the upper atmosphere. In order to bring out these properties, a coefficient of proportionality a_i may be introduced into formula (3) to characterize the physical properties of particles as manifested in various processes (in the scattering of light, this would be the scattering power of dust particles in comparison with that of molecules; in the case of ionization, a coefficient characterizing the "cross section" for the ionization process for dust particles of various masses and velocities, $a_i = f(m, v)$, and so on).

$$K_i = \frac{a_i}{D} nv'. \tag{4}$$

As a typical example, let us consider the phenomenon of scattering in the atmosphere at high altitudes.

Scattering of Light by Cosmic Dust Particles*

The intensity of scattered light in the upper atmosphere produced by meteoric material depends on the size and number of the individual dust particles causing the scattering. Generally speaking, the larger the size of a particle, the more strongly it scatters light. However, in our case, where the total mass of the scattering meteoric material is fixed in advance, an increase in the sizes of particles leads to a sharp reduction in their number, which under certain conditions will not be offset by the growing "scattering coefficient."

Thus, for example, for particles with $r > 0.3$-$0.4\,\mu$, i.e., for all particles encountered in cosmic dust, the scattering power increases as r^2, but their number decreases as r^3, hence with increasing r the total intensity of scattered light produced by cosmic dust will fall. Because of this, the scattering of light by large particles in the upper atmosphere may be neglected, and it may be assumed that the principal amount of scattered light is determined by scattering from particles of 0.3-$0.8\,\mu$, the smallest cosmic dust particles, which enter the earth's atmosphere in the greatest numbers. The sizes indicated include meteors ranging over three or four stellar

*First order scattering is considered here. Higher-order scattering may be neglected at such high altitudes.

magnitudes (from +29 to +26), and therefore, given a Watson mass distribution, they account for 500 tons of material per day. From the assumptions made concerning the quantity of material and the sizes of particles entering the earth's atmosphere, and taking the density of meteoric matter to be $\delta = 4$, we find that the flux D of particles of the four stellar magnitudes with mean radius 0.4μ is 10^{-3} cm^{-2} sec^{-1}.

In order to determine the character of the scattering in the atmosphere, the coefficient a_i in (4) must be determined. This coefficient depends on the ratio α_μ/α_n, where α_n and α_μ are the scattering coefficients for dust particles and air molecules, respectively. However, in order to decide the character of scattering in the atmosphere, it is not sufficient merely to compare integrated intensities; it is also necessary to take into account the indicatrices of scattering, which sharply distinguish light scattered by dust particles from light scattered by molecules. Dust particles, having an elongated indicatrix of scattering, concentrate most of the scattered light in a narrow beam, while molecules, having Rayleigh indicatrices, scatter light approximately equally in all directions. Therefore, in making a comparison of the intensity of light scattered by dust particles and by molecules in the direction of the principal beam, one should consider in the case of molecular scattering only the small portion of the scattered luminous flux contained within the solid angle in which the light scattered by the dust particles is propagated. In other words, if the dust particle scatters light in a solid angle γ, then for appropriate comparison with molecular scattering, the fraction of the total luminous flux defined by the coefficient $\gamma/4\pi$ should be considered.

Returning to equation (4), we may now write

$$\alpha_i = \frac{\alpha_\mu}{\alpha_n}\frac{\gamma}{4\pi},$$
$$K_i = \frac{1}{D}\frac{\alpha_\mu}{\alpha_n}\frac{\gamma}{4\pi}nv'. \qquad (5)$$

Representing the combination of all constant terms in equation (5) by

$$\sigma = \frac{1}{D}\frac{\alpha_\mu}{\alpha_n}\frac{\gamma}{4\pi}, \qquad (6)$$

we have

$$K_i = \sigma n r'. \quad (7)$$

Let us determine the constant σ in (7). The scattering coefficient α_μ, calculated for one air molecule[11,12] at $\lambda = 0.5\mu$, is $\alpha_\mu = 6.7 \times 10^{-27}$ cm^2. The scattering coefficient α_n for a dust particle of 0.4μ radius may be calculated by using the well-known expression

$$\alpha_n = \pi r_n^2 K(\rho), \quad (8)$$

where r_n is the radius of the dust particle, $\rho = (2\pi r_n/\lambda)$, $K(\rho)$ is the scattering function tabulated in Refs. 13 and 14. The value of the function $K(\rho)$ will depend on the index of refraction and varies, for $\rho = 5$ ($\lambda = 0.5\mu$, $r_n = 0.4\mu$), from a value of 2 for opaque absorbing particles to 4 for transparent particles. Since the index of refraction for cosmic dust particles has not been sufficiently studied, we shall adopt a mean value $K(\rho.) = 3$ in these calculations. This appears all the more justified since, in our rough calculations where we are concerned with correct order of magnitude estimates, a change in $K(\rho)$ even by a factor of 1.5-2 will not affect the result obtained.

Taking into account what has been said above, we obtain in accordance with (8)

$$\alpha_n = 1.5 \cdot 10^{-8} \text{ cm}^2, \quad \frac{\alpha_\mu}{\alpha_n} = 4.5 \cdot 10^{-19}.$$

The value of this ratio will increase with decreasing λ.*

It may be assumed[13] that a dust particle of 0.4μ radius ($\rho = 5$) sends out its main stream of scattered radiation in a cone of angle not less than 20-25°. For this case $\gamma/4\pi = 10^{-2}$. Substituting all of the values obtained into expression (6), we have

$$\sigma = 4.5 \cdot 10^{-18},$$

and (7) assumes its final form:

$$K_i = 4.5 \cdot 10^{-18} n v'. \quad (9)$$

*For $\lambda \approx 0.4\mu$, $\frac{\alpha_\mu}{\alpha_n} = 10^{-18}$.

Formula (9) permits an estimate of the character of the scattered light in the atmosphere. The quantity K_i ultimately represents the ratio of the intensity of scattered light due to the molecules present in unit volume to the intensity due to dust particles in the same volume. Therefore the condition $K_i = 1$ expresses the equality of these intensities in the direction of the principal angle of scattering by a dust particle; in all other directions, for $K_i = 1$, molecular scattering (blue light) will exceed by 2 to 3 orders of magnitude the intensity of white light scattered by the dust particle. For the case where $K_i = 1$, the over-all indicatrix of scattering in the atmosphere will still differ strongly from the Rayleigh indicatrix; its asymmetry (the intensity of scattered light in the direction of the source as compared with the intensity of scattered light in the opposite direction) will be 2, instead of 1 as for the Rayleigh indicatrix. Choosing the condition $K_i = 6$, we find that the asymmetry of the over-all indicatrix differs little from 1 (Refs. 4 and 15); in this case, it may be assumed without serious error that Rayleigh scattering is observed at the given height in the atmosphere. Thus, a certain criterion is established for determining the character of scattered light: if $K_i \geq 6$, it is Rayleigh scattering; if $K_i < 6$, it is a non-Rayleigh type of scattering.*

In order to obtain concrete values of K_i at different heights in the atmosphere, the values of nv' listed in the fifth column of Table 3 are to be substituted into (9). These values are collected in Table 4, where in addition to K_i calculated for the maximum fallout of cosmic matter, 5000 tons/day, (K_{i5000}) figures are given which characterize K_i for other amounts: 1000, 500, and 200 tons/day.†

From the data given in Table 4, the following conclusions may be drawn:

1. At an altitude between 92 and 96 km in the atmosphere there must be a sharp transition between scattering of non-

---

*Depending upon the precision with which the character of the scattering must be determined, the boundary may be moved either way.

†It should not be forgotten that of all the amounts indicated, in accordance with the Watson distribution, only that fraction of the material included within a corresponding interval of three or four stellar magnitudes is actually used in studying the scattering of light.

Rayleigh type ($K_i < 6$) and Rayleigh scattering ($K_i \geq 6$). The stability of this boundary is worthy of note, for it hardly varies in height with variation in the amount of meteoric dust entering the atmosphere.

TABLE 4

	H, km										
	220	200	150	140	130	120	110	100	95	90	80
$K_{i\,5000}$	0.03	0.2	1.3	2.6	5.4	11	31	30	11	0.2	0.2
$K_{i\,1000}$	0.1	1	6.5	13	27	55	155	150	55	1	1
$K_{i\,500}$	0.3	2	13	26	54	110	310	300	110	2	2
$K_{i\,200}$	0.7	5	32	65	135	275	755	750	275	5	5

2. Above 92-96 km there is a region of Rayleigh scattering which extends upward to an altitude that depends on the amount of incoming meteoric material we assume. As is evident, the boundary of the upper transition, from the condition $K_i \geq 6$ to

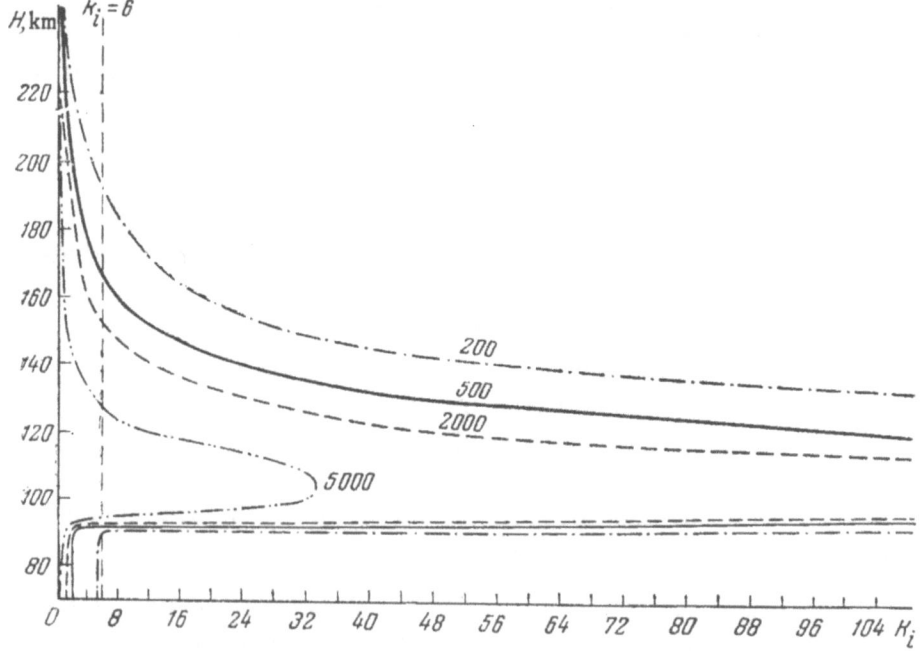

FIG. 1. Dependence of K_i on altitude. The different curves characterize $K_i = f(H)$ for different amounts of meteoric material coming into the earth's atmosphere (the figures on the curves give the number of tons per day).

to the condition $K_i < 6$, occupies different heights for different cases.

From Fig. 1 it is evident that with a change in the amount of material from 5000 tons to 200 tons the upper boundary of the region of Rayleigh scattering is shifted from an altitude of 130 km to about 190 km.

3. Above the upper boundary, where K_i again becomes less than 6, the scattering must be of a non-Rayleigh type.

Thus, within the limits of the assumptions made, there must exist in the atmosphere a layer which is characterized by a Rayleigh indicatrix of scattering. The layer is sharply bounded at the bottom and less sharply bounded at the top. Some diffuseness of the upper boundary is produced by the initial velocities of the incoming dust particles (Fig. 2).

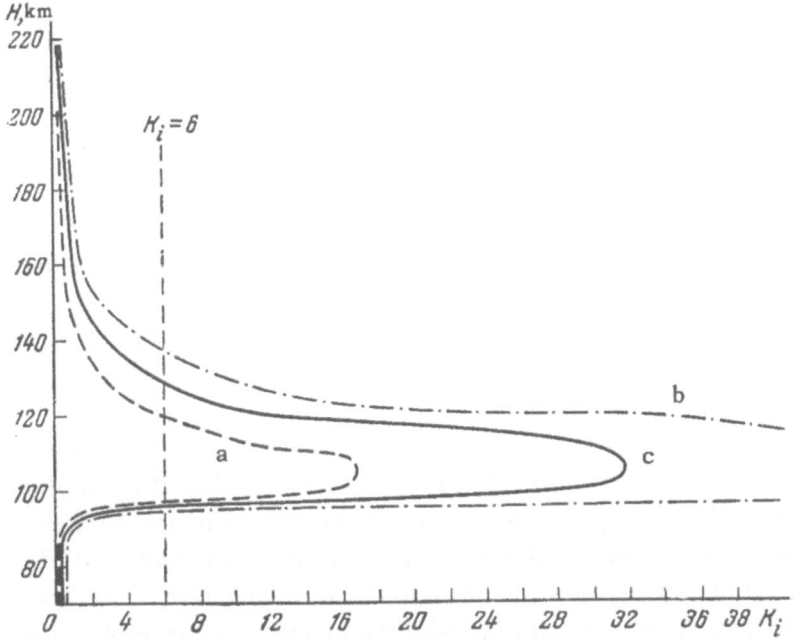

FIG. 2. Dependence of K_i on altitude for various initial velocities of the meteor particles and for M = 5000 tons/day: a) velocity $v_0^v = 1.2 \cdot 10^6$ cm/sec; b) $v_0 = 3 \cdot 10^6$ cm/sec; c) $v_0 = 6 \cdot 10^6$ cm/sec.

The thickness of the "Rayleigh scattering layer" will vary according to the amount of cosmic dust entering the earth's atmosphere. The layer may disappear entirely only if the amount of meteoric dust should exceed by one and a half orders of magnitude the maximum possible amount of 5000 tons per day assumed by us, which is unlikely. On the other hand, if the amount of meteoric matter in the atmosphere should be considerably smaller, on the order of 20-25 tons/day, then only the upper boundary of the layer lying at a height of 400-420 km would be observed in the atmosphere. If we ignore the presence of terrestrial dust in the lower layers of the earth's atmosphere, as well as the mixing processes which determine the dustiness of these layers, then the region of Rayleigh scattering will occupy the entire thickness of the atmosphere from the surface of the earth to a level of 420 km. However, in reality this does not happen because the dust of terrestrial origin, necessarily present in the layers near the ground, forms at the bottom its own boundary for the Rayleigh scattering layer.

METEORIC MATERIAL AND THE LUMINOSITY OF THE NIGHT SKY

The energy of meteor particles entering the earth's atmosphere is very high. Because of this it is possible for a particle to alter the state of air molecules it encounters: to impart additional kinetic energy to them, to cause dissociation, ionization, excitation, and so forth.

Apparently, the principal role in the interaction of a micrometeor with the surrounding medium is played by molecules of the medium itself, which rebound from its surface in the act of direct collision. These molecules move relative to the slow molecules of the undisturbed gas with velocities somewhat greater than the velocity of the meteor itself, and as a result of elastic and inelastic collisions produce the changes described.

The peculiar process of energy loss by the meteor in the upper layers of the earth's atmosphere, which manifests itself in an abrupt retardation over a small portion of its path, is responsible for the fact that its energy is spent in different ways at different altitudes. Figure 3 shows the variation of velocity with altitude for two typical kinds of micrometeor

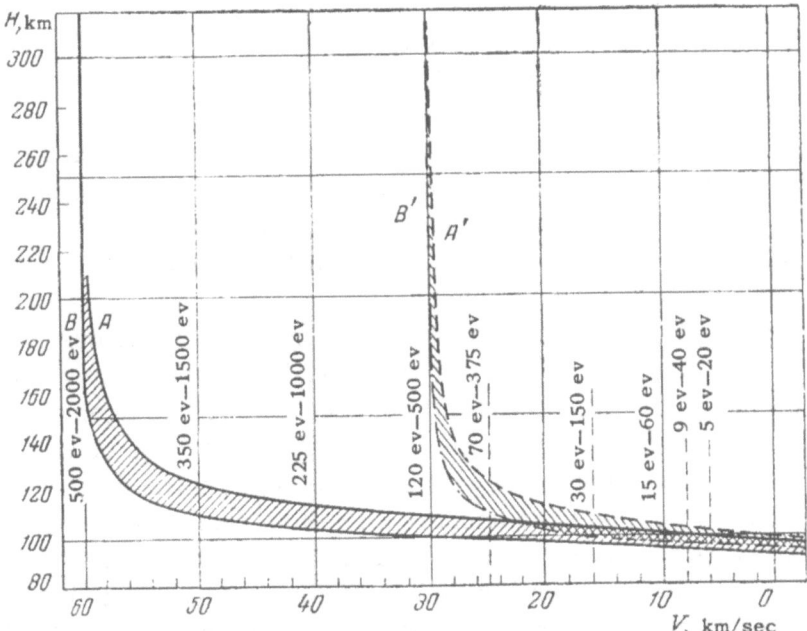

FIG. 3. Velocity variation of micrometeors in passing through the earth's atmosphere. Curves A and B are for meteors of radius 0.4 and 1 μ, respectively, with initial velocity $v_0 = 6 \times 10^6$ cm/sec. Curves A' and B' likewise, but with $v_0 = 3 \times 10^6$ cm/sec. Energies (in electron-volts) are given for N_2 and O_2 molecules rebounding from the meteor.

($r = 0.4 \mu$ and $r = 1\mu$) with two typical initial velocities (3×10^6 and 6×10^6 cm/sec) on the assumption that the density of the material is $\delta = 4$.* All the curves indicate that the principal energy loss of the most numerous meteor particles occurs in a thin layer of the atmosphere between 115 and 95 km in altitude. It must be emphasized that the bearers of this "lost" energy are fast molecules, which are capable of vigorous reactions even at altitudes of 95-100 km.

In calculations of the amount of energy brought into the earth's

*New estimates of meteoric density made by Whipple (0.1–0.01 g·cm⁻³) refer to comparatively large particles. The finest particles of meteoric dust (stellar magnitude +25 to +30) may not have a friable structure and hence such low density.

atmosphere by micrometeors, different authors give different values: thus, in Ref. 15, 0.02 erg $\times$ cm^{-2} $\times$ sec^{-1} is given for this quantity, in Ref. 16, 0.004 erg $\times$ cm^{-2} $\times$ sec^{-1}, in Ref. 17, 0.1 erg $\times$ $\times$ cm^{-2} $\times$ sec^{-1}.* In our own rough calculations, we shall assume the mean value given in Ref. 15. Taking into account the energy of larger meteors which also enter into the energy balance of the layer of micrometer retardation (115-95 km), it may be assumed that an energy of 0.04 erg is liberated per second in a column of 1 cm^2 cross section. This energy is not very large; however, its contribution to the energy producing the luminosity of the night sky may prove to be very substantial.

The luminosity of the night sky is made up of various emissions concentrated at different heights in the atmosphere. These emissions are apparently caused by various agents. The emission of the green line at λ = 5577.3 A may be explained by the influence of sporadic meteors on the upper layers of the earth's atmosphere. The hypothesis that the night sky glow may be caused by meteoric bombardment is not new. It will be recalled that Humphreys suggested such an idea as early as the Twenties. It is now possible, on the basis of broad experimental data, to explain the complex effects of night sky emission (in the region of the green emission at λ = 5577.3 A) as due to the influence of sporadic meteors.

Such an assumption seems quite probable: the mean intensity repeatedly measured for the green line of the night sky leads to a value of the order of 4×10^{-4} erg $\times$ cm^{-2} $\times$ sec^{-1}, i.e., it constitutes about 1% of the retardation energy of meteors in the atmospheric layer between 115 and 95 km.

At the present time it is difficult to point to a concrete mechanism for the transformation of a meteor's kinetic energy into radiant energy at λ = 5577.3 A, since the cross sections for dissociation, ionization, and excitation processes involving collisions of molecules with one another at relatively low energies (tens and hundreds of electron volts) have been studied little. It may be assumed that these cross sections are small, but apparently the emission of the green line

*If we assume that approximately one third of the total energy of meteors is due to micrometeors.

is supported by various processes which take place following the meteor's passage.*

In spite of the vagueness of the mechanism of energy transfer, such a transfer does occur with appreciable intensity; this has definitely been shown in Ref. 18. In photographing the emission spectra of high meteor trails, Halliday observed in them, without any doubt, the bright green line of the night sky (λ = 5577.6 A). This line behaves quite differently from all the other lines in the meteor spectrum: in spite of the fact that the photography was carried out with a rotating shutter and all the usual lines of the spectrum were interrupted, the line at 5577.6 A was continuous, which is indicative of a prolonged afterglow. Extremely important also is the fact that the observed line appears earlier than the other meteoric lines and, remaining for a short time, disappears before the usual lines of the meteor spectrum itself fade out. The latter property indicates the atmospheric nature of the green line.

Any hypothesis concerning the luminosity of the night sky must explain not only the theoretical possibility of the existence of this emission, but also all the complexity and diversity of the processes which characterize the behavior of the green line with time.

Until very recently, the height of the luminescent layers had been very uncertainly determined, and only the latest measurements, made with the aid of rockets[19-22] have permitted a reliable determination of the height of the 5577 A emission layer; it turned out to be 90-118 km,† i.e., just that altitude in the neighborhood of which micrometeors give up their maximum energy to the air molecules.‡ The height of the 5577 A luminescence observed in Ref. 18 was somewhere

*The basic processes responsible for emission in the 5577.3 A line are those mentioned by Chapman: 1. $O_2 \to O + O; O + O + O \to O_2 + O^*$;
2. $N_2^+ + O^- \to N_2^* + O^*$.

†Recent experiments[23-25] to determine the altitude of the 5577 A emission layer, performed "from the ground," have also yielded an altitude of about 100 km.

‡The center of the layer's luminescence is at an altitude of 95-105 km.

in the interval 110 to 97 km, in spite of the fact that the meteor penetrated considerably deeper into the atmosphere. This means that at altitudes of 110-97 km in the atmosphere, conditions are favorable for the occurrence of the forbidden transition of the oxygen atom $'S \rightarrow {}'D$. Below 97 km, conditions become unfavorable for the emission of the green line because of the sharply rising probability of de-excitation of the excited oxygen atoms by collisions. This phenomenon should set a rather sharp lower boundary for the 5577 A emission layer, and this is in fact observed.[22] The atmospheric property described will cause the luminescence produced by any large meteor to be concentrated near an altitude of 100 km, as if the meteor had not penetrated deeply into the earth's atmosphere. The upper boundary must be considerably less sharp. It is formed by the decrease in luminescence due to the falling density of the medium itself.

The observed trend of intensity for the green line in the course of the night has not found satisfactory explanation in existing theories. As is known, the intensity of the green line changes continuously in the course of a single night, as well as from night to night. These variations are, as a rule, not systematic. However, amid all this diversity there is one outstanding regularity, the character of which has been noted by all observers: the maximum luminescence occurring about 1 or 2 hours after midnight. If the daily trend in the numbers of sporadic meteors is examined,[16] it can be seen that it displays precisely the same form, the post-midnight maximum being just as distinctive as in the case of the green line. Figure 4 shows a comparison of the characteristic trends of these two phenomena. The correlation goes even further: the relative increase in the intensity of the 5577 A line during the night is typically 2-3 times; approximately the same increase is observed in the daily run of sporadic meteor rates.

Up to now, not one hypothesis concerning the luminosity of the night sky has even attempted to explain the annual variation in the intensity of the 5577 A line, which is displayed in the fact that the intensity of the green line reaches a maximum in the winter months and a minimum in the spring. Some observers[25] point out the presence of two maxima, in November and February, others only one, in November.[26] The minimum of emission always falls in the period March-June. If

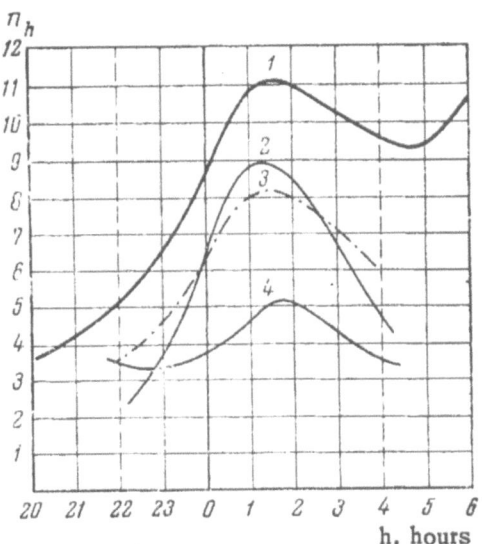

FIG. 4. Comparison of the nightly variation of sporadic meteor rates: 1) with the variation in the intensity of the green line; 2, 3) from Ref. 23; 4) from Ref. 26. Intensities of emission have been reduced to relative units, the scale being different for the various curves.

this trend is compared with the annual run of sporadic meteor rates, good agreement is observed (Fig. 5). An exception in the run of green line intensities is shown only by the February maximum, which is not reflected in the meteor trend. True, this maximum is considerably smaller than the principal one and is not recorded by all observers. Possibly it is of a random, fluctuating character.

Besides the "regular" variations spoken of above, there are well-known irregular changes in the intensity of the green line[27] which indicate an erratic occurrence of the processes responsible for the 5577 A emission. It should be noted that the generally erratic character of the luminosity in the region of the green line, a typical "raggedness" in the luminosity of the night sky, are in the best possible harmony with the idea of a meteoric agent as the cause of this glow, because it need not be expected that the stream of meteoric particles constantly entering the earth's atmosphere is spatially and temporally stable. In Refs. 28-31, the conclusion is drawn that a

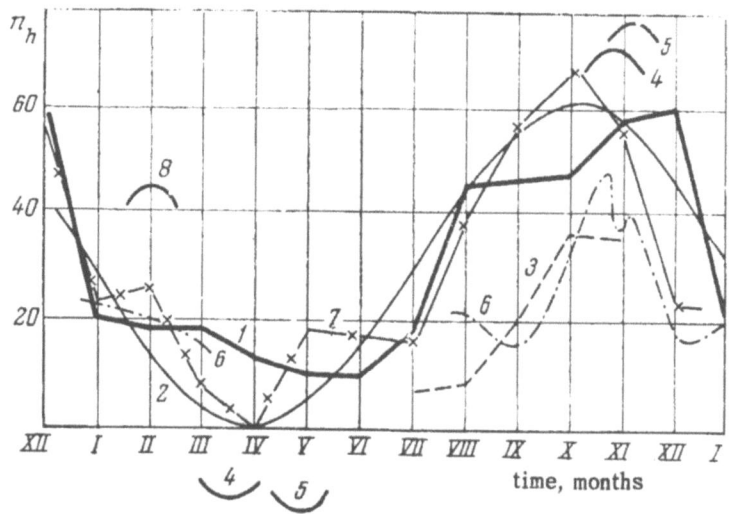

FIG. 5. Comparison of the annual variation of sporadic meteor rates: 1) with the annual variation in the intensity of the green line; according to Rayleigh and Jones (1923-1934); 3) Barbier (1941); Garrique (1946), maximum and minimum emission; 5) Dufay and Mao-lin (1946), maximum and minimum emission; 6) Roach (1951); Dufay (1941-1943); 8) secondary maximum noted by Dufay and Rayleigh. Intensities have been reduced to relative units which differ for different observations. All data are for the northern hemisphere.

characteristic feature of the luminosity of the night sky is the presence of very irregularly shaped regions which differ in brightness and move about on the celestial sphere in the course of the night. The average size of these optical nonuniformities is of the order of 2500 km, their speed of motion is 70-90 m/sec, and the predominant direction from northeast to southwest, although other directions are encountered.

The authors of these papers[28-31] note that, in the first place, the phenomenon observed by them does not agree with Chapman's theory and, in the second place, the observed motion of the nonuniformities may be related to macroscopic motions of the radiating oxygen atoms caused by winds presumably originating at heights of the order of 100 km.

Nonuniformity of the luminosity of the night sky in the emission line at 5577 A is a direct consequence of the hypothesis of the meteoric origin of this phenomenon. According to this hypothesis, the motion of the nonuniformities that arise should have a predominant direction from northeast to southwest.* Under certain conditions, other directions might also be observed.† According to our supposition, the motion of the regions observed by Roach is not related to atmospheric movements: the meteor "beam" sweeps over the atmosphere and produces a glow in the parts of the sky which it "illuminates." Thus, what has been recorded in the papers referred to is apparently some internal structure of the streams of sporadic (and other) meteors which cause the sky glow above a given station.

Among the peculiarities in the behavior of the green line, some relationship between it and solar activity, and with latitude, has been noted. In the first case, an increase in intensity is observed with the growth of solar activity, and in the second case, an increase in intensity from the equator to the pole.[26,27] Both of these effects find their place in the meteoric hypothesis of the luminosity of the night sky, if it is assumed that cosmic particles carry an electric charge. At the present time, the existence of such a charge, which is bound to appear on dust particles as a result of a number of processes taking place outside the atmosphere (photoelectric effect, attachment of free electrons, etc.), can hardly be doubted. Problems concerning the charge on particles of cosmic matter have been discussed in Refs. 32 and 33. Recently these problems were considered in greater detail by Singer,[34] who showed that at times of increased solar activity dust particles may attain high velocities in the direction away from the sun. Such "hard" particles, striking the earth's atmosphere, should increase

*The resultant motion of the region measured by a terrestrial observer is made up of motions in which the earth participates and motions of groups of meteoric particles which cause the luminescence of a particular optical nonuniformity.

†In general, curvilinear motions in the directions stated should predominate.

the luminescence in the green line, which has probably been noted by some observers.

In the case of charged particles, the latitude effect also becomes easy to explain: the dust particles, falling into the earth's magnetic field, will enter the earth's atmosphere in large numbers in the polar regions, where they cause more intense luminescence.

The list of "coincidences" between the character of the 5577 A emission and meteoric phenomena could be continued to include the relationship of this emission to the earth's magnetic field, to disturbances in the E and F layers, to the Hoffmeister bands, to the presence of sodium lines in spectra of night sky emission,* and so on.

Thus the data available at the present time provide a basis for advancing the hypothesis of the meteoric (micrometeoric) origin of the luminosity of the night sky in the 5577.3 A line as one of the most likely hypotheses for this phenomenon.

LITERATURE

1. O. D. Komissarov, T. N. Nazarova, L. N. Neugodov, S. M. Poloskov, and L. Z. Rusakov, Artificial Earth Satellites, Volume 2, Plenum Press, New York, 1960.
2. M. Dubin, Cosmic Debris of Interplanetary Space, Second OSR Astronautics Conference, Denver, 1958.
3. H. E. LaGow, D. H. Schaefer, J. C. Schaffert. Micrometeorite impact measurements on a 20" diameter sphere at 700 to 2,500 km altitude, U. S. Naval Res. Lab., Washington, 1958.
4. A. Kh. Khrgian, Atmospheric Physics (in Russian), Moscow, GTTI, 1953.
5. J. W. Townsend, C. J. Johnson, J. C. Holmes, E. B. Meddows, J. G. J. Rocket Rep. Series, 1958, p. 131.
6. B. A. Mirtov, Uspekhi Fiz. Nauk 63, 181 (1957).
7. B. Yu. Levin, The Physical Theory of Meteors and Meteoric Material in the Solar System (in Russian), Izd AN SSSR, 1956.
8. F. Watson, Between the Planets (Russian translation), Gostekhizdat, Moscow, 1947.

*Sodium lines in the spectrum of the night sky apparently have a meteoric origin. They are observed in the layer where the most intense ablation of meteors occurs (80–95 km).

9. V. V. Mikhnevich, B. S. Danilin, A. I. Repnev, and V. A. Sokolov, see Volume 3, p. 119.
10. Mémoires de la Société Royale des Sciences de Liège, 1952.
11. E. O. Hurlburt, J. O. S. A., 43, No. 2, 1953.
12. K. Ya. Kondrat'ev, The Radiant Energy of the Sun (in Russian), Gidrometizdat, 1954.
13. K. S. Shifrin, Scattering of Light in a Turbid Medium (in Russian), Gostekhizdat, Moscow, 1951.
14. G. Gertner, Transparency of a Cloudy Atmosphere to Infrared Light (in Russian), GTTI, Moscow, 1949.
15. F. Whipple, Bull. Am. Net. Soc., 33, 13, 1952.
16. A. C. B. Lovell, Meteor Astronomy (Russian translation), Fizmatgiz, Moscow, 1958.
17. I. S. Astapovich, Meteoric Phenomena in the Earth's Atmosphere (in Russian), Fizmatgiz, Moscow, 1958.
18. J. Halliday. Astrophys. J., 128, 441, 1958.
19. O. E. Berg, M. Koomen, M. Meredith, R. Scolnik, J. Geophys. Res., 61, 302, 1956.
20. M. Koomen, R. Scolnik, R. Tousey, J. Geophys. Res., 61, 304, 1956.
21. M. Koomen, R. Scolnik, R. Tousey, Chem. Aeronomy, London, 1957.
22. J. P. Heppner, L. H. Meredith, U. S. Naval Res. Lab., 1957.
23. H. Elsässer, H. Siedentopf, J. Atm. Terr. Phys., 8, 222, 1956.
24. F. E. Roach, A. B. Meinel, Astrophys. J., 122, 530, 1955.
25. F. E. Roach, Annalen Geophys., 11, No. 2, 1955.
26. I. A. Khvostikov, The Night Sky Glow (in Russian), Izd AN SSSR, 1948.
27. E. K. Mitra, The Upper Atmosphere (Russian translation), IL, 1955.
28. F. E. Roach, H. Pettit, J. Geophys. Res., 56, 325, 1951.
29. F. E. Roach, L. R. Megill, M. H. Rees, J. Atm. Terr. Phys., 12, 171, 1958.
30. F. E. Roach, E. Tandberg-Hanssen, L. R. Megill, J. Atm. Terr. Phys., 13, 113, 1958.
31. F. E. Roach, E. Tandberg-Hanssen, L. R. Megill, J. Atm. Terr. Phys., 13, 122, 1958.
32. H. Alfen, On the Origin of the Solar System, Oxford, 1954.
33. L. Spitzer, M. P. Savedoff, Astrophys. J., 111, 593-608, 1950.
34. S. F. Singer, Scientific Uses of Earth Satellites, 1956.

MAGNETOMETERS IN THE THIRD SOVIET EARTH SATELLITE

S. Sh. Dolginov, L. N. Zhuzgov, and V. A. Selyutin

INTRODUCTION

Measurements of the magnetic field of the earth in the upper atmosphere, specified by the program of the International Geophysical Year, were carried out in May and June of 1958 with the third Soviet earth satellite.[1]

The first data to be obtained were detailed data on the spatial distribution of the magnetic field of the earth at 230-1880 km. Although obtained principally over the European-Asiatic region, to south of the 65th parallel, the experimental data allow us to resolve general problems of geomagnetism, since the magnetic field of the territory of the Soviet Union contains all the structural features of the magnetic field pertinent to the global sphere: the world-wide anomaly and regional and local anomalies.

The period during which the magnetic measurements were carried out (15th of May to 5th of June) was both turbulent and quiet as far as the magnetic field was concerned and thus serves for investigation of problems touching on the variable part of the magnetic field of the earth.

The experimental data are being processed and analyzed. The results will be published as the work is completed.

In the present paper we describe the magnetometers which were used, the problems associated with installation and operation of the magnetometers in the satellite, data characterizing the stability and accuracy of operation of the instruments, the nature of the experimental data which were obtained, and methods of taking account of the magnetic deviation.

GENERAL APPARATUS REQUIREMENTS

The method of measurement and the characteristics of the magnetometer were determined by the geophysical problems of the experiment as well as conditions in the satellite.

The measurements of the magnetic field in the upper atmosphere were carried out for the purpose of obtaining experimental information concerning the spatial distribution of the magnetic field of the earth, the homogeneity of this field, and the nature and absolute values of the gradients.

Another problem was understanding the nature of changes in the earth's magnetic field at high altitudes during magneto-ionospheric disturbances.

The magnetometer specifications (sensitivity, measurement range, time delay) were determined from the basic problems of the experiment.[2] To evaluate these requirements one can use the values of the normal gradients of the field as determined by calculation and from terrestrial charts. As is well known, the vertical gradient of the total vector for the dipole part of the earth's field is determined from the formula $(\Delta T/\Delta R) = (3T/R)$ where R is the distance from the center of the earth. The latitudinal gradient of the normal field at the surface of the earth is approximately 3.5 γ/km ($\gamma = 10^{-5}$ oe). For a given orbit one would expect that the minimum values of the field (above the equatorial region) could reach 15,000γ. The maximum value of the field allowed for measurements by the magnetometer was chosen from geophysical and practical considerations to be 60,000γ.

The estimates of possible effects due to field of ionospheric currents on the system were somewhat indefinite because an analysis of the terrestrial variations in the magnetic field does not uniquely determine the intensity of the equivalent current systems. At the surface of the earth the amplitude of the diurnal variations on a quiet day (S_q) averages 40-50 γ, that is to say approximately 0.1% of the constant field of the earth. The current systems, the thickness of which is small compared with the transverse dimensions (current systems at the lower latitudes), may be compared to a fine, uniform current layer.

For this reason the field close to such a layer should not differ appreciably from that which is observed at the surface of the earth. However, in magnetometer measurements at altitudes of 90-105 km effects of the order of 400γ due to current systems have been observed.[3]

The magnitude of the turbulent diurnal variations S_d varies from 40 to 300γ depending on the magnetic activity and the latitude.

According to a spherical analysis highly non-uniform currents of considerable intensity flow in relatively narrow bands in the polar regions. The intensity of the field due to these current systems is inversely proportional to the distance to the current system. Consequently, at high latitudes one may expect the effects due to current systems to reach several thousand gamma.

The amplitudes of the oscillations of the magnetic elements during the time of magnetic storms vary over extremely wide limits. According to an estimate made by Chapman[4] the maximum effect of a current system in a magnetic storm may reach 27,000 γ.

Thus, it is apparent that the magnetometer in a satellite must be capable of a wide range and must have high sensitivity. With a satellite velocity of 8 km/sec the device must be capable of measuring variations in magnetic field occurring at the rate of 300 γ/sec.

Another important characteristic of the apparatus is the stability of the zero reading. A comparison of magnetic data obtained at the beginning and end of the period of operation of the magnetometer can be made only if the zero setting is stable or if its variation is regular and subject to investigation under laboratory conditions. An acceptable limit may be taken as approximately 50γ per day; with an instability of this kind in the zero reading the uncertainty in the measured field values (after the introduction of appropriate corrections) is no worse than the errors in terrestrial magnetic charts. The error in present-day magnetic charts is, as is well known, on the average 200-300 γ.

CHOICE OF THE TYPE OF MEASUREMENT APPARATUS

As is well known, at each point in space the magnetic field of the earth is characterized by a magnitude and direction. It will be apparent that the most fruitful possibility for further analysis and comparison with terrestrial values of the magnetic field will be an apparatus which measures the magnitude and direction, or the components, of the magnetic field. Most magnetic sensing units or transducers measure the field components directly.

In its motion in orbit the satellite continuously changes its orientation in space (therefore with respect to the earth's field) so that measurement of an individual component (for instance, the vertical component) would entail great technical difficulties.

For this reason, the most feasible approach lies in magnetometers which measure the scalar magnitude of the field. Such devices are the nuclear-induction magnetometer and magnetometers which make use of magnetic-saturation transducers.

Nuclear-Induction Magnetometer

The nuclear-induction magnetometer measures the intensity of a magnetic field by using the free precession of protons in an external magnetic field. The proton free-precession frequency in a magnetic field is given by the Larmor relation $\omega = \gamma_p H$, where γ_p is the gyromagnetic ratio and H is the strength of the magnetic field.

The intensity of the earth's field can be determined by measuring the frequency of free precession of protons in this field. The method of observing free precession of protons in the earth's magnetic field (Packard and Varian[5]) is extremely simple. To a sample (a liquid with a high proton content located in a field H_0) for a short time we apply a strong field H (the order of 100 oe), which is then shut off suddenly. The field H is set up by an excitation coil which encloses the sample and is approximately perpendicular to the field H_0. Under the effect of the field H the sample experiences a macroscopic magnetization of strength $I_n = \chi_n H$, where χ_n is the nuclear susceptibility. When the field H is switched off the macroscopic magnetic moment produced in the sample starts to precess freely about the field H_0 at a frequency $\omega = \gamma_p H_0$. The magnitude of the macroscopic magnetic moment of the sample falls off gradually but the relaxation time is approximately three seconds and is sufficient for a measurement of the frequency of the alternating voltage which is induced in the signal coil by the precessing moment of the sample; this coil is actually the excitation coil, which is now connected to an amplifier. The alternating electromotive force induced in the coil is

$$E = K\chi_n\gamma_p H_0 H \sin^2\theta e^{-\frac{t}{T_2}},$$

where K is a constant which depends on the coil parameters; θ is the angle between the field H and H_0; t is the time after the field is switched off; and T_2 is the relaxation time.

Measurement of a field by the nuclear-induction method has several advantages as compared with other methods of measurement: 1) the field measurement is essentially a frequency measurement; 2) the accuracy of the measurement is independent of the coil parameters and the apparatus is used to amplify the signal; 3) the magnetometer readings are given in absolute values; 4) the pick-up unit and the amplifying channels for the signal are, in principal, free from zero drift; the accuracy of the measurement is determined only by the accuracy of the frequency standard; 5) the results of the measurement for a fixed position of the pick-up unit are independent of the orientation of the pick-up unit in the field being measured.

The signal from the pick-up coil, which is amplified by an audio-frequency amplifier, can be transmitted by telemetering and measured accurately at remote distances (the signal can be used directly to modulate the transmitter). The accuracy of measurements by proton magnetometers in an observatory was estimated as 1γ (10^{-5} oe), and is limited only by the accuracy of the determination of γ_p.

When this method is used in moving objects certain complications arise which make the measurement difficult and which increase the measurement errors: 1) although the precession frequency is independent of the orientation of the excitation coil with respect to the field being measured the amplitude of the signal is proportional to $\sin^2 \theta$ and can become vanishingly small at small values of θ; 2) if the coil is fixed in a moving object which rotates with an angular velocity $\dot{\varphi}$ around an axis perpendicular to the coil axis, the field is measured with an error $\Delta H = \pm 3.7 \dot{\varphi}$ (ΔH is measured in gammas while $\dot{\varphi}$ is measured in radians per second)[6]; 3) the non-uniformity of the magnetic field over the volume of the magnetometer pick-up unit reduces the relaxation time T_2: even for field gradients of $3-4\gamma$/cm the signal decays so rapidly that it is essentially impossible to measure the field.

Another limitation on this method is its extremely high sensitivity to noises due to external ac voltages. As far as this problem is concerned the most favorable conditions obtain for toroidal pick-up units; in the ideal case these should be free from effects due to external noises which are uniform over the volume of the pick-up unit. In practice, however, toroidal pick-up units are not free from noises due to ac voltages at both low and high frequencies.

In terrestrial magnetometers, which are designed for field measurements over relatively narrow ranges (in this case the pick-up units are located at points which are free from noise), the required signal-to-noise ratio can be easily achieved at relatively small polarization fields. For example, for an amplifier band width of 100 cycles a signal-to-noise ratio of 20 can be easily obtained with a polarization power of approximately 20 w.

In a rocket magnetometer[7] with a relatively narrow measurement range (and thus a rather narrow amplifier band width), in order to achieve a signal-to-noise ratio of 12 the polarization power must be increased to 70 w.

In order to achieve this same ratio in a magnetometer designed for measurements of the magnetic field with a satellite in a field range of 15,000-55,000 γ, given an amplifier band width of the order of 1600 cycles it would be necessary to increase the polarization power to 200 w.[7]

In the case of satellite measurements the associated apparatus for the proton magnetometer is complicated by the fact that with a limited number of terrestrial stations there is a need for standard frequency marker generators in the memory system and in the associated apparatus; these markers must be recorded together with the magnetometer signals. Finally, it is necessary that there be no sources of spurious magnetic fields with gradients of 3-4 γ/cm in the immediate vicinity of the pick-up coil.

Almost all the requirements indicated above imply the need for independent measurements and special capsule design.

Total Field Magnetometer with Saturation Pick-up Unit

The first measurements of the magnetic field of the earth made with a rocket[8] were carried out by means of a magnetometer with saturation pick-up units. Use was made of a scheme in which signals from three mutually perpendicular pick-up units were fed to the input while the signal at the output was proportional to the square of the total field. The size and power requirements for this device were rather small. However, this method can be used only for making measurements over a limited range of field variation; moreover, it is more or less limited to a given orientation in space. In a rocket flight this condition can be satisfied. In measurements carried out with a satellite, however, a simple scheme of this kind can be used only to obtain very rough results.

The absolute accuracy of the squaring elements is less than 5% over a wide signal range.[9]

Measurements of the magnetic field over a wide range with a high sensitivity and the transmission of these data by telemetering with an accuracy which exceeds the accuracy of the telemetering system are possible if a compensation method is used in the measurement.

The compensation method can be used in a three-component magnetometer in a rotating satellite only if some means is provided for continuous automatic orientation of one (measurement) pick-up unit of the magnetometer in the direction of the total magnetic field vector.

Analysis of the actual conditions of operation of the magnetometer apparatus in the third artificial earth satellite has shown that the experiment on measurement of the magnetic field was completely realized with the use of a magnetometer with saturation pick-up units which were automatically oriented along the total field. The following circumstances were considered:

1) in the presence on board of a large number of scientific and technical instruments it is impossible to avoid magnetic and electrical noises, so that the observation of signals from a nuclear magnetometer would be extremely difficult;

2) in order to achieve a measurement accuracy of 0.05-0.1% over a wide range of measured fields it is necessary to use a compensation method of measurement;

3) in moving over its orbit the satellite changes its orientation in space continuously, so that the compensation method cannot be used without automatic orientation of the pick-up unit of the magnetometer in the direction of the field;

4) if use is made of transistors and appropriate circuits it is possible to obtain automatic orientation along the total magnetic field in a magnetometer of small weight with small power requirements which is capable of operation at any orientation of the satellite and at any magnetic latitude;

5) a magnetometer of this kind can be used to measure the earth's magnetic field and to determine the change in the orientation of the satellite in space and the nature of its motion with respect to its own center of mass; these data are necessary for the interpretation of data of other experiments.

MAGNETOMETER IN THE THIRD ARTIFICIAL EARTH SATELLITE

General Description of the Principles of Measurement

The magnetometer in the third satellite was an automatic instrument which carried out the following two functions continuously:

1) orientation of a special measurement pick-up unit in the direction of the earth's magnetic field at any satellite orientation;

2) measurement, by means of this pick-up unit, of the intensity of the magnetic field.

Orientation in the direction of the field and the measurement of the intensity were realized by means of sensitive elements the operating principles of which are based on the same physical principle. The signals from the pick-up units were used separately in the orientation and measurement channels of the magnetometer.

A schematic diagram of the magnetometer is shown in Fig. 1. It consists of four basic parts: 1) measurement channel; 2) orientation channels; 3) mechanical orientation unit; 4) automatic control elements and power supplies.

The measurement and orientation channels contain a number of similar functional elements: pick-up units, selective amplifiers, phase-sensitive detectors. These elements serve for transformation of the signals due to the constant magnetic field into electrical signals (direct current) of the required power.

The elements used as sensing elements for the earth's magnetic field were magnetic-saturation pick-up units of the second-harmonic type. In the simplest case a typical saturation pick-up unit consists of plates or wires made from a material with a high magnetic permeability (permalloy) and the primary and secondary windings which are mounted on it.

When the wire is magnetized by a field $H = H_0 + H_m \sin \omega t$ (here H_0 is the magnetic field of the earth which is being measured, while $H_m \sin \omega t$ is the auxiliary sinusoidal field

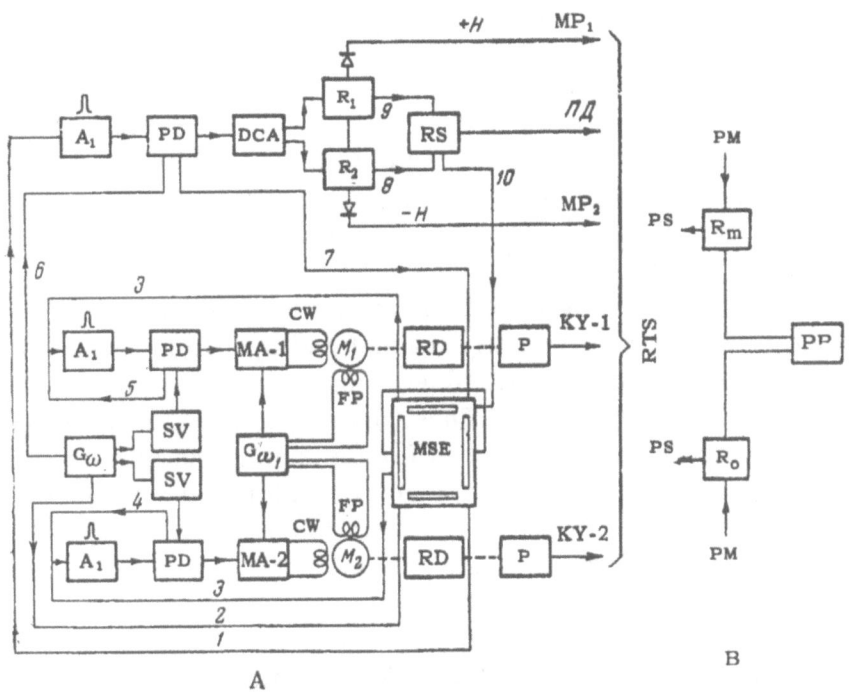

FIG. 1. Block diagram of the SG-45 magnetometer

A) Orientation and measurement channels: A_1—amplifier for frequency 2ω; PD—phase-sensitive detectors, $G\omega$—generator for frequency ω; SV—synchronization voltage at frequency ω; DCA—dc amplifier; MA—magnetic amplifiers for frequency ω; $G\omega_1$—generator for frequency ω_1; R_1 and R_2 — RP-5 relays; M_1—motor for the orientation shaft; M_2—motor for the orientation platform; RS—solenoid range switch; RD—reducers; P—potentiometers; MSE—platform with magnetic-sensitive elements; CW—control windings of the motors; FP—fixed-phase windings of the motors; 1) signal from the measurement element (frequency 2ω); 2) excitation for the magnetic-sensitive elements at frequency ω; 3) signal from the magnetic-sensitive element in the orientation channel (frequency 2ω); 4) suppresser for oscillations in the platform; 5) suppresser for oscillations in the shaft; 6) synchronization voltage; 7) feedback; 8) compensation for the field $-H$; 9) compensation for the field $+H$; 10) step compensation, frequency $\omega = 2000$ cps, $\omega_1 = 400$ cps. B) Automatic control elements and power supply: R_m—relay for switching the measurement channel; R_o—relay for switching the orientation channels; PM—circuit from the programming mechanism; PS—power supply.

at the excitation frequency ω, which magnetizes the wire to saturation twice in each cycle), a voltage at the double frequency $U_{2\omega}$ is induced in the secondary coil. The amplitude of the double-frequency voltage $U_{2\omega}$ is, within certain limits, proportional to the strength of the field being measured H_0. When $H_0 = 0$ the voltage $U_{2\omega}$ vanishes. When there is a change in the sign of H_0 the phase of the second harmonic voltage changes.

A thin, long wire made from permalloy has one further useful property: it is easily magnetized by a field along its axis and essentially impossible to magnetize when the field is perpendicular to its axis. Thus, the magnetic saturation pick-up units are ideal transducers for the signal due to a fixed magnetic field, producing an electrical signal at a frequency of 2ω. These pick-up units have two important properties: sensitivity to the magnitude and sign of the magnetic field and selectivity with respect to the direction of the field.

In the total field magnetometer of the self-orienting type three mutually perpendicular pick-up units are used; these are fastened to the mounting plate of a special orientation unit. One of these pick-up units is the measuring unit, while the other two, located in the plane of the plate, serve to orient it perpendicular to the total field.

An excitation voltage of frequency 2 kc, which serves for periodic magnetization of the cores, is obtained from a special 2-kc generator. The pick-up signals (second harmonic) are amplified in selective amplifiers which are tuned to a frequency of 4 kc, and then applied to the inputs of phase-sensitive detectors. The constant component of the output voltage of the phase-sensitive detector in each channel is, within certain limits, a linear function of the magnitude of the constant field being measured; the sign depends on the sign of the field.

Beyond this point the measuring and orientation channels are different. The dc signals from the phase-sensitive detectors in the orientation channels are applied to the inputs of magnetic amplifiers of the servo system of the orientation channels, where they are converted into 400-cycle ac voltages.

After power amplification these voltages are applied to the control windings of small inertialess motors in the mechanical orientation unit.

The motors rotate the plate in the plane of the orientation pick-up units as long as this plate is not perpendicular to the total magnetic field vector. When this orientation is achieved the second-harmonic signals from the orientation pick-up units vanish and the motors stop.

The measurement pick-up unit is then oriented along the total field vector and the signal from this unit is used for measurement of the intensity of the magnetic field. A measure of the magnetic field and its variations is the constant current applied in the compensation winding mounted on the pick-up unit in the measurement channel.

The main part of the magnetic field is compensated by a special high-stability current source. The compensation current which flows through the pick-up winding can be varied discretely by means of an automatic range switch. Continuous variation of the field over limits somewhat greater than the individual steps in the automatic range switching device is realized automatically by the introduction of heavy negative feedback; the compensation is accomplished by means of a current which is obtained from the phase-sensitive detector in the measurement channel from a signal which is produced by the uncompensated part of the magnetic field. The negative feedback current compensates for the entire field except for a very small fraction which serves for maintenance of the compensation current. The use of negative feedback serves to increase the stability of the magnetometer.

After power amplification in a dc amplifier the automatic compensation current is applied to the winding of polarized relays which are connected in series with the load resistance in the measurement channel. When the automatic compensation current exceeds the limiting value for a given range of continuous field compensation the polarized relays close an appropriate contact in the automatic range switching device. By this method it is possible to extend the total measurement range in such a way that the field can be measured over a wide range of values with a constant high sensitivity.

The position of the moving platform with respect to the magnetic field vector is determined by the moving contacts of two potentiometers (the frames of these potentiometers are attached to the frames of the platform), a gear drive, and shafts which rotate the platform relative to an axis fastened in a yoke, and the yoke itself with respect to a perpendicular axis. One turn of a potentiometer coincides exactly to one turn of the platform or yoke.

DESCRIPTION OF THE OPERATION OF THE INDIVIDUAL UNITS AND THE DESIGN OF THE SG-45 MAGNETOMETER

Circuit diagrams for the magnetometer are shown in Figs. 2 and 3.

Pick-up Units of the Magnetometer

A harmonic analysis of the output voltage from the pick-up winding (when the core is magnetized by an excitation field $H_m \sin \omega t$ and a fixed field H_0) shows that the output voltage contains a discrete spectrum of odd and even harmonics; the amplitudes of these harmonics are functions of the fields H_0 and H_m. The even harmonics are odd functions of the field H_0 and even functions of the auxiliary excitation field H_m; a change in the phase of H_m does not change the phase of the even harmonics. On the other hand, the odd harmonics are even functions of the field being measured H_0 and odd functions of the field H_m.

An actual pick-up unit for measurement of the earth's magnetic field consists of two cores (permalloy ribbons) located in frames. On each of the frames there are excitation windings W_e which are connected in series and wound in opposite directions (Fig. 4). Over the winding W_e in the pick-up units in the orientation channels there are signal windings W_s which are connected in series and wound in the same direction. With this method of connection the odd harmonics must be compensated in the W_s winding (if the elements of the pick-up unit are identical). In the pick-up unit of the measuring channel there is one winding W_s which encloses both ribbons.

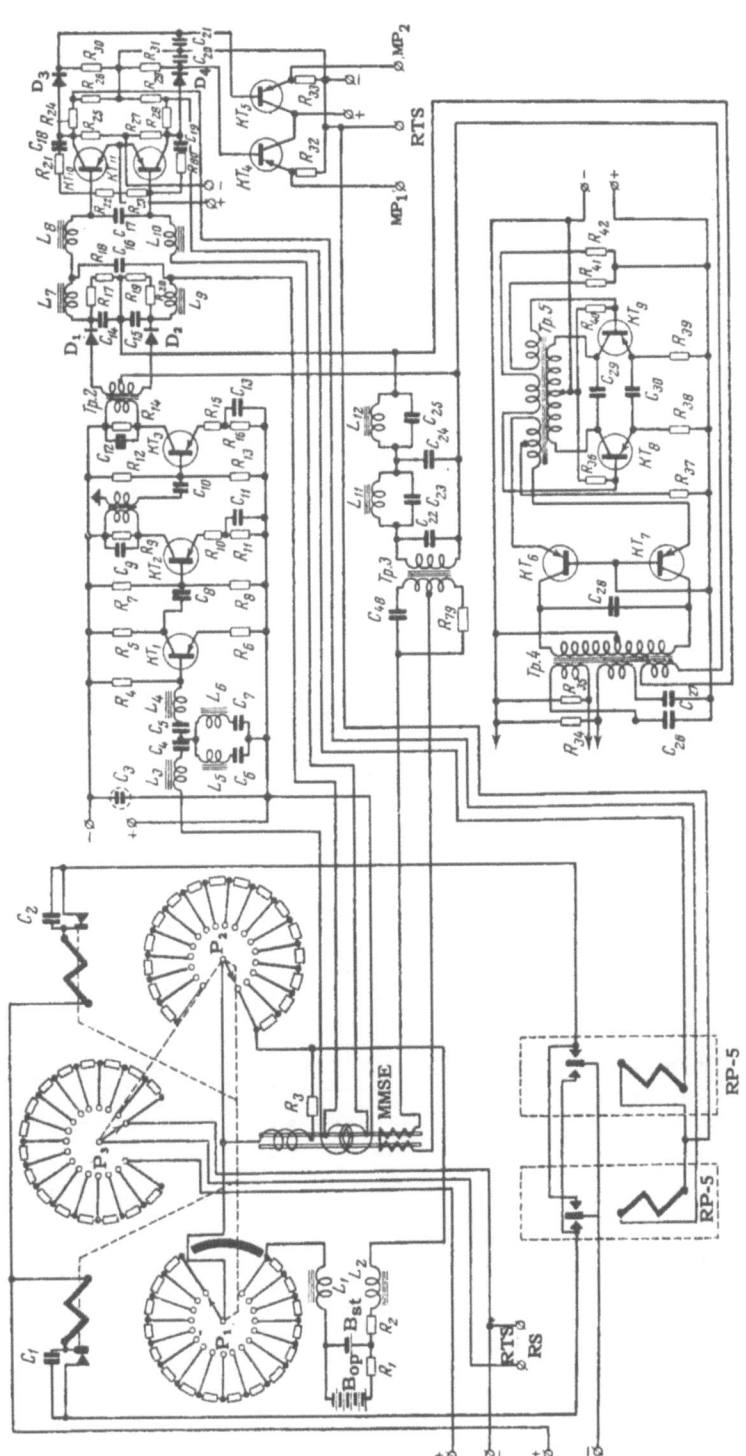

FIG. 2. Schematic diagram of the measurement channel of the SG-45 magnometer.

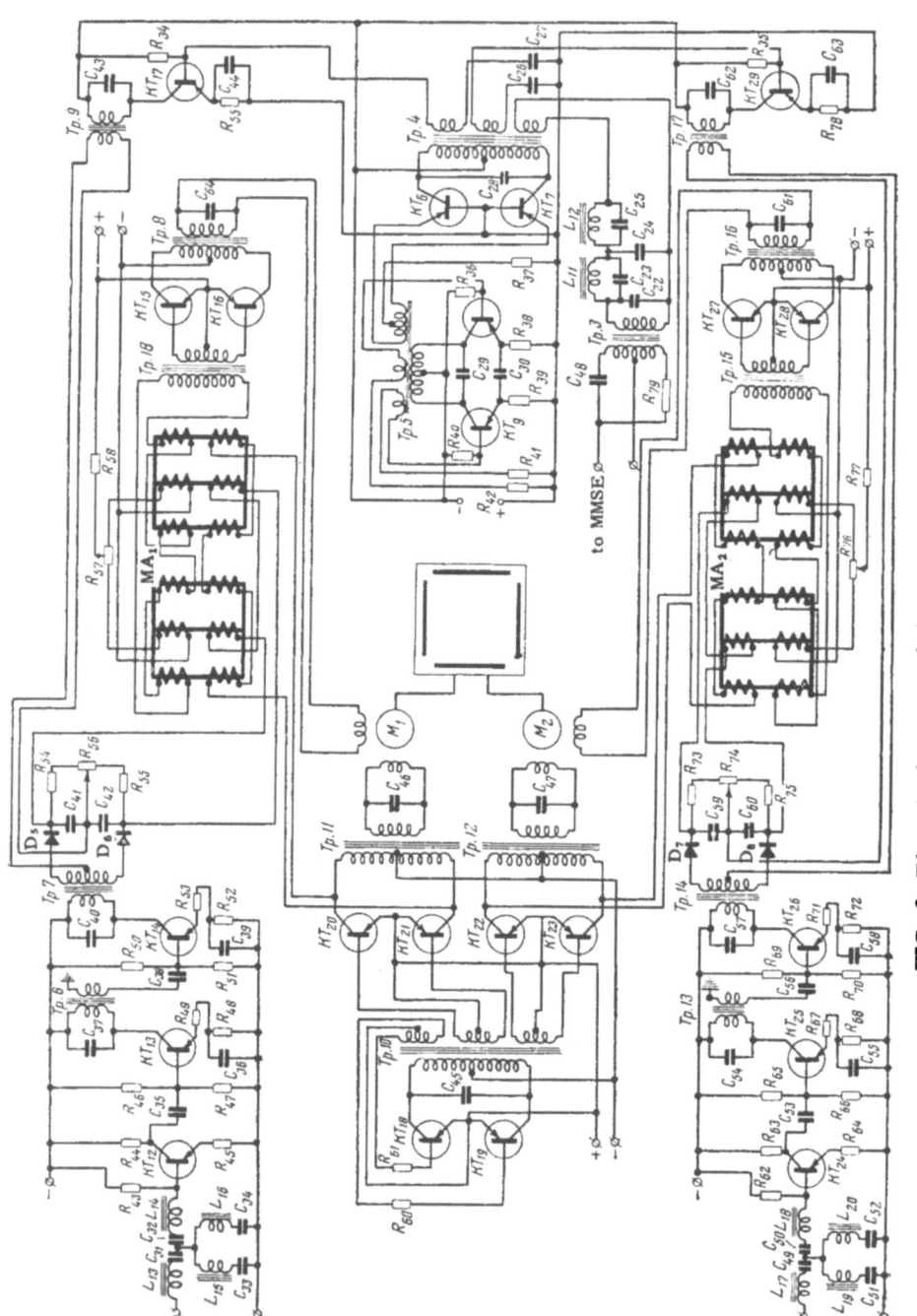

FIG. 3. Electrical circuit of the orientation channel.

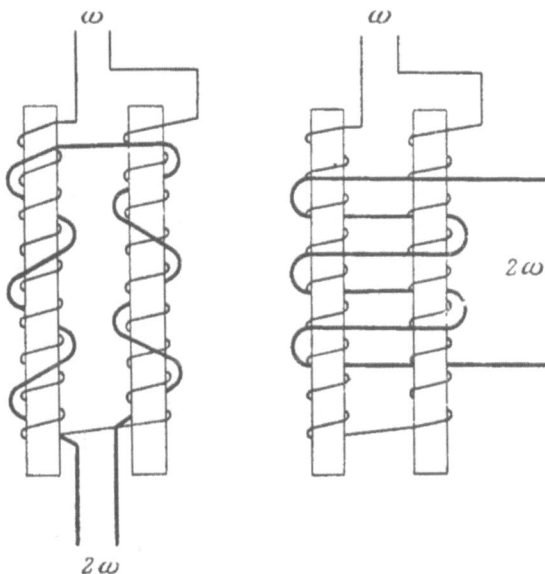

FIG. 4. Diagram showing the connections of the magnetometer coils. The thin lines show the W_e windings, while the thick lines show the W_s windings.

In addition to the indicated windings on the frames of the pick-up units in the orientation channels there are feedback windings W_{fb}, while the pick-up unit in the measuring channel has a compensation winding W_c.

Since the cores are fabricated in the form of thin plates with lengths considerably greater than the perimeter of the cross section, or in the form of a thin wire, the demagnetization factor for the cores in the direction of the major axis is a thousand times smaller than the demagnetization coefficient in the transverse direction. A core of this kind is practically magnetized not by the entire field but only by the component of the field in the direction of the major axis. For this reason the second-harmonic voltage in the winding of the pick-up unit which is perpendicular to the field is zero. If the pick-up unit deviates in direction a voltage of appropriate phase is produced.

Magnetometer Generator

The auxiliary voltage used for excitation of the pick-up unit is obtained from a 2,000-cycle sinusoidal generator.

In order to achieve stable operation of the magnetometer it is very important that the generator voltage be free of even harmonics, in particular the second harmonic. For this reason the generator makes use of a push-pull circuit, while the supply voltage for the pick-up unit passes through low-frequency filters (L_{11}, L_{12}, C_{23}, C_{24}, C_{25}).

The generator consists of two stages: the master oscillator, which uses P2B germanium transistors (KT_8 and KT_9) and a buffer power amplifier which uses P3V germanium transistors (KT_6 and KT_7) in a common base circuit. The circuit of the master oscillator (Tp. 5) is made on an alsifer core. The voltage from the buffer transformer stage (Tp. 4) is applied to the input of a low-frequency filter and excites two power amplifiers which make use of P2B transistors (KT_{17} and KT_{29} in Fig. 3), the voltage of which is used for synchronization of the phase-sensitive detectors in the orientation channels. The synchronization voltage of the phase-sensitive detector of the measurement channel is obtained from the second winding of transformer Tp. 4. The magnetometer pick-up units require a current of 75 ma at a voltage of 4 v. The 2,000-cycle generator uses a battery current of 60 ma at a voltage of 25 v.

Magnetometer Amplifier

With a core length of 40 mm and a practically convenient number of turns for the W_s winding the sensitivity of the pick-up units is approximately 10 microvolts/γ. Because of a difference in the core thickness and differences in the winding frames or differences in capacity distribution and fringing flux at the signal winding the odd-harmonic voltages may reach 50-100 mv.

In order to suppress odd harmonics and obtain voltage amplification and power amplification of the desired signals, the system makes use of three identical tuned amplifiers which are tuned to a frequency of 4,000 cycles. The input element of each amplifier is a passive band pass filter which has maximum attenuation at frequencies of 2,000 and 6,000 cycles (first and third harmonics of the excitation voltage). The filters are intended to provide a good signal-to-noise ratio and thus to prevent the possible appearance of spurious even harmonics because of non-linear effects in the amplifier.

The amplifiers contain three stages which use type P6D transistors with a common emitter circuit. The circuits are made on oxyfer cores, "Oxyfer 200."

The total voltage gain is 20,000. The band width at the 0.7 point is 350 cycles. The first-harmonic attenuation in the filter is 54 db. The relatively wide band width is necessary for stable operation of the magnetometer. The amplifier supply voltage is 11 v. All three amplifiers are adjusted to achieve strictly uniform phase characteristics.

Phase-Sensitive Detectors

The principle of operation of the phase-sensitive detector used in the magnetometer is based on the fact that a combination of odd and even harmonics which are shifted in phase gives rise to a distorted voltage wave; the heights of the positive and negative peaks of this wave are different (Fig. 5). The direction of the higher peak depends on the phase of the even harmonic, while the difference in the peaks depends on the magnitude of the signal and the phase difference between the harmonics. By changing the phase of the excitation voltage of the pick-up unit it is possible to shift the phase between the synchronization voltage (2,000 cycles) and the signal voltage (4,000 cycles) so that this difference is 90°.

The difference in amplitude of the half-waves, which, within certain limits, is proportional to the field being measured, is measured by the detector, which acts as a differential peak voltmeter.

The circuit (cf. Fig. 2) consists of an input transformer (Tp. 2) the primary winding of which serves as a resonance circuit for the last stage of the amplifier, and a secondary winding (with center-tapped output) connected through two germanium diodes to a resistance-condenser system (C_{14}, C_{15}, R_{17}, R_{18}, R_{19}, R_{20}). In the center arm of this balanced circuit there is a transformer winding through which the synchronization voltage of 2,000 cycles is applied.

Under these conditions the average values of the voltage at the condenser are equal if the amplitudes of the voltage half-waves from the transformer are equal, and are different if the amplitudes of the half-waves are different.

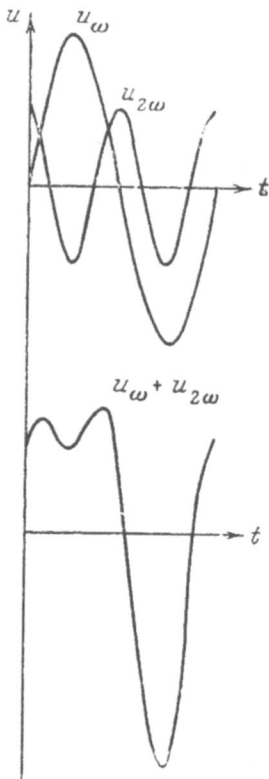

FIG. 5. Addition of even and odd harmonics shifted in phase.

In the latter case a current flows through the load resistance of the phase-sensitive detector; the sign and magnitude of this current depend on the voltage difference, which, in turn, depends on the sign of the magnetic field being measured.

Compensation Circuit for the Measurement Channel of the Magnetometer and Telemeter Output

As has already been indicated, the measurement pick-up unit works as a null detector; the main part of the field is compensated by means of a stable field source, while the small continuously varying part of the field is compensated by the introduction of heavy negative feedback.

The voltage from the phase-sensitive detector is applied through filters (L_7, L_9, and C_{16}) to the feedback winding. By changing the number of turns in the feedback winding, it is possible to vary the sensitivity of the instrument over wide limits. The automatic compensation current changes sign when the sign of the magnetic field changes.

The telemetering system accepts voltages up to 6 v of only one sign. In order to satisfy this requirement the output to the telemetering system is as follows. The automatic compensation current is amplified by a dc amplifier which makes use of a balanced circuit with P6V transistors (KT_{10}, KT_{11}). The load for the dc amplifier is the control winding of two type RP-5 polarized relays, whose purpose will be clear from the following explanation. The windings of the relay are connected in series and each of the windings is shunted by resistances in such a way that the relay contacts close at a current of 1 ma. Resistances R_{24} and R_{28} are connected in series with the windings of the relay; these resistances are chosen so that at a current of 1 ma the voltage drop across the entire circuit is 12 v. This voltage also changes sign when the field changes sign.

The voltage from the ends of the circuit consisting of the relay winding and the resistances R_{24} and R_{28} is applied to a circuit consisting of two diodes D_3 and D_4 and load resistances R_{30} and R_{31}, connected as shown in the diagram. The voltage from resistances R_{30} and R_{31} is applied through the emitter followers KT_4 and KT_5 to the alternately operating telemeter channels MP_1 and MP_2, one of which measures positive field variations, while the other measures negative field variations. Under these conditions a voltage of one sign which varies from 0 to 6 v is applied to channels MP_1 and MP_2.

The automatic compensation of the main part of the field ($40,000 \pm 24,000\gamma$) is provided by a circuit which is supplied by an external stabilized current source. A system of parallel resistances is connected to the coil of the compensation winding in the measurement pick-up unit through a special switching system; these resistances are chosen so that the compensation field in the coil can vary by steps of $3,000\,\gamma$ (Fig. 6). In order to keep the resistance of the entire system

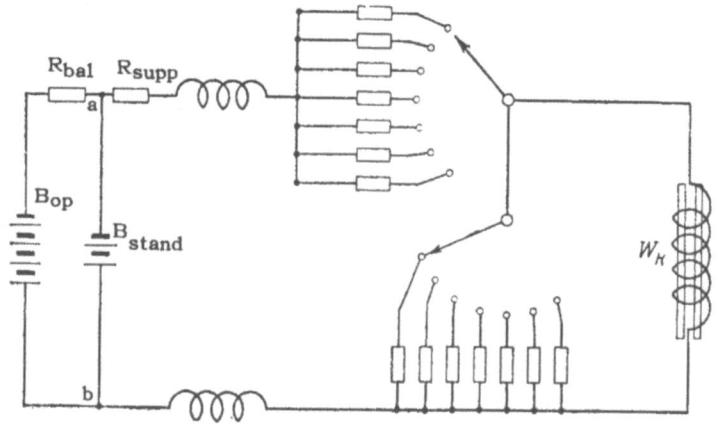

FIG. 6. Electrical diagram of the range switch.

constant as the compensation winding is shunted, there is an additional deck on the switching unit which simultaneously switches in series-connected resistors whose values are such that the total resistance of the compensation circuit remains fixed.

The current in the compensation coil is applied through a ballast resistance R_{bal} from a 60-v mercury oxide battery with a capacity of 3 ampere-hours. Two filtering chokes are connected in series with the compensation coil together with an additional resistance which is chosen so that with the nominal current of 5 ma through the coils the potential difference between points a and b is equal to the voltage of a standard battery which, as indicated in the circuit, is connected as a buffer for the load. The reference battery is a mercury oxide battery (2.7 v) with a capacity of 30 ampere-hours. This battery supplies power to the compensation coil only when the voltage of the operating battery starts to drop. An analysis of the operation of this circuit[10] shows that the stability of the current in the compensation circuit is higher, the higher the ratio R_{op}/R_{stand}, where R_{stand} is the internal resistance of the reference battery, and the higher the stability of this battery voltage.

This circuit makes it possible to maintain the compensation current constant with a definite degree of accuracy without any regulation.

The entire automatic compensation system operates as follows: when the voltage drop at the relay windings reaches 12 v (while the resistances on which the signal is applied to the telemeter channel reaches 6 v) the left or right contact of relay RP-5 closes, depending on the sign of the current. The sign of the current depends on whether (for a given intensity of the magnetic field) there is over-compensation or under-compensation of the field in the coil of the pick-up unit in the measurement channel.

Closure of the right or left pairs of contacts in relay RP-5 causes the closure of one of the solenoids in the range switch, which shifts the moving contacts of both decks of the range switch over the required number of steps up to the point at which relay RP-5 is opened, corresponding to compensation of the field within the limits of the discontinuous variation range. In order to avoid hunting effects in the range switch, differentiating circuits R_{21}, C_{18}, and R_{30}, C_{19} are used to introduce feedback which is proportional to the rate of change of the signal.

Mechanical Orientation Unit

A diagram of the orientation unit is shown in Fig. 7. On the platform are mounted three mutually perpendicular pick-up units; one of these is the measurement unit, while the other two, which lie in the plane of the platform, serve as orientation units.

The platform can be rotated about an axis fastened in the yoke. In turn, the yoke can be rotated about a perpendicular axis. The rotation is realized by means of two gear drives. The driving gears in the drive are fastened to drive shafts which transmit the rotation through a reducer from two low-inertia DID-0.5 motors. One of the motors (M_2) provides

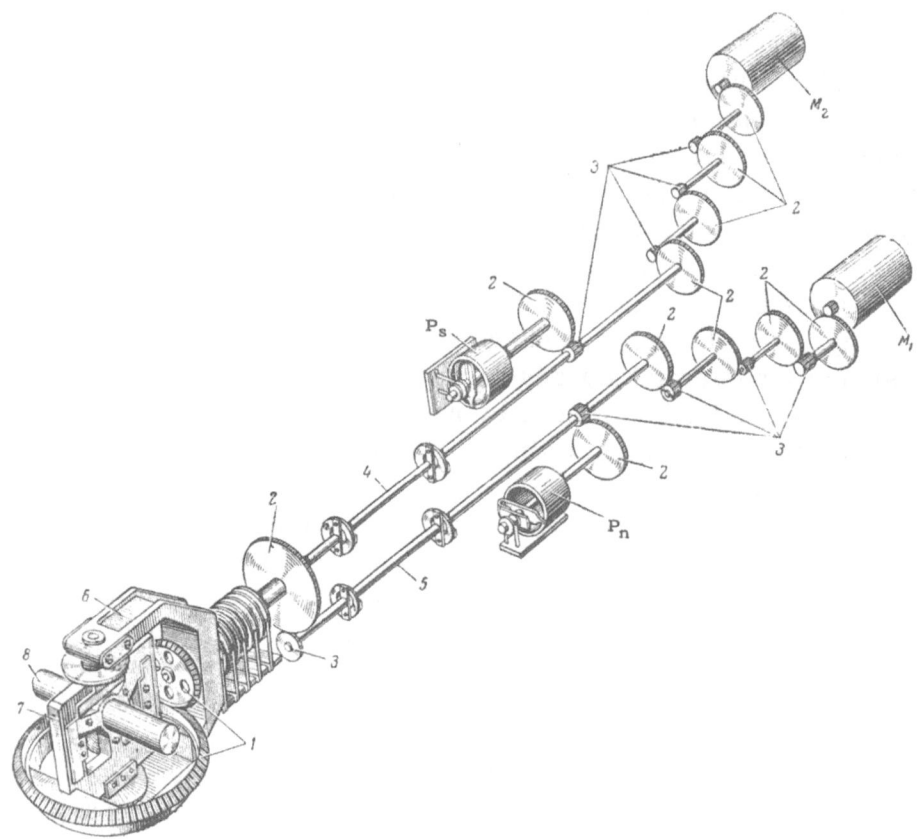

FIG. 7. Diagram of the orientation unit. M_1 – DID-0.5 motor (shaft rotation); M_2 – DID-0.5 motor (platform rotation); P_s – slant potentiometer; P_n – inclination potentiometer; 1) bevel gears; 2) gear wheel; 3) gear (small driving gears).

rotation of the platform about an axis perpendicular to the plane of the magnetic meridian while the other (M_1), the azimuthal motor, brings the system into a position in which the platform with the orientation pick-up units is perpendicular to the total field vector.

The voltage to the windings of the pick-up units is applied through sliding contacts. Because of this design of the orientation unit it is possible to carry out measurements while the platform executes more than one revolution. On the drive shafts of the orientation units there are gears which mesh with the gears of two circular potentiometers to which is applied a voltage from a fixed voltage source (6 v). The voltages taken from the moving contacts of the potentiometers depend on the positions of the moving contacts and indicate the position of the moving platform and the plate on which the pick-up units are mounted.

In Fig. 8 we show the error-angle dependence of the signal at the orientation pick-up units and the error-angle dependence

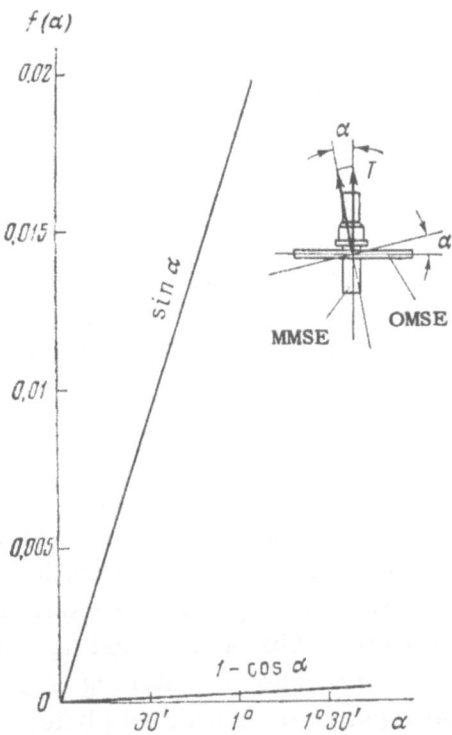

FIG. 8. The measurement error as a function of platform orientation.

of the signal in the measurement pick-up unit. The error signal produced in the orientation pick-up units for deviation of the normal to the platform from the direction of the total field vector by an angle α is proportional to $\sin \alpha$ and the corresponding change in signal in the measurement pick-up unit is proportional to $(1 - \cos \alpha)$. With an error of 1° in the orientation platform the error in the measurement pick-up unit is less than 4γ (for measurements of the field at the surface of the earth).

Servo System of the Magnetometer

The electronic units in the orientation unit are designed for power amplification and voltage amplification of the error signals produced in the windings of the orientation pick-up units; the electronic units also serve to shape the control signals for the motors in the mechanical orientation unit.

As has already been mentioned, there are similar elements in the orientation and measurement channels: pick-up units, selective amplifiers, and phase-sensitive detectors. Beyond this point the circuits are different.

The signals from the phase-sensitive detectors in the orientation channels are applied to the input of magnetic amplifiers MA_1 and MA_2 which are connected in push-pull. The magnetic amplifiers operate at 400 cycles. The 400-cycle voltage is obtained from a special generator consisting of a driving stage made up of P2B germanium transistors (KT_{18} and KT_{19}) and two power amplifiers in a push-pull circuit which uses P3V germanium transistors in a common emitter circuit (KT_{20}, KT_{21} and KT_{22}, KT_{23}). The output windings of the transformers of these amplifiers are connected to the fixed-phase windings of the motors in the orientation units. These same amplifiers provide voltage for the excitation windings of the magnetic amplifiers. The output winding of each magnetic amplifier is connected through a matching transformer (Tp. 15 and Tp. 18) to the input of the power amplifiers which supply the control windings of the motors. The power amplifiers of the controlling phases make use of P3V transistors in a push-pull circuit. The voltages applied to the fixed-phase and control windings of the motors are shifted by 90°. Thus, the low-inertia motors of the orientation unit operate at a frequency which differs from the excitation frequency of the pick-up units and the frequency of the signal; in this way it is possible to avoid effects on the measurement system due to the motors which are in direct proximity to the pick-up units.

Since the servo system does not contain narrow-band elements its operation is not affected by instabilities in the supply voltage. The response rate of the servo system is of the order of 45°/sec.

The magnetometer is provided with a separate switching system for the orientation channel and the measurement channel; this is achieved through the use of two independent DP-10 relays. The switching device has a special programming mechanism in which the signal is connected from the orientation system to the measurement channel, thus avoiding the possibility of overloading the latter.

Construction of the Magnetometer

The magnetometer consists of three units which are connected by means of socket connectors (Fig. 9); 1) orientation unit; 2) electronics unit; 3) power supply.

FIG. 9. Photograph of the magnetometer. 1) Orientation unit; 2) electronics unit.

The orientation unit is light and consists of three basic parts: the front part with the sensitive magnetic elements and the current-sensing units, a center part with the reducer, potentiometers, and motors, and a rear flange which holds the connector socket (Fig. 10).

The low-inertia DID-0.5 motors which drive the measurement pick-up unit system to align it along the total vector can rotate at 13,000 rpm when running free and 7,000 rpm under a torque load of 3 g-cm, providing rotation of the platform or

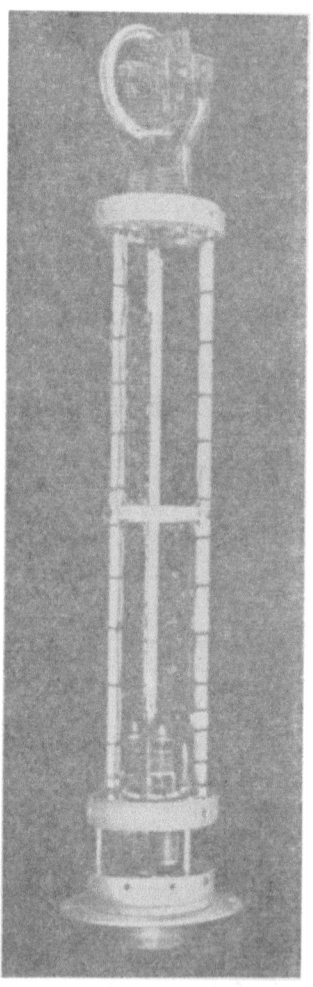

FIG. 10. Photograph of the orientation unit without cover.

drive shafts at 1/4 revolution per second. Because of the use of conical gears, rotation in the two mutually perpendicular directions can be realized in clockwise and counterclockwise directions.

In order to avoid any effect due to the motors (the only

magnetic element in the unit) these are mounted at considerable distances from the pick-up units. The rotation drive unit in the front part has long drive shafts which are connected to the appropriate axles by elastic couplings.

The orientation unit is provided with a reliable shock absorber system so that it can operate under conditions of extreme vibration.

The electronics unit is rack-mounted. The lowest level contains the 400-cycle generator, the servo system, and the power amplifiers. In the center of the lower deck are mounted the 2,000-cycle generator, the low-frequency filter, and the power amplifier; the center of the upper deck contains the three selective amplifiers and the phase-sensitive detectors. The upper deck contains the range-switching unit, the two DP-10 switching relays, the RP-5 relay, the voltage distribution deck, the chokes, and other elements of the circuit.

The range switching unit makes use of a gear differential so that it is possible to uncouple the two rotations. For high electrical reliability, reduction of the transmission resistance, and reduction of the variation in resistance, the contacts and commutators of the switching unit are tandem-units and are palladium-plated.

On the ceiling of the electronic unit are mounted all the connecting cables of the magnetometer. The decks are mounted by four rods. If the rods are withdrawn, it is possible to remove the decks of the electronic unit without unsoldering. The electronic unit is covered by one case. This unit is provided with four pins in its lower base so that it can be mounted.

The weight of the electronics unit is 12.5 kg.

The power supply contains twelve batteries, the outputs of which are connected to a socket connector. The separate supply for the generators, amplifiers, and servo system makes it possible to avoid interference between the channels and noise between the individual units without causing an increase in the weight or dimensions. In this respect it is especially important to separate the power supplies for units which require high power but do not require high stability from those of the circuit elements which require low power but for which a high stability in voltage is necessary.

METROLOGICAL CHARACTERISTICS OF THE MAGNETOMETER, RECORDING AND DECODING

All models of the magnetometer are carefully checked for their basic measurement characteristics: sensitivity, zero stability as a function of time, temperature coefficient, rate of processing of signals, and dynamic error.

The readings of the test models of the magnetometer were reduced to absolute value for comparison with magnetometers from the observatory. The zero point of the flight model of the magnetometer was determined absolutely by means of a proton magnetometer in measurements made with both devices in the same place on a special platform.

Data from the magnetometer on board the satellite were transmitted by five telemeter channels, two of which (KY_1 and KY_2) transmitted information on the position of the orientation pick-up units, while the other three provided information on positive and negative variations in the magnetic field (MP_1 and MP_2) and on the position of the range switching unit (RS). The sensitivity of the magnetometer was such that the total measurement range of field over two channels of the telemetering system was $\pm 2400\,\gamma$. Channel MP_1 transmitted positive changes in the field with respect to the original level, while channel MP_2 transmitted negative changes. When there was a reading in channel MP_1 the reading in channel MP_2 was zero and vice versa. The total range of both channels ($4800\,\gamma$) exceeded the range of one step in the discrete compensation range switch ($3000\,\gamma$).

A sketch of the voltages in the telemetering channels for linear variation of the magnetic field is shown in Fig. 11. As is apparent from these curves (Fig. 11a) an increase in field is indicated by an increase in the voltage in channel MP_1. When the limiting value is reached the range switch operates and this is indicated by a simultaneous discontinuity in the voltage variation in channel RS. Simultaneously there is a voltage in channel MP_2 which corresponds to the difference between the magnitude of the range channel MP_1 and the magnitude of the range of one step in the discrete compensation

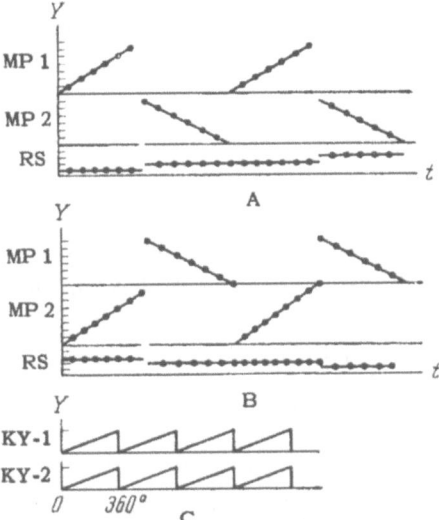

FIG. 11. Voltages in the telemetering channels. A) Increasing field; B) decreasing field; C) voltages in the pick-up units for the turning-angle control.

(this difference is 600 γ). With a further increase in field the signal in MP_2 falls to zero and then appears in channel MP_1, and so on.

When the field falls off the nature of the variation in the recording is similar to that described above. The succession of variations in channel RS is reversed, while the increase in voltage takes place in channel MP_2, and the drop in channel MP_1 (Fig. 11b).

Thus, a changeover in the range switch is indicated by readings in the three telemetering channels.

The intensity of the magnetic field of the earth from the readings of the three telemetering channels is given by the following formula:

$$T = T_0 - (n - n_0) K + \varepsilon_1 (\alpha - \alpha_0) - \varepsilon_2 (\beta - \beta_0),$$

where T_0 is the zero point of the magnetometer on the standard stage as compared with the proton magnetometer; n_0, α_0, and β_0 are the readings in channels RS, MP_1, and MP_2 on the standard platform; n, α, and β are the readings in channels RS, MP_1, and MP_2 in the measurements: ε_1 and ε_2 are the sensitivities of channels MP_1 and MP_2; K is the range setting.

In Fig. 11c we show sketches of the voltage from the pickup units of the angular control element at a constant rotational velocity. The construction of the orientation unit is such that rotation along the longitudinal axis (rotation of the "shaft") causes a change in the position of the platform on which the pick-up units are mounted. Hence, under these conditions the following relation obtains between the potentiometers which monitor the angular position (KY_1 and KY_2) and the turning angles:

$$\Delta V_{KY_1} = k_1 \Delta \varphi,$$

$$\Delta V_{KY_2} = k_2 \Delta \psi - k_3 \Delta \varphi.$$

The stability of the zero setting of the magnetometer depends on the stability of operation of the electronic units and the stability in time of the current which compensates the main part of the field. In Fig. 12 is shown a curve which indicates the typical variation of the compensation current over a

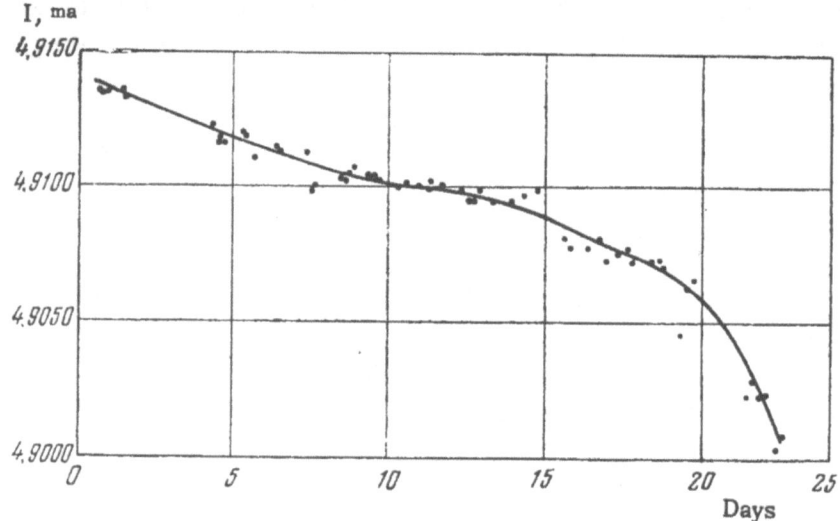

FIG. 12. Variation in compensation current as a function of time.

long period of time. This curve refers to a circuit in which the reference battery (B_{stand}) has a capacity of 12 ampere-hours at 4 v. The average current variation for this circuit is equivalent to a zero deviation of 135 γ in 21 days. As has already been indicated the stability properties of the circuit depend on the internal resistance of the standard battery. The capacity of the flight standard battery was increased to 30 ampere-hours, while the voltage was reduced to 2.7 v.

In Fig. 13 we show the variation in compensation current for a circuit with parameters completely analogous to the parameters of the compensation circuit in flight types. These

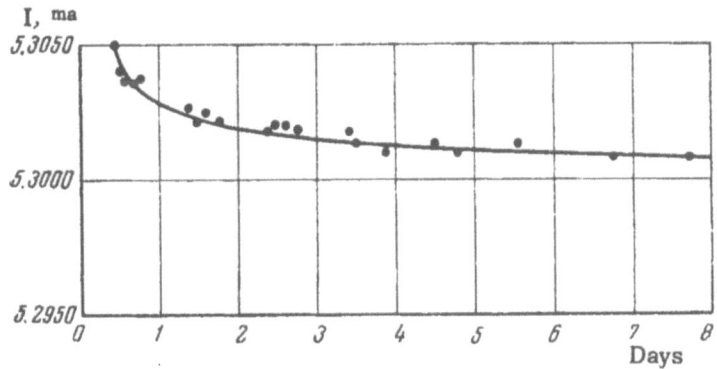

FIG. 13. Variation in compensation current as a function of time at fixed temperature.

observations are part of a series of observations carried out to determine the temperature coefficient. The temperature was changed rather frequently. The readings are reduced to one temperature. In spite of the fact that in this case the internal resistance of the standard battery was almost four times smaller, the mean deviation of the zero is 10γ per day. This is explained by the considerable temperature coefficient of the batteries and the thermal inertia of the circuit which actually leads to a somewhat higher instability.

In Fig. 14 the compensation current is shown as a function of temperature. In order to reduce the effect of thermal inertia of the circuit the measurements were carried out over a period of 80 hours. The temperature coefficient was found to be variable. In the first approximation it may be assumed in estimating the temperature errors that the temperature coefficient has two values: 1) between 0 and 20° the value is $6\gamma/\text{deg}$; 2) between 20 and 50° the value is $2\gamma/\text{deg}$.

The temperature in the satellite varies over a rather

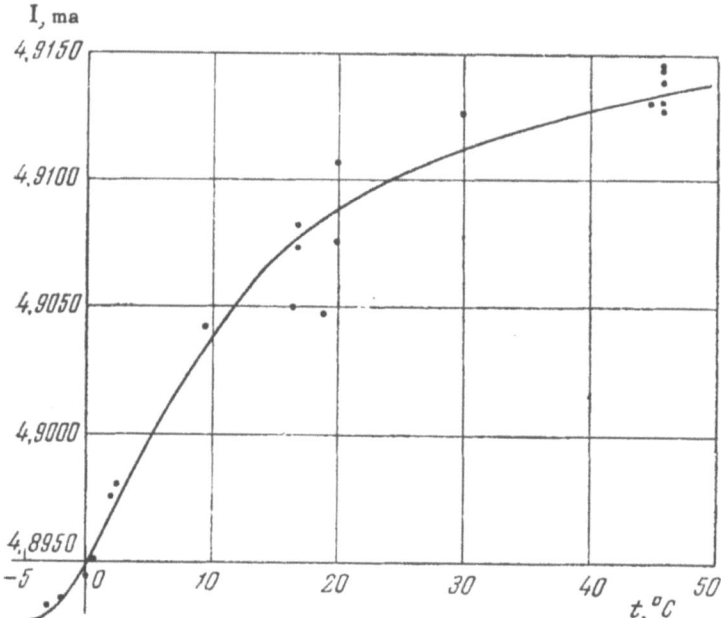

FIG. 14. Compensation current as a function of temperature.

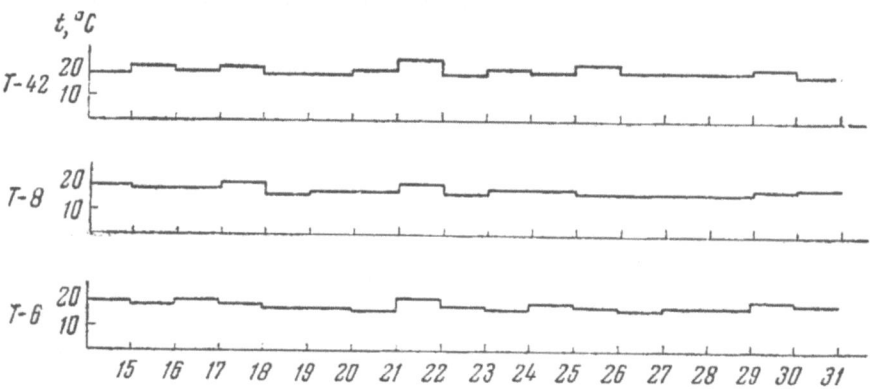

FIG. 15. Graph showing the change in temperature from May 14 to May 31 for three temperature pick-up units.

small range because of the reliable temperature-controlling system; this is apparent from Fig. 15.

The stability of the zero point also depends on the stability of operation of the electronic units. The nature of the variation of the zero point of the entire magnetometer in continuous operation can be ascertained from consideration of Fig. 16 in which we show the variations in magnetic field from the data of observatory magnetometers and from the data of the magnetometer in the third satellite. The solid curve in Fig. 16 shows the variation in field ΔZ and ΔH from readings on observatory magnetographs. The upper dashed curve shows the

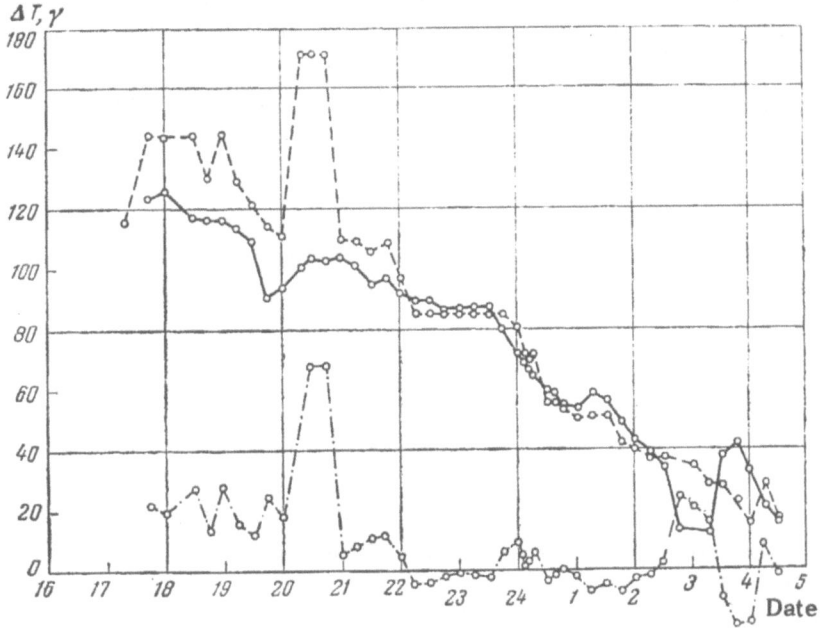

FIG. 16. Variation of magnetic field as a function of time.

readings of the magnetometer of the satellite for this same time. The accuracy in the reading from the visual device was approximately 10γ in the latter case. The lower curve shows the difference between the first two curves. It characterizes the stability of the magnetometer over the first twelve hours of continuous operation.

It is apparent from the curves that an appreciable instability appears in the first hours after the instrument is turned on. Later the spread in the points for the difference curve shows a random character and the absolute value lies within the limits of accuracy of readings for a visual device of the 1.5% class.

For a magnetometer which is placed on a rotating satellite a very important characteristic is the rate of correction of the signal. This refers to the servo system as well as the elements in the measurement channel. The correction rate of the servo system was determined by measuring the time required to establish the stage in a position perpendicular to the meridian when it was set at a given angle. The correction rate for small error angles was determined by similar measurements with the application of a uniform sinusoidal field of known frequency perpendicular to the earth's field. The current source in this case was a sub-audio oscillator. These measurements show that the servo system is capable of correcting an error signal at a rate of 40-45°/sec with angular accelerations up to $130°/sec^2$. The range switch in the measurement channel makes it possible to correct an error signal of 48,000 γ in 1-1.5 sec.

The time constant of the output telemeter circuit is approximately 0.08 sec. The hunting which is sometimes found in the range switch is due to this time constant in the output circuit; fast action of the range switch was eliminated by the introduction of negative feedback proportional to the rate of change of the signal.

For the specified voltage and quiescent conditions the device requires a power of the order of 12 w. At the maximum correction rate the required power is 20 w.

INVESTIGATION OF DEVIATION AND ELIMINATION OF DEVIATION

The orientation unit with the sensitive pick-up units was placed in a special compartment furthest removed from the remainder of the apparatus and power supplies. However, even in this case it was not possible to remove all effects of the magnetic elements in the apparatus used for other experiments. The units which created the greatest trouble were the permanent magnets in the magnetic ionization gauges, the electrostatic fluxmeters, and the magnetic parts of other instruments. The position of the sensitive pick-up units of the magnetometer and the elements which create the greatest trouble is apparent from Fig. 17.

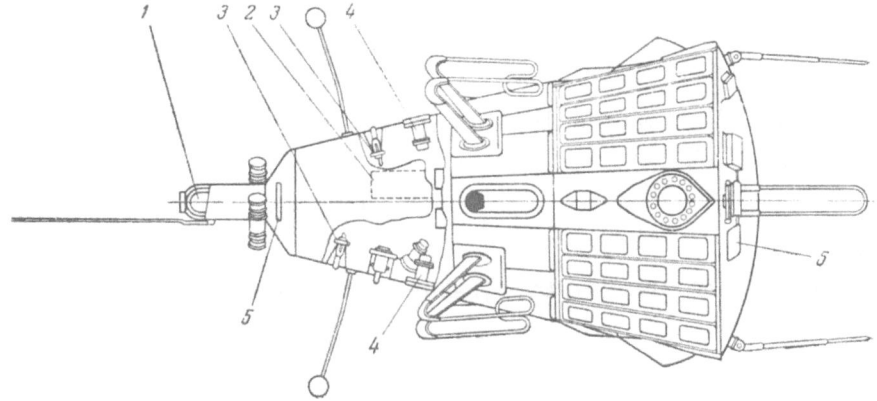

FIG. 17. Diagram showing the location of the instruments on the third artificial satellite. 1 and 2) magnetometer; 3) magnetic and ionization pressure gauges; 4) electrostatic fluxmeter; 5) solar batteries.

Special measurements were carried out in order to determine the magnitude of the total magnetic deviation and its characteristics. The satellite with all its apparatus was placed on a holder with which the satellite could be rotated around a longitudinal horizontal axis and around a vertical axis. An example of these measurements is shown in Fig. 18. The maximum value of the magnetic deviation was found to be 3500γ.

It was found possible to eliminate magnetic deviations from the results of the measurements because of the following circumstance: for the indicated magnetic deviation the magnetic interference causes a change in the intensity of the field but has essentially no effect on the direction of the field; the satellite executes regular precession motion with a large precession angle and the velocity of precessional motion is so large that the magnetic interference creates a variation in field, the gradient of which is considerably different from the gradient of the earth's magnetic field at great heights.

If H_e is the earth's magnetic field and H_i is the field due to the magnetic interference, while φ is the angle between them, the scalar magnitude of the total vector is given by the expression

$$|T_s| = |T_e + T_i| = \sqrt{T_e^2 + T_i^2 - 2T_e T_i \cos\varphi}. \tag{1}$$

Because T_i is so small this expression can be expanded in powers of T_i

$$T_s = T_e - T_i \cos\varphi + \frac{1}{2}\frac{T_i^2}{T_e}\sin^2\varphi. \tag{2}$$

Limiting ourselves to the first term we have

$$T_s = T_e - T_i \cos\varphi.$$

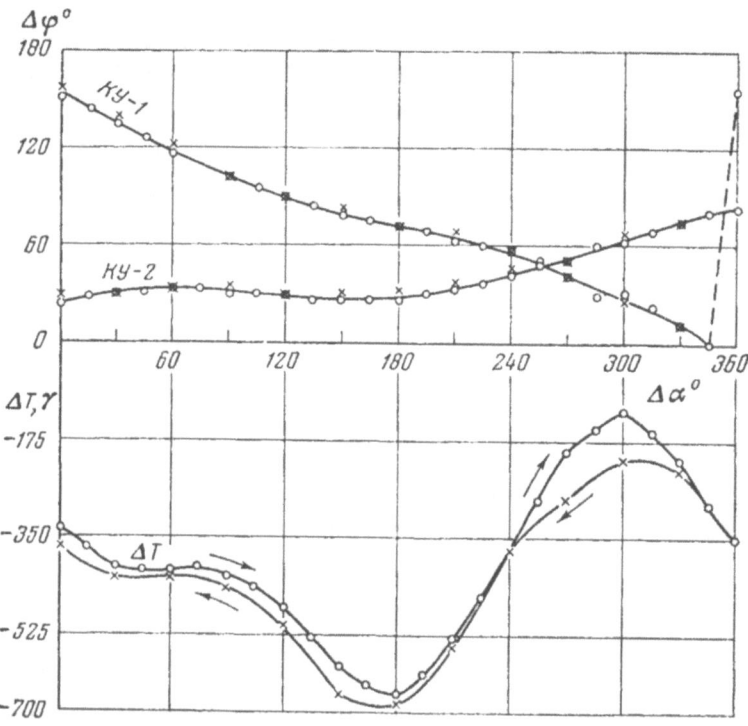

FIG. 18. Deviation curve and reading in the orientation channels for simultaneous rotation around longitudinal and vertical axes. The circles show the forward motion and the crosses show the backward motion.

Thus, the magnetometer in the satellite actually measures the intensity of the magnetic field of the earth and the projection of the vector of the magnetic interference in the direction of the magnetic field. Because of the precession of the satellite the angle φ changes continuously in magnitude and sign.

In Fig. 19 we show a typical magnetometer recording. The slow smooth variations of the earth's magnetic field are modulated by the magnetic interference, which varies periodically in magnitude and sign. By graphical or analytical averaging of the curves, effects due to magnetic interference can be eliminated to any desired degree of accuracy. As is apparent from Equation (2), the third term in the expansion is not eliminated. Neglecting the third term leads to a maximum error of the order of 100γ.

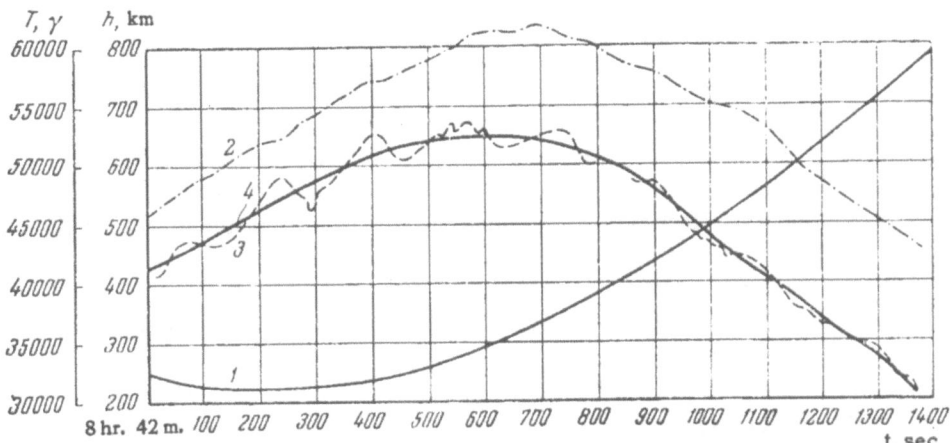

FIG. 19. Typical magnetometer recording. 1) Variation in satellite height (h) as a function of time; 2) magnetic field strength according to the data of terrestrial measurements; 3) magnetic field strength from the data of satellite measurements; 4) magnetic field strength measured on the satellite after the introduction of corrections for deviations due to the satellite itself.

ACKNOWLEDGMENT

The construction, adjustment, and testing of this apparatus has been carried out by a number of our colleagues in the magnetic laboratory of the NIZMIR and the OKB of the Ministry of Geology and Mineral Conservation as well as the authors. Great contributions to this work have been made by A. V. Klimovskii, V. Ya. Kulagin, L. O. Tyurmina, E. G. Eroshenko, P. V. Chernov, N. V. Lavrent'eva, and L. I. Ignatova.

The authors wish to express their gratitude to all persons who have participated in the construction of the apparatus, in the carrying out of the experiment, and in the first analysis of the observational data.

LITERATURE

1. S. Sh. Dolginov, L. N. Zhuzgov, and N. V. Pushkov, Artificial Earth Satellites, Volume 2, Plenum Press, New York, 1960.
2. N. V. Pushkov and S. Sh. Dolginov, Usp. Fiz. Nauk, 63, 645 (1957).
3. S. F. Singer, Rocket Investigations of the Upper Atmosphere, IL (1957).
4. S. Chapman, Rocket Investigations of the Upper Atmosphere, IL (1957).
5. M. Packard and R. Varian, Phys. Rev., 93, 941 (1954).
6. J. Laurence et al., J. Geophys. Res., 61, 547 (1956).
7. A. L. Bloom and L. E. Johnson, Electronic Industr. and Tele-Tech., August, p. 76, 1957.
8. S. F. Singer et al., J. Geophys. Res., 55, 115-126 (1950); Phys. Rev., 82, 957 (1951).
9. S. A. Ginzburg, Nonlinear Amplifiers and Their Functional Characteristics (in Russian), Gosenergoizdat, 1958.
10. E. Cherepanov, Reports NIIZM, 1955.

METHOD OF DETERMINING THE ELECTRICAL POTENTIAL OF A BODY IN A PLASMA

Ya. M. Shvarts

When rockets and artificial earth satellites are used for geophysical investigations it is frequently necessary to measure the electrical potential of these bodies since this quantity has an effect on the readings of certain instruments installed in the rockets and artificial satellites.[1,2] In this connection a number of methods have been developed and used for the determination of the potential of objects which are used to probe the upper atmosphere.[1,3,4]

The potential of a rocket was determined by Johnson and Meadows[4] by the effect of the potential on the readings of a radio-frequency mass spectrometer for determining the ionic composition of the atmosphere, which was on board the rocket. Because of the potential there is a shift between the true values of the mass numbers of the ions and the values obtained by the mass spectrometer. It should be noted, however, that a shift can also be produced for other reasons, for example the effect of the motion of the probing body, etc. Without giving a detailed analysis of the operation of the radio-frequency mass spectrometer (an analysis of this kind is given, for example, by Mirtov and Istomin[2]) we may note that the relative error in the measurements of the electrical potential of a probing body made by this method is directly proportional to the error in the measurements of the values of the mass numbers of the ions (neglecting the effect of electrical potential) and inversely proportional to the shift in the mass numbers of the ions due to the effect of the electrical potential of the probing body. Since the size of the measurement error can be significant while the shift can be small, when the method discussed is used it is possible only to make an estimate of the potential.

It has been proposed[1,3] that the potential can be determined by simultaneous measurements of the electrostatic field at the surface of the satellite and the concentration of positive ions in the immediately surrounding space. This method, as follows from Imyanitov,[3] is based on the fact that the quantities measured by the two instruments (electrostatic fluxmeter and a device for measuring the concentration of positive ions) are closely related. Actually, the concentration N_+ calculated from the readings of the instrument for measuring the concentration of positive ions is a function of the satellite potential, i.e., $N_+ = f(V)$. At the same time the intensity of the electrostatic field at the surface of the satellite (in the absence of external electrostatic fields) is a function of the potential and the concentration of positive ions, i.e., $E = f(N_+, V)$.

It is apparent that in order to determine the potential by this method one must assign the functional relations $N_+ = f(V)$ and $E = f(N_+, V)$. The nature of the relation $E = f(N_+, V)$ will to a great extent depend on the velocity distribution of the charged particles of the plasma in the region being investigated and on the orientation of the portion of the surface of the probing body where the measurement of the electrostatic field intensity is carried out, with respect to the trajectory of the body.[3] The complexity and, in a number of cases, the fact that it is impossible to find the functional relation $E = f(N_+, V)$, are the most important shortcomings of this method and limit its application.

The method which is simplest in concept and easiest to carry out has been given by Gringauz and Zelikman.[1] We shall not dwell on a description of this method since it has been described in detail but note only that the accuracy of the determination of the potential depends strongly on how clearly defined the discontinuity in the probe characteristics is; in turn this is determined by the ion velocities and the velocity of the probing body and the nature of the mass spectrum of all the ions. Moreover, in order to determine potential by this method, one must know the mass number of the heaviest ions in the region being studied.

There is, however, another possible method for solving the problem of determining the potential of a body in a plasma (rocket or satellite). It follows from the theory of space charge[5] that the intensity of the electric field at the surface of an isolated body is given by the equation

$$E_w = \frac{k}{m} \frac{V}{\delta}, \qquad (1)$$

where E_w is the intensity of the electric field at the surface; V is the potential of the body; δ is the thickness of the layer of space charge at the surface of the body; and k and m are coefficients which assume different values depending on the density of the gas.

The thickness of the layer of positive charges at the surface of the body* is determined by the fact that the ion current j_p traverses this layer at a potential difference V and that the initial energy of the ions is small compared with V. The relation between j_p, V, and δ is[5]

$$j_p = A \frac{V^m}{\delta^k}. \qquad (2)$$

It is obvious that if we know the numerical values of the coefficients A, m, and k, by measuring the electrostatic field at the surface of an isolated body (satellite or rocket) in a plasma and the density of the ion current at the location of the device for measurement of the intensity of the electric field we can determine both the electrical potential of the body V as well as δ, the thickness of the space charge layer at which the instrument for measuring the electric field intensity is located. For example, the expression for the potential of the body is given by the following:

$$V = A^{\frac{1}{k-m}} \left(\frac{m}{k}\right)^{\frac{k}{k-m}} E_w^{\frac{k}{k-m}} \cdot j_p^{\frac{1}{m-k}}. \qquad (3)$$

Carrying out an experiment by this method involves no special difficulty since methods for determining the ion current at a surface have been fairly well developed. There is considerably less work on measurements of the electrostatic fields at the surface of a body in a plasma, but, as has been

*We may note that a body located in a plasma is generally charged negatively; for rockets and satellites this has been verified by experiment.[4,6,7]

shown,[3] there is no fundamental difficulty in carrying out such measurements. It is important to note that the method being considered is based on the use of boundary conditions at the surface of a body located in a plasma and is free, to a considerable extent, from errors associated with the necessity for an exact knowledge of the plasma parameters (concentration, temperature) or the velocity distribution of the particles.

The determination of the numerical values for the coefficients k, m, and A, specifically the determination of the coefficients k and m, should not be difficult. Under those conditions which obtain in ionospheric measurements by means of rockets and satellites, the coefficients k and m are respectively 2 and 1.5. The value of the coefficient A depends on the ionic composition of the plasma and on the ratio of the layer thickness δ to the radius of the probe used for measuring the ion current.[5] However, for the case of a plane probe the value of A is independent of the layer geometry. For a plasma produced under laboratory conditions the determination of A is not difficult since the composition of this plasma is known beforehand. For ionospheric measurements things may be somewhat more complicated but the fact that oxygen and nitrogen ions, which are the basic components of the ionosphere[6,7] at heights greater than 230 km, have close lying mass numbers (16 and 14 respectively) facilitates the solution to the problem.

If we assume that $A \sim \sqrt{1/M}$, where M is the mass number of the ion, it is apparent that the error made as a consequence of assuming one of the molecular weights indicated above is small. At heights less than 250-230 km the ionosphere undoubtedly contains a more complicated mixture of different ions. The accuracy in the determination of the numerical value of A at these heights is determined by our knowledge of the ionic composition of the atmosphere at these heights. However, the data for these heights are contradictory. The absence of negative ions at these heights may also complicate the interpretation of measurements of the electrical potential made by this method. It is apparent that in each individual case a different analysis of the measurement conditions is required. There is a condition which fundamentally, however, limits the application of this potential measurement method. This condition is the satisfaction of the relations in Equations (1) and (2) for that portion of the surface of the body where the electrical field measurement device and the positive ion current measuring device are located. For a fixed probing body these conditions are observed over the entire surface of the body. However, in the case of a body which moves with high velocity, for example a satellite, there is a limitation on the application of (1) and (2). They are not satisfied over the rear portion of the surface of the moving body where there is a region which is struck primarily by electrons.[1] Hence, in order to determine the potential it is necessary to use only the data of instruments for measuring the intensity of the electrostatic field and the density of the positive ion currents which relate to the case in which the normal to these surfaces does not deviate by an angle of greater than 90° with respect to the trajectory of the moving body.

The method described for determining the electrical potential of satellites or rockets can be applied to determining the potential of any body which is located in a plasma.

LITERATURE

1. K. I. Gringauz and M. Kh. Zelikman, Usp. Fiz. Nauk 63, 239 (1957).
2. B. A. Mirtov and V. G. Istomin, Usp. Fiz. Nauk 63, 227 (1957).
3. I. M. Imyanitov, Usp. Fiz. Nauk 63, 267 (1957).
4. C. J. Johnson and E. Meadows, J. Geophys. Res. 60, 193 (1955).
5. V. L. Granovskii, Electrical Currents in Gases, GTTI, Mos.-Len., 1952.
6. Pravda, October 5, 1958.
7. V. I. Krasovskii, Priroda, No. 12, 71 (1958).

STUDY OF METEORIC PARTICLES THROUGH INSTRUMENTS ON THE THIRD SOVIET ARTIFICIAL SATELLITE

T. N. Nazarova

Our solar system contains, in addition to the planets and their satellites, asteroids and comets, a host of more minute bodies referred to, in their totality, as meteoric matter.

The bulk of the available information on meteoric bodies intruding into the earth's atmosphere from interplanetary space has been obtained to date by astronomical methods.

Visual, photographic, and radar tracking of meteors, observation of the zodiacal light and the Fraunhofer component of the solar corona, and different techniques of studying meteoric debris filtering down to the earth's surface have led to the establishment of a physical theory of meteors, and are being employed to determine the spatial density of meteoric bodies in the solar system.[1-7] The first three methods named make it possible to find out the number of particles encroaching upon the earth's atmosphere, as well as their masses and particle-size distribution. However, the scope of these techniques is limited by the presently prevailing technical capabilities of recording particles with masses of 10^{-4} g or larger.

In order to secure information on still smaller particles, recourse may be had to the integral characteristics of meteoric matter, found from observation of the zodiacal light, and extrapolation may also be resorted to.

As we may learn from Table 1, photometric investigations have yielded sharply divergent values for the spatial density of dust material in the vicinity of the earth's orbit.

The study of this meteor matter—its cosmic nature (rate of incidence, space density, mass of the meteoric bodies, etc.),

TABLE 1

Author	Density ρ, g/cm^3
V. G. Fesenkov (1947)	$6 \cdot 10^{-23}$
Allen (1947)	$4 \cdot 10^{-23}$
Van de Hulst (1947)	$3 \cdot 10^{-21}$
Behr and Siedentopf (1953)	10^{-23}
Elsässer (1954)	$2 \cdot 10^{-23}$
Minnaert (1955)	$6 \cdot 10^{-22}$
Siedentopf (1955)	$2-4 \cdot 10^{-22}$

its place and role in the system—is of great interest to astronomy, geophysics, and also for the solution of problems concerning the evolution and origin of planetary systems, since it could throw light on a number of questions which are decisive for modern cosmogonic hypotheses.

One paper in the present symposium[8] examines some new aspects of the effect of micrometeoric particles of geophysical processes in the atmosphere.

The investigation of meteoric matter is also necessary for the solution of some problems of an applied nature, particularly for problems related to the study of the conditions governing motion of rockets and artificial earth satellites in outer space.[9-11]

It is common knowledge that the danger of an encounter between a space ship and a meteor body capable of destroying the former or of damaging the hermetic sealing of a ship is rather remote. The spatial density of meteoric bodies capable of inflicting substantial damage on a rocket ship has been studied at length. Bodies of that type cause glowing and ionization, or fall to the earth's surface in the form of meteorites, when they plunge into the earth's atmosphere. Information on the spatial density, the mass of the meteoric bodies, and any number of other important data may be ascertained through analysis of data derived from observations of meteors as they come into collision with the earth.

Protracted exposure of the skin of a rocket ship or satellite to bombardment by most minute meteoric particles will lead to a gradual erosion of the surface, and to deterioration of solar batteries and optical equipment. We may thus see that the study of micrometeoric particles is of both great scientific interest and purely practical interest, and is therefore a highly pressing problem since, as indicated above, the only data we have for particles of minute dimensions, in contrast to the situation regarding particles with masses 10^{-4} g or larger, are integral characteristics or extrapolation data.

Data characterizing an individual particle may be obtained apparently only by direct methods, employing rockets, and particularly artificial earth satellites remaining aloft a long time in the upper atmosphere.

When a meteor body collides with an obstacle, several physical phenomena occur which may prove useful to the design of recording equipment intended for installation on rockets or satellites. The range of equipment serving this purpose includes special crystal microphones, photomultipliers, capable of recording the glow induced by a meteor impact, transducers consisting of a set of thin wires bursting upon impact by a particle, transducers operating on the accelerometer principle, and piezoelectric transducers of various designs. Erosion of the satellite skin surface owing to collisions with micrometeoric particles is detectable via changes in the resistance of a thin metal layer, and by variation in the intensity of beta radiation.[12-18]

The equipment employed on the third Soviet artificial earth satellite was capable of recording the number of particle impacts and the momentum acquired by the material of the impact sensor, as the material was torn loose from the surface by the collision with the meteoric particle.[19] Several hypotheses have been proposed to account for the relation between the "reaction" momentum and the particle mass, the particle speed, diameter of the particle, etc.

Theoretical calculations carried out by K. P. Stanyukovich[20] have shown that the momentum recorded is proportional to the energy of the incident particle, for high speeds:

$$I \approx \bar{\theta} \sqrt{\frac{2\alpha}{\varepsilon_k}} E_0,$$

where $\bar{\theta}$ is a factor dependent on the transient nature of the process and particularly on the angular distribution of the mass of material dislodged (for a typical meteoric body burst, $\bar{\theta} \approx 1/3$), ε_k is the energy density of the crystal lattice of the vaporized body, and the value of the dimensionless factor depends on the properties of the vaporized medium. Thus, $I \sim F(E_0)$.

According to a contribution by M. A. Lavrent'ev[21]

$$I = Aa^3 v_0^{1.6} T_0^{-0.3}$$

where A is a constant, a^3 is the mass of the body, v_0 is the speed of the body, T_0 is the minimum density of the body necessary to convert its matter to the gaseous state.

The mass of the particles being recorded may be determined from the relation $I = AE_0$ if we take as point of departure the simplest theoretical formulation of the dependence of the momentum acquired by the impact sensor on particle energy, assuming in the process that the mean particle velocity is 40 km/sec.

The proportionality factor A between the reaction momentum and the kinetic energy E_0 of the particle has been determined more precisely in the time elapsed since the 5th Assembly of IGY. A change in the factor A quite naturally entailed a change in the determination of the amount of mass attributed to the recorded particle. At the present time, the factor A has been determined to a precision of within one half of an order of magnitude.

Measurement of the "reaction momenta" on the third satellite was performed by means of a ballistic piezoelectric transducer.[22] Electrical signals appearing in the transducer as a consequence of the impacting of the meteoric particle were divided in amplitude into 4 ranges (channels) by a converter-amplifier designed to count the number of pulses in each channel.

The equipment was calibrated to record particles with masses within the ranges 8×10^{-9} to 2.65×10^{-8} g, 2.65×10^{-8} to 1.5×10^{-7} g, 1.5×10^{-7} to 5.6×10^{-6} g, and larger than 5.6×10^{-6} g. A telemetering signal was broadcast after 32 impacts were accumulated in the range of particles of smallest mass, after accumulation of 16 and 4 impacts in the following two ranges, and after each impact in the last range.

Vibration and noise originating inside the satellite were ignored by the recording equipment, since the sensitivity threshold was set higher than the noise level.

It is fitting to note that the piezoelectric transducer mounted on the satellite was not restricted to recording particles colliding with its surface. It was also somewhat sensitive to impacts on the body of the satellite, as was ascertained in processing the experimental data.

Four piezoelectric transducers occupying a total area of 3410 cm^2 (including the body of the transducer) were mounted to record meteoric particles impacting on the underside of the satellite.

If, as stated above, we proceed from the theoretical formulation giving the dependence of the momentum acquired by the impact transducer on the energy of the meteoric particle, and assume that the mean velocity of the particle is 40 km/sec, we find that impacts by particles with masses ranging from 8×10^{-9} to 2.65×10^{-8} g, and possessing an energy of the order of 10^4 to 10^5 ergs, were recorded during the experiment.

On May 15, while data on our parameter was being telemetered, the frequency of impacts ranged from 4 to 11 m$^{-2} \cdot$ sec^{-1}, and on May 16 and 17 it dropped to 4×10^{-3} m$^{-2} \cdot$ sec^{-1}, later dropping further to 5.3×10^{-4} m$^{-2} \cdot$ sec^{-1}. Not a single response was recorded during the following 8 days, i.e., the amplifier-transducer did not store the necessary number of impacts required for telemetering, in other words the intensity of the flux of meteoric particles fell below 10^{-4} g $\cdot$ m$^{-2} \cdot$ sec^{-1}.

Let us take note of some of the peculiarities of the phenomenon observed on May 15.

1. Despite the large spread of points (particularly at apogee), it is possible to discern the existence of minima having a period of the order of 150 sec (Fig. 1).

The minimum number of impacts was recorded by the im-

STUDY OF METEORIC PARTICLES 407

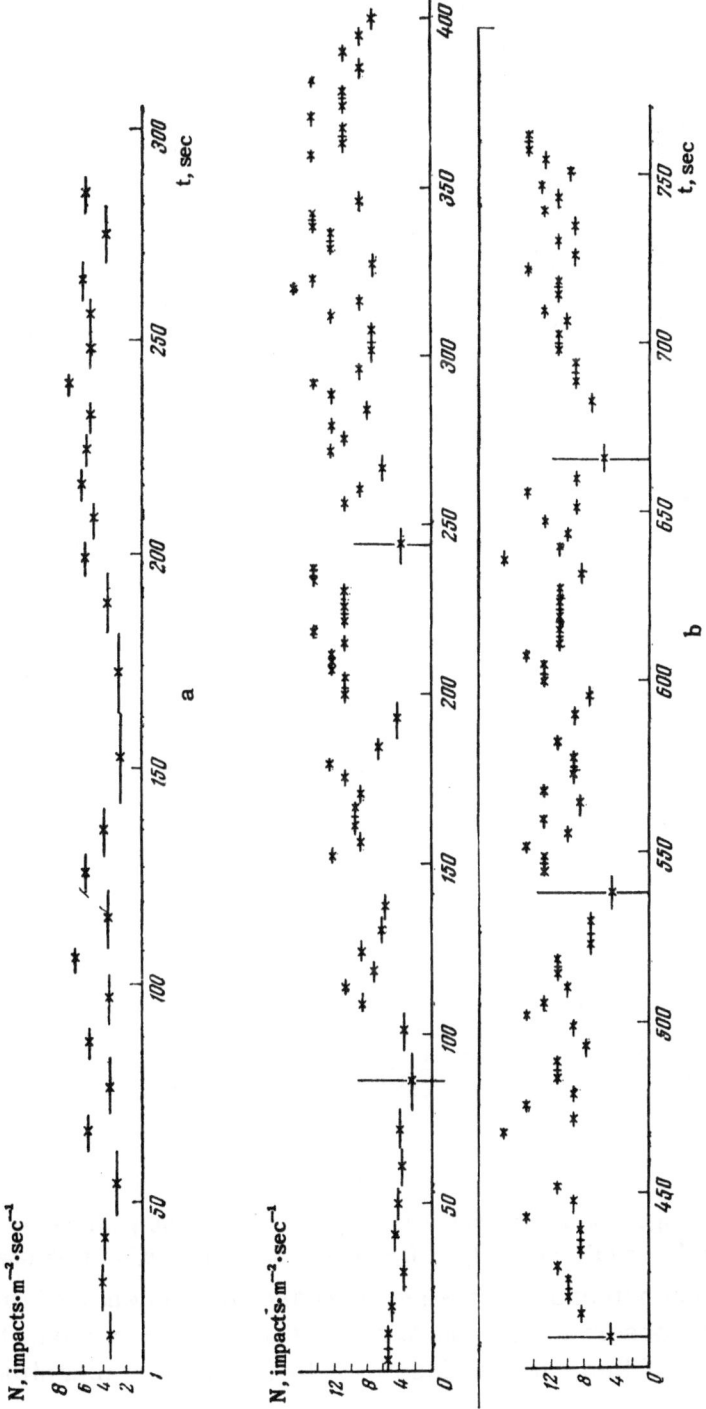

FIG. 1. Number of impacts recorded by sensors at the following heights: a. 1300–1500 km; b. 1700–1800 km.

pact sensor at moments when the bottom of the satellite was facing the earth. At that time the nose cone was exposed most, with the bottom only partially exposed (owing to its convex geometry).

2. The frequency of impacts varied with time as the satellite proceeded on its orbit (Table 2).

TABLE 2

Number of pass	Height, km	Number impacts per m² per sec
2	1300—1500	5
2	1700—1880	11
3	400—600	7
5	400—700	4

The arrangement of the segments marked off on the orbit may be seen from Fig. 2. The diagram makes it clear that the number of impacts varied with changes in the position of the satellite in orbit, but not with variations in height.

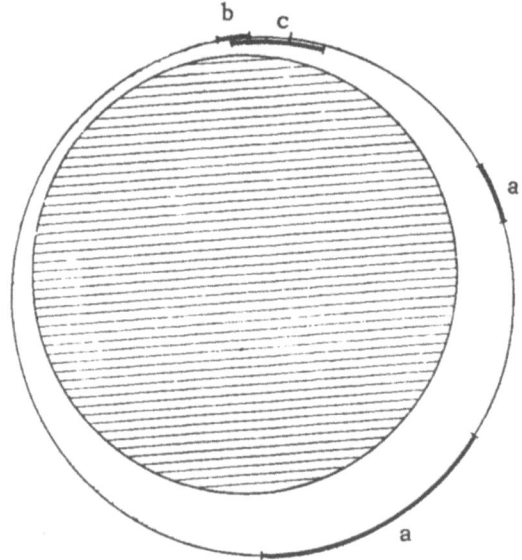

FIG. 2. Arrangement of intervals when impacts of meteoric particles were recorded, on the satellite's orbit; a. first pass; b. third pass; c. fifth pass.

The phenomenon of intense increase in frequency of impacts was observed only on May 15. On subsequent days, the frequency of impacts was less by 3 to 4 orders of magnitude.

At the present time, it is difficult to say just what might be the cause of this step-up in the frequency of impacts. This phenomenon may not be attributed with full confidence to the fact that the satellite was at that time sweeping through a meteor swarm (because of the much too high frequency of impacts[3]), although it is known from observational data that the space density of meteoric bodies in the Draconid shower is thousands of times larger than the density in swarms of active showers, where it in turn varies over a range of two orders of magnitude. Investigations yet to come will shed more light on this question.

ACKNOWLEDGMENT

In conclusion, the author would like to express her profound gratitude to O. D. Komissarov, L. N. Neugodov, A. A. Trukhachev, L. Z. Rusakov, A. K. Bektabegov, and G. M. Kurtev for their kind and active participation in the work.

LITERATURE

1. P. G. Watson, Between the Planets, Harvard University Press, Cambridge, 1956.
2. F. L. Whipple, Physics and Medicine of the Upper Atmosphere, The University of New Mexico Press, New Mexico, 1958.
3. B. Yu. Levin, Fizicheskaya teoriya meteorov i meteornoe veshchestvo v solnechnoi sisteme (Physical Theory of Meteors and Meteoric Matter in the Solar System), U.S.S.R. Academy of Sciences Press, Moscow, 1956.
4. V. G. Fesenkov, Meteornaya materiya v mezhplanetnom prostranstve (Meteoric Matter in Interplanetary Space), U.S.S.R. Academy of Sciences Press, Moscow, 1947.
5. A. C. B. Lovell, Meteor Astronomy, Oxford University Press, London, 1954.
6. I. S. Astapovich, Meteornye yavleniya v atmosfere zemli (Meteoric Phenomena in the Earth's Atmosphere), Physics and Mathematics Press, Moscow, 1958.
7. L. A. Katasev, Fotograficheskie metody meteornoi astronomii (Photo-

graphic Techniques in Meteor Astronomy), State Technical Press, Moscow, 1958.
8. B. A. Mirtov, This volume, p. 334.
9. M. W. Ovenden. J. of the British Interplanetary Society, 10, No. 6 (1951).
10. G. Grimminger. J. Applied Physics, 19, 947 (1948).
11. F. L. Whipple, Vistas in Aeronautics, Pergamon Press, N.Y., 1958.
12. M. Dubin, Rocket Exploration of the Atmosphere, Pergamon Press., London, 1954.
13. O. E. Berg and L. H. Meredith, J. of Geophys. Res., 61, No. 4 (1956).
14. E. Manring and M. Dubin, Some Preliminary Reports of Experiments in Satellites 1958 alpha and 1958 gamma, IGY World Data Center A, Rockets and Satellites, National Academy of Sciences, Washington 25, D. C.
15. M. Dubin, Cosmic Debris of Interplanetary Space, Proc. of the Second OSR Astronautics Conference, Denver, Colorado, 1958.
16. H. E. LaGow, D. H. Schaefer, and I. C. Schaffert, Mircometeorite Impact Measurements on a 20" Diameter Sphere at 700 to 2500 Kilometers Altitude, Report to V Assembly Special IGY Committee, 1958.
17. S. F. Singer, Jet Propulsion, 26, No. 12 (1956).
18. K. Edward and E. Manring, Planet Space Science, vol. 2, Pergamon Press, 1959.
19. M. A. Isakovich and N. A. Roi, Artificial Earth Satellites, Volume 2, Plenum Press, New York, 1960.
20. S. A. Baum, S. A. Kaplan, and K. P. Stanyukovich, Vvedenie v kosmicheskuyu gazovuyu dinamiku (Introduction to Cosmic Gas Dynamics), Physics and Mathematics Press, Moscow, 1958.
21. M. A. Lavrent'ev, see Volume 3, p. 85.
22. O. D. Komissarov, T. N. Nazarova, et al., Artificial Earth Satellites, Volume 2, Plenum Press, New York, 1960.

SOME RESULTS OF MEASUREMENT OF MASS SPECTRA OF POSITIVE IONS ON THE THIRD SOVIET ARTIFICIAL EARTH SATELLITE

V. G. Istomin

The radio-frequency mass spectrometer installed on board the third satellite recorded positive ions with mass numbers 32, 30, 28, 18, 16, and 14,* which were identified with the singly charged ions of molecular oxygen, nitric oxide, molecular nitrogen, monatomic oxygen, and monatomic nitrogen, respectively. Data were obtained over a range of altitudes from 225 to 980 km, and a latitude interval of 27-65° N. Certain regularities were observed in the variation of ionosphere composition with height and geographic latitude.

INTRODUCTION

A 7-5 cycle variant of the Bennett radio-frequency mass spectrometer was used in investigating the mass spectra of positive ions in the ionosphere, on the third satellite. The design of the instrument, its basic parameters, and the procedure followed in taking the measurements are all described in an earlier paper.[1]

In line with the program laid out for the operation of the equipment mounted in the satellite and the power supplies available, the mass spectrometer was operated from May 15 to May 25. During that time, a wealth of data was accumulated: about 15,000 mass spectra at heights ranging from 225 to 980 km. Measurements were performed in the Northern

---

*18 and 16 for two isotopes of monatomic oxygen, see below—Translator's note.

hemisphere only, over the latitude interval of 27 to 65°N.

By virtue of the fact that the height and geographic latitude of the satellite in orbit are interrelated in some definite fashion, with this relation subject to variation in time (on account of regression of the orbit), at a fairly slow rate in the case of the third satellite, the data on ionosphere composition obtained in all the passes give pretty much the same complex height vs. latitude cross sections of the atmosphere.

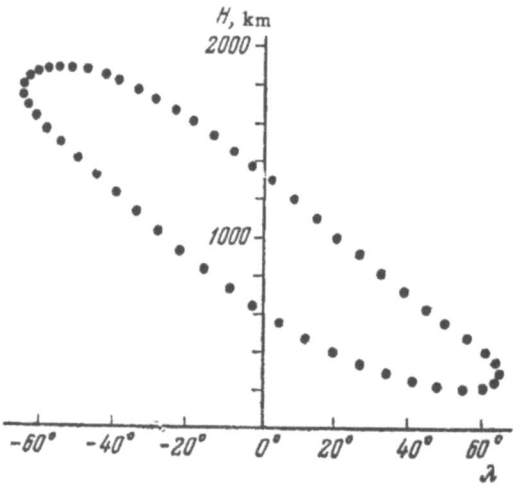

FIG. 1. Orbit of third satellite in height vs. geographical latitude coordinates approximately for the first ten days in orbit.

Figure 1 depicts the orbit of the third satellite during its first ten days in orbit. It is evident that there are only two values of the satellite orbit corresponding to any given value of geographic latitude over the range −65° to +65°, and vice versa: only two latitude values corresponding to any given height of orbit. This adds complications to any interpretation of the results obtained, making it difficult to obtain relationships for the atmosphere parameters under study which might be referable purely to height or purely to latitude.

In addition to the above drawback, note the fact that the entire observational material has reference to the daytime, i.e., the mass spectra obtained are of ions in the sunlit atmosphere.

Most of the data were obtained during morning hours, 7 to 11 A.M. Moscow time. Local time will quite naturally differ from this, depending on the geographic longitude of the corresponding point on the satellite's trajectory.

THE BASIC FEATURES OF THE SPECTRA OBTAINED

A characteristic feature of the mass spectra of ions obtained on the third satellite is the presence, in addition to the fundamental (true) mass peaks, of so-called harmonic (false) peaks which complicate interpretation of the tracings.

The presence of the harmonic peaks is attributed to the fact that the effective stopping potential of the mass spectrometer is too low for two reasons: 1) owing to the fact that the mass spectrometer is mounted in a moving vehicle, all of the atmospheric ions have, in addition to the thermal velocities relative to the instrument, a directional (ordered) velocity equal to the speed of the satellite; and 2) owing to the presence of this ordered velocity $v = 8 \times 10^5$ cm/sec, an ion of mass number M and charge $q = 4.8 \times 10^{-10}$ in cgs units will possess a certain energy equivalent to the presence of an accelerating voltage

$$\Delta V_v = 300 \frac{M \cdot m_0}{2q} v^2 \text{ volts} \tag{1}$$

where $m_0 = 1.67 \times 10^{-24}$ g is a quantity equal to 1/16 the weight of the oxygen atom. For ions of mass number 16, the velocity of 8×10^5 cm/sec is the equivalent to an energy of 5.35 ev, and for ions of mass number 30, the energy is 10^5 ev.

By reason of that fact, the stopping potential of the instrument for ions of different mass proves to be too low for different values numerically equal to the ion energies expressed in electron volts. Furthermore, as shown by an analysis of the spectra obtained, the satellite had a negative potential of several volts, which contributed to a further lowering of the effective stopping potential, by an amount which was uniform for all atmospheric ions.

As a result of the reduction in effective stopping potential for those two reasons, the sensitivity of the mass spectrometer was found to be overestimated several times, more so for heavy ions than for light ions, while the mass resolution correspondingly decreases to about half that achievable under laboratory conditions.

The instrument was adjusted to a resolution $R = (M/\Delta M) \approx 20$ in the region of mass numbers 20. Here, as usual, M is the mass number of the peak and ΔM is its base width expressed in amu. The resolution at the base of the peak ranges from 7 to 10 in the region of mass numbers 16-14, in the spectra of atmospheric ions obtained by the satellite.

This means that, in the given experiment, mass peaks in the region of light masses ($M \sim 16$) were completely resolved if their mass numbers differed by about 2 or more amu. In the region of heavy masses ($M \sim 30$), the peaks whose mass numbers differed by 2 amu were not completely resolved.

The dynamic range of the spectrometer turned out to be measurably extended in the direction of high intensities, on account of the presence of harmonic peaks in the spectra. Since the level of a light harmonic of mass 16 is on the average equal to 0.1 of the amplitude of the fundamental peak, when the peak at 16 was so intense that the low-sensitivity output of the amplifier became saturated (current exceeding 10^{-8} amp), its value could be estimated from the value of the light-harmonic peak.

The mass numbers of all the peaks in the spectra, as might be anticipated,[2] were found to be shifted on the spectrometer mass scale in the direction of light masses, on account of the effect of the satellite's velocity and negative charge.

Remember that the sweep over the mass spectrum in the radio-frequency mass spectrometer employed is executed by varying the negative accelerating voltage in a sawtooth-wave pattern.

Under laboratory conditions, the mass number of the peak is determined from the formula

$$M = \frac{V}{k}, \qquad (2)$$

where M is the mass number of the peak in amu, V is the sweep voltage in volts at the moment the peak appears, and k is an instrument constant equal to 7.2 v/amu.

When the spectrometer was used on the satellite, an increment equal numerically to the energy of a corresponding ion moving at a speed of 8×10^5 cm/sec, expressed in electron volts, was added to the negative sweep voltage V, with the negative potential φ of the satellite superposed additively. Equation (2) must then be transformed to

$$M = \frac{1}{k}(V + \Delta V_v + \varphi) = M_{dec} + \Delta M_v + \Delta M_\varphi. \tag{3}$$

Here M_{dec} is the mass number of the peak when determined in decoding or interpreting the tracings on the basis of the sweep voltage value at the moment the peak appeared, ΔM_v is a correction accounting for satellite velocity, and ΔM_φ is a correction accounting for satellite charge.

The quantity ΔV_v, as is evident from Equation (1), itself depends on the ion mass number, and the equation for determining the mass number of the peak may consequently be written as follows:

$$M = \frac{1}{k}\left(\frac{V+\varphi}{1-300m_0v^2/2qk}\right) = \frac{M_{dec}+\Delta M_\varphi}{1-300m_0v^2/2qk} \tag{4}$$

or, finally, by substitution of the numerical values

$$M = \frac{1}{k}\left(\frac{V+\varphi}{1-0.334/k}\right) = \frac{M_{dec}+\Delta M_\varphi}{0.954}. \tag{5}$$

As is evident from Equation (5), the shift in mass peaks due to the velocity effect must be 0.74 amu, for this spectrometer, for mass 16, and 1.39 amu for mass 30.

INTERPRETATION OF THE SPECTRA

The basic difficulty in interpreting the mass spectra of the ions resides in the separation of the fundamental (true) mass peaks from the harmonics (false peaks). Figure 2 gives some idea of the nature of tracings of the spectra, showing reproductions of four photographs.

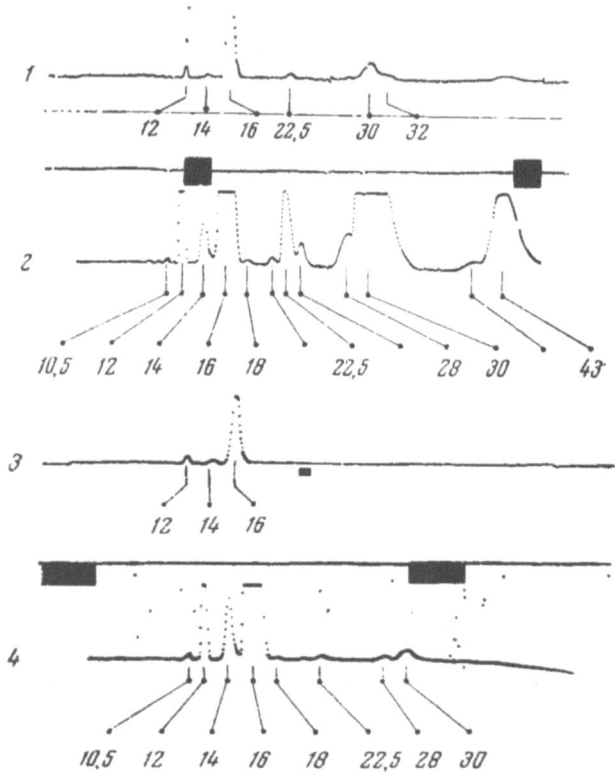

FIG. 2. Tracings of ion mass spectra taken on May 23, 1958, at about 9 A.M. Moscow time. 1 and 2 are tracings from the low- and high-sensitivity channels of the mass-spectrometer amplifier. Height ~230 km, latitude 55°N; 3 and 4 are the same, with height ~350 km, latitude 64°N.

The first and second reproductions are of tracings of ion spectra in the region of the perigee, obtained via the low- and high-sensitivity channels of the mass spectrometer amplifier, respectively, while the third and fourth reproductions are similar tracings relevant to a height level of the order of 350 km.

The tracings date from May 23, 1958, at about 9 A.M. Moscow time. The geographic latitude of the satellite orbit at that time was 55° N for the first two tracings and 64°N for the latter two.

Conspicuous in the ion spectra from the perigee region are the abundance of peaks and the poor resolution, particularly in the region of large mass numbers.

Separation of the fundamentals and harmonics in the spectra was achieved by comparing their amplitudes and the pattern of variation observable, as a function of the height of satellite orbit and as a function of satellite orientation in space.

The positions of the harmonics in the spectra were determined in prior laboratory tests for the particular spectrometer used. An adequate yardstick substantiating the correctness of the identification of fundamentals and harmonics was the circumstance that the relative width of the harmonic peaks was less than that of the fundamentals. This fact was also given substantiation in laboratory experiments.

After introduction of corrective terms accounting for satellite velocity in conformity with Equation (5), the mass numbers of peaks assumed to be fundamental (true) peaks were distinguished from integral even values by some fraction of an amu. This difference was attributed to the negative charge of the satellite. For the two days May 22 and May 24, for instance (passes 96 and 122), the difference averaged over the interval of the satellite pass relevant to the measurements was, respectively, 0.3 and 0.5 amu.

Taking into account the value of the instrument constant $k = 7.2$ v/amu, the negative potential of the satellite is found from these data to be -2.2 and -3.6 v.

In all of the spectra, we find one peak towering above the others in intensity whose mass number, after corrections introduced for satellite velocity, is found to be close to 16. On that basis, this peak was identified with the peak of monatomic oxygen O^+.

The "light" peak which ranked second in intensity, also present in spectra taken at all altitudes, was the peak with mass number 14. It seems most natural to identify it with the peak of monatomic nitrogen N^+. Peaks with mass numbers 12 and 10.5 were attributed to "light" harmonics of peaks 16 and 14, respectively.

On the heavy-mass side, peak 16 is encroached upon by a very weak peak of mass number 18 which is clearly detectable in many of the spectra. Its relative intensity was found to be

$i_{18}/i_{16} = (0.15 + 0.05)\%$. This peak is definitely not a harmonic of any of the fundamental peaks present in the spectra, and its appearance is to be explained solely by the presence of some corresponding ionized molecule or atom in the atmosphere.

This ion may be identified either with the water ion H_2O^+ or the ion of another hydrogen compound, e.g., ammonium NH_4^+.

In the light of the identification so performed, the presence of the ion with mass number 18 must serve as an indirect indication of the presence of appreciable quantities of neutral or ionized hydrogen in the upper atmosphere. However, another possibility appears more realistic: the mass peak 18 should apparently be attributed to the ion of the oxygen isotope of atomic number 18, whose relative abundance is 0.2%.

The question will be settled definitively only by subsequent and careful measurements.

In addition to the group of "light" peaks 18, 16, 14 in the spectra obtained at heights in the perigee region, a group of "heavy" peaks with mass numbers 32, 30, 28 which were also construed as fundamental peaks (true peaks) stood out in the spectra. The most intense of these was the peak of mass number 30, which must be identified with the peak of the nitric oxide ion NO^+. The peaks of mass number 32 and 28 are naturally attributed to the ions of molecular oxygen O_2^+ and molecular nitrogen N_2^+. Despite the fact that these heavy peaks were not fully resolved and the fact that their shoulders, especially in the case of peak 32, often run together with the peak 30 intermediate between them, their presence in the spectra is established beyond question.

All of the other peaks found in the spectra seen in Fig. 2 must be viewed as harmonic (false) peaks.

Accordingly, the group of three peaks, of which the most intense is the peak of mass number 22.5, is a component of the "light" harmonics of peaks 32, 30, and 28. On some spectra, it is possible to also single out a heavy harmonic of peak 16, having a mass number 22.8. The peaks with mass numbers 40 and 43 are "heavy" harmonics of peaks 28 and 30.

VARIATION IN COMPOSITION OF THE IONOSPHERE AS A FUNCTION OF HEIGHT AND GEOGRAPHIC LATITUDE

In viewing the results reported in this section, we must bear in mind the fact that all of the data are referable solely to the intensities of the mass peaks measured in current units.

The intensities of the peaks must naturally be related to the relative and absolute concentrations of the corresponding ions, but this relation may be a rather complicated one, and the question has not been elucidated in definitive manner at present.

The difficulty arising in correlating the relative intensities of the ion peaks to secure information on the relative concentrations of the relevant ions resides in the attendant need to take into account discrimination with respect to mass, which must inevitably occur in the experiment described.

Discrimination of ions of different masses may occur first of all in the field of the satellite itself, since the satellite does not have zero potential relative to the undisturbed plasma. The presence of a retarding field set up by the first grids of the mass spectrometer tube is conducive to additional discrimination of ions. Finally, ions of different mass will experience different conditions in the analyzer itself, depending on the initial energy they had on entering the analyzer. As stated in the discussion of the features of the spectra obtained, the initial energy of ions with mass numbers 16 and 30 differs by roughly a factor of two under the experimental conditions obtaining on board the satellite. In line with this, the level of the stopping potential for these ions will also differ, and the intensity ratio of the peaks will not be equal to the ratio of the concentrations of the ions involved as they gain admittance to the analyzer.*

*The first of the two factors, discrimination in electric fields, leads to a reduction in the relative intensities of the heavy ions compared to light ions, while the second factor exerts an opposite action and may offset this reduction to an extent, or even bring about the reverse effect.

However, we must also bear in mind the fact that the effects of mass discrimination, whatever may be the causative factors responsible for it, may be relatively large only for those ions whose masses differ considerably from each other, while they will be small for ions closely similar in mass.

In this last case, the intensity ratio of the mass peaks is close to the ratio of the concentrations of the ions involved. For example, the intensity ratio of the peaks of monatomic nitrogen and monatomic oxygen, i_{14}/i_{16}, or that of molecular nitrogen and nitric oxide i_{28}/i_{30}, must be close to the ratio of their concentrations in the ionosphere, while the ratio of the intensities of the peaks of nitrogen oxide and monatomic oxygen will quite possibly differ considerably from the ratio of their relative concentrations.

Both the absolute and the relative intensities of the mass peaks in the spectra disclose appreciable variation in time. Since the operation of the spectrometer was monitored and found to be steady over the entire period of time concerned, variations in the spectra must be ascribed to: a) variation in the orientation in space of the analyzer entrance relative to the direction of flight of the satellite, brought about by the satellite's rotation in space; b) variation in the coordinates of the satellite: height, geographic latitude and longitude; c) variation in the external environment; factors affecting the ionization of the atmosphere.

The first set of variations, related to the rotation of the satellite, was brought to light by the fact of a periodicity coinciding with that found from data furnished by other experiments conducted on the third earth satellite.[3,4] In examining rather extended intervals of the tracings (of the order of tenths of a second), we can make out groups of spectra obtained at the most favorable orientation of the analyzer entrance relative to the direction of flight, i.e., the orientation where the angle formed by the normal to the entrance of the mass spectrometer tube and the velocity vector of the satellite is at a minimum. As demonstrated by a theoretical treatment of the problem, discrimination of ions of different mass must be minimal under such conditions. Confirmation of this deduction is also brought by an analysis of the tracings obtained. Data based on the reduction of just these spectra were utilized predominantly in analyzing the composition of the ionosphere and its changes as a function of satellite height and geographical coordinates.

As reported earlier,[5] the monatomic oxygen ion O^+ was found to be a constantly present ionospheric constituent prevailing at all heights studied in the range of 226 to 980 km. For that reason, the intensities of all the remaining mass peaks may be conveniently compared to the intensity of the peak of monatomic oxygen.

The second monatomic constituent of the ionosphere is the nitrogen ion N^+. The ion current (peak intensity) of monatomic nitrogen, relative to monatomic oxygen, varies as a function of height and geographic latitude over a range extending from 1.3 to 8-10%. The trend toward increased content of ions of monatomic nitrogen, reported earlier,[5] is substantiated by all of the spectra reduced.

This relationship shows up in bold relief in the graph in Fig. 3. Height in kilometers is plotted on the horizontal axis, with relative peak intensity of monatomic nitrogen in terms of percentage of the monatomic oxygen peak plotted on the vertical axis. As indicated above, this quantity is close to the ratio of concentrations of monatomic nitrogen and oxygen ions. Each point on the graph was obtained by averaging the measurements of several spectra (from 5 to 20 spectra). The different symbols refer to different passes of the satellite executed during the period covering May 18 to May 24, 1958. Worthy of note is the rather large spread of points, amounting to ± 3% at some heights; however, it is readily seen that this spread is caused not by random errors in measurement, but is a reflection of real changes in ionospheric composition diurnally and from one satellite pass to the next. This may be seen in Figs. 4 and 5, in which we have the height dependence of the relative concentration of monatomic nitrogen ions, but data from one single day are used for each graph.

It is obvious that the points corresponding to a single pass fit well onto a smooth curve, with a small spread, while points

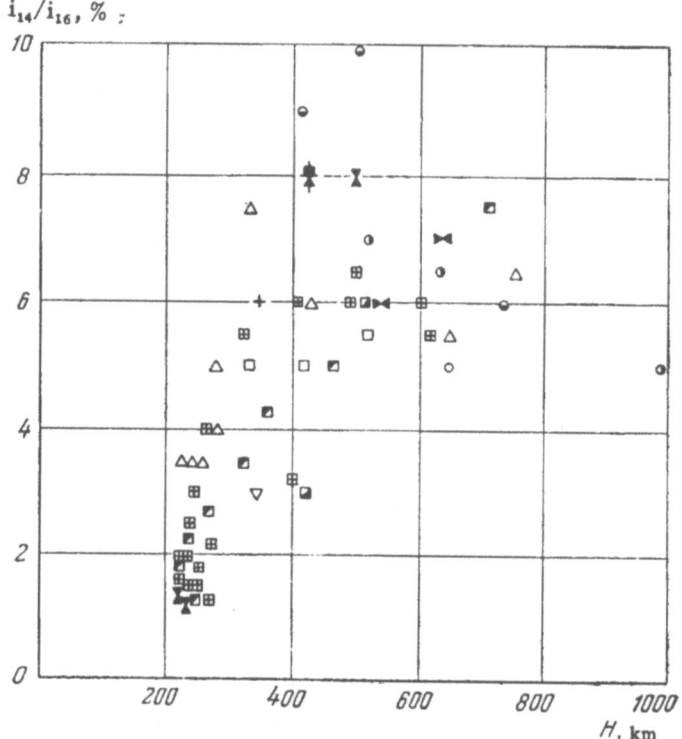

FIG. 3. Variation of concentration of ions of monatomic nitrogen with respect to ions of monatomic oxygen, as a function of height (data based on 12 satellite passes executed during May 18-24, 1958).

obtained on the preceding or succeeding pass depart noticeably from the curve plotted.

In considering Figs. 3-5, one must bear in mind that these graphs provide no basis for inferring a variation of ionospheric composition with height, on account of the aforementioned specific nature of the experiment conducted aboard the satellite, since the geographic latitude also varies with the change in height of satellite orbit. The latitude dependence of the ionospheric composition stands out in more salient fashion over the range of heights 225 to 350 km, for which the latitude varies extensively, viz., from 25° to 65° N.

Figure 6 provides data on the latitude dependence of the ratio of concentrations of monatomic nitrogen ions and

MEASUREMENT OF MASS SPECTRA 423

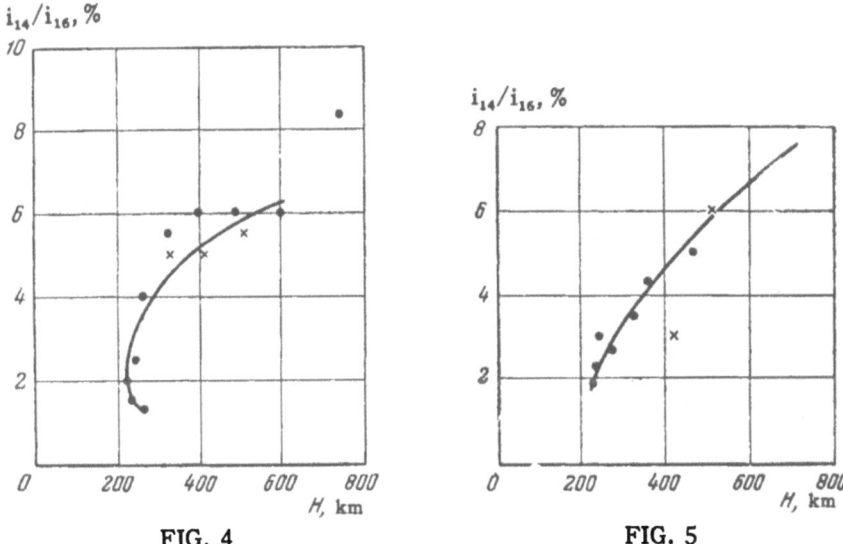

FIG. 4 FIG. 5

FIG. 4. Variation of concentration of ions of monatomic nitrogen with respect to ions of monatomic oxygen, as a function of height (data based on two satellite passes on May 21, 1958).

FIG. 5. Variation in concentration of ions of monatomic nitrogen with respect to ions of monatomic oxygen, as a function of height (data based on two passes on May 23, 1958).

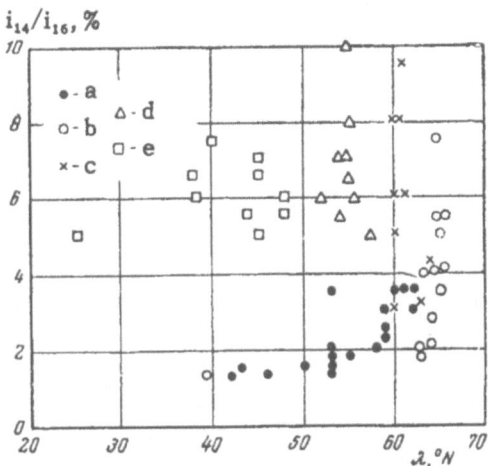

FIG. 6. Variation in concentration of ions of monatomic nitrogen with respect to ions of monatomic oxygen, as a function of geographic latitude (data based on 12 satellite passes in period covering May 18 to May 24, 1958).

monatomic oxygen ions. The entire interval of heights is broken down into five subintervals, and the points relating to different intervals are distinguished by appropriate symbols.

It is evident that the relative concentration of ions of monatomic nitrogen increases appreciably, for the ranges of height 225 to 250 km and 251 to 350 km, as the satellite proceeds from the latitude belt 30° to 50° into the belt 55° to 65° N. For the interval of heights 351 to 450 km, no latitude dependence is revealed, since the latitude varies little over that range. As for still higher altitudes, in the 451-600 km and 601-980 km ranges, we may draw the conclusion that the relative concentration of ions of monatomic nitrogen either loses its dependence on latitude, or else that this dependence loses much of its force although retaining the same qualitative features that distinguish it at lower altitudes.

With respect to the spread of points observed on the graph of Fig. 6, we may note that this spread of points is to some extent a reflection of the variation in composition from one pass to the next, and from day to day. The graphs in Figs. 7 and 8, each of which reflects the latitude variation of the relative concentration of ions of monatomic nitrogen for a single day, confirm this surmise. It is clear then that all of the

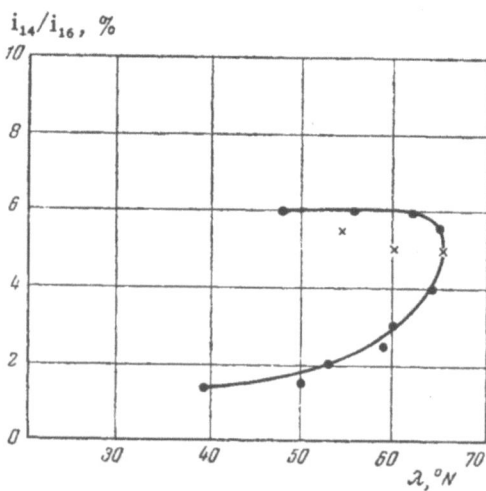

FIG. 7. Variation in concentration of ions of monatomic nitrogen with respect to ions of monatomic oxygen, as a function of geographic latitude (data based on two satellite passes on May 21, 1958).

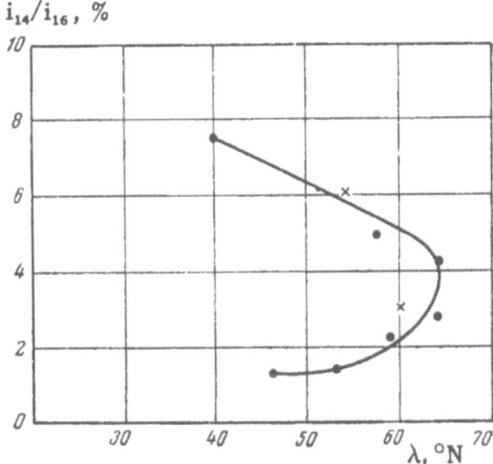

FIG. 8. Variation in concentration of ions of monatomic nitrogen with respect to ions of monatomic oxygen, as a function of geographic latitude (data based on two satellite passes on May 23, 1958).

points show good fit onto a smooth curve, with a relatively small spread.

The height dependences of the relative concentrations of molecular oxygen and nitrogen ions, and of nitric oxide ions, are roughly of the same nature. At perigee, the relative intensities of the corresponding ion peaks are at the maximum, and fall off rather rapidly with height. The molecular ion of highest intensity is the nitric oxide ion. The intensity of the nitric oxide peak at a height of about 230 km amounts to 25-35% of the intensity of the peak of the monatomic oxygen ion.

Ranking second in intensity, among the molecular ions, is the oxygen ion O_2^+. Its intensity at perigee is about 7-12% that of monatomic oxygen. The intensity of the peak of molecular nitrogen at perigee comes to 1.5-3%.

The characteristic curves of the height dependence of the relative intensities of the molecular peaks NO^+, O_2^+, and N_2^+ (relative to O^+), based on data from a single pass (no. 96, on May 22, 1958), are shown in Figs. 9-11. A striking feature of these curves is the fact that all of them show a clearly marked bifurcation into two branches: a southern and a northern branch. Points plotted on the diagrams are enumerated in the

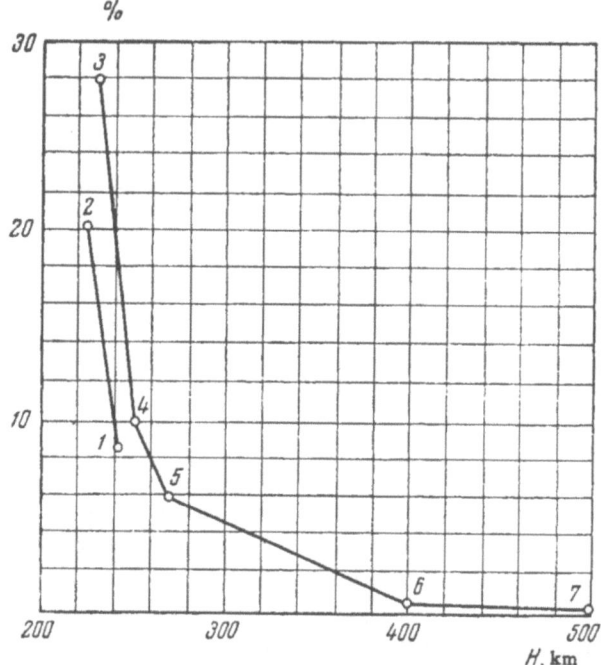

FIG. 9. Variation in relative intensity of ion peak of nitric oxide, as a function of height and geographic latitude (data based on single pass executed on May 22, 1958). Points plotted are enumerated in the order of increasing geographic latitude: 1, 2. southern latitudes (prior to perigee); 3 to 7. northern latitudes (subsequent to perigee).

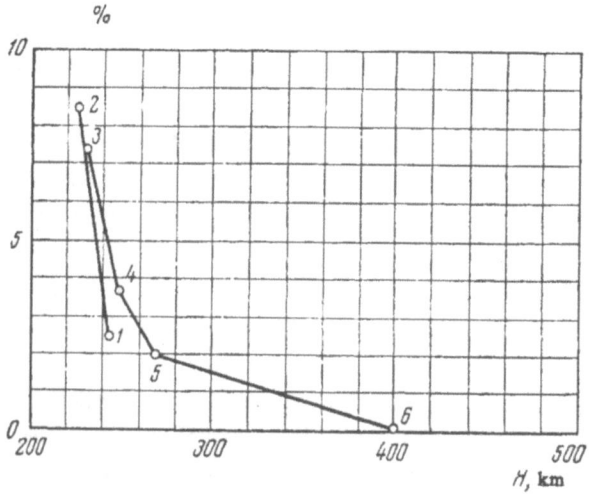

FIG. 10. Variation in relative intensity of ion peak of molecular oxygen as a function of height and geographic latitude (data based on single pass executed on May 22, 1958). Notation same as in Fig. 9.

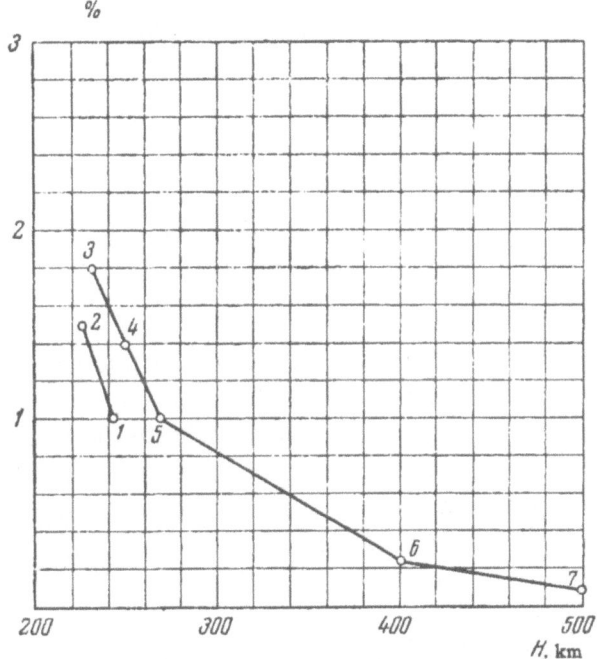

FIG. 11. Increase in relative intensity of ion peak of molecular nitrogen as a function of height and geographic latitude (data based on a single pass executed on May 22, 1958). Notation same as in Fig. 9.

order of increasing geographic latitude: points 1 and 2 refer to southern latitudes (prior to perigee), and points 3 to 7 refer to northern latitudes (subsequent to perigee). The latitude consequently increases from point 1 progressively to point 6. The last point, point 7, corresponds to approximately the same latitude as point 4.

On all three diagrams Figs. 9-11, the southern branch of the curve passes below the northern branch, indicating a dependence of the relative concentrations of the molecular ions NO^+, O_2^+, and N_2^+ on geographic latitude: as latitude increases, the concentration of ions of nitric oxide, molecular nitrogen, and molecular oxygen relative to the concentration of ions of monatomic oxygen also increases.

The ion peak of molecular oxygen trails out to the maximum height at 400 km (relative intensity of peak $O_2^+/O^+ \sim 0.1\%$), while the peaks of the ions of nitric oxide and molecular nitrogen go out to a height of 500 km. The intensities of the NO^+

and N_2^+ peaks relative to O^+ at heights 400-500 km become approximately the same, equal to 0.2-0.1%. Only above 500 km, then, do molecular ions fail to be detected, and the ionosphere is found to be purely monatomic in composition, with nitrogen and oxygen atoms, to an accuracy of 0.1%.

SUMMARY

Measurements of the mass spectra of positive ions taken at heights ranging from 225 to 980 km at daytime showed that ions of monatomic oxygen O^+ predominate at those heights. In addition to the ions of monatomic oxygen, ions of mass number 14—monatomic oxygen N^+; of mass number 18—oxygen isotope $(O^{18})^+$; of mass number 28—molecular nitrogen N_2^+; of mass number 30—nitric oxide NO^+; and of mass number 32—molecular oxygen O_2^+, were recorded.

It was revealed that the composition of the ionosphere in the region investigated varies as a function of height. The relative content of heavy molecular ions NO^+, O_2^+, and N_2^+ (i.e., relative to O^+) falls off with increased height, and the relative content of the light ion N^+ increases with increased height. The relative content of ions of molecular oxygen drops to less than 10^{-3} at heights above 400 km, while the content of molecular nitrogen ions and of nitric oxide ions drops to less than 10^{-3} at heights above 500 km.

It was disclosed that the composition of the ionosphere varies as a function of geographic latitude. The relative content of ions of monatomic nitrogen at heights of 225-350 km increased markedly as the satellite passed from the 30-50° N belt of latitudes to the 55-65° N belt. The relative content of ions of molecular nitrogen, molecular oxygen, and nitric oxide is also much higher in the 55-65° N latitude belt than at more southerly latitudes.

ACKNOWLEDGMENT

The author avails himself of this opportunity to express his acknowledgment to B. A. Mirtov for the latter's unflagging interest in the work and in discussion of the results, and also to S. V. Vasyukov, A. A. Perno, and R. P. Shirshov, for their kind and valuable assistance in decoding the telemetered tracings and in processing the experimental data.

LITERATURE

1. V. G. Istomin, Artificial Earth Satellites, Volume 2, Plenum Press, New York, 1960.
2. B. A. Mirtov and V. G. Istomin, Uspekhi fiz. nauk, 63, 227 (1957).
3. S. Sh. Dolginov, L. N. Zhuzgov, N. V. Pushkov, Artificial Earth Satellites, Volume 2, Plenum Press, New York, 1960.
4. V. V. Mikhnevich, Artificial Earth Satellites, Volume 2, Plenum Press, New York, 1960.
5. V. G. Istomin, Artificial Earth Satellites, Volume 2, Plenum Press, New York, 1960.

COSMIC RAY MEASUREMENTS BY GEOPHYSICAL ROCKETS

Yu. G. Shafer and A. V. Yarygin

In 1958 measurements of the total intensity of cosmic rays were completed with the help of self-contained Geiger counters and of an ionization chamber placed aboard a geophysical rocket.

These investigations were intended to find the height dependence of the intensity up to 200 km and to determine the average specific ionization of particles of the primary radiation. In addition we were also concerned with the problem of testing the equipment under the high load and particular temperature conditions characteristic of rocket flights.

The apparatus was placed in the hermetic nose section of the rocket, whose steel walls did not exceed in thickness 3 mm (~ 2.5 g/cm^2); the axis of the apparatus coincided with the axis of the rocket. About 8% of the total solid angle of the upper semisphere was shielded by a large amount of material (> 50 g/cm^2). The solid angle in which the equipment was shielded by the rocket body on the side of the lower semisphere was 0.16π. Figure 1 shows schematically the arrangement of the measuring apparatus.

The particle recording was done by halogen counters. The geometrical dimensions of the counters were 18×150 mm, the thickness of its steel wall was 0.1 mm. Important properties of a counter of this type are its stability of work for temperature variations from +50 to −40°C, its long working life, and its low working voltage (380-400 v). The electric supply was provided separately by semiconductor converters. At the input of one of these converters was given a voltage of 6.5 v with a current of 7.5 ma; the output was 400 v with a

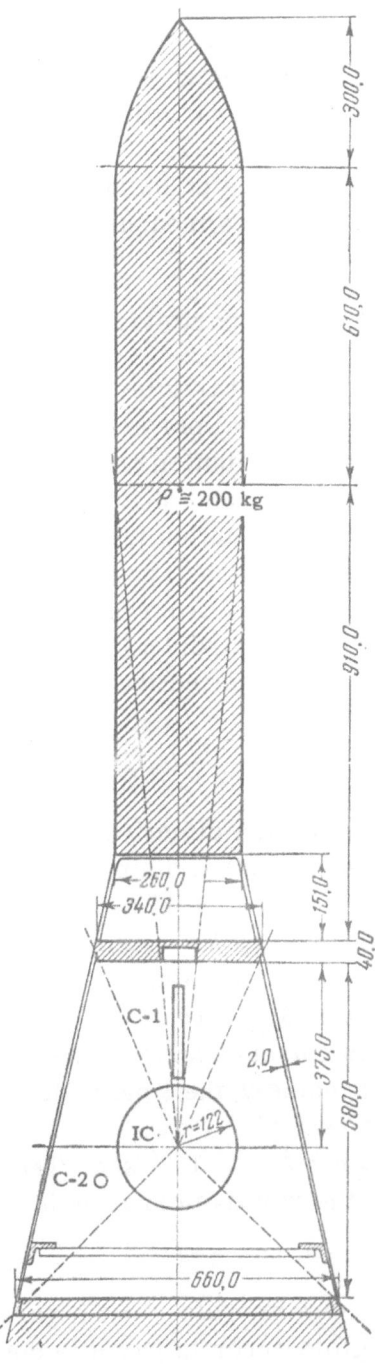

FIG. 1. Schematic arrangement of the horizontal counter (C-2), of the vertical counter (C-1), and of the ionization chamber (IC) in the head section of the geophysical rocket.

current of 50 μa. A gas stabilizer was connected at the output in order to decrease the voltage oscillations. The stabilized current was independent of variations in the load current and of the input voltage within 1-2%.

The global intensity was calculated on the basis of the number of charged particles measured by a single counter from the relation

$$J = \frac{N}{G} \text{ particles cm}^{-2} \text{ sec}^{-1},$$

where N is the number of particles recorded by the counter in 1 sec, G is a geometrical parameter which is a function of the angular radiation distribution, of the counter orientation, of its effective length l_{eff} and of its effective radius r_{eff}.

Starting from the assumption that the angular distribution of the primary radiation on the upper semisphere was isotropic, and that the intensity from the lower semisphere was zero, we obtained, for a cylindrical counter oriented arbitrarily in space, $G = S_{eff}$ (effective counter section):

$$S_{eff} = \frac{\pi r_{eff}}{2} (r_{eff} + l_{eff}). \tag{1}$$

It is well known that the effective diameter of a Geiger counter differs from its actual diameter by at most 0.1 mm. The effective length of the type of counter used by us was determined experimentally by Yu. I. Logachev and was equal to 100 ± 1 mm. Substituting the values of r_{eff} and of l_{eff} in formula (1), we obtain $S_{eff} = 15$ cm^2.

It was important for our experiments to know the dead time of the counter, namely, the time τ during which the counter was not able to record.

With our counter $\tau \approx 2 \times 10^{-4}$ sec. The value of τ determines the value of the relative error σ in the counting speed:

$$\sigma \cong N\tau.$$

For the maximum counting rate N = 40 particles per sec, σ = 0.8%. This negligible error in particle counting will not be taken into account in our further calculations.

The evaluation of the angular displacement of the rocket body during its flight is very important. Since the apparatus was not spherically symmetric, variations due to the irregular motion of the rocket could arise during the measurements.

A theoretical calculation of these variations would be extremely complicated even if the intensity of the primary radiation remained constant. Therefore we placed in the rocket two counters: one was situated so that its axis coincided with the axis of the rocket; the other was situated perpendicularly to the rocket axis. With this arrangement of the counters the unpredictable angular motion of the rocket would influence in the different ways their counting speed.

The systematic difference between N_{vert} and N_{hor} could make it possible to determine the influence of the counter shielding by the rocket mass and the lowest accuracy limit. Moreover, the indications of two counters with independent radio circuits and power supplies was a useful means for checking their operations.

In our apparatus one counter was located vertically above the ionization chamber, the other was located horizontally at a side of the chamber (see Fig. 2). The pulses coming from the counters passed through emitter-repeaters at the input of the counting device with a counting rate of 1:256. Each trigger unit of the counting circuit could count up to 5×10^4 pulses per second. With such a resolving power all arriving pulses would be recorded by our apparatus.

The last trigger of the counting circuit commanded a polarized relay (P_1) through the emitter-repeater. The change in position of the nucleus of this relay took place after each 128 pulses (Fig. 3). The contacts of the relay P_1 short-circuited the corresponding potentiometric circuits of the rocket telemetric system. As a result of the switch-over of the last trigger when 128 pulses have been counted a voltage value characteristic for each counter was transmitted by radio from aboard the rocket and was recorded on a film in a recorder in the receiving station. The same film received a continuous time signal which permitted measuring time intervals with an accuracy up to 0.02 sec. An example of this photographic recording is given in Fig. 4.

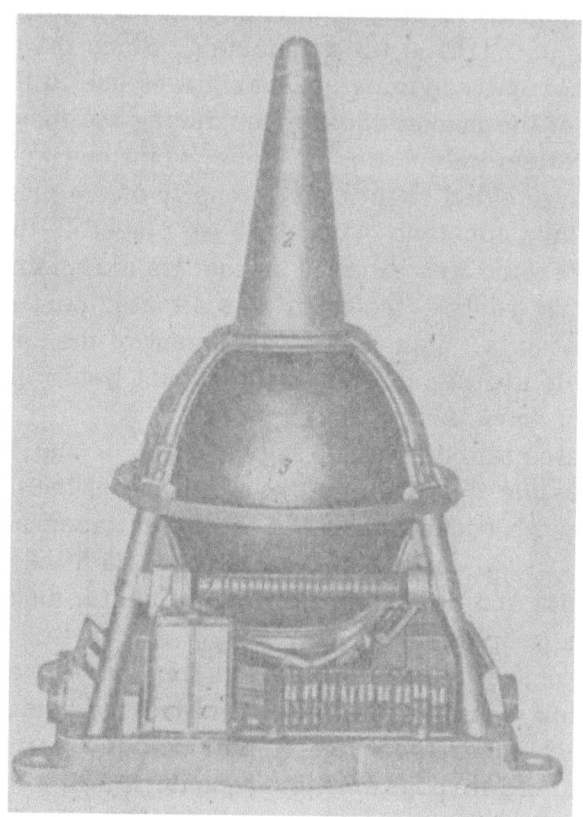

FIG. 2. Apparatus for cosmic-ray recording placed aboard the geophysical rocket. 1, Horizontal counter; 2, vertical counter (in a thin-walled duralium container); 3, ionization chamber.

The number of particles recorded by the counter in one second could be calculated from the ratio $N = n/t$, where n denotes the counting rate, t represents the time measured on the film between two changes of the voltage of a given counter.

In our apparatus the horizontal counter had also the function of master counter. When it had counted 256 pulses, a relay P_4 applied the voltage to the chamber relay P_5 (Fig. 5). This relay started operating and took the charge accumulated because of ionization on the chamber collectors after 256 particles had passed through the master counter.

The spherical ionization chamber had a diameter of 24 cm; it was made of stainless steel 0.4 mm thick. A chamber having such a volume has a statistical accuracy of not more than 5%,

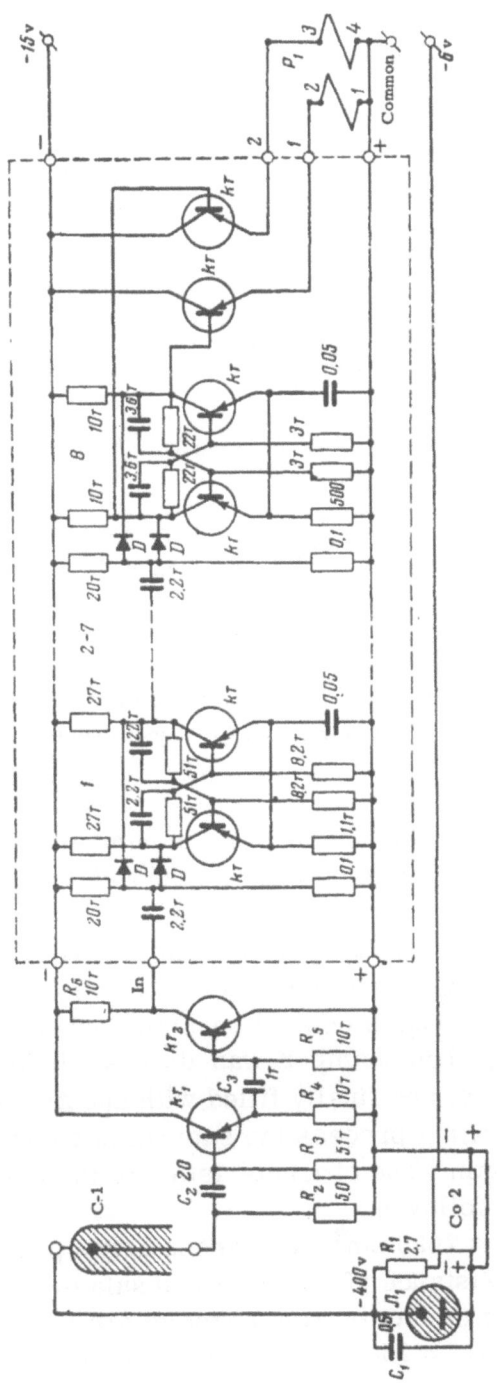

FIG. 3. Basic counting circuit.

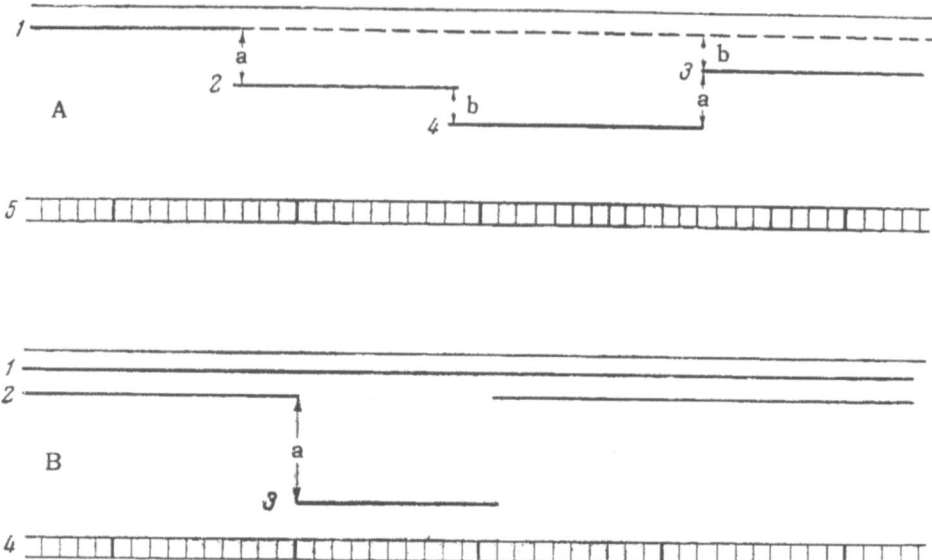

FIG. 4. Examples of the photographic recording of the apparatus signals at the receiving stations. A, Counter signals: 1, zero level; 2, level of the vertical counter signal (amplitude a); 3, level of the horizontal counter signal (amplitude b); 4, total level of the signals from both counters (a + b); 5, time markers. B, Signals of the ionization chamber: 1, calibration level (100%); 2, carrying level; 3, chamber signal (a, signal amplitude); 4, time markers.

which corresponds to the fluctuations in number of the primary relativistic charged particles passing through the chamber during one second.

A vibrostable collecting electrode having the shape of a thin-walled semisphere with a diameter of 28 mm was fastened to a jaw passing through a central ceramic insulator provided with a guard ring. The chamber was deaerated, evacuated, flushed many times, and finally filled with spectrally pure argon. During the filling process the saturation curves were recorded for each atm. The characteristics of the ionization camera were the following:

1) Volume $V = 7020$ cm^3; 2) Excess argon pressure $P = 7$ atm; 3) Resistance of the central insulator $R = 1.5 \times 10^{15}$ ohms; 4) Conversion factor from argon to air $K = 1.4$; 5) Saturation potential $U_s = 250$ v; 6) Working voltage $U_w = 380$-400 v; 7) Ionization current I of the order of magnitude of

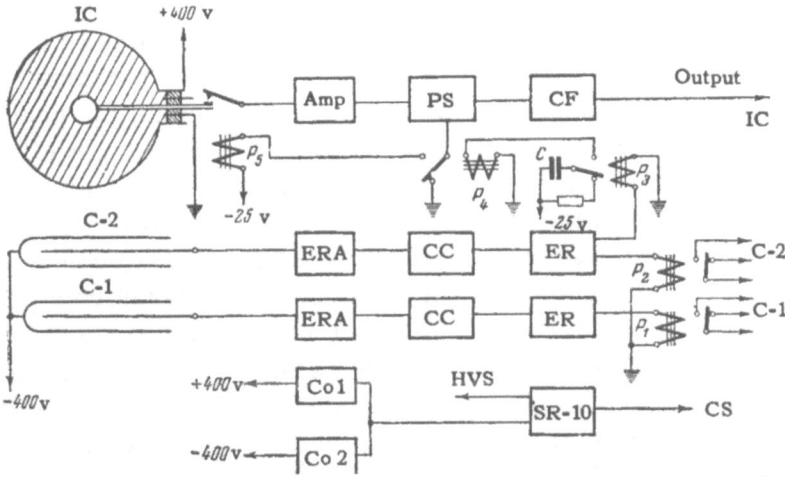

FIG. 5. Block scheme of the apparatus for measuring the intensity of cosmic rays. ERA, emitter-repeater and amplifier; CC, counting circuit; ER, emitter-repeater; Co 1, Co 2, converters; SR-10, power-supply automatic switch; IC, ionization chamber; C-2, horizontal counter (master counter); C-1, vertical counter; P_5, chamber relay; Amp, amplifier; PS, pulse stretcher; CF, cathode follower; P_1, P_2, P_3, P_4, polarization relays; HVS, electronic circuit supply; CS, current supply.

10^{-11} amp; 8) Electric capacity $C = 6.5 \times 10^{-12}$ F.

In order to calculate the possible statistical error due to fluctuations in the number of many-charge particles of the primary current we chose the following composition of the primary radiation: protons, 80% of the total number of particles arriving from all directions; helium nuclei, 18%; Li, Be, B, 0.3%; C, N, O, 1.2%; other nuclei with $Z \geq 9$, about 0.5%.

As is well known, the ionization produced by one particle is proportional to the square of its charge. Therefore, the contribution of ionization of primary nuclei to the total ionization current of the chamber as calculated by us, will be

$$I_p = 24\%; \quad I_\alpha = 21.5\%; \quad I_{\text{Li, Be, B}} = 2.2\%;$$

$$I_{\text{C, N, O}} = 23\%; \quad I_{z \geq 9} = 29\%.$$

As a result of these calculations the mean value of the statistical accuracy was found to be

$$\bar{\sigma} = \sqrt{\overline{\Sigma\left(\frac{\Delta I_z}{I_z}\right)^2}} = 8.3\%,$$

where $(\Delta I_z/I_z)$ is the relative statistical error due to the fluctuations of the number of primary nuclei with charge Z.

The best method for ionization measurements by means of a chamber consists in removing the electrical charge from the collector electrode by means of a relay. This method was first elaborated and applied by A. E. Chudakov.[1]

In our apparatus the chamber relay P_5 (Fig. 5) connected the collector to the amplifier input (Amp). Therefore a positive pulse of the duration of about 1 μsec would arise on the load resistance of the chamber. This pulse passed to the input of an amplifier with a strong feedback and having an amplification coefficient roughly equal to 6. Subsequently the signal was stretched by the diode stretcher PS which had a time constant of the order of 100 sec. This pulse was broken by the polarized relay P_4. Therefore, square signals u_{output} having a duration of 1 sec and graduated in volts arrived at the input of the telemetric system through the cathode follower CF. These signals were finally transmitted by radio and recorded on a photographic film at the ground observation point.

As has been already mentioned the chamber relay was controlled by the horizontal master counter. This took place when the reed of the relay P_3, whose coil was connected to the output of the emitter repeater ER, changed position. The charge current of the condenser C closed the relay P_4. The stretcher circuit was thus opened and the chamber relay was activated.

Since the pulse at the output of the cathode follower had a negative polarity, the operating conditions of the final cascade of the chamber electronic system were chosen so that a strictly constant positive reference voltage level would be obtained at its output. In our experiments this reference voltage level was equal to +5.65 v. The negative signal from the chamber was subtracted from this voltage. The chamber electronic system was made with direct-heating dim-filament miniature tubes.

Figure 4 shows an example of the photorecording obtained at the ground station of the chamber signals u_{ch} transmitted by radio from aboard the rocket. In order to determine the value of u_{output} in volts it was necessary to measure the amplitude of the square pulse according to scale determined by the position on the film of the calibrating level of the known voltage u_{st} of a standard battery placed on the rocket, and of the known level of the reference voltage u_{ref} arriving from the output of the chamber cathode follower in the intervals between the chamber signals.

In our experiments $u_{st} = 6.18 \pm 0.05$ v, and $u_{ref} = 5.60 \pm 0.08$ v; u_{output} was calculated from the formula $u_{output} = u_{ref} - u_{ch}$, where u_{ch} was measured from the photorecording.

The potential of the collector, u_{in} was determined from the calibrating curve $u_{in} = f(u_{output})$, which was a construction characteristic of the apparatus (Fig. 6).

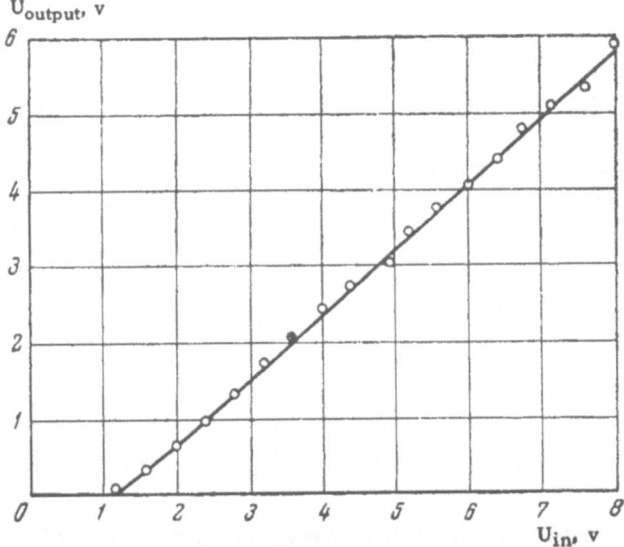

FIG. 6. Calibration curve $u_{in} = f(u_{output})$

The total ionization I produced by primary nuclei passing through the chamber during the exposure time t was calculated according to the formula:

$$I = \frac{u_{in} C}{Pte},$$

where P is the pressure in the chamber, C is the capacity of the chamber, and e is the electron charge.

The apparatus was supplied by mercury oxide batteries. The total power required by the apparatus was 2.1 w.

The flight equipment underwent a whole series of tests during which all circuits were accurately checked, their insulation was tested, the voltages at the output of the cathode follower and of the standard battery were measured, the leakage resistance of the collector was tested as well as the gas pressure in the chamber and the conditions of the scales.

The counters and the chamber were calibrated by means of a γ preparation of Co^{60} and of the natural background noise. The latter was equal to 70 pulses for the counters, and to 0.48 v per min for the chamber. During the testing the calibration curve was repeatedly recorded for different power supply voltages within 10% of their nominal values.

The ionization level and the counting rate during the tests were transmitted by radio from aboard the rocket and recorded on a film. The fact that all tests gave identical results proved that the apparatus was working correctly.

The rocket was fired on the 2nd of July, 1958 at 9:00 A.M. Moscow time. During the flight the counters and the chamber were followed visually on an oscillograph screen at the radio receiving station during 264 sec. These observations showed that the vertical counter worked normally during the whole flight. The signals from the horizontal counter and the ionization chamber started to arrive at the receiving station after 213 sec. The whole apparatus worked normally till the end of the flight.

At the vertex of the trajectory, when a height of 210 km had been reached, after the separation of the rocket nose from its body the radio transmission of the signals was interrupted. However, the apparatus remained connected and was safely dropped by parachute in working condition.

No abnormalities were discovered when the apparatus was tested in field conditions at the landing point and during the accurate laboratory tests.

The data obtained during the flight do not allow us to decide whether during the powered flight of the rocket any appreciable deviations of the intensity of the cosmic rays from its average value took place. From measurements made at the NIZMIRe (Moscow) station it was found that during the period from June 30 to July 7 the intensities of the hard and of the neutron components remained constant; the number of solar spots was comparatively small (198 against a monthly average of 238). The perturbations of the earth's magnetic field and of the ionosphere during this time remained at the June level. Only at the end of July four perturbation periods apparently connected with some increase of the activity of the sun were observed.

These observations indicate that the day of the rocket flight was comparatively free from perturbations.

One of the important results of our measurements consisted of the fact that we obtained data for the height dependence of the total intensity of the particle current. These results are presented in Fig. 7 and in Table 1.

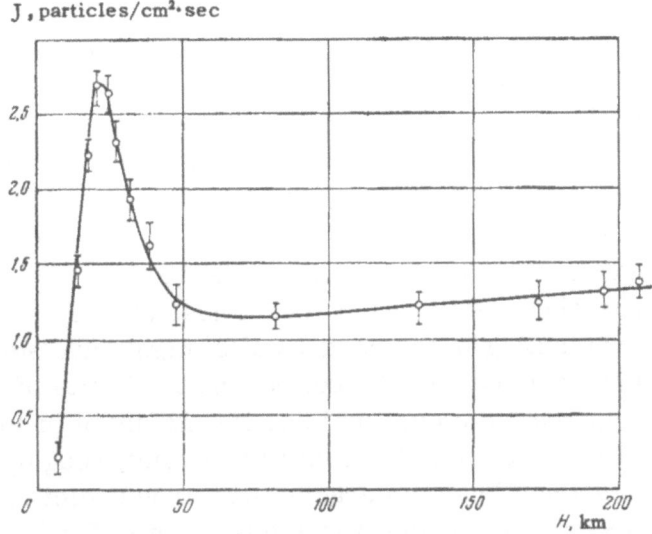

FIG. 7. Dependence from the height of the total intensity of cosmic rays.

Figure 7 shows that the maximum counter rate, at the height of 20.5 km, was higher than that at 70 km roughly by a factor 2.3. Starting with 70 km the intensity increased monotonically with the height (approximately by 17% from 70 km to 210 km).

The considerable mean square error for heights around 200 km is probably explained by the irregular changes in position of the rocket in free flight when its velocity decreased at the uppermost point of its trajectory.

TABLE 1

Dependence of the Total Intensity of Cosmic Rays upon Height

H, height, km	J, cosmic-ray intensity, particles/cm^2·sec	H, height, km	J, cosmic-ray intensity, particles/cm^2·sec
6.18	0.21±0.01	38.18	1.63±0.10
14.15	1.46±0.09	47.73	1.25±0.07
17.64	2.21±0.13	82.00	1.17±0.06
20.50	2.69±0.16	130.82	1.25±0.07
24.00	2.63±0.16	171.70	1.27±0.12
27.00	2.30±0.12	195.10	1.34±0.12
31.75	1.94±0.12	207.40	1.40±0.16

The intensity measured by us at the maximum of the height curve was 20% higher than according to A. E. Chudakov's data,[1] and 30% higher than the intensity measured by A. N. Charakhch'yan in 1958 at the latitude of Simferopol.* This increase was presumably due to the fact that our apparatus had counted particle showers produced by cosmic rays in the rocket material, or to the influence of X-rays.

Powerful showers from the rocket nose and secondary particles originating in the rocket walls might have some influence in our experiment. Since, beyond the limits of the atmosphere, the particle current consists mainly of protons whose flux is approximately equal to 67 g/cm^2, roughly 4% of the particles should have interacted with iron nuclei of the rocket walls, whose thickness was d ≈ 3 mm (2.5 g/cm^2). Each of these particles would produce about 5 new particles, of which not less than 30% would be neutral. Therefore the particles recorded which must have been secondary particles

*Private communication of A. N. Charakhch'yan.

originated in the rocket walls should have been $4\% \times 5 \times 2/3 \approx 13\%$.

The effect of showers from the rocket nose ($d \approx 150$ g/cm^2) was related to interaction of high-energy particles ($\sim 10^{11}$ ev) which accounted for 1% of the primary radiation. Each of these particles would generate about 100 new particles. Considering that the solid angle corresponding to the rocket head on the upper semisphere was about 0.05π, one can find that this effect must have been roughly equal to $1 \cdot 0.05\pi \cdot 100 \approx 16\%$. This figure can be reduced if one takes into account the comparatively small distance from the center of mass of the rocket head. The angle of divergence of the shower particles was not large in our experiment, and therefore not every particle of the shower should be considered separately. Taking this into account we suggest that the particle count due to showers must be reduced to 8%.

The above calculations allow us to evaluate tentatively the upper limit of the contribution due to "excess" secondary particles. Its value is about 20%. Therefore our measurements give a value for the total intensity beyond the atmosphere at a height of 50-70 km of 0.97 ± 0.06 particles $\cdot$ cm^{-2} $\cdot$ sec^{-1}.

As a result of our experiment the mean specific ionization of primary particles in 1 cm^3 of argon at normal pressure was determined as $j = 260 \pm 19$ ion pairs $\cdot$ cm^{-3} $\cdot$ sec^{-1}.

This value is 1.8 times higher than the mean ionization of charged relativistic particles.

The specific ionization value measured by us for a path of 1 cm in argon at normal pressure was 189.1 ± 8.2 cm^{-1}.

Another important result obtained by us has been the fact that we have shown the high reliability of the ionization chamber with its electronic system, and the fact that the supply voltage was not critical for variations up to 20%. This was confirmed by a repeated verification, before and after the flight, of the form of the pulses obtained from the chamber collected. These pulses were each time strictly identical in form, purely exponential, without peaked traces, without breaks, and of the same duration. The same also held for the output pulses from the chamber. All relays and the chamber relay starting system connected to the master counter worked wholly satisfactorily.

ACKNOWLEDGMENT

We express our gratitude to S. N. Vernov and N. L. Grigorov for their constant guidance, to A. E. Chudakov, A. I. Kuz'min, and to Yu. I. Logachev for their valuable advice and help.

The authors also thank the radio technician V. A. Belomestnykh, who gave an important practical contribution to this work.

LITERATURE

1. S. N. Vernov and A. E. Chudakov, Study of cosmic rays with the help of rockets and satellites in the Soviet Union, Communication to the V International Congress of the IGY, 1958.

AN ARTIFICIAL COMET AS A METHOD FOR OPTICAL TRACKING OF COSMIC ROCKETS

I. S. Shklovskii

Knowledge of the coordinates of a space rocket is necessary for determining its orbit. The coordinates may be secured by either radiophysical or optical observations. However, optical tracking runs up against serious difficulties. Because of the great distances involved, the flux of solar radiation reflected by a space rocket will be very small. Simple calculations show that when the space rocket is 100,000 km distant from the earth, its apparent stellar magnitude (with phase taken into account) will be fainter than $+14^m$. At a distance of 400,000 km, its stellar magnitude will diminish to a still fainter $+17^m$. Taking into account the fact that such observations may be carried out at small angular distances from the moon, where the brightness of the night sky is considerable, we can easily hazard the conclusion that serious difficulties will be encountered. Add to this the fact that telescopes of modest aperture and large focal distance (necessary to mitigate the effect of the bright sky background noise) have a small field of vision. Since the preliminary ephemerides worked out for a space rocket are very crude, it is virtually impossible to successfully train such telescopes on the object. We may venture the conclusion that astronomical observations of space rockets, even at "lunar" distances are, if possible at all, accessible to a very limited number of unique telescopes (in the 100-inch and 200-inch reflector class). There is no point, however, in relying on one or two observatories in possession of such telescopes for the observational work, since the probability of failure due to unfavorable meteorological conditions may be quite large.

In such circumstances, we are naturally confronted with the question: is it possible to enhance the brightness of the space rocket by some artificial means? The simplest approach is one proposed[1] for artificial earth satellites. The gist of this approach is to orbit a satellite-launched balloon whose surface would be made of a thin, lightweight, and fairly rugged synthetic material. This balloon would be transported in a folded configuration by the carrier rocket and would be inflated by some pneumatic device upon ejection. This would result in a sphere 30 or even 50 meters in diameter. The sphere would be an excellent reflector of sunlight if coated with a layer of some appropriate substance.

The obvious principle behind this project is to increase the brightness of a cosmic rocket by increasing its surface area. The practical execution of this approach nevertheless runs up against serious difficulties of a technical nature. For example, when the diameter called for is 70 cm, the weight of the entire balloon system will be about 300 g.[1] This leads us to the conclusion that the balloon will weigh no less than 100 kg when the diameter is 30 meters. The stellar magnitude of a balloon of this type (30 meters diameter) will be weaker than 11^m, seen at a distance of 100,000 km. It will still be difficult to observe such a faint object against the rather bright sky background in integral light. Summing up, we might state that this concept obviously fails to provide any radical solution to the problem of optical tracking of a cosmic rocket. Another question then arises: wouldn't it be possible to obtain a substantial increase in the brightness of a space rocket for at least a brief interval? In other words, wouldn't it be possible to produce a bright burst of light on the rocket? From the very start, we can exclude the possibility of producing such a burst of light by conventional means of illumination, since the weight penalty of the equipment required would be exorbitantly high when we consider the enormous power needs (of the order of 10,000 kw, see below) to make the burst visible.

Nevertheless it is possible to produce such a burst. Nature itself shows us how. We have in mind the phenomenon of the comet.

It is well known that the radiation of the head of a comet is predominantly due to resonant scattering of sunlight by various molecules (e.g., C_2, OH, CN, etc.), and also by Na atoms. The light emitted by the gaseous cometary tails is also due to resonant scattering of sunlight by CO^+ and N_2^+ molecules (cf., for example, the review paper by Swings[2]). There is no problem in estimating the total number of molecules responsible for emission of light in the head and tail of a comet. An elementary estimate was first made for the 1911 comet by Wurm in 1943.[3] The results of Wurm's calculations reduce to the following.

The concentration of CO^+ molecules is about 1 cm^{-3} in the tail of the 1911 comet at a distance of $0°.5$ from the head (assuming the tail thickness to be 3×10^{10} cm). The oscillator strength for the resonant band of this molecule is taken equal to unity.

For the head of this comet, assuming the reduced value of the apparent stellar magnitude at perihelion passage ($r_p = 0.59$ a. u.) to be 3.6^m, and the oscillator strength for the C_2 molecule (the principal molecule responsible for radiation in the visible portion of the spectrum) $f_{C_2} = 2 \times 10^{-2}$, Wurm found the total number of molecules $N_{C_2} = 1.5 \times 10^{32}$.

If the oscillator strength of a radiating molecule or atom were close to unity, the number of molecules in the head of the comet would be reduced to 3×10^{30}. Assuming the mass of the imagined molecule to be 3×10^{-23} g, we find that the total mass of the gas in the head of the comet required to produce an optic effect equivalent to that of the 1911 comet would be of the order of 10^8 g or 100 tons. This is a very low figure (on a cosmic scale, that is). Now the distance to that comet was approximately 1 a.u. If the distance were only 100,000 km, roughly the same optic effect would be provided by a cloud of some suitable gas, of about 300 g in mass. In our calculations, we took into account the fact that the 1911 comet was much closer to the sun than to earth at its perihelion.

This naturally gives rise to the idea of ejecting a cloud of suitable gas from an interplanetary rocket to yield resonant fluorescence of solar radiation. The solar photons scattered by the cloud would then be observable on earth. A cloud of this type could properly be termed an "artificial comet."

Starting out from these considerations, we proposed a project for creating an artificial comet as a technique of optical tracking of a rocket in interplanetary space.

We performed the calculations relevant to the artificial comet by a method different from that used for a "real" comet.[3] We chose sodium in our case as the substance to be vaporized. As will become clear from what follows, sodium is not the ideal substance for making an artificial comet. It should be noted, however, that an upper-atmosphere experiment involving sodium vaporization had already taken place in the U.S.A. by the time our project began to take shape, so that the principal technical avenues to solving the problem were fully elucidated. Assuming isotropic scattering (this restriction is not overly important), the number of photons scattered in a unit volume of the cloud during unit time in unit solid angle will be

$$\varepsilon = \frac{1}{4\pi} n_1 B_{12} u_{12}, \qquad (1)$$

where u_{12} is the spectral density of solar radiation in the vicinity of the earth (on the frequency scale), and B_{12} is the Einstein coefficient.

The value of u_{12} may be arrived at by direct observation of the intensity distribution of solar radiation over the spectrum.

According to observations by Abbott and Wilsing, the average radiation over the solar disk, computed on a basis of unit length of the spectrum in the neighborhood of the D lines, is[4]

$$F_\lambda = 33 \cdot 10^{13} \text{ ergs} \cdot \text{cm}^{-3} \cdot \text{sec}^{-1}.$$

The residual intensity at the centerline r = 0.055.[5] The spectral density of solar radiation near the earth, on a frequency scale, is

$$u_{12} = \frac{\lambda^2}{c^2} \pi F_\lambda \left(\frac{r_\odot}{R}\right)^2 r, \qquad (2)$$

where $r_\odot = 6.99 \times 10^{10}$ cm is the radius of the sun, $R = 1.495 \times 10^{13}$ cm is the distance separating the earth and the sun.

Using the well-known relationships between Einstein coefficients, we may derive the following expression from Equations (1) and (2) to give the surface brightness of a sodium cloud:

$$I = \varepsilon l = \frac{l}{4\pi} n_1 \frac{g_2}{g_1} \frac{A_{21} \cdot c^3}{8\pi h \nu^3} u_{12} = \frac{l}{4\pi} \frac{A_{21} g_2}{g_1} \frac{n_1 \lambda^5}{8hc^2} F_\lambda \left(\frac{r_\odot}{R}\right)^2 r. \quad (3)$$

We here assume that I is the total intensity of both components of the yellow sodium doublet. For the D_1 line, the oscillator strength $f = 2/3$, $A_{21} = 5 \times 10^7$ sec; $g_2/g_1 = 2$; for the D_2 line, $f = 1/3$, $A_{21} = 2.5 \times 10^7$ sec, $g_2/g_1 = 1$.

Performing the indicated operations, we arrive at

$$I = 5.6 \cdot 10^{-2} n_1 l \text{ photons} \cdot \text{cm}^{-2} \cdot \text{sec}^{-1} \cdot \text{steradians}^{-1} \quad (4)$$

We learn from Equation (4) that each sodium atom present in the field of solar radiation at a distance of 1 a.u. from the sun scatters ~ 0.7 quantum at D resonance lines in all directions, each second.

We now assume that q g of sodium in the vapor phase is ejected at the distance $R_1 = 100,000$ km. After expanding over a time interval t into a complete vacuum, the sodium cloud will acquire a radius ρ. The rate of expansion of the cloud will be of the order of the thermal velocities of sodium atoms, V_T. As a consequence

$$\rho = V_T t.$$

When q exceeds a certain (infinitesimal) value, the optical thickness τ of the cloud at the resonance lines will be appreciably greater than unity in the initial expansion period. Consider first the radiation of an optically thin cloud, which will always be the case when $\tau \ll 1$. The number of sodium atoms in the cloud is

$$N_{Na} = \frac{q}{m_{Na}} = 2,6 \cdot 10^{22} \cdot q. \quad (5)$$

The radiation flux from the cloud

$$F_{cloud} = \frac{5.6 \cdot 10^{-2}}{R_1^2} \cdot N_{Na} \approx 15 \cdot q \text{ photons} \cdot \text{cm}^{-2} \cdot \text{sec}^{-1}$$

The radiation flux from the sun is $F_\odot = 1.32 \times 10^6$ ergs $\cdot$ cm$^{-2} \cdot$ sec^{-1}, and its bolometric apparent stellar magnitude

$m_b = -26.95$. From this, we see that the bolometric stellar magnitude of an optically thin sodium cloud weighing 1 g and located at 100,000 km distance from the earth will be 14^m. Taking into account the fact that ~ 15% of the energy incident falls in the visible portion of the solar spectrum, that the visual apparent magnitude of the sun $m_v = -26.86$, and that the sensitivity curve of the eye in the neighborhood of the D lines reaches 0.7 of its maximum value, we find that the visual stellar magnitude of our cloud will be 12.5^m. If the mass of the optically thin cloud is 1 kg, we shall have a visual magnitude of 5^m at the same R_1.

Consider the case where the optical thickness of the cloud $\tau > 1$ at the resonance lines. In that case, the cloud may be simulated to a first approximation by a surface reflecting sunlight in a narrow spectral region $\Delta\lambda_1$ with an albedo equal to the residual intensity of the D line in the solar spectrum, i.e., roughly 0.055. Note that the albedo of the moon is approximately the same (0.07).

When τ is not very large, we have $\Delta\lambda_1 \approx \Delta\lambda_D$, where $\Delta\lambda_D$ is the Doppler width of the spectral line scattered by the cloud. When $\tau \gg 1$, we have $\Delta\lambda_1 = k\Delta\lambda_D$, where $k = 2-3$.

Proceeding further with our analogy, we estimate the surface brightness of the cloud, whose optical thickness is not very large, although greater than unity. Assuming the temperature of the sodium vapor in the cloud to be $T = 1500°$, we find that

$$\Delta\lambda_D = \frac{\lambda}{c}\sqrt{\frac{2kT}{m_{Na}}} = 2 \cdot 10^{-2} \text{A}. \tag{6}$$

On the other hand, assuming the effective length of the spectrum to be ~ 1000 A, we find that the surface brightness of the cloud in the visual region is approximately 75,000 times less than the surface brightness of the full moon.* Assuming the latter value to be -14^m for a square degree, we find that the surface brightness of the cloud is -1.5^m for a square de-

---

*We took into consideration the curve of the spectral response of the eye in the region of the D line.

gree, i.e., hundreds of times greater than the brightness of the nighttime sky.

When $\tau \gg 1$, $\Delta\lambda_1$ will be roughly equal to $(2-3)\Delta\lambda_D$. Furthermore, we must take into account the fact that there are two sodium lines (D_1 and D_2), which doubles the width of the spectral region over which the albedo is fairly high. The surface brightness of a very dense sodium cloud may therefore reach -3.5^m for a square degree.

As the cloud expands, its optical thickness will drop off rapidly (as ρ^{-2}), and the surface brightness will fall off slowly at first (as long as $\tau > 1$), and then rapidly later, as ρ^{-2}. In the final stages of the expansion of the cloud into the vacuum, when $\tau \ll 1$, the cloud, as we might infer from the discussion, will resemble a ring-shaped nebulosity with highly smeared edges.

On a moonlit night, when the sun has not sunk too far below the horizon, the surface brightness of the sky may well exceed the surface brightness of the sodium cloud. However, use of light filters may succeed in greatly enhancing contrast of the cloud against the sky background, since the radiation emitted by the former is monochromatic. For example, if we use an interference filter with a transmission bandwidth of 30 A, the contrast of the cloud against the background will be increased about 50 times, i.e., by more than 4 stellar magnitudes.

Let us now estimate the maximum linear dimensions of an optically thick cloud.

ρ_2 will denote the radius of the cloud at which the optical thickness will be 2. We shall then obviously be confronted with the condition

$$\frac{S_\nu \cdot N_{Na}}{4/3 \pi \rho_2^3} \cdot 2\rho_2 = \frac{S_\nu \cdot 2.6 \cdot 10^{22} q}{2\rho_2^2} = 2, \tag{7}$$

where $S_\nu = \frac{\sqrt{\pi} e^2}{mc} \cdot \frac{1}{\Delta\nu_D} \cdot e^{-\left(\frac{\nu - \nu_0}{\Delta\nu_D}\right)^2}$ is the absorption coefficient.

At $T = 1500°$, $\nu = \nu_0$ ($S_\nu \approx 10^{-11}$ cm^2) and q = 1000 we will have $\rho_2 \approx 10^7$ cm. In general, $\rho \sim q^{1/2}$, i.e., it evinces a relatively weak dependence on the mass of the cloud. At the distance of 100,000 km, the maximum angular dimensions of an optically thick cloud at q = 1000 will be 6'.7, and at a distance of 400,000 km, i.e., near the moon, such a cloud will have angular dimensions of 1'.7. The cloud will be observed as an object possessing finite area even in a shot taken with a short-focus camera.

The time during which the cloud remains optically thick is an important parameter in this case. At $V_T \sim 10^5$ cm/sec (corresponding to thermal velocities of sodium at T = 1000° C), the cloud will remain optically thick for about 100 sec, after which its surface brightness will begin to weaken rapidly as the cloud dissipates further. The rate of expansion of the cloud in space and the speed of the translational motion of the cloud, equal to the speed of the carrier rocket, is a factor limiting the length of the exposure. It is an obvious prerequisite that the time of exposure of the cloud not be displaced over the plane of the picture by more than, say, half its value. It is also obvious that longer exposures at $\tau < 1$ will not result in increased grain darkening. If, for example, the transverse component of the rocket's velocity with respect to the earth is not large, while the cloud is expanding outward at a speed of ~ 1.5 km/sec, the optimum exposure time should not exceed 25-30 sec. This condition then obviously predetermines the focal ratio of the camera used to take shots of the artificial comet. The requirement that our "comet" be photographed as an object possessing finite area sets the lower bound to the focal length of the camera.

The considerations adduced here enable us to determine the parameters of the camera used in photographing the comet. With this as a point of departure, we succeeded in working out the technical problems affecting the equipment used in observing the artificial comet.

Ionization of sodium atoms by ultraviolet rays emitted by the sun places a limit on the expansion of the cloud. The ionization potential of sodium is 5.12 ev. The flux of solar photons of energy less than 5.12 ev ($\lambda < 2400$ A) capable of ionizing sodium is, in accord with direct data, $F(\lambda < 2400) \sim 10^{14}$ $cm^{-2} \cdot sec^{-1}$.

According to Bates' calculations[6] the effective ionization cross section for Na is very small, $\sigma_{Na} \approx 3.7 \times 10^{-19}$ cm^2 (at absorption cutoff). The lifetime of neutral sodium atoms in interplanetary space will be

$$t_{Na} \approx \frac{1}{\sigma_{Na} F\,(\lambda < 2400)}\,. \qquad (8)$$

We see from this that $t_{Na} \approx 2.7 \times 10^4$ sec or 7.5 hrs. This means that the photoionization effect may be safely disregarded as a factor impeding observations of a sodium artificial comet. The comet's brightness diminishes long before that point owing to expansion, rendering the comet unobservable.

The concept of vaporizing sodium atoms in the upper atmosphere to study the physical and chemical properties of the latter was proposed by Bates in 1950.[7] In 1955, Edwards, Bedinger, and Manring performed an experiment involving vaporization of 3 kg of sodium in the earth's atmosphere, by means of rockets sent to heights of 52-113 km under crepuscular conditions.[8] Highly intense glow was observed. There was no abrupt change in the intensity of the glow on passage through the boundary of the earth's shadow.

In 1956, Bedinger and Manring carried out a similar experiment, this time under nighttime conditions.[9] Two kilograms of sodium vapor were introduced into the atmosphere at heights ranging from 60 to 140 km. The intensity of the sodium glow varied sharply as a function of height and reached high values at times. These observations demonstrated that sodium rapidly engages in reaction with the atoms and molecules present in the earth's atmosphere. The lifetime of sodium atoms in the atmosphere at those heights is several minutes. This shows that, to produce a burst of sodium glow under purely twilight conditions it is necessary to bring the rocket up to rather considerable heights. Special equipment for vaporizing sodium (cf. Ref. 8) with the aid of ignited chemically pure thermite was devised for the experiments described above. According to Ref. 8, almost all of the sodium present was converted to the atomic state of aggregation in the vaporization process.

Vaporization of 1 kg of sodium required 3.2 kg thermite. The Na_2 molecule, as indicated in Ref. 8, must be very small, since the dissociation energy of this molecule is small. On the other hand, ionization of sodium does not occur either, since the temperature of the vaporization was relatively low (T = 1000°C). Note however that a sizable portion of the sodium may become involved in the formation of fine solid and liquid particles during the vaporization process. This effect was not considered in Ref. 8.

Equipment for vaporizing sodium to the atomic state was developed as part of the project for an artificial comet in the U.S.S.R. To test the efficiency of the vaporizer's performance, an experiment was conducted on September 19, 1958, and resulted in a cloud of sodium vapors illuminated by the sun at a height of 440 km. A high-altitude geophysical research rocket was employed in the experiment. The absolute photometry of the sodium cloud formed enabled the experimenters to determine the number of sodium atoms vaporized. This in turn made it possible to find the vaporization coefficient, which turned out to be rather high. At the same time, observations of the diffusion of the cloud being formed led to more accurate determinations of the density of the earth's atmosphere at a height of 440 km.[10]

The first functioning artificial comet was released during the flight of the first Soviet cosmic rocket, on January 3, 1959. At the previously programmed instant, $3^h56^m20^s$ (Moscow local time), 1 kg of sodium was vaporized on board the rocket. A network of tracking stations equipped with specially designed small-size wide-angle cameras (f = 250 mm, d/f = 1:2.5, and f = 500 mm, d/f = 1:5) were organized in good time to perform observations on the artificial comet. The camera parameters were governed by the expected conditions under which the artificial comet would be observed, for which see above. All of the cameras used were equipped with multilayered interference filters with transmission bandwidth of about 30 A. The transparency of the filters varied over the range 30-50%. A special procedure was worked out beforehand to make it possible to secure the coordinates of the artificial comet by reference to background stars.

Unfortunately, poor visibility prevailed at most of the tracking stations. A reliable photographic plate of the artificial comet was obtained at theKislovodsk High-Altitude Station of the Main Astronomical Observatory (cf. Fig. 1).

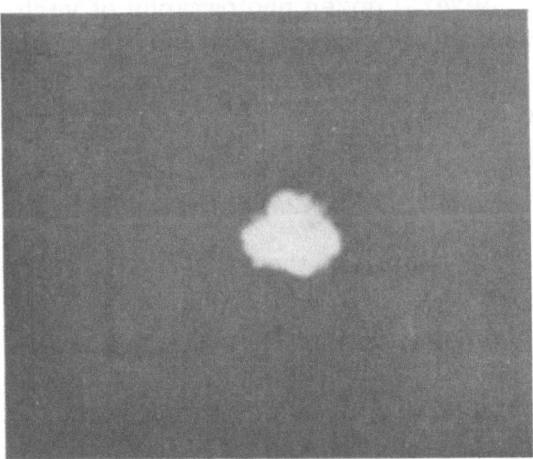

FIG. 1. Plate of artificial comet formed on January 3, 1959, in the flight of the first cosmic rocket.

The artificial comet was formed at the instant when the space rocket was at a distance of 119,500 km from the earth.

According to preliminary estimates, the stellar magnitude of the comet was about 7.5^m. This means that approximately 10% of the sodium was vaporized, i.e., about 100 g. For this reason, the "lifetime" of the artificial comet came to only about 30 sec. Despite the relatively low efficiency of the vaporizer, the first artificial comet was able to fulfill its function. It made it possible to determine the angular coordinates of the space rocket to an accuracy greatly in excess of radio déterminations.

The topocentric coordinates for Kislovodsk (for the instant of time indicated) are

$$\alpha = 14^h 08^m.4 \quad \delta = -5°04'.$$

There is every ground for assuming that techniques for forming and observing artificial comets will be greatly improved in the future.

During the flight of the second Soviet space rocket on September 12, 1959, a sodium artificial comet was formed at $21^h49^m30^s$ Moscow official time. Successful observations of this comet were carried out at several Soviet and foreign observatories. Several dozen photographs of various stages of the development of the comet were obtained. On the basis of these plates, it proved possible to determine the coordinates of the rocket with great accuracy. We give here the topocentric coordinates for the town of Alma-Ata:

$$\alpha = 20^h36^m00^s.24, \quad \delta = -9°7'12''.2.$$

The geocentric distance of the artificial comet was 156,000 km. The apparent photovisual magnitude was about 4.5^m. The angular diameter of the sodium cloud extended to 30' at the limit of accessibility to observation. The measured rate of expansion of the artificial comet was 1.7 km/sec. The artificial comet could be observed for 6 min in the cameras used during the observations. Several phases in the development of the artificial comet are visible on the accompanying photographs, which (Fig. 2a-m) were obtained by V. F. Esipov at Stalinabad on an electronic image-converter telescope camera.

Let us now consider briefly one possible variant in the artificial-comet technique which appears to us highly promising.

Although sodium possesses a number of prominent advantages favoring its use as material for an artificial comet, it nevertheless suffers from one important drawback. The scattered solar radiation is highly attenuated, owing to the presence of the intense D_1 and D_2 Fraunhofer lines in the solar spectrum. Lithium is free from this disadvantage. As far as is known, there is practically no lithium in the sun, so that the resonance Fraunhofer line $\lambda 6708$ is absent. The intensity of emission obtainable through this means alone would be an increase of almost twenty times. Lithium is moreover almost three times as light as sodium. Note too that the Einstein coefficient A_{21} for the $\lambda 6708$ line is 1.6 times smaller than for the sodium D line.

The net result is that, for the same number of photons scattered, an optically thin lithium cloud will contain approximately 40 times less mass than an optically thin sodium cloud (we here disregard the mass of the thermite; it is by no means excluded that new vaporization techniques will be developed). The surface brightness of an optically thick lithium cloud will be -5^m (bolometric) for a square degree.

It is important, however, to note that the sensitivity of the human eye is not very great in the spectral region of 6708 A, particularly with regard to twilight vision. Lithium artificial comets would therefore hardly serve the purpose of visual observations.

The ionization potential of lithium is 5.38 ev. According to Ref. 5, the effective photoionization cross section for Li at the absorption cutoff is 3.2×10^{-18} cm^2.

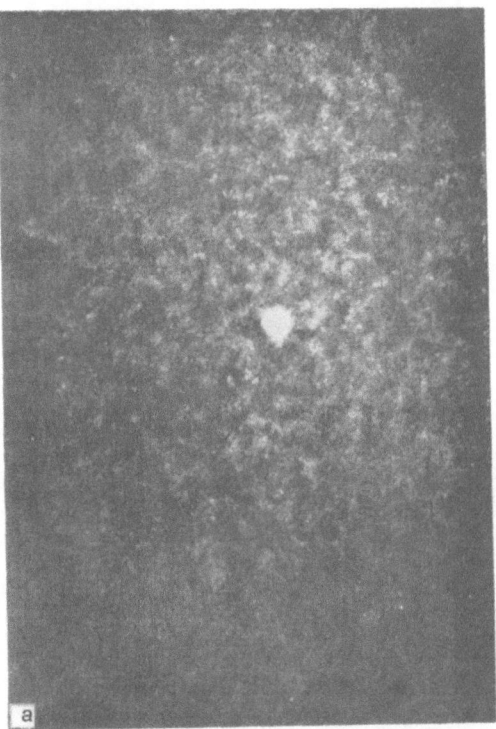

FIG. 2. Plates of artificial comet formed at 21h49m30s on September 12, 1959, in the flight of the second cosmic rocket. (a) 21h50m03s.

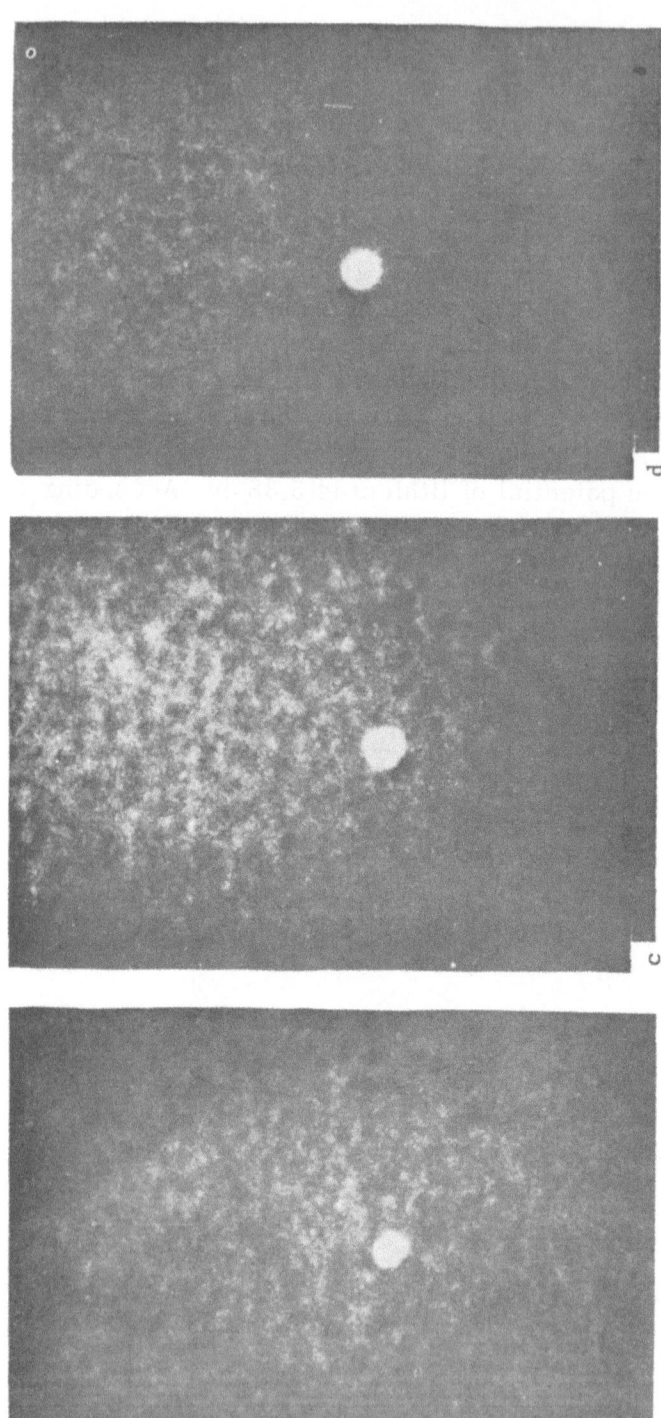

FIG. 2 (continued). (b) 21ʰ50ᵐ11ˢ; (c) 21ʰ50ᵐ20ˢ; (d) 21ʰ50ᵐ32ˢ;

AN ARTIFICIAL COMET FOR OPTICAL TRACKING 459

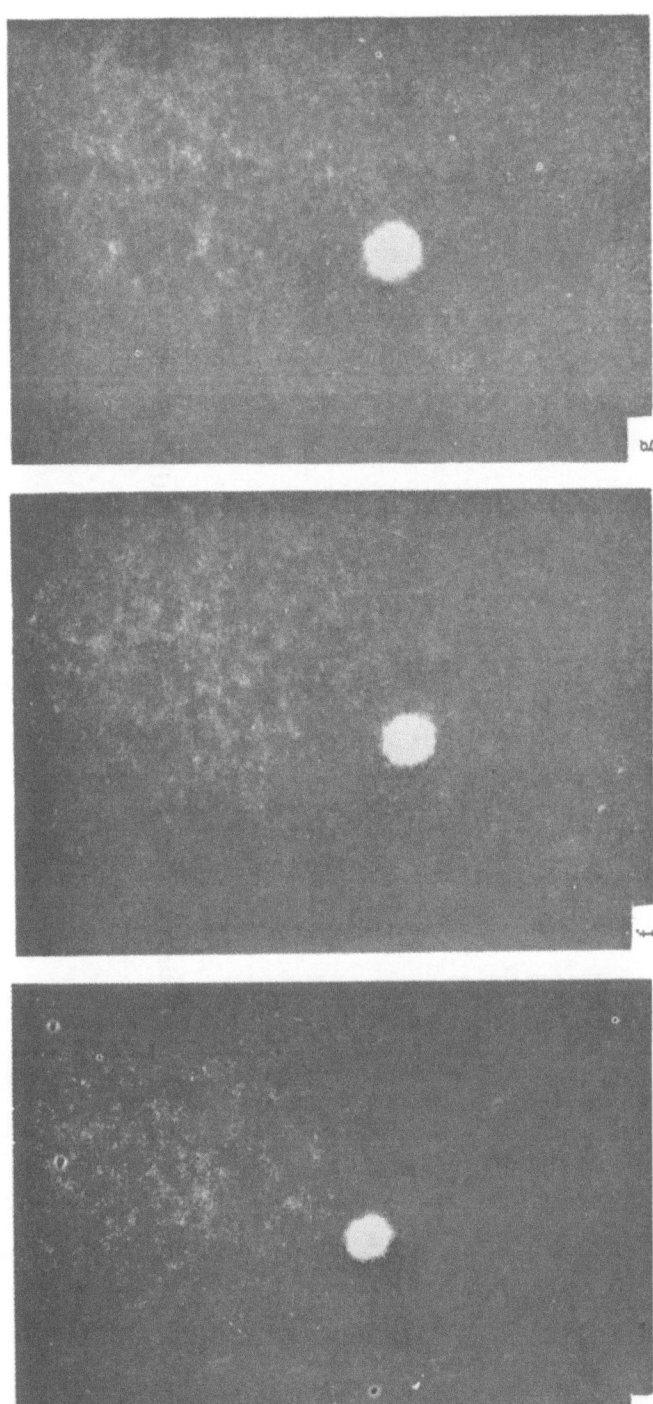

FIG. 2 (Continued). (e) 21ʰ50ᵐ43ˢ; (f) 21ʰ30ᵐ51ˢ. (g) 21ʰ51ᵐ03ˢ;

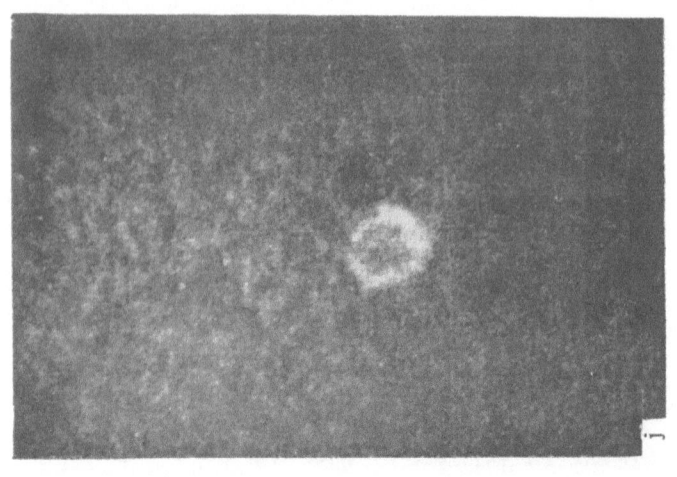
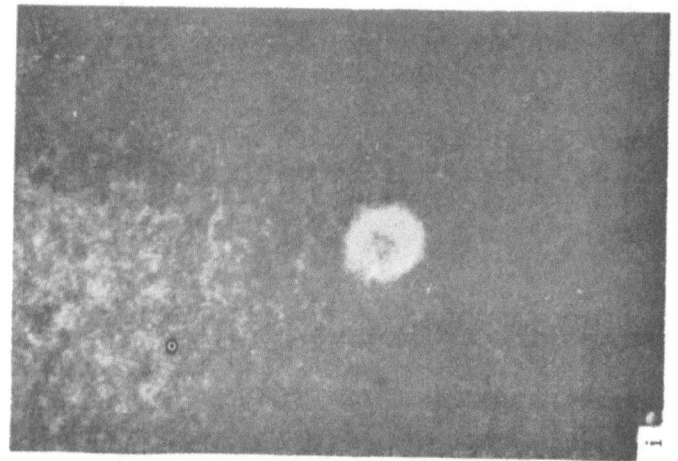
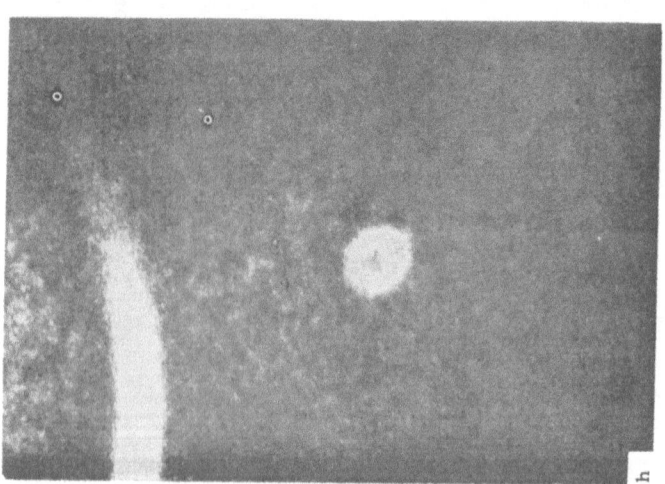

FIG. 2.(Continued). (h) $21^h51^m13^s$; (i) $21^h51^m25^s$; (j) $21^h51^m38^s$;

AN ARTIFICIAL COMET FOR OPTICAL TRACKING 461

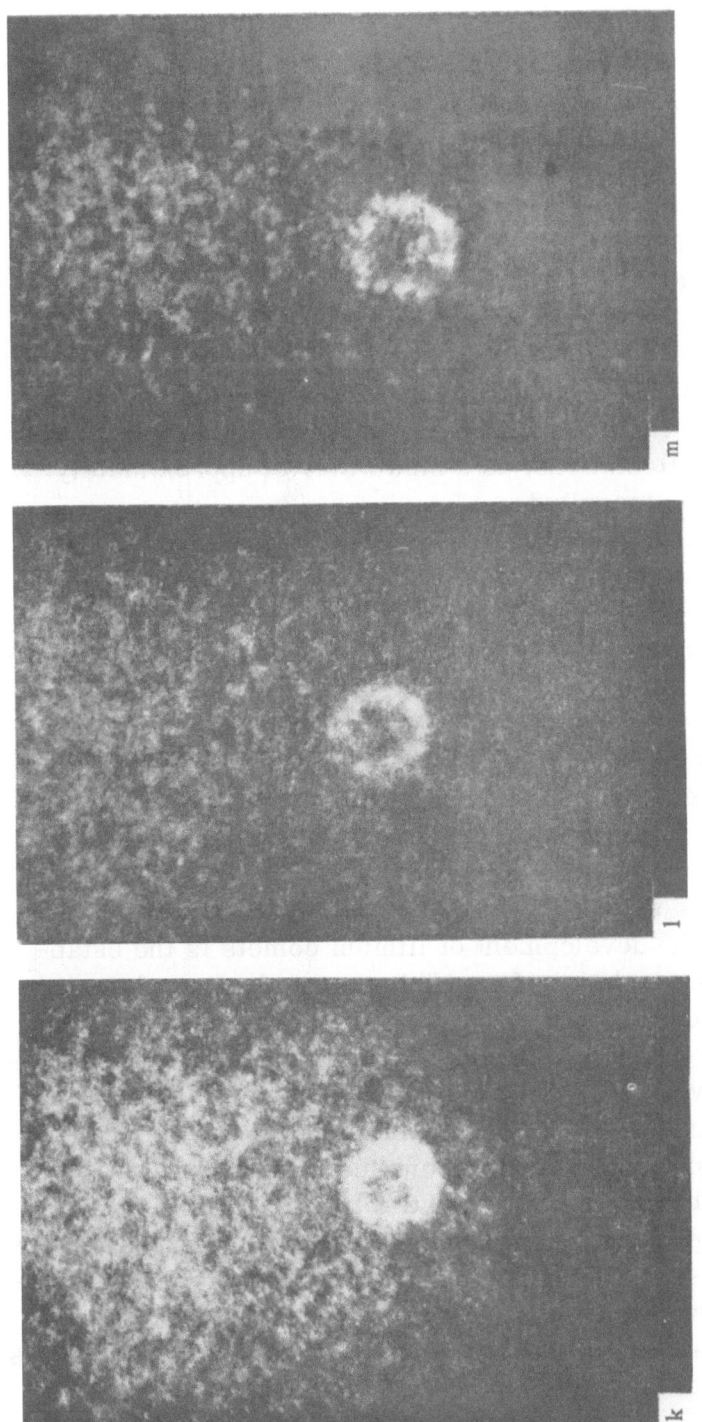

FIG. 2 (Continued) (k) $21^h51^m56^s$; (l) $21^h52^m09^s$; (m) $21^h52^m28^s$.

Taking this higher ionization potential of lithium into account, we find that the lifetime of neutral lithium atoms in interplanetary space is approximately 5 times shorter than that for sodium.

Once the problem of the lithium artificial comet has been solved in practice, we shall have solved the problem of creating a "skywriting" interplanetary rocket, at least in the vicinity of the earth at distances measuring several hundred thousands of kilometers. The duration of a flight passage through this region of cosmic space is of the order of several hundred hours. By vaporizing 1 g of lithium every 1-2 sec, the "comet" will become observable, as may easily be demonstrated by calculations, as a small star of approximately 10th bolometric magnitude.

The total lithium charge required would be several tens of kilograms (assuming, of course, ideal conditions with thermite and other "parasitic" weight factors left out of consideration). Various uses may be found for "skywriting" interplanetary rockets. In particular, this approach may present a radical solution to the problem of observing and tracking translunar rockets traversing interplanetary space.

Lithium comets open up possibilities for optical observations of space rockets at distances out to several millions and even tens of millions of kilometers.

It should be stressed once again, however, that one prerequisite for the development of lithium comets is the establishing of some technique for efficient vaporization of the lithium. The associated equipment and igniter fuel must be of as low weight as possible. The possibility of the future use of atomic energy to solve the intriguing problem of designing a lithium artificial comet is not at all excluded.

In conclusion, let us take up one more question.

One possible variant in the artificial comet project is that of a dust comet. Dust, as is well known, does a fairly good job of scattering sunlight. This phenomenon is observed in a number of cases within the confines of the solar system. For example, dust tails are observed on some comets. Dust particles concentrated toward the plane of the ecliptic are responsible for the optical effects of the zodiacal light and the so-called Fraunhofer component of the solar corona.

Comparisons of the efficiency of dust as against gaseous artificial comets are therefore not a matter devoid of interest.

Suppose we have 1 kg of some suitable substance which we can completely pulverize, with the aid of appropriate equipment, to spherical particles of radius a. We shall forego discussion of the technical side of the question involving the practical possibility of such a complete pulverization process.

The number of particles forming the dust cloud will be

$$N = \frac{1000}{4/3 \pi a^3 \delta}, \tag{9}$$

where δ is the specific weight of the substance to be pulverized.

Each dust grain will scatter sunlight towards the earth with an effective scattering cross section

$$\Sigma = K(\theta) \chi \pi a^2, \tag{10}$$

where $K(\theta)$ is the scattering indicatrix, and the value of χ depends on the nature and grain-size of the scattering particles.

For a concrete example, let us take $a = 0.3\mu$ (dust grains of smaller dimensions could hardly be produced in practice; even grains with $a = 0.3\mu$ would be a difficult technical accomplishment).

Then the value of χ, governed by diffraction phenomena, will be 2 (for dielectrics), according to Ref. 11. Suppose now that the indicatrix is spherical, i.e., $K(\theta) = 1$. Then the overall surface of all of the grains taken collectively will be $N\Sigma = 1.6 \times 10^7$ cm^2, which is equivalent to the surface of a sphere 44 meters in diameter.

On the other hand, 1 kg of sodium vapor contains 2.6×10^{25} atoms. Assuming an effective resonant fluorescence cross section of 10^{-11} cm^2, and taking into account the fact that the residual intensity in the sodium lines is $r = 0.055$, while the equivalent width of the scattering coefficient is 0.02 A (which amounts to 1/75,000 of the effective extent of the visible portion of the spectrum), we now find the "equivalent surface" of the sodium cloud "smeared out" over the visible portion of the spectrum: 1.9×10^8 cm^2, i.e., 12 times greater than that for a dust cloud of the same mass. The flux of sunlight scattered from a dust comet may be reduced considerably by taking into account the scattering indicatrix, since the indicatrix is greatly extended in the direction of scattering for such particles, as a rule.

In summary, we might say that the efficiency of a dust comet is far below that of a sodium comet (not to mention a lithium comet). The inestimable advantage of gaseous comets over their dust counterparts is the monochromaticity of their emissions, which makes it possible to carry out fruitful observations against an enhanced sky background brightness, by use of appropriate light filters.

The unique strong feature of the dust artificial comet is its protracted lifetime, since the dust grains will disperse in space at a rather leisurely pace (at a rate of the order of 10^4 cm/sec).

On the whole, a dust artificial comet might be said to occupy an intermediate position between a gas comet and a balloon, with respect to efficiency.

Meriting special attention is the question of the use of several molecules exhibiting resonance bands in the optical portion of the spectrum (CN, for example) as artificial-comet material. But we shall not go into an analysis of "molecular" comets in this article.

LITERATURE

1. Science News Letter, 72, No. 16, 1957, 244.
2. P. Swings, Vistas in Astronomy, v. 11, London, 1956, p. 358.
3. K. Wurm, Mitt. Hamburger Sternwarte, 8, 51, 1943.
4. A. Unsöld, Physics of Stellar Atmospheres, Moscow, 1947 (Russian translation).
5. C. W. Allen, Astrophysical Quantities, London, 1955, p. 137.
6. D. Bates, Monthly Notices R. A. S., 106, 432, 1946.
7. D. Bates, J. Geophys. Res., 55, 347, 1950.
8. Edwards, Bedinger, Manring, and Cooper, The Airglow and the Aurorae, London, 1957, p. 123.

9. Bedinger, Ghosh, and Manring, J. Geophys. Res., 62, 170 (1957).
10. I. S. Shklovskii and V. G. Kurt, see Volume 3, p. 92.
11. Shifrin, Rasseyanie sveta v mutnykh sredakh (Light Scattering in Turbid Media), Moscow, 1950.

Volume 5

THE ORBITS OF COSMIC ROCKETS IN THE DIRECTION OF THE MOON*

L. I. Sedov

The realization and application of interplanetary flight is based on the theoretical analysis of the laws of motion of aircraft in cosmic space and on the numerical computations resulting from these relationships. The fundamental requirements imposed on the power and guidance program for rocket-carriers, as well as the optimum and feasible conditions for take-off, are established by analytic means.

As the initial reference system in a description of the motion of the aircraft we can choose a Cartesian coordinate system with origin at the center of the earth, where the system is moving transversely with respect to the stars. In many practical problems it is necessary to use spherical coordinate systems which are rigidly connected with the earth and for which the origin is located at the center of the earth or at different points on its surface. It is sometimes necessary to consider the motion of objects relative to the moon or the other planets.

In the differential equations of motion of celestial ballistics describing free flight in cosmic space it is necessary to introduce only the interaction forces defined by Newton's law of universal gravitation. In solving the problem of landing on the moon and circling the moon, it is necessary to consider the motion of a body in the known gravitational field set up by the sun, moon and stars (taking into account the oblateness of the earth).

---

*This article represents the text of a review paper presented at the annual meeting of the American Rocket Society in November, 1959.

In delving into the optimum conditions for launching a rocket to the moon, we can make use of approximation methods and regard the motion of the body as Kepler motion relative to the earth when the distance to the moon is greater than 66,000 km, and as Kepler motion relative to the moon when this distance is less than 66,000 km.

For specified points on the earth's northern hemisphere we can to a first approximation (taking into account only the gravitational forces of the earth) obtain the optimum take-off conditions with the aid of the following considerations. Let us consider hitting the moon. In Fig. 1 we use A to denote the

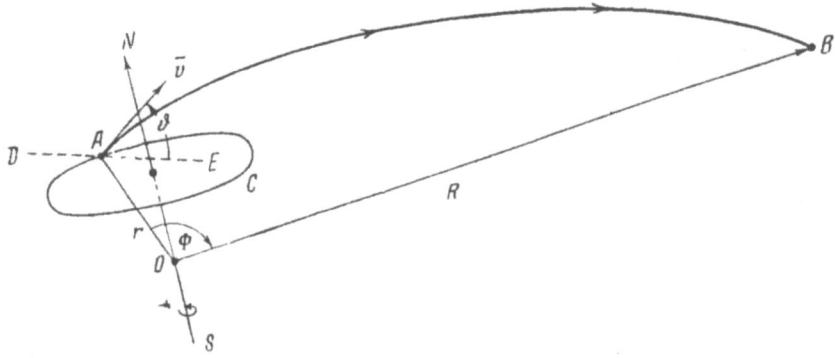

FIG. 1. Schematic path of a cosmic rocket. SN) axis of rotation of the earth; AC) parallel on which the point A lies; DE) horizontal in the plane of the orbit.

position of the end of the active participation of the rocket, B as the position of the center of the moon at the instant of contact, O as the center of the earth. Every orbit that corresponds to a hit from the point A to the point B lies in the plane defined by the three points A, O, B. For the complete determination of the orbit and magnitude of the initial velocity v it is sufficient to specify the angle θ of incline of the initial velocity v with respect to the horizontal.

Solving the two-body problem, it is easy to arrive at the functional relation

$$\frac{v^2}{v_p^2} = f\left(\frac{r}{R}, \theta, \Phi\right), \tag{1}$$

where v_p is the velocity along the parabola. The angle between the directions OA and OB, designated by Φ, is called the angular range. The point A, which is rigidly fixed in the earth, on rotation of the earth describes a parallel. Here the angle Φ varies within limits easily determined from Fig. 1.

Since the ratio r/R is small and is for all practical purposes constant, according to (1) the ratio v^2/v_p^2 depends mainly on θ and Φ.

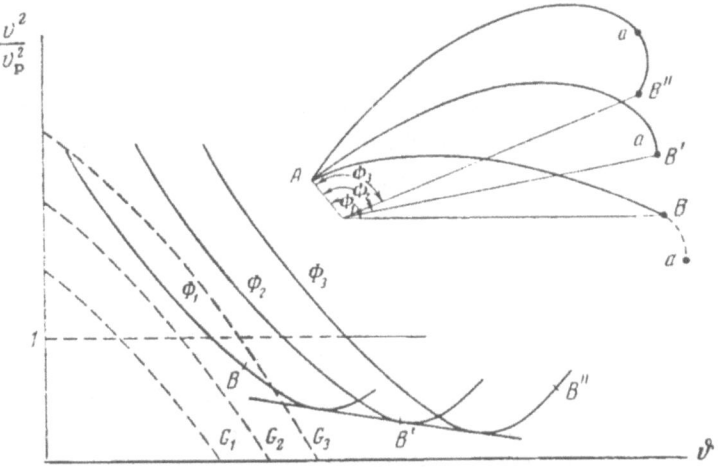

FIG. 2. Dependence of v^2/v_p^2 on the angle of incline θ for various values of Φ (solid curves: $\Phi_1 > \Phi_2 > \Phi_3$) and for several values of the rocket weight (dashed curves: $G_1 < G_2 < G_3$). Points B, B', B" correspond to a hit by the projectile on the moon. Above right are shown the corresponding trajectories; a) position of apogee, A) point of emergence into the trajectory.

In Fig. 2 this dependence on θ for Φ = const is represented graphically by solid lines. In the same graph dashed lines are used to show a typical form for the characteristics of the rocket for various constant values of the rocket weight at the end of active flight. From examination of Fig. 2 it follows that the optimum launching time, for which the useful weight is greatest, corresponds to maximum value of the angular range Φ. It is obvious that in this case the plane of the orbit is perpendicular to the plane of the equator. The indicated maximum value of the angle Φ depends on the position of the moon in its orbit around the earth. It is clear that for points of the northern hemisphere the greatest value of Φ corresponds to the lowest position of the moon beneath the plane of the equator. The indicated conditions define for the given point A the most favorable launching time during the lunar month and the instant for launching during the earth day.

The foregoing conclusions were made without taking account of the earth's rotation and the suitability of applying the additional component of the rocket's velocity obtained due to this consideration. Allowance for the earth's rotation reduces the inclination of the optimum orbit with respect to the plane of the equator. It is necessary to introduce also certain corrections connected with allowance for the influence of the moon, sun, oblateness of the earth, and characteristics of the rocket's active acceleration. An additional substantial increase in the useful weight can be obtained by choosing the take-off point at an appropriate point on the earth's surface. The most favorable points for take-off are found near the equator.

For the Soviet cosmic rockets a flight variant was chosen with the angle of inclination of the plane of the initial part of the orbit with respect to the plane of the equator equal to 65°, which is nearly the most ideal. The fundamental calculations of the laws of motion were performed by high-speed computers. The optimum trajectories were found, and the group of trajectories near the optimum ones were studied.

In choosing a specific orbit, one may begin by establishing the total specific energy of the launched object at the end of the acceleration interval. The value of the total specific energy is determined by the details of the flight conditions and the requirement of obtaining favorable conditions for observing the object from the U.S.S.R. at the instant of impact or during the time when it is near the moon. This also determines the time of travel to the moon. From an analysis of the optimum conditions it follows that the flight time from the earth to the region of the moon should be about 0.5, 1.5, 2.5, or 3.5 days.

In order to realize launching it is necessary (after having determined the total energy) to organize the data on the possible useful weights and on the conditions necessary for take-off during the few days near that most favorable, which corresponds for northern latitudes to least inclination of the moon at the time of closest approach to it. This problem is solved by means of ballistic analyses with account for the characteristics of the rocket.

The calculations show that when the time of take-off deviates from the optimum time by one or two days in either direction, the losses in useful weight are found to lie within permissible limits. Consequently, it is convenient to launch only during a few days in each lunar month. A substantial deviation in the launching day from the most favorable involves large losses in the size of the possible payload.

For the detailed investigation of the influence of the initial parameters on the characteristics of the trajectory it has proved useful to establish certain important properties of a cluster of trajectories whose center trajectory passes through the moon's center. The properties enumerated below for such a cluster of trajectories are rather difficult to realize; but the application of these properties makes it convenient to solve many practical problems.

1. Near the moon and in a small circular region about the center trajectory of the type chosen, which can be regarded as a straight line, the nearby trajectories at distances up to 20,000 km from the moon form a set of lines having rotational symmetry relative to the center axis of the trajectory.

2. In the vicinity of the moon the trajectories have nearly conical cross sections lying in the meridian planes.

3. The various trajectories of this family are obtained for small deviations of the six independent parameters from their principal values. For these six parameters one might choose the coordinates of the rocket and the components of the absolute velocity at the end of the acceleration interval.

4. On variation of only a single one of any of the parameters or with a proportional change in any set of six parameters different trajectories of the cluster lying in the same meridian plane are obtained.

5. In the meridian planes it is possible to introduce the

universal dimensionless variable λ, which is the same for all planes defined by the relation

$$\lambda = \frac{\Delta \xi}{\Delta \xi^*},$$

where $\Delta \xi$ is the increment in the parameter ξ at the end of the acceleration interval, and $\Delta \xi^*$ is an increment corresponding to the specified value of the minimum distance of the perturbed trajectory from the center of the moon.

Consequently, in any meridian plane, for various trajectories there exists the universal relation

$$\rho_{min} = f(\lambda),$$

where ρ_{min} is the minimum distance of the trajectory from the moon's center.

On the basis of the properties named above the six-parameter cluster becomes reduced to a two-parameter cluster, which can be reflected uniquely into an arbitrary plane π in which the points are defined by polar coordinates: the radius λ and polar angle ω (ω denotes the angle that determines the meridian plane in the space of the trajectory).

This circumstance makes it possible to investigate and describe the set of perturbed motions in a well-outlined manner.

In the plane π it is convenient to represent isolines for the various numerical characteristics of the trajectories. This is particularly significant when investigating a circling trajectory and for finding the initial data corresponding to orbits with the required and most suitable characteristics. An example of two families of isolines used in choosing a circling trajectory for a robot space station is shown in Fig. 3.

In this graph are plotted families of isolines of the maximum and minimum distances from the object to the earth in the first loop after nearing the moon. For the cluster of trajectories cut out, the initial energy corresponds (not considering the influence of the moon) to a distance of the apogee from the center of the earth equal to 550,000 km. For trajectories of the cluster passing to the left of the isoline "apogee at 550,000 km" retardation occurs, and the height of the apogee is reduced; for trajectories passing to the right of this line acceleration occurs, the height of the apogee increases, and the energy of the object relative to the earth after nearing the moon increases. The corresponding quantitative effects are well visualized in the graph.

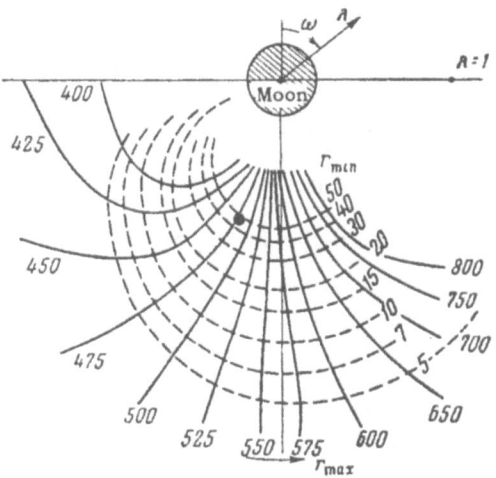

FIG. 3. Isolines of the maximum distance from the earth's center (solid curves) and minimum distance (broken curves) in the plane of λ, ω. The circle represents the RSS. The numbers near the curves indicate the constant value of r (10^3 km) corresponding to the given curve.

The value of ρ_{min} = 7900 and the values of λ and ω that are realized in the motion of a robot space station (RSS) are determined by calculation. After plotting these data in graphical form (Fig. 3), we obtain the distance of the apogee from the center of the earth equal to 480,000 km and the distance to the perigee in the first loop equal to 475,000 km.

The perturbing action of the moon on the motion of the RSS is very great; it can be characterized quantitatively by the data given in Table 1.

We may note that the magnitude of the momentum vector of the amount of movement increases approximately three times while the direction changes almost to opposite.

Consideration of the plane π simplifies investigation of the influence of dispersion in the initial parameters on the characteristics of the orbits. The introduction of the plane π permits the problem of hitting the moon and the problem of passing near and circling the moon to be united into one. Obtaining the circling trajectories with specified characteristics reduces to the problem of hitting in the plane π at a specified point corresponding to the values sought for the characteristics of the orbit.

TABLE 1

	Total geocentric energy, $km^2 \cdot sec^{-1}$	Geocentric angular momentum, $10^3\ km^2 \cdot sec^-$			
		C_x	C_y	C_z	C
Before moon's sphere of influence	−0.68	55	−30	21	6
After entering moon's sphere of influence	−0.74	−177	65	32	19

Note: C_x is the projection on the axis of vernal equinox x; C_y is the projection on the axis perpendicular to x in the plane of the equator; C_z is the projection on the axis of rotation of the earth: $C^2 = C_x^2 + C_y^2 + C_z^2$.

The moon can be circled along various types of trajectories.

A distant circling trajectory (at distances of the order of 40,000-100,000 km from the moon, with little influence by the moon) approaches an ellipse with focus at the center of the earth. For such a trajectory and with the take-off point located on the northern hemisphere the return loop (after passing near the moon) on approaching the earth will pass below the equator. This makes it impossible to observe the object from the territory of the U.S.S.R. for distances of close approach to the earth. Moreover, for these trajectories the distance of the perigee from the center of the earth is less than the radius of the earth, so that the object enters the earth's atmosphere in the first loop. The lifetime will thus be small.

Research has shown that the most favorable trajectory is of another type, in the motion along which the object passes at a distance of the order 5,000-20,000 km from the moon. In this case the moon turns out to be a strong perturbing factor, as a result of which it is possible to have trajectories returning to the earth's vicinity, after which the object can revert to a satellite with a high perigee, of the order of 40,000 km and an apogee of the order 500,000 km. Now the return to earth after nearing the moon will occur over high latitudes of the northern hemisphere of the earth, making it very convenient for observations and for obtaining radio-transmitted information. Besides, such trajectories are very convenient for solving the problem of orienting the object and photographing from distances in the 40,000-150,000 km range. The calculations show that for this range of distances, which is covered in the period of about half of a twenty-four hour day, the direction of the "object-moon" vector in absolute space is almost completely conserved. This is conducive to normal action of the orientation and photographing system. This feature of the orbit makes it possible to turn on the orientation system and subsequent operations connected with photographing by means of a programming apparatus operating according to a preset timed program.

In order to photograph the nonvisible side of the moon, it is necessary to make the launching during days near the new moon. At this time the hidden side of the moon is illuminated by the sun. Combination of the conditions of most favorable power with the conditions for illumination of the moon determines the months and days best suited for obtaining photographs of the moon's hidden side.

The above considerations served as the basis for choosing the circling trajectory for the third cosmic rocket, where for the best solution of the problem posed, a strong influence of the moon on the motion of the rocket was applied.

Actual data on the orbits of the first three Soviet cosmic rockets are given below.

1. The last stage of the first cosmic rocket without fuel carried 1472 kg, the weight of the container, scientific apparatus, and supply sources amounted to 361.3 kg. The rocket was launched January 2, 1959 in the direction of the moon; it passed to the north of the moon at a distance of about 5000 km from its surface. The flight time from the earth to closest approach with the moon was 1.5 days.

In the launching of this rocket the parabolic velocity was at first exceeded, as a result of which after passing near the moon the rocket continued to move away from the earth, left the earth's gravitational field, and became an artificial satellite of the sun.

This rocket is moving relative to the sun in an orbit having the following characteristics:

a) incline of the plane of the orbit with respect to the ecliptic plane i = 1°;
b) minimum distance from the sun 146×10^6 km;
c) maximum distance from the sun 197×10^6 km;
d) period of rotation T = 450 days.

The minimum distance from the sun was attained in the middle of January, 1959. The shortest distance of the orbit of the artificial planet to the orbit of Mars is equal to 15×10^6 km.

Radio communication with the first cosmic rocket was realized up to distances of 400,000-500,000 km.

2. The second cosmic rocket was launched for a hit. The weight of the last stage without fuel was 1511 kg, the weight of the container with scientific apparatus was equal to 390 kg. The pattern of motion of the rocket is shown in Fig. 4.

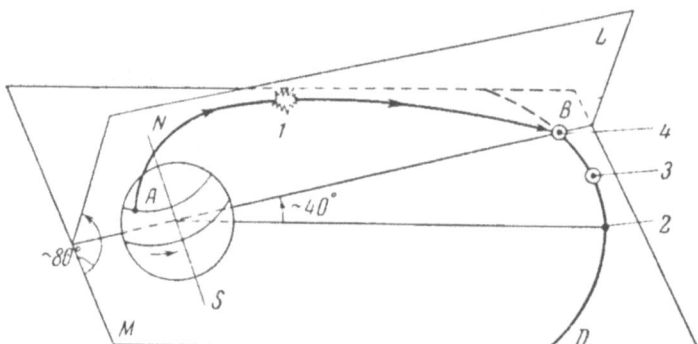

FIG. 4. Pattern of motion of the second cosmic rocket. M) plane of moon's orbit; DB) moon's orbit; L) plane of rocket's trajectory; AB) rocket trajectory; 1) formation of artificial comet; 2) point of least inclination of the moon's orbit; 3) position of moon at time of take-off of the rocket; 4) position of the moon when hit by the rocket.

The following data characterizes the orbit of this rocket. The trajectory lies almost exclusively in one plane, forming an angle of 65° with the equator. Flight from the earth to the moon required about 1.5 days. The initial velocity in free flight exceeded the local parabolic velocity.

From the standpoint of requirements for accuracy of the initial data a flight time of about 1.5 days proved to be better for hitting than the longer flight times (2.5 or 3.5 days), for which a smaller initial energy is required. For flight to the moon during a period of about 0.5 days the expenditure of energy demanded is too high.

The minimum geocentric velocity in the orbit somewhat exceeded 2 km/sec. The velocity of impact with the surface of the moon was equal to 3.3 km/sec. The velocity vector formed an angle of about 60° with the surface of the moon.

The point of impact with the moon is located to the north of the center of the moon's visible disk at a distance of about 800 km and situated south of the craters of Archimedes, Aristillus, and Autolycus (Fig. 5). The corresponding value of ρ_{min} for the projected trajectory is equal to 500 km.

The instant of contact was determined from the sudden cessation of reception of radio signals on September 14, at 0 hours, 2 minutes, 24 seconds, Moscow time. This time agrees with the data of trajectory measurements performed both during the course of the flight and in the time interval immediately preceding the instant of contact.

3. The third cosmic rocket was launched October 4, 1959, to circle the moon. The weight of the last stage without fuel was 1553 kg. The weight of the RSS together with apparatus and supply sources placed inside the last stage amounted to 435 kg.

During the flight of the RSS from the earth to the moon the inclination of the orbit with respect to the equatorial plane was equal to 65°. After perturbation by the moon the further motion, under the influence of the earth's gravitation, followed a nearly elliptical orbit with an inclination with respect to the equator of about 80°. Calculation of the latter motion shows that the sun and moon influence the orbit of the RSS so that the inclination of the orbit varies irregularly and gradually diminishes. In the tenth loop the inclination attains 48°. In the eleventh loop under the moon's influence the inclination of the orbit once again increases to 57°. It is remarkable that the minimum distance to the earth as a consequence of the influence of the sun and moon decreases from loop to loop. Calculation shows that after circling eleven times the RSS will enter the earth's atmosphere at the end of March, 1960, in the northern hemisphere and its existence will come to an end.

The indicated circumstance is connected with the form of the orbit and the nature of its disposition relative to the earth

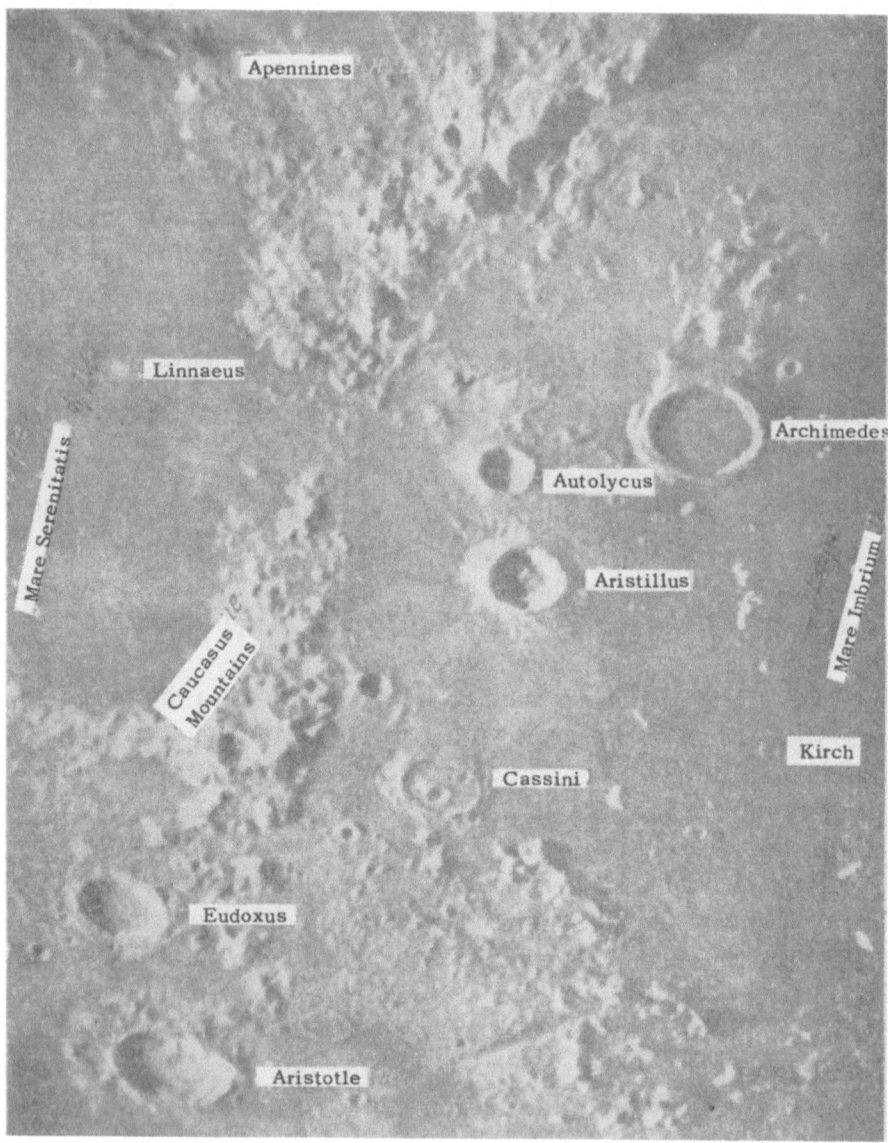

FIG. 5. Portion of the moon's surface where the second cosmic rocket came in contact with the moon.

and sun. This effect, at first glance unexpected, is due simply to Newtonian forces. It is obvious that such effects must necessarily be taken into account in the theoretical analysis of problems in the construction of planetary systems and of the peculiarities of the orbits of different planets and their satellites in the solar system. As a result of perturbations created by the sun, there occurs an evolution of the orbit, which can lead to collision of the satellite with the parent planet; thus after an extended period of time the only satellites that can "live" are those having certain specific types of orbits.

A projection of the first loop of the RSS orbit on the plane

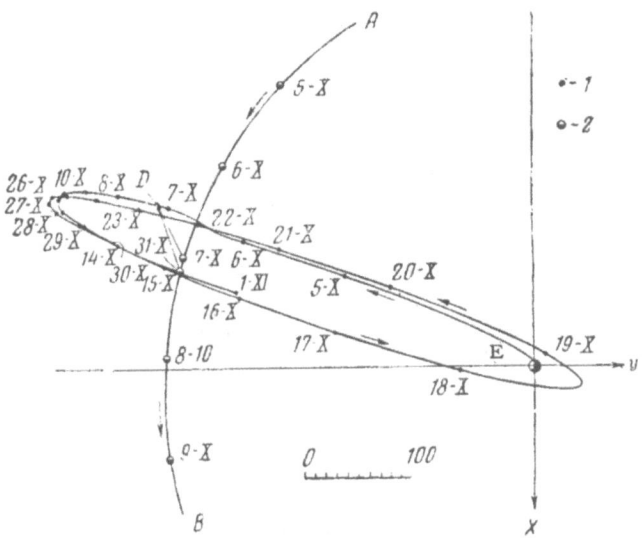

FIG. 6. Projection of the first loop of the RSS orbit on the plane of the earth's equator. E) earth; AB) orbit of moon; the x-axis is directed to the point of vernal equinox; at the point D the hidden side of the moon was photographed; the numbers 1 and 2 represent the positions of the RSS and the moon, respectively, at 0 hours, universal time (3 A.M., Moscow time) during the days indicated in the drawing; the distance scale is given at the bottom (10^3 km).

of the equator is shown in Fig. 6, while the projection on the plane perpendicular to the direction from the earth's center to the point of vernal equinox is shown in Fig. 7.

A projection of the first loop of the orbit on the surface of the earth is given in Fig. 8. The lines of projection describe the loop at the instant of nearness to the earth's axis of rotation, both in approaching the earth and in leaving it. The data

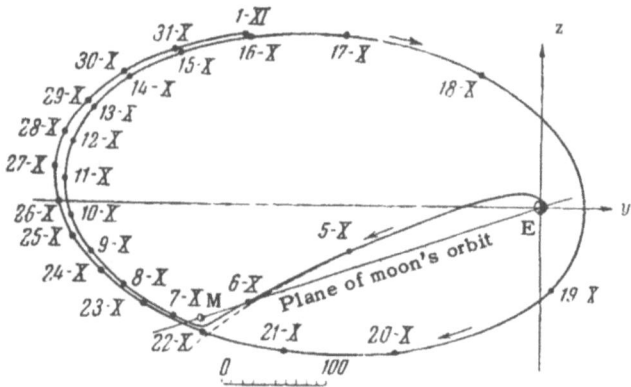

FIG. 7. Projection of the first loop of the RSS orbit on the plane perpendicular to the direction from the earth's center to the point of vernal equinox. The z-axis coincides with the earth's axis of rotation; the y-axis represents the intersection of the equatorial plane with the plane of projection; M) moon; the line ME is the intersection of the plane of the moon's orbit with the plane of projection; the remaining notation is the same as in Fig. 6.

on the characteristic points of the RSS orbit are compiled in Table 2.

TABLE 2

Characteristic points of the RSS trajectory	Date	Moscow time, hours and minutes	Distance from earth's center, 10^3 km	Distance from moon's center, 10^3 km
Minimum distance from moon	6.10.1959	17.21	368.08	7.94
Photographing	7.10.1959	6.30—7.10	399.50—400.50	65.5—68.5
Maximum distance from earth	11.10.1959	0.44	480.50	442
Minimum distance from earth	18.10.1959	19.49	47.49	363
Maximum distance from earth in second loop	26.10.1959	20.59	489	697

A projection of the orbit on the surface of the moon is shown in Fig. 9, where the points corresponding to minimum

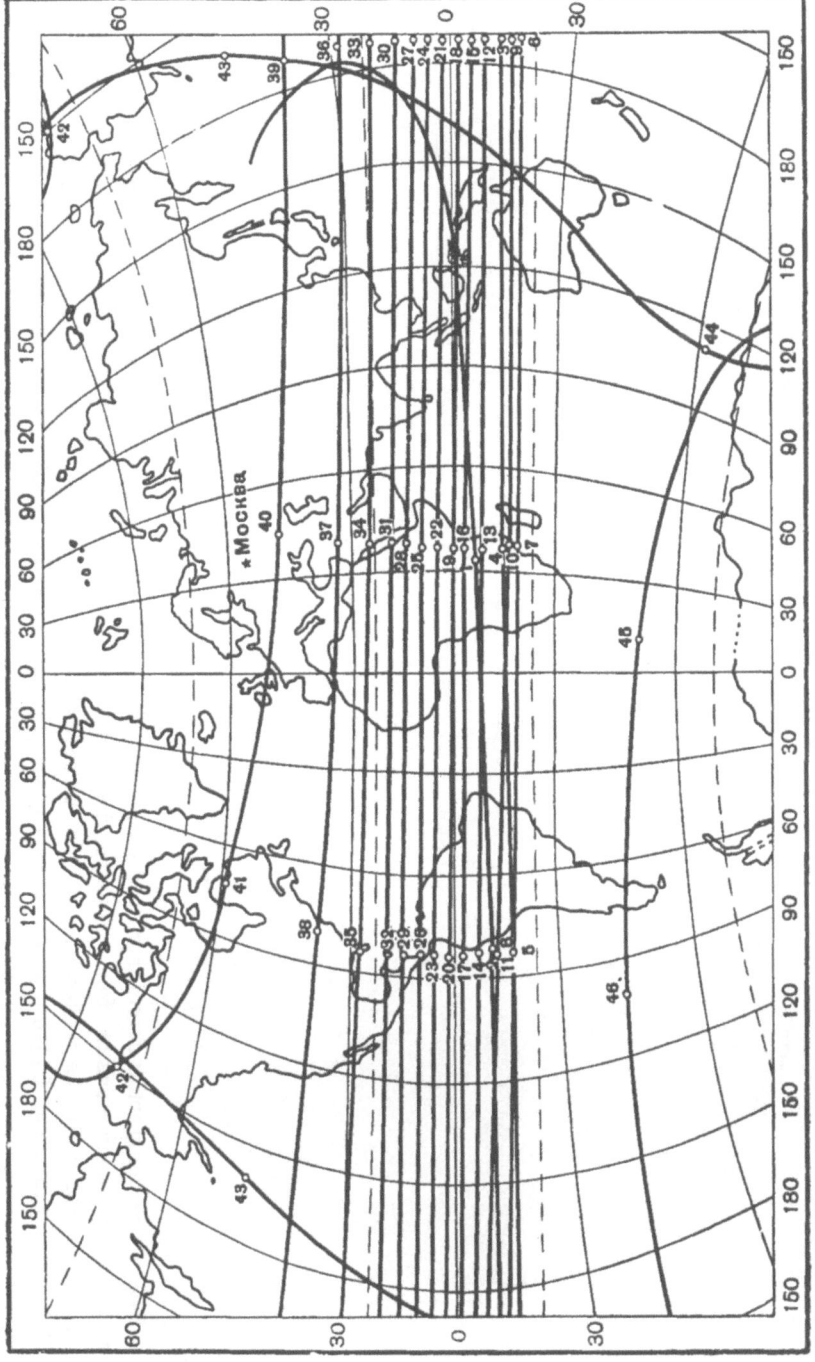

FIG. 8. Projection of RSS orbit on the earth.

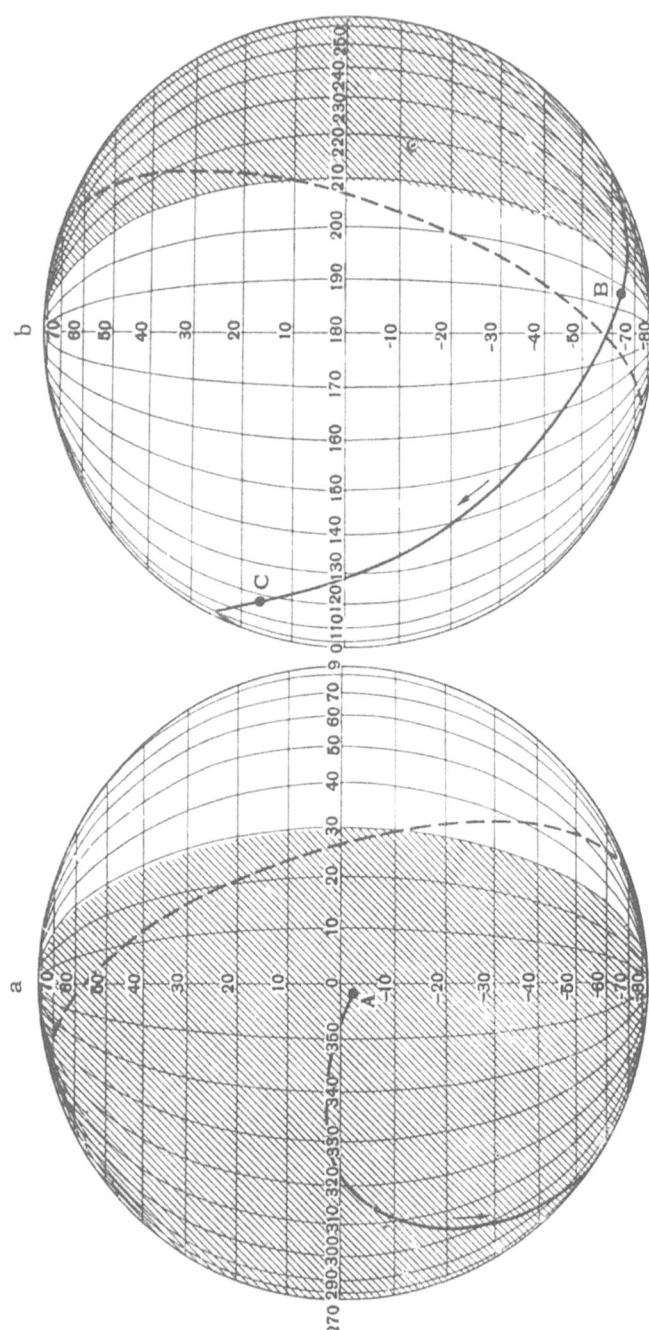

FIG. 9. Selenographic coordinates and projection of the RSS orbit on the moon's surface. a) Hemisphere facing the earth; b) hemisphere unseen from the earth; the striated region is unilluminated by the sun; the solid line is the projection of the RSS orbit; the broken line bounds the area photographed on the moon; A) beginning of flight; B) shortest distance to the moon; C) point at which photographing was made (65,000-68,000 km from the moon's surface).

distance to the moon's center and time of photographing are marked.

The conditions for photographing were favorable from the point of view of illumination of the moon's hidden side, where about 2/3 of the unknown part of the moon was photographed. An unfavorable situation in the photographing was the presence of direct illumination of the surface, as a result of which there were shadows from nonuniformities on the moon's surface, and the photographic image was obtained only due to differences in the reflecting power of its various portions.

Prior to the photographing, an orientation system was put into operation, which directed one end of the RSS axis toward the sun, the other end with photographic objectives pointed in the direction of the moon. After the RSS had been oriented, the hatch covering the illuminator and sensitive elements for the lunar orientation and photographing apparatus was opened. In photographing the moon a special photoelement was used to switch off automatically the photoelements reacting to the light of the sun; after this a more refined orientation of the axis of the photoinstruments toward the center of the moon was carried out. After the orientation process the center of the moon was photographed, beginning on October 7 at 6 hours, 30 minutes, Moscow time; at this time the distance of the RSS to the center of the moon was equal to 65,600 km. To obtain favorable temperature conditions inside the RSS, after completion of the photographing, rotation about a transverse axis with an angular velocity equal to 2 deg/sec was imparted to it. Then the film was developed, fixed, and dried automatically, and the photographic image was sent to earth by means of a special television system.

Samples of the photographs obtained and processed so far are given in Figs. 10, 11, and 12.

In the photographs which were transmitted to earth from the RSS, besides the part of the moon unseen from the earth, a small part of the surface of the moon visible from the earth was also recorded. This made it possible to link some of the objects on the moon's surface not viewed before with those already known and to determine their coordinates on the moon.

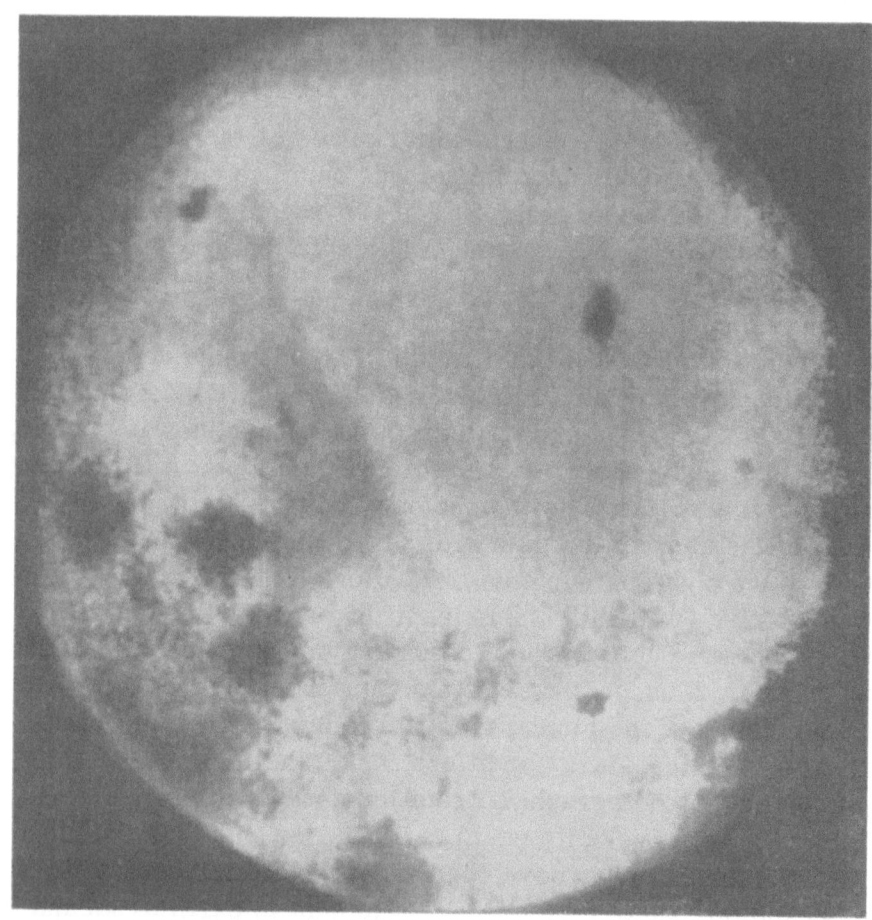

FIG. 10. Photograph of the moon's surface obtained from the RSS.

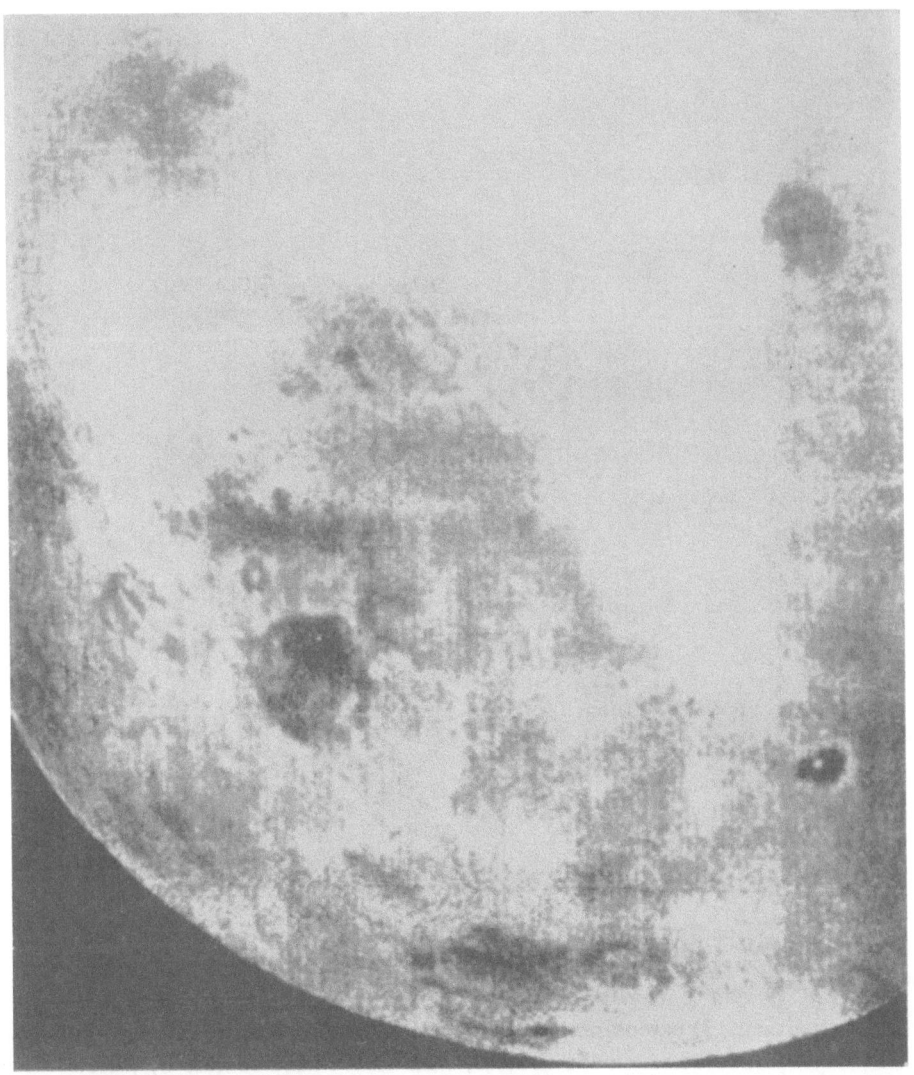

FIG. 11. Photograph of the moon's surface, obtained from the RSS.

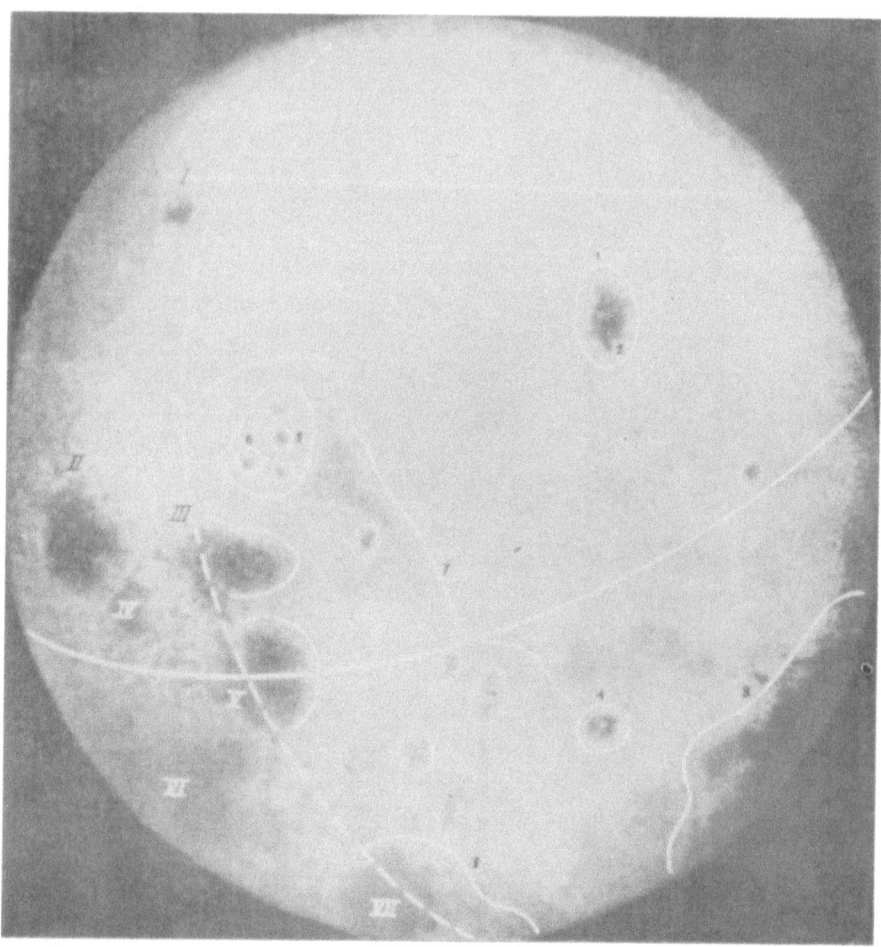

FIG. 12. Location on the moon's surface of a number of objects as obtained following preliminary processing of the photographs transmitted from the RSS. 1) Sea of Moscow (crater sea with a diameter of 300 km); 2) Bay of Astronauts; 3) extension of the Southern Sea to the unseen side; 4 and 5) Tsiolkovskii and Lomonosov Craters, with hills in the center; 6) Joliot-Curie Crater; 7) Soviet Mountain Range; 8) Sea of Mechta; the solid line is the lunar equator; the broken line is the boundary of the moon not visible from the earth; the objects contained within the solid curves have been reliably established; those objects within broken curves require more refined description; classification of the objects encircled by dotted curves are being refined. The objects located on the visible part of the moon are indicated by Roman numerals: I) Humboldt Sea; II) Sea of Crises; III) Rim Sea; IV) Sea of Waves; V) Smith's Sea; VI) Sea of Fertility; VII) Southern Sea.

For objects located at the edge of the visible disk, the form of which was poorly understood due to distortions of perspective, more distinct features have now been obtained. In other instances the objects were known only partially; the photographs obtained made it possible to define more precisely the form of their parts located on the visible side of the moon and to determine the form of their unknown continuations into the nonvisible side.

In the nonvisible part the presence of a mountain chain, named the Soviet Range, was established, along with the presence of seas and craters. One of the seas was named the Sea of Mechta in honor of the first cosmic rocket, launched on January 2, 1959.

The photographs in Figs. 10 and 11 are original photographs, from which the clearly pronounced interference from electronic noise has been removed. The photograph in Fig. 12 was obtained as the result of preliminary processing of a number of photographs and represents a first attempt at composition on the basis of the furnished details exposed in a number of frames.

For a complete description of the nonvisible part of the moon the photographing operation will have to be continued further. It is obvious that subsequent photographs with side illumination can display new formations on the hidden side of the moon.

It seems to us that the experiment carried out (that of photographing in space with radio transmission of the images to earth) not only is significant in informing us on the nature of the moon's hidden side but also is of utmost importance as the approbation of a new methodology in modern experimental astronomy.

MAGNETIC MEASUREMENTS WITH THE SECOND COSMIC ROCKET

S. Sh. Dolginov, E. G. Eroshenko, L. N. Zhuzgov, N. V. Pushkov, and L. O. Tyurmina

The magnetic measurements with the second cosmic rocket were carried out for the following purposes: 1) to discover whether or not the moon has a magnetic field (dipole); 2) if there is such a field, to determine its magnitude; 3) to obtain new information on the magnetic field in the outer corpuscular zone (radiation belt) of the earth, the first information concerning which was obtained by measurements with the first cosmic rocket.[1]

These problems are of fundamental importance for an understanding of the magnetism of the earth and cosmic bodies and for an explanation of magnetic turbulence and its connection with solar activity, the northern lights and cosmic rays.

In the capsule of the rocket launched to the moon on September 12, 1959 there were three independent one-component magnetometers with even-harmonic magnetic saturation pick-up units. The pick-up units measured the components of the magnetic field along three mutually perpendicular axes, X, Y, Z fixed in the rocket capsule.

For measurements of the weak fields which might be expected close to the moon the sensitivity of the magnetometers in the second rocket was increased by a factor of four as compared with that of the magnetometer in the first rocket. This required a reduction of the range of the magnetometer measurements and is the reason that the measurements of the magnetic field of the earth carried out by the second rocket were started at a distance of approximately 18,000 km from the center of the earth whereas in the first rocket the field measurements were started at a distance of approximately 14,700 km.

The magnetometers which were used were relative instruments. The zero points of the magnetometers were determined primarily from the data of measurements at large distances (outside the magnetic field of the earth). The stability of the zero points of the X, Y, and Z magnetometers are shown by the curves in Fig. 1. Each of the points repre-

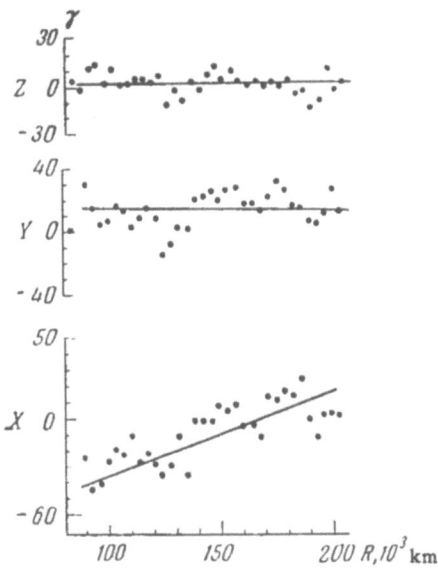

FIG. 1. Readings of the X, Y and Z magnetometers at distances of 75,000–200,000 km from the earth which characterize the stability of the zero points.

sents the average value of readings taken over twenty minutes of flight.

It is apparent from these curves that the zero points of the Y and Z channels remained within $\pm 20\gamma$ close to the earth. The zero point of the X channel varied to some extent. For the portion of the trajectory close to the earth it may be assumed that the displacement of the zero point varied at a constant rate of the order of $7\gamma/hr$. For the portion of the trajectory outside the magnetic field of the earth the mean error in the determination of the channel zero points of the magnetometers was $\pm 30\gamma$.

In addition measurements were made of the possibility of checking the zero points of the channels of the magnetometers along portions of the trajectory in the terrestial magnetic field by the measured value of the field components and the value of the total vector T computed on the basis of these measurements.

We consider the rotation of a pick-up unit fixed at a given angle ψ with respect to the horizontal which rotates about a vertical axis with an angular velocity $\dot{\varphi}$. The pick-up unit is located at a point at which the vector T_0 forms an angle α with respect to the horizontal. The readings of the pick-up unit will vary periodically:

$$H_{p.u.1} = T_0 (\sin\alpha \sin\psi + \cos\alpha \cos\psi \sin\varphi),$$

where the mean value is determined by the first term in the expression given above.

The readings of another pick-up unit which lies in the same plane as the first but at an angle of 90° with respect to the first will be given by

$$H_{p.u.2} = T_0 (\sin\alpha \sin\psi + \cos\alpha \cos\psi \cos\varphi),$$

i.e., it will also be a periodic function with the same mean value but at each position between $H_{p.u.1}$ and $H_{p.u.2}$ there will be a phase shift of $\pi/2$.

The difference in the levels relative to which the readings of the two such pick-ups vary (i.e., the variation in the mean values of $H_{p.u.1}$ and $H_{p.u.2}$) indicates the shift in the zero in one of the channels. In a coordinate system fixed with respect to the vector T_0, the readings on the pick-up units of a three-component magnetometer will be

$$Z = T_0 \cos\theta, \; X = T_0 \sin\theta \sin\varphi, \; Y = T_0 \sin\theta \cos\varphi,$$

where θ is the angle between the axis of the Z unit and the vector T_0, φ is the angle of inclination of the axis of the X unit with respect to the meridian plane.

It is apparent that the square of the modulus of

T_0 ($T_0^2 = X^2 + Y^2 + Z^2$) is independent of orientation. However, if there are zero shifts in channels X, Y and Z of amounts a, b, and c, respectively, the expression for the square of the total vector will contain two additional terms:

$$T^2 = T_0^2 + (a^2 + b^2 + c^2) + 2T_0(a\cos\theta + b\sin\theta\sin\varphi + c\sin\theta\cos\varphi).$$

The second term in this expression leads to a general displacement of the experimental curve and is independent of orientation.

The third term depends on the magnitude of the field and on the orientation of the capsule. It can be found from the experimental curve for the total field vector T by modulation of the amplitude at a period equal to the period of rotation of the capsule. From the total data it is then possible to determine the magnitude and sign of the displacement of the zero points for those pick-up units on which capsule rotation has an effect.

Thus, by virtue of the rotation of the capsule there is an additional possibility for eliminating the effects of zero points displacement; to a considerable degree this feature reduces the limitations imposed by the relative nature of the apparatus which is used.

The errors due to the nonsimultaneity of the interrogation of the channels of the magnetometer is of the order of 30γ only at the initial portion of the measurements. The error in the determination of the total vector is approximately 50γ.

MEASUREMENTS OF THE MAGNETIC FIELD OF THE EARTH

The results of measurements of this field are shown in Fig. 2. The values of T computed from the measured values of X, Y, and Z are shown here by dots. The solid curve 2 gives the values of the total intensity of the earth's magnetic field computed from the data of a spherical harmonic analysis of the measurements of the magnetic field at the surface of the earth.

By analyzing the results of measurements near the earth's surface and the difference between the measured and computed values of the total field strength we may draw the following conclusions.

The measured and computed values of the total field component are in rather good agreement at distances greater

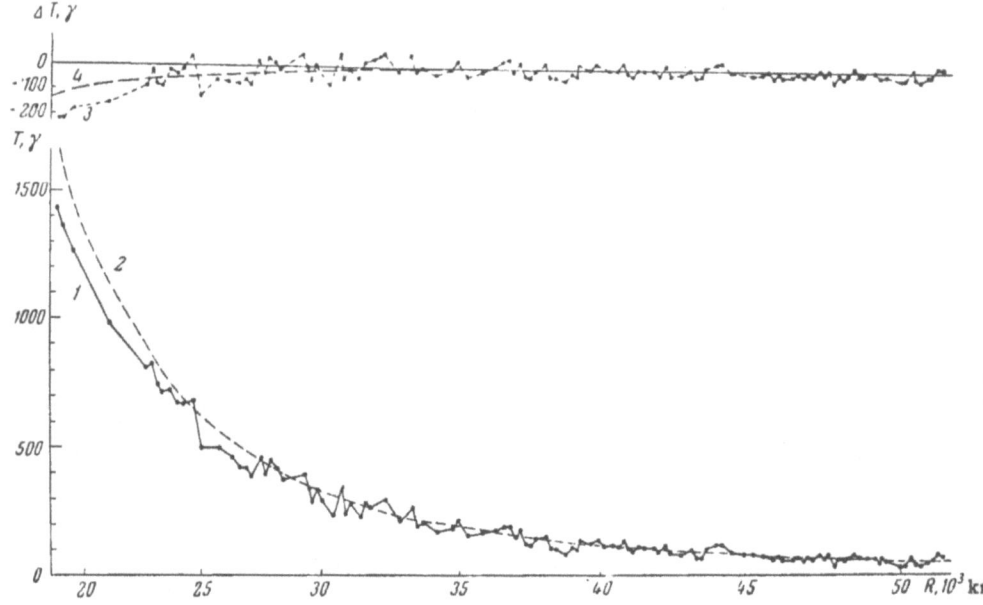

FIG. 2. The total magnetic field vector of the earth as a function of distance from the center. 1) Measured data; 2) calculated dipole field dependents; 3) the difference $\Delta T = T_{meas} - T_{cal}$ between the measured and calculated intensities; 4) smoothed curve for the difference ΔT.

than 23,000 km from the center of the earth. At smaller distances there are discrepancies similar to those which were observed in measurements with the first rocket; the former are considerably smaller in magnitude.

The intensity of the corpuscular radiation in the outer radiation zone has been measured by American investigators in the Pioneer III and Pioneer IV rockets[2,3] and by Soviet workers with the first and second cosmic rockets.[4-6]

The rocket Pioneer III, which revolved about the earth, made two measurements in the radiation zone: on leaving the earth and on returning to the earth. The results of the measurements[2-5] are shown in Fig. 3a. Curves 1, 2 and 3

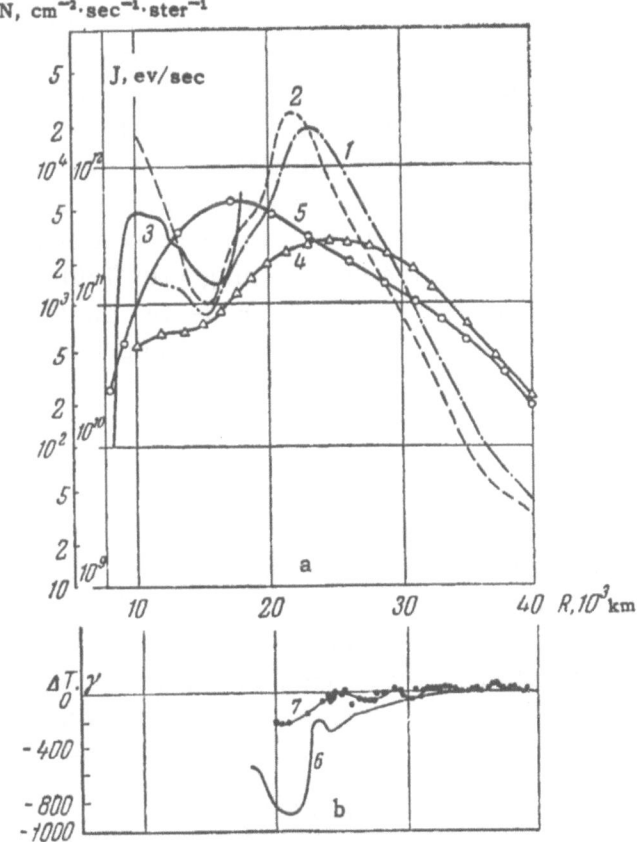

FIG. 3. The radiation intensity (a) and the difference $\Delta T = T_{meas} - T_{cal}$ (b) as a function of distance from the center of the earth in the outer radiation belt. 1 and 2) radiation measurements on the Pioneer III rocket (ascending and descending branches); 3) measurements on the Pioneer IV; 4 and 5) measurements of the total ionization with the first and second cosmic rockets (scintillation counter measurements); 6 and 7) the difference ΔT in the magnetometer measurements for the first and second cosmic rockets.

show the variation with distance from the center of the earth in the intensity of the corpuscular radiation as recorded in Refs. 2 and 3 (the intensity of the radiation is characterized by the number of particles incident on 1 cm² in 1 ster per second). Curves 4 and 5 are plotted from the data of Refs. 4 and 5 (the intensity of the radiation is characterized by the total energy evolved in the crystal of a scintillation counter). The rising and falling branches of the trajectory of Pioneer III intersected the corpuscular zone at different distances from the geomagnetic equator and this explains the difference in the positions of the maxima for the radiation recorded in the return and launching of the Pioneer III rocket.

The magnetic field in the outer radiation zone was meassured twice—with the first and second cosmic rockets. In Fig. 3b curves 6 and 7 show the difference between the values of total intensity of the earth's magnetic field measured with the first and second cosmic rockets and the theoretically calculated values based on data of a spherical harmonic analysis.

Consideration of the available data indicates that the dimensions of the outer zone, the intensity of the radiation in the zone and the separation of the region of maximum radiation intensity from the earth vary with time (from measurement to measurement).[2,5,6] These variations can be correlated with magnetic storms and the northern lights and this indicates that the outer corpuscular region is formed by solar corpuscles which are captured by the magnetic field. An analysis allows us to establish certain relations between the position of the radiation maxima and magnetic activity.

TABLE 1

Rocket	Distance from the center of the earth to the zone maximum, 10^3 km	Reduction in the average values of the horizontal components, $\Delta H, \gamma$
First cosmic rocket (USSR)	25	50
Pioneer III (USA)	22—24	70
Second cosmic rocket (USSR)	18	78

In Table 1 we show the distances from the center of the earth for the maxima of the intensity of the corpuscular radiation taken from Fig. 3 and the reduction in the average daily values of the horizontal components (ΔH). As is apparent from the table, the maximum intensity of the radiation falls closer to the earth the stronger the magnetic storm responsible for the formation of the corpuscular region.

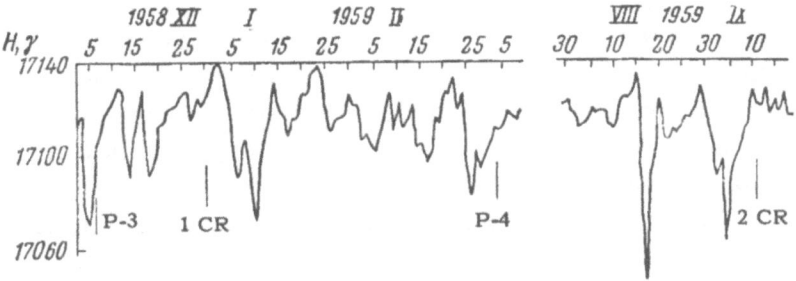

FIG. 4. The time variation of the averaged values of H by measurements from terrestrial stations during the flights of the cosmic rockets. Day of firing the rocket: P-3) Pioneer III; 1 CR) first cosmic rocket; P-4) Pioneer IV; 2 CR) second cosmic rocket.

The behavior of the average diurnal values of the horizontal component is shown in Fig. 4. Comparing the curves in Figs. 3 and 4, S. Sh. Dolginov has proposed that the radiation maximum is probably found farther from the earth the quieter the day preceding the experiment. The following model of the effect is proposed. Under the effect of the corpuscular flux the zone approaches closer to the earth. After completion of the chief phases of the magnetic storm the zone is drawn out by the inhomogeneous magnetic field of the earth into a region of smaller fields and gradients (similarly to the ionized gas of a gas burner in the inhomogeneous field of an electromagnet). Thus it may be assumed that the outer zone of corpuscular radiation has diamagnetic properties. The last finding does not contradict a fact observed in the flight of the first cosmic rocket, namely that there is a discrepancy between the maximum for cosmic radiation and the maximum of the magnetic field.

A comparison of the variation with distance of the intensity of the corpuscular radiation and that of the magnetic field recorded in the flight of the first cosmic rocket shows that the region in which magnetic effects connected with the outer radiation zone are observed is several thousand kilometers closer to the earth than the maximum in the intensity of the corpuscular radiation. If a similar situation obtained in the flight of the second cosmic rocket and if during this time the region of magnetic effects was closer to the maximum of the intensity of the corpuscular radiation, all those magnetic effects could not be recorded since they were in a region in which the intensity of magnetic field was large and beyond the limits of measurement of the magnetometer.

The features indicated above lead to the conclusion that the outer corpuscular region plays the role of an intermediate reservoir of solar particles which is filled during magnetic disturbances and which gradually loses particles in the time periods between disturbances. This then, serves to explain why the northern lights are observed daily at high latitudes and not only in days of magnetic disturbances.

In the first report concerning the experimental observation of the magnetic field of the outer corpuscular region[1] it was proposed that one of the most probable causes for the magnetism of the radiation zone could be the electrical drift currents which are produced by virtue of the drift of charged particles in the magnetic field of the earth. The possibility of drift of charged particles in the magnetic field of the earth has been verified by observations of nuclear explosions at high altitudes carried out in the U.S.A.[7]

In accordance with contemporary theories of the origin of magnetic storms and the northern lights it may be suggested that the outer corpuscular zone is formed by the injection into the magnetic field of the earth of solar corpuscular streams which consist of neutral particles and of an equal number of positively and negatively charged particles (protons, positive ions and electrons). In the flow around the earth of these streams, some of the charged particles are captured by the magnetic field and enter the inner corpuscular zone of the earth.

The charged particles in the radiation zone make numerous trips from one hemisphere to the other, moving along the lines of force of the field, and simultaneously drift in a direction perpendicular to the field and its gradient. In this motion the positive particles move to the west while the negative particles move to the east. It can be shown[8] that because of the small radius of curvature of the electrons, their contribution in the magnetic field due to drift currents in the corpuscular zone is insignificant. Consequently, the magnetic phenomena observed by magnetometers in rockets must be due to drift currents of protons and positive ions.

The existence of protons in the outer corpuscular region is also indicated by observations of northern lights, the spectrum of which contains hydrogen lines. At the present time there are still no experimental data on the existence of protons with energies of several tens of kilo-electron volts but indirect evidence concerning the existence of protons in the corpuscular zone can be obtained from the data of magnetic measurements in this region and from the observation of hydrogen lines in the spectra of the northern lights.

On the basis of the data of magnetic measurements with the first cosmic rockets one of the authors of the present paper, N. V. Pushkov, has suggested that two regions of reduction of magnetic field observed by the magnetometers on this rocket (one of these at a distance of 23,000 km from the center of the earth and the other at a distance of 25,000 km) are related to the two maxima in the intensity of protons of which one was formed during the time of an intense magnetic storm (December 3-6, 1958) and the other during two moderate magnetic storms observed between December 12 and 20, 1958.

An analysis of the variation of the field components shows that in the flight the capsule precessed in such a way that the X and Y magnetometers rotated about the Z axis, while the Z axis itself rotated about another axis, the precession axis, whose direction in space remained fixed. The nutation angle (the angle between the precession axis and the Z axis) was approximately 89°. The period of the natural rotation of the capsule appearing in the periodic variation of the X and Y components was approximately 900 sec, while the precession period appearing in the periodic variation of the Z component was approximately 86 sec.

The values of the X, Y and Z components varied periodically as a consequence of the rotation of the capsule and the consequent rotation of the magnetometer pick-ups (rigidly fastened in the capsule) in the magnetic field of the earth. This rotation continued after the capsule left the magnetic field of the earth and it can give additional criteria for evaluating the existence of a field close to the moon and in the interplanetary space: if a field exists in these regions there should also be a periodic variation of the field components similar to that observed close to earth.

MEASUREMENTS OF THE MAGNETIC FIELD OF THE MOON

The magnetometers on the second cosmic rocket were also designed for observations of the magnetic field of the moon. They operated normally up to the point at which the container struck the moon. The last measurements were made at approximately 50 km from the surface of the moon. The measurements showed no magnetic field at the moon.

An analysis of the accuracy of the measurements of the magnetic field of the earth and the accuracy of the telemeter recordings of the measurements allows us to conclude that if there is a magnetic field at the moon with intensity at the surface which is greater than 50-100γ, it would have been observed.

In Table 2 we show the measured values of the total intensity of the magnetic field at various distances from the center of the moon.

It is apparent from Table 2 that the measured values lie within the error limits of the measurements and show no significant increase in field with approach to the moon. It is well known that if the moon has a total field, as the earth, similar to the field of a uniformly magnetized sphere, there would be an increase in field as the moon is approached inversely proportional to the cube of the distance to the center of the moon.

TABLE 2

Distance from the center of the moon, km	T, γ	Distance from the center of the moon, km	T, γ	Distance from the center of the moon, km	T, γ
5150	30	4000	70	2800	50
5100	**80**	3800	**70**	2700	110
4950	50	3700	30	2600	40
4850	50	3400	90	2290	40
4700	20	3300	40	2160	**170**
4550	50	3200	70	2055	60
4400	70	3100	70	1935	50
4300	40	3000	60	1795	120

Remarks: The values are rounded off to tens of gamma. The most reliable values from an estimate of the quality of the telemetered recordings are given by the heavy numbers.

In view of the data given above it may be concluded that either the moon in general has no magnetic field or the field, even at the surface, is so small that it is completely masked by the error limits of the measurements. The largest possible value of the average intensity of magnetization of the moon is of the order of 0.0002 cgs units. In any case it may be asserted that the intensity of the magnetic field is at least 400 times smaller than the intensity of the magnetic field at the surface of the earth (the mean value of the intensity of magnetization of the moon is less than 0.25% of the mean value of the intensity of magnetization of the earth).

LITERATURE

1. S. Sh. Dolginov and N. V. Pushkov, Dokl. Akad. Nauk SSSR 127 (1959).
2. J. A. Van Allen and L. A. Frank, Nature 183, 430 (1959).
3. J. A. Van Allen, The geomagnetically trapped corpuscular radiation; paper presented at conference on cosmic rays (Moscow, July 6-11, 1959).
4. S. N. Vernov, A. E. Chudakov, P. V. Vakulov and Yu. I. Logachev, Dokl. Akad. Nauk SSSR 125, 305 (1959).

5. S. N. Vernov, A. E. Chudakov, P. V. Vakulov, Yu. I. Logachev, and A. G. Nikolaev, This volume, p. 503.
6. L. V. Kurnosova, V. I. Logachev, L. A. Razorenov, and M. I. Fradkin, This volume, p. 512.
7. S. F. Singer, Missiles and Rockets 5, Nos. 13, 14, 15, 16 (1959).
8. S. F. Singer, Trans. Amer. Geophys. Union 38, 175 (1957).

RADIATION MEASUREMENTS DURING THE FLIGHT OF THE SECOND LUNAR ROCKET

S. N. Vernov, A. E. Chudakov, P. V. Valukov, Yu. I. Logachev, and A. G. Nikolaev

Radiation survey equipment was installed in the second Soviet lunar rocket, launched September 12, 1959, in order to obtain new data concerning the earth's outer radiation belt, to record cosmic radiation on the way from the earth to the moon, and to detect the moon's radiation belt if the latter should exist.

The second rocket carried a larger number of instruments and made more measurements than did the first lunar rocket.[1] Furthermore, part of the radiation survey instrumentation was installed outside the hermetically sealed casing (56 cm from its surface), thus decreasing the shielding of these instruments by other equipment to a considerable degree.

The instrumentation consisted of six Geiger counters and three scintillation counters.

The following instruments were located inside the casing:

1. Scintillation counter A (whose detector was a sodium iodide crystal 39.5 mm in diameter and 40 mm high). This instrument recorded the total ionization created in the crystal by ionizing radiation, and also the pulse-counting rate corresponding to the following energy liberation in the crystal: greater than 60 kev (first threshold); greater than 600 kev (second threshold); greater than 3.5 Mev (third threshold).

2. Geiger counter No. 4, 1 cm in diameter and 5 cm long, surrounded by 1.5 mm of supplementary copper shielding.

3. Geiger counter No. 5, 1 cm in diameter and 5 cm long,

surrounded by supplementary shielding of 3 mm lead and 1 mm aluminum.

These three counters were located within an aluminum jacket 1 g/cm^2 thick. In addition about 20% of the entire solid angle was covered by approximately 10 g/cm^2 of material.

The following instruments were located outside the casing:

4. Scintillation counter B (whose detector was a cylindrical sodium iodide crystal 39 mm in diameter and 40 mm high). This instrument recorded the total ionization created in the crystal, and also the pulse-counting rate corresponding to the following energy liberation in the crystal: greater than 45 kev (first threshold); greater than 450 kev (second threshold). The crystal of this counter was shielded with 1 g/cm^2 aluminum, while up to 5% of the entire solid angle was covered by a larger amount of material (up to 10 g/cm^2).

5. Scintillation counter C (whose detector was a cesium iodide crystal 3 mm thick and 30 mm in diameter with the sides of the free space covered by a layer of 1.2 mg/cm^2 aluminum). This instrument recorded the total ionization created in the crystal.

6. Geiger counter No. 1 with shielding of 3 mm lead plus 1 mm aluminum and having a window of 0.28 cm^2 area.

7. Geiger counter No. 2 with the same shielding and a 1.6 cm^2 window covered with 0.2 mm copper foil.

8. Geiger counter No. 3 with the same shielding and a 1.6 cm^2 window covered with 0.5 mm copper foil.

In addition, the windows of all three counters were covered on the outside with aluminum foil 0.2 mm thick.

The wall thickness of all counters was 50 mg/cm^2 of stainless steel. Counters No. 2 and No. 3 operated only in the high-intensity zone. After leaving the high-intensity zone, the corresponding telemetering channels were used to transmit information on the counting rate of the scintillation counters B (first and second thresholds).

The indicated change of measuring program was executed at a designated radiation intensity; an additional counter (without supplementary shielding) was located inside the casing for this purpose. The program shift was based on a counting rate of about 500 pulses/sec by this counter.

Semiconductors were used in all instrument circuits. The resolving time of the counter and discriminator was 10^{-5} second.

The present paper gives a partial account of the results of preliminary processing of measurements taken at 8000 to 120,000 km from the center of the earth, and also in the vicinity of the moon, beginning at 40,000 km from the moon's surface.

DATA CONCERNING THE SPATIAL LOCATION OF THE OUTER RADIATION BELT

Figure 1 shows the trajectories of the first and second lunar rockets with respect to the earth's magnetic field, and also results of ionization measurements.

The two rocket trajectories do not differ significantly; the trace of the second lunar rocket traverses the high-intensity zone 200 to 300 km closer to the plane of the geomagnetic equator than does the trace of the first rocket. The indicated trajectory shift cannot be responsible for the change of shape and displacement of the maximum of the curve showing the dependence of intensity on flight altitude; instead, it rather accentuates this difference.

The general picture of the deformation of the high-intensity zone on September 12 with respect to its position on January 2, 1959, amounts to a displacement of the zone toward the inner region of the magnetic field.

The maximum intensity on September 12 was observed at 17,000 km from the center of the earth on the 59° line of force,* while on January 2 it was 27,000 km away (on the 63° line of force).

What causes the observed deformation of the outer radiation belt? It should be noted that the flights of the first and second lunar rockets were along trajectories which were quite close together with respect to geographic coordinates

---

*The line of force is designated by the geomagnetic latitude on which it intersects the earth's surface.

but significantly different in their direction with respect to the sun; this may have produced a systematic deformation of the earth's magnetic field. It is however more probable that the deformation of the outer radiation belt is related to the variable character of the solar corpuscular flux and correspondingly to the variable character of particle injection into the high-intensity zone. This hypothesis is supported by the diversity of particle energy spectra observed during the January 2 and September 12 experiments as well as by comparison of the over-all intensity pattern with that obtained during the flight of the American rocket Pioneer III.[2]

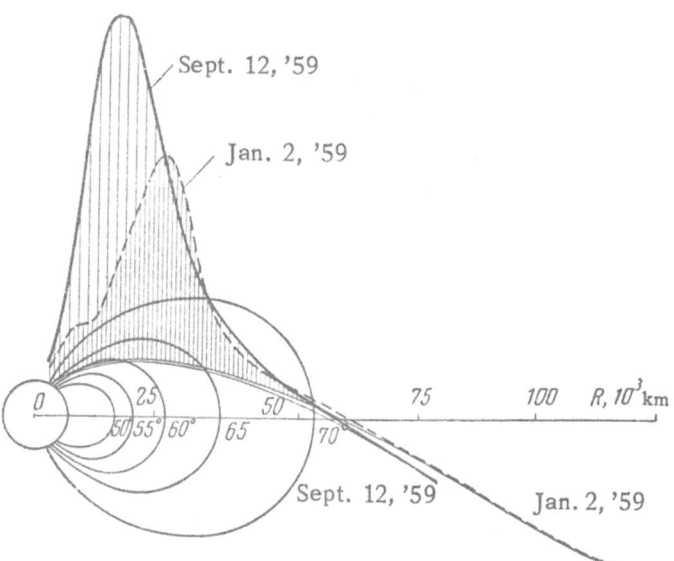

FIG. 1. Trajectories of the first and second lunar rockets with respect to the earth's magnetic field. The lengths of the vertical lines show the radiation intensity at a given point of the trajectory. The drawing shows magnetic lines of force intersecting the earth's surface at latitudes 50, 55, 60, 65, and 70° (the magnetic field is assumed to be a dipole field with geomagnetic pole coordinates 78.5° N, 67° W).

For this American rocket, the flight trajectory with respect to the direction of the sun was close to that of the first Soviet lunar rocket. Nevertheless, the intensity maximum occurred at a distance of 22,000 km from the center of the earth on the 57° line of force; this is in better agreement with the data of the second Soviet lunar rocket than it is with the first Soviet rocket. From this, we assume that the data of the second rocket give a more typical picture of the spatial location of the outer radiation belt.

COMPOSITION OF THE EARTH'S OUTER RADIATION BELT

Figure 2 shows the readings of several instruments in the second lunar rocket related to distance from the center of the earth.

The counting rate of the 3.5 Mev threshold scintillation counter (curve 1) offers confirmation with significantly better accuracy (than in the first lunar rocket) that the outer radiation belt contains no particles having a range of several grams per square centimeter. The small (about 30%) counting rate increase in the region of the maximum, which occurs even in this case, might be explained by the superposition of pulses of smaller amplitude.

Thus the flux of electrons with energy above 5 Mev (or protons with energy above 30 Mev) comprises less than one particle/cm^2 · sec even at the zone maximum.

Significantly new data were disclosed by the readings of Geiger counters Nos. 4 and 5, which were located inside the casing and shielded with supplementary copper and lead filters (curves 2 and 3). The data from the 3.5 Mev threshold scintillation counter show that the counting rate increase in counters Nos. 4 and 5 could not be due to penetration of the casing by charged particles. This means that both counters were recording photons. Since the counting rates of counters Nos. 4 and 5 differed only by a factor of 1.5, these photons must be assigned a relatively high energy (over 400 kev).

In principle one can offer two explanations for the appearance of photons with the observed energies:

1) x-rays from electrons of 2-3 Mev; it must be noted

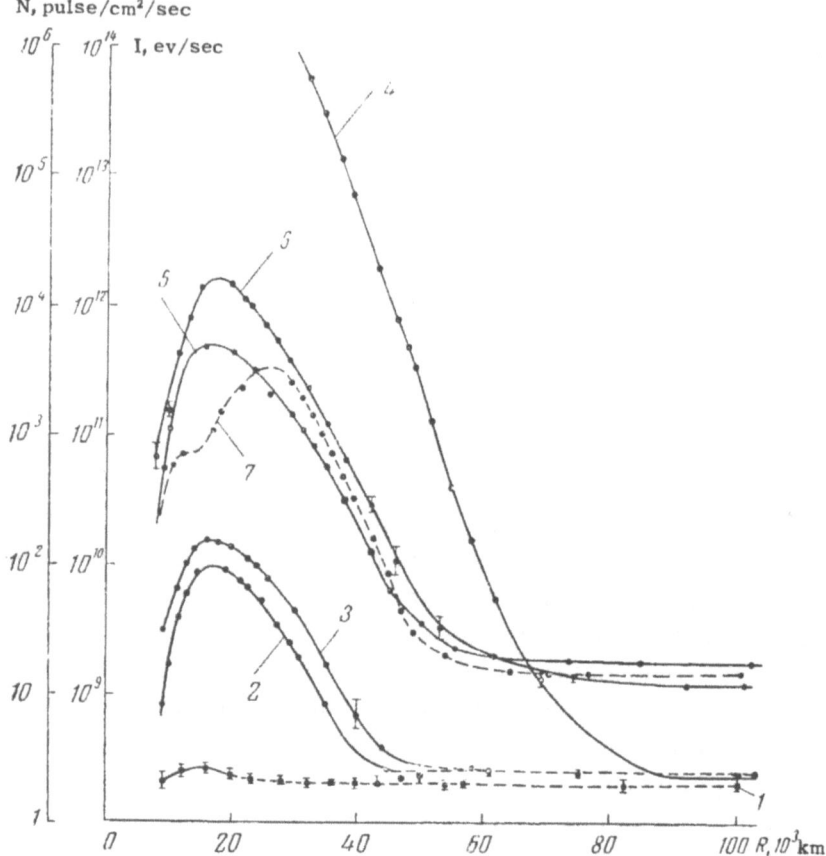

FIG. 2. Dependence of intensity on distance from the center of the earth. The vertical axis shows the ionization occurring in the crystal (ev/sec) and the pulse-counting rate per unit area of perpendicular cross section. In the case of counter No. 1, only the area of the window was considered for measurements within the zone. 1) Pulse counting rate of scintillation counter A (third threshold); 2) pulse counting rate of counter No. 4 (1.5 mm copper shielding, inside casing); 3) pulse counting rate of counter No. 5 (3 mm lead and 1 mm aluminum shielding, inside casing); 4) pulse counting rate of counter No. 1 (0.2 mm aluminum window, outside casing); 5) ionization in crystal A; 6) ionization in crystal B; 7) ionization in crystal of the first lunar rocket.

that the possible electron energy range is quite narrow;

2) radioactivity induced in the casing wall by bombarding protons of about 10 Mev energy. Just as in the case of the electrons, the spectrum of these protons must cut off sharply

on the high-energy side (there are practically no protons with energy over 30 Mev).

The first explanation seems to be more plausible at the moment. But in this case the particle energy spectrum (of the electrons) is very surprising. The readings of counters Nos. 4 and 5 give an estimate of about 5×10^5 particles/cm^2 · sec for the flux of ~ 2 Mev electrons in the region of maximum intensity; the flux of 0.1-1.0 Mev electrons is of the same order of magnitude or less; as has already been mentioned, the flux of 5 Mev electrons is less than 1 particle/cm^2 · sec. On the other hand the experiments of the first lunar rocket showed a quite high flux of electrons with energies 20-50 kev (10^{10} particles/cm^2 · sec). This soft component of the electron spectrum was also detected at the edge of the zone by the scintillation counters of the second lunar rocket. In the region of the maximum it was observed more weakly here than it was on January 2, but it nevertheless gave a noticeable contribution to the readings of counter No. 1 (curve 4, Fig. 2).

Thus we have obtained evidence of two distinct particle energy groups: electrons of about 20 kev and electrons of about 2 Mev (or else protons of about 10 Mev). Evidently the two groups are formed by quite different mechanisms. The energy of the first group is close to the average energy of the protons in the solar corpuscular flux; this permits us to assume that thermodynamic equilibrium is established between the protons and electrons as they are introduced into the earth's magnetic field. On the other hand the second group must evidently be formed in a non-equilibrium process. One cannot help but note that the magnitude of the second-group electron pulses is close to that of the corpuscular proton flux pulses.

THE SEARCH FOR INCREASED RADIATION IN THE VICINITY OF THE MOON

During the flight toward the moon and to within 1000 km of its surface, the radiation intensity was not observed to rise more than 10% above cosmic background. It was difficult to obtain accurate data in the distance range 0 to 1000 km from the lunar surface because of the short time of flight through this region, but no significant intensity increase was found here either.

If we compare the radiation intensity of a hypothetical lunar radiation belt with the maximum intensity of the earth's outer radiation belt, using the readings of the detectors most sensitive to soft radiation (the scintillation counters), then the relative intensity at altitudes above 1000 km is 10^{-6} or less, and at altitudes between 0 and 1000 km, it is 10^{-4} or less. Thus we can say that a lunar radiation belt is practically nonexistent.

If we assume that the existence of the outer radiation belt and the intensity of its particles are both determined by the magnetic field strength, then the magnetic field strength which exists at the edge of the earth's outer radiation belt gives an upper limit for the magnitude of the magnetic field at the lunar surface. This limit is 10^{-3} of the field at the earth's surface.

MEASUREMENTS OF COSMIC RAY INTENSITY

After leaving the earth's outer radiation belt, beginning at 70,000 km from the center of the earth and continuing to the vicinity of the moon, all instruments showed a constant intensity. Placing part of the instruments outside the casing produced an appreciable effect in the sense that the contribution of secondary radiation arising by the action of cosmic rays on surrounding material was decreased. Figure 2 shows the results of ionization measurements within the casing (curve 5) and outside the casing (curve 6).

Inside the radiation belt, curve 6 rises significantly higher than curve 5 because of absorption of the relatively soft radiation; outside the zone of soft radiation, the effect is then reversed, as is to be expected.

Other parameters indicate an analogous effect. Table 1 shows a summary of radiation intensity data obtained by the first and second rockets.

TABLE 1*

Date	Location of instruments	Geiger counter intensity, particles cm² · sec	Scintillation counters		Ionization in the 180 g NaI crystal, 10⁹ ev·sec
			Threshold energy	Intensity*****	
Jan. 2, 1959	Inside casing	2.3±0.1	4.5 Mev 450 kev 45 kev	1.9 ±0.1 3.0 ±0.15 6.75±0.3	1.42±0.05
Sept. 12, 1959	Inside casing	2.46±0.1 ** 2.46±0.1 ***	3.5 Mev 600 kev 60 kev	2.12±0.1 2.77±0.15 6.7 ±0.3	1.55±0.05
Sept. 12, 1959	Outside casing	1.98±0.1 ****	450 kev	2.02±0.1	1.15±0.05

*Errors are characteristic of maximum dispersion of counter areas.
**Counter with supplementary shielding of 1.5 mm Cu.
***Counter with supplementary shielding of 3 mm Pb
****Counter with supplementary shielding of 3 mm Pb.
*****Indicates number of pulses per second per unit area of the crystal (19 cm²).

The data obtained is quite self-consistent. The last line of the table (instruments outside casing) gives the most clear data concerning primary cosmic radiation.

LITERATURE

1. S. N. Vernov, A. E. Chudakov, P. V. Valukov, and Yu. I. Logachev, Dokl. Akad. Nauk SSSR 125, 304 (1959).
2. J. A. Van Allen and L. A. Frank, Nature 183, 430 (1959).

COSMIC RADIATION STUDIES DURING THE FLIGHT OF THE SECOND LUNAR ROCKET

L. V. Kurnosova, V. I. Logachev, L. A. Razorenov, and M. I. Fradkin

Cerenkov counters were placed in the second lunar rocket to record the nuclear components of cosmic radiation. One of these recorded α particles; another recorded particles with charges greater than 5 and greater than 15. Both Cerenkov counters had plexiglass detectors of cylindrical shape with diameter 2.6 cm and height 2.6 cm. The detectors were covered on the front face with a layer of aluminum 2 mm thick, and on the sides with 3 mm (0.8 g/cm^2). The counter used to record nuclei with $Z \geq 5$ and $Z \geq 15$ was placed inside the hermetically sealed container, whose wall thickness did not exceed 1 g/cm^2 of aluminum. The α-particle counter was located outside the casing. Other instruments located within the casing shielded the inside Cerenkov counter somewhat more than the outside one, but for neither counter did the coverage exceed 20% of the total solid angle; this consisted generally of coverage of the side surfaces of the detectors with a layer of substance of the order of 30 g/cm^2.

Charged particles cause light flashes in such Cerenkov counters if their speed exceeds 0.67 of the speed of light (this is the condition for light emission in the detector); for a nucleus, this corresponds to a total energy of 1.3×10^9 ev/nucleon. Since the intensity of the Cerenkov light flash is proportional to the square of the nuclear charge (Z^2), the electronically registered pulses of various thresholds, which are taken from the collector (or from the diode) of the photomultiplier, permit us to determine the flux of various groups of nuclei.

The thresholds, as computed from registration of various nuclear groups, were established on the basis of average value of pulses created in the counters by cosmic ray mesons. Curves of the amplitude distribution of μ mesons were taken in the laboratory with each instrument before the flight; a Geiger-counter telescope was used to select those mesons which were in the direction of the detector axis. The average pulse amplitude occurring as photomultiplier output upon passage of a μ meson through the detector, was equal to

$$\overline{J}_\mu \approx k\left(1 - \frac{1}{n^2}\right),$$

where n is the refractive index of the detector material; k is a proportionality constant.

The channel for α particles had a minimum amplitude of pulse registration equal to $2\overline{J}_\mu$. For the channels for recording nuclei of charge greater than 5 and greater than 15, this value was approximately $20\overline{J}_\mu$ and $200\overline{J}_\mu$ respectively.

The pulse amplitude caused by a nucleus with charge Z and velocity $v = \beta c$ was equal to

$$J_z = k\left(1 - \frac{1}{\beta^2 n^2}\right)Z^2$$

(for a nucleus passing parallel to the detector axis). The value of $J_{Z=2}$ can only exceed the value of $2\overline{J}_\mu$ for those velocities corresponding to the total energy of the α particle, equal to 1.6×10^9 ev/nucleon.

In the channel used for recording nuclei of $Z \geq 5$, the threshold energies were 1.6×10^9 ev/nucleon, 1.5×10^9 ev/nucleon, and 1.3×10^9 ev/nucleon for nuclei with charges 6, 7, and 15 respectively. These threshold energies were determined bearing in mind the fact that the particle could traverse a path of over 2.6 cm within the detector. For nuclei of higher charge the energy threshold approaches that required to produce Cerenkov radiation (1.3×10^9 ev/nucleon).

In the channel recording nuclei of $Z \geq 15$, the threshold energies were 2.1×10^9 ev/nucleon, 1.7×10^9 ev/nucleon, and 1.55×10^9 ev/nucleon, for nuclei of charge 15, 17, and 20 respectively; for nuclei of higher charge, this likewise approaches 1.3×10^9 ev/nucleon.

The nuclear count was made with the aid of a conversion scheme employing semiconductor triodes and diodes. The data on the number of pulses entering the amplifier was transmitted back to the earth through a telemetering system operating constantly during the period reception was possible in USSR territory. The minimum number of pulses for which information was received on the channels recording nuclei of $Z \geq 2$, $Z \geq 5$, and $Z \geq 15$ was 32, 4, and 2, respectively.

In addition to the channels recording nuclei, there was also a channel earmarked for observation of the intensity of all charged particles in the radiation belts. For this purpose, we used the fact that the Cerenkov counter photomultipliers were sensitive to x-rays created in the casing walls by charged particles. For sufficiently sensitive electronic counting equipment (threshold $1/3\,\bar{J}_\mu$) the instruments were able to record x-rays through the photoeffect arising in the diodes, diaphragms, and photomultiplier glass, and also through ionization of residual gas inside the FÉU. In the case where the equipment does not operate as a Cerenkov counter, it can record γ-quanta of 15-20 kev and above. Preliminary irradiation of the equipment with x-rays of varying hardness showed that the counting rate was proportional to the intensity of irradiation within wide limits; furthermore, the proportionality coefficient depends on the irradiation energy. The threshold of $1/3\,\bar{J}_\mu$ was established in one of the channels of the same instrument used to record nuclei with $Z \geq 5$ and $Z \geq 15$. We shall henceforth call this channel the radiation indicator.

The radiation indicator, besides recording charged particles produced by x-ray and γ radiation in the casing wall, could also record electrons penetrating the casing wall and having kinetic energy over 2 Mev. These were detected by the Cerenkov radiation they produced in the detector. The radiation indicator also recorded by Cerenkov radiation those protons and nuclei with total energy above 1.5 and 1.3×10^9 ev/nucleon respectively.

INSTABILITY OF THE OUTER RADIATION BELT, ESTABLISHED WITH THE AID OF THE RADIATION INTENSITY INDICATOR

During the flight of the second lunar rocket, the radiation indicator showed the presence of a zone of increased radiation intensity with a maximum at a distance of 17,000 km from the center of the earth. Figure 1 shows the curve depicting the variation of radiation intensity with distance. This fig-

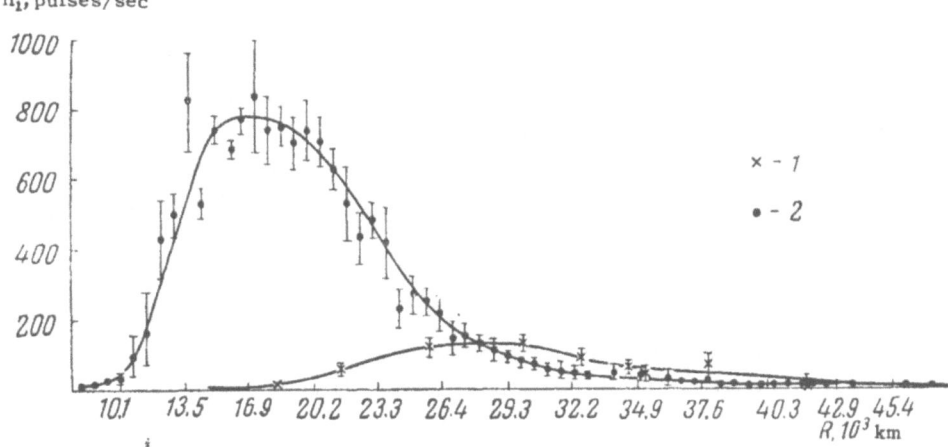

FIG. 1. Dependence of radiation intensity n_i in the outer radiation belt on distance from the center of the earth. 1) Data of radiation indicator in the first lunar rocket (January 1959); 2) data of radiation indicator in the second lunar rocket (September 1959).

ure also shows the curve obtained with a similar indicator during the flight of the first lunar rocket.* Comparison of these two curves shows that the radiation-intensity maximum, and indeed the whole radiation belt, as registered by the second lunar rocket, had shifted toward the earth by a distance equal to one-and-a-half earth radii (approximately by 10,000 km) with respect to the position registered by the first lunar rocket.

The trajectories of the first and second lunar rockets were so similar that it is not plausible to explain the differ-

*The radiation indicator of the first lunar rocket had a threshold corresponding to $\frac{1}{2}J_\mu$. Therefore, as calibration measurements show, the results should be increased by a factor of approximately 2–3 for comparison with the data of the second lunar rocket radiation indicator. However, in view of the poor accuracy of the indicated comparison, we can only indicate qualitatively the increase of intensity in the September maximum as compared with the January maximum.

ence in positions of the maxima on the basis of trajectory differences. Thus on January 2 the radiation-belt intensity maximum was at 27,000 km from the earth's center, while on September 12 it was at 17,000 km. This fact is interesting in itself because it indicates an instability in the location of the zone of increased radiation. The change in intensity of the radiation maximum registered by the indicators also points to this instability of the radiation belt.

This shift in the maximum intensity of the radiation belt toward the earth was also detected during the same flight by the luminescent counters.[1]

DATA FROM THE RADIATION INTENSITY INDICATOR DURING THE LUNAR FLIGHT

Beyond the earth's radiation belt the indicator registered generally cosmic ray protons. The flux recorded was 2-4 particles/$cm^2 \cdot$ sec. This value did not change significantly during the entire path of the rocket outside the belt (according to our present evaluation of the data, for the portion of trajectory up to 150,000 km from the earth's surface, and for the trajectory portion near the moon, beginning at 40,000 km from its surface).

The accuracy is not high for flux determinations at large distances, since the measured values are related to the initial portion of the indicator scale. In any case, one can state that there is no evidence of an increase in radiation intensity in the vicinity of the moon within the limits of a two-fold increase in cosmic radiation. This agrees with the results of Dolginov et al.[2] concerning the absence of a lunar magnetic field.

THE NUCLEAR COMPONENT OF COSMIC RADIATION

At great distances from the earth the average flux of α particles and of nuclei with charge greater than 5 and greater than 15 does not change with distance. The average counting rate in the channels registering these nuclei (according to the basic material evaluated to date) for distances 80,000-160,000 km from the earth's center was respectively 23.5 ± 0.02, 1.91 ± 0.06, and 0.08 ± 0.01 per minute.

In order to obtain the flux values for these groups of nuclei from the basic data, and to obtain the relationship among these flux values, it is necessary to calculate the "geometry factor" of the Cerenkov counters;* one must further determine its dependence on the charge and energy of the registered nuclei. As has been previously indicated, for each threshold nuclei of higher charge can be registered even if they have less energy than the threshold for nuclei of lesser charge. Therefore a significant number of nuclei with higher Z which have passed through only the edge of a detector will be registered. Thus the geometry factor increases with increasing Z for a given threshold of the electronic counting system.

For particles of given Z the geometry factor will likewise depend on the energy, since a low-speed particle having a sufficiently long path in the detector can produce a lightflash exceeding the threshold. At the same time particles of higher speed produce a similar lightflash in traversing a shorter path.

All these factors play a significant role in the registration of nuclei having a charge near the threshold value (Z_T). With increase of charge, the dependence of the geometry factor on Z and on the energy becomes insignificant, and for a given threshold the geometry factor approaches 50 cm^2 · steradian for all particles with $Z \gg Z_T$.

Geometry factors have been calculated for various groups of nuclei. In order to make the calculations, it has been assumed that the Cerenkov counter registers the distribution of isotropic radiation only from the forward hemisphere. Such an assumption is based on the fact that due to the directional character of the Cerenkov radiation, the pulses produced by

*We designate by "geometry factor" a quantity having dimensions of cm^2 · steradians, which when multiplied by the flux magnitude (particles/cm^2 · sec · sr) gives the counts recorded per second by the counter.

charged particles going in the direction of the cathode are several times (3-5) larger than for similar particles traveling in the opposite direction. Therefore for values of Z not differing greatly from the threshold value Z_T, such particles of "reverse" direction will not be registered. In computing the geometry factor, we did not take into account the corrections relating to the registration of particles of reverse direction for $Z \gg Z_T$, nor those relating to incomplete light gathering for particles passing perpendicular to the detector axis.

The geometry factor for α particles was computed for various assumed values of γ, which is the exponent in the energy spectrum expression $N(>E) = AE^{-\gamma}$. Table 1 shows the geometry factors and their relation to the nuclear flux magnitudes.

TABLE 1

Threshold	Nuclei making main contribution to count rate		Geometry factor Γ, $10^{-4} m^2 \cdot sr$		Value of Γ used to determine flux	Flux, particles/m^2 ·sec· sr
	Symbol	Z	$\gamma = 1.2$	$\gamma = 1.6$		
$Z > 2$	He	2	26	24.5	26	150
$Z > 5$	C	6	24	22.5		
	N	7	29	28		
	O	8	33.5	32	30	10.6
	Ne	10	38	36.5		
$Z \geqslant 15$	Ca	20	26.5	25.5		
	Cr	24	33	32		
	Fe	26	33.5	32.5	33	0.4
	Ni	28	35	34		

As is shown by Table 1, the values of the geometry factors obtained under various assumptions as to the shape of the spectrum do not differ much. For each of the registered groups, the fluxes were computed with the geometry factors which correspond to those nuclei which predominate quantitatively in the stratosphere. We do not show the statistical errors in the flux values because the main source of error is inaccuracy in the geometry factor. For the relative flux of nuclei with $Z \geq 2$, $Z \geq 5$, and $Z \geq 15$, we obtained 1000, 75, and 3.

There is some difference in the flux values shown in the literature for various nuclear groups. It is possible that this is connected with extrapolation to the atmospheric boundary, which is carried out by various authors under varying assumptions as to the probability of nuclear dissociation in the layer of atmosphere remaining above the instrument. It is also possible that the observed differences in flux intensity for several nuclear groups can be attributed to solar activity.[3]

Data on the flux of various nuclear groups is given in the

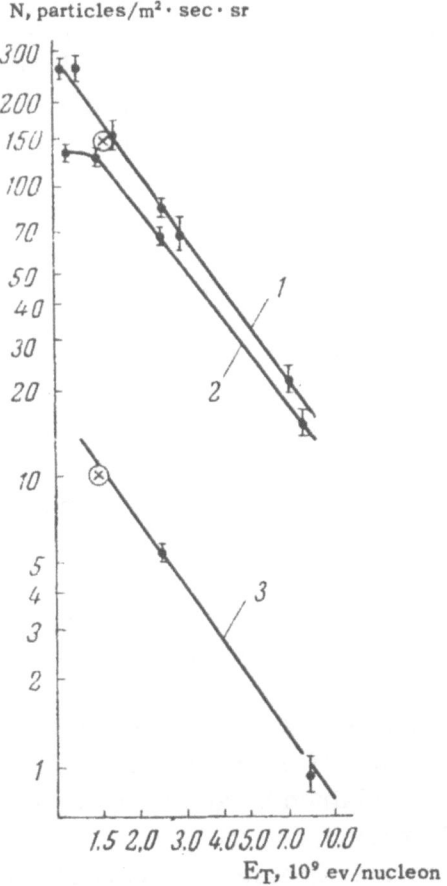

FIG. 2. Integrated energy spectra of α particles (curves 1, 2) and of the nuclear group C, N, O, F (curve 3) from the data of Ref. 3. Curve 1 is for the period of minimum solar activity, while curve 2 is for the maximum. The x's indicate results measured in the second lunar rocket.

summary article.[4] More recent data are given in Ref. 3. Figure 2 shows the integrated energy spectra for α particles and C, N, O, F nuclei, taken from Ref. 3. The crosses indicate flux values obtained in measurements of the second lunar rocket for nuclei of charge from 2 to 5 and from 5 to 15.

As has been mentioned earlier, the average flux values for various nuclear groups did not change with distance at large distances from the earth. The fluxes registered during various portions of the path corresponding to short time intervals did deviate from the average values. We can point to a case in which all three channels showed an increased count rate for a short period of time (17 minutes) between 14 hrs 27 min and 14 hrs 44 min (Moscow time) at a distance of 72,000–74,000 km from the earth's surface. The normalized counting rates during this period in the channels for $Z \geq 2$, $Z \geq 5$ and $Z \geq 15$ were 32.0 ± 1.4, 3.3 ± 0.6, and 0.75 ± 0.25 counts/minute respectively, i.e., 1.4, 1.8 and 9.4 times higher than the average. Further analysis is needed to determine whether other cases of deviation from the average are simply statistical or represent true variations of the nuclear component with time.

There was a large increase of counting rate in the α-particle channel at a distance 10,000–30,000 km from the center of the earth. It is hard to designate the maximum point accurately, since there was saturation of the counting scheme between two successive interrogations of the telemetering system (count rate above 4 pulses/sec). Figure 3 shows the dependence on distance of the counting rate in the α-particle channel.

Two possible explanations can be offered for the observed increase in counting rate of the α particle channel:

1. The effect is explained by an actual increase of α particle flux in this region.

The Cerenkov counter registers α particles possessing energies over 1.6×10^9 ev/nucleon and therefore having a radius of curvature of several thousand kilometers in the earth's magnetic field as found at a distance of 10,000–30,000 km.

The observed counting rate increase could possibly be attributed to the accumulation of α particles on orbits about the earth (the limiting case of such an orbit is a circle lying in the equatorial plane).*

2. The effect is explained by the influence of the intense radiation in the radiation belt on the instruments. During preliminary irradiation of the instruments with x-rays of energy above 15-20 kev before the flight, there was no observed counting in the α-particle channel. However, this irradiation was of somewhat lower intensity than that at the maximum of the radiation belt. After the rocket flight, similar instruments were irradiated with γ quanta of various energies. It was found that several instruments gave a noticeable number of counts in the channel earmarked for α-particle counting when irradiated. Thus, on irradiating one of the instruments with γ rays of 98 kev, 280 kev, and 1.17 − 1.33 Mev, we observed a number of readings all smaller by a factor of at least four than was established during the time of flight through the zone in which the α-particle channel showed a high counting rate. During this irradiation the intensity was so selected as to produce the same energy release in the crystal of a luminescent counter, similar to one used aboard the lunar rocket,[1] as was found during the flight.

The readings observed on irradiation evidently cannot be explained by photoeffects in the FÉU diode, since the pulse amplitude arising through photoeffects should be several times smaller than the amplitude $\bar{J}_\mu$ (the pulse of a μ meson passing through the detector); furthermore, the photoeffect amplitude is less than the threshold, established at $2\bar{J}_\mu$. The readings from the α-particle channel cannot be explained as the superposition of individual pulses, since the value of these readings depends almost linearly on the irradiation intensity. It is possible that they can be explained by an increase of noise pulses under the influence of irradiation which evokes gas ionization within the amplifier.[6]

*One possible mechanism for accumulation of high-energy particles in such an orbit was examined in Ref. 5, and was independently proposed by L. V. Johanson.

The increase of counting rate in the α-particle channel cannot be explained by multiple superposition of pulses from the electrons having several Mev energy which were detected[1] with the aid of Geiger counters surrounded with lead and cop-

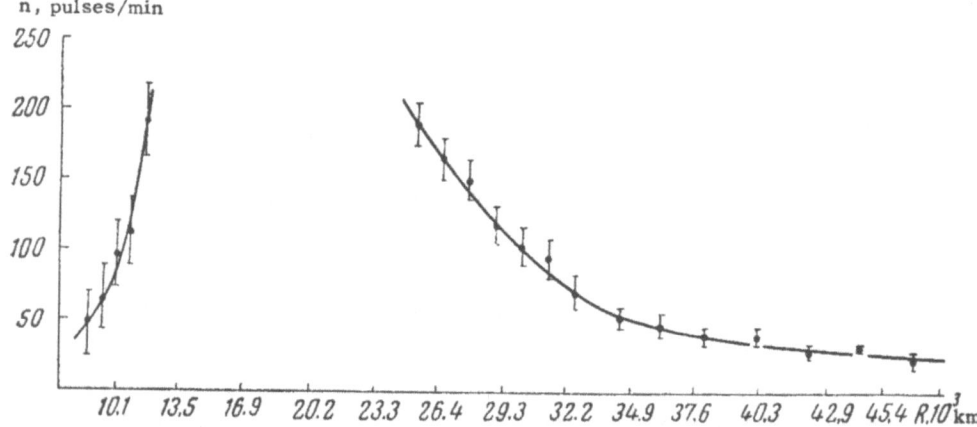

FIG. 3. Dependence on distance from the center of the earth of the counting rate n in the channel for registering particles with $Z \geq 2$.

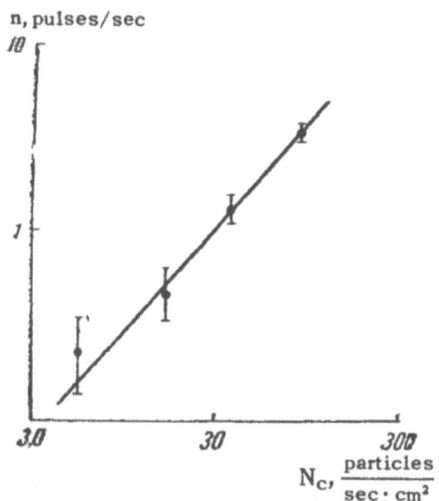

FIG. 4. Dependence of α-particle channel counting rate (n) on the indicated Geiger counter shielding (N_c).

per. Figure 4 shows (in log-log form) the dependence of counting rate in the α-particle channel on the indicated shielding of the Geiger counter while in the high-radiation zone. This dependence is represented by a straight line inclined at 45° to the abscissa; this means that the counting rate in the α-particle channel is proportional to the counting rate of the Geiger counter. If the increase in counting rate had been caused by superposition of pulses, then this dependence would have been at least quadratic in nature.

At present we cannot make a final choice between the possible explanations of this effect that we have expounded above, although it is possible that it is caused by the high intensity of the radiation field. At the same time, the accumulation of α particles and other nuclei into orbits surrounding the earth is also possible in principle. Further investigation should show whether this possibility is realized in actuality.

ACKNOWLEDGMENT

In conclusion the authors offer their deep thanks to Professor V. L. Ginzburg for his guidance, and to Professor N. A. Dobrotin for his constant interest in the work. The authors are grateful to G. S. Dragun, V. V. Marevskii, V. D. Razhin, I. A. Sirotkin, and V. S. Tishkina for participation in various stages of the work. The authors thank all their associates whose work made it possible to set up the experiments in the lunar rocket.

LITERATURE

1. S. N. Vernov, A. E. Chudakov, P. V. Vakulov, Yu. I. Logachev, and A. G. Nikolaev, This volume, p. 503.
2. S. Sh. Dolginov, E. G. Eroshenko, L. N. Zhuzgov, N. V. Pushkov, and L. O. Tyurmina, This volume, p. 490.
3. C. J. Waddington, Progress in Nuclear Physics, 8, Pergamon Press, 1960.
4. S. N. Vernov, V. L. Ginzburg, L. V. Kurnosova, et al., Uspekhi Fiz. Nauk 63, 131 (1957).
5. R. Gall and D. Livshits, Conference on Cosmic Rays, Moscow, 1960.
6. N. O. Chechik et al., Élektronnye umnozhiteli (Electronic Amplifiers), GTTL, Moscow, 1954.

RESULTS OF A STUDY OF IMPACTING OF METEORIC MATTER BY MEANS OF INSTRUMENTS MOUNTED ON SPACE ROCKETS

T. N. Nazarova

The space rockets launched on January 2, September 12, and October 4, 1959 were instrumented to record meteoric particles, the equipment installed being similar to that used on the Third Soviet Artificial Earth Satellite.[1]

When a meteoric particle impacted on the sensor, the latter recorded the momentum of the sensor material* dislodged by the bursting of the particle on the surface of the sensor.[2]

In interpreting the experimental data, we took as point of departure the dependence of the momentum imparted to the sensor on particle energy[3] subject to the relationship $I = AE_0$, where I is the "reaction momentum," E_0 is the particle kinetic energy, A is a proportionality factor. Assuming that the speed of the meteoric particle averages 40 km/sec, we can determine the mass of the particles recorded.

The equipment installed in the first space rocket was calibrated to record particles with masses falling within the ranges 2.5×10^{-9} to 1.5×10^{-8} g, 1.5×10^{-8} to 2×10^{-7} g, and heavier than 2×10^{-7} g, a telemetry signal being sent out after 16 impacts were accumulated in the most sensitive range, 4 impacts in the next most sensitive range, and for each impact in the least sensitive range.

Vibrations and noise originating inside the satellite were ignored by the instruments, since the sensitivity threshold

---

*The momentum acquired by the material of the sensor, ejected in the impact, far exceeds the momentum of the particle itself.

was set higher than the noise level. In contrast to the instrumentation of the third earth satellite, the sensors on board the space rockets were insensitive to impacts of meteoric particles on the skin of the rocket vehicle.

In the January 2, 1959 experiment, during the 10 hours of operation of instruments recording impacting of meteoric particles, sensors occupying a total area of 0.2 m^2 failed to register a sufficient number of impacts to produce a telemetering signal, i.e., particles having masses within the range 2.5×10^{-9} to 1.5×10^{-8} g executed less than 2×10^{-3} impacts per m^2 per sec, particles of mass 1.5×10^{-8} to 2×10^{-7} g executed less than 5×10^{-4} impacts per m^2 per sec, and particles of mass greater than 2×10^{-7} g executed less than 10^{-4} impacts per m^2 per sec.

On the second and third space rockets, the equipment installed was designed such that no storage of data was necessary, each impact being recorded.

The instrumentation of the second space rocket was calibrated to record meteoric particles with masses falling within three ranges: 2×10^{-9} to 6×10^{-9} g, 6×10^{-9} to 1.5×10^{-8} g, and heavier than 1.5×10^{-8} g.

In the September 12, 1959 experiment, during the 30 hours sensors occupying a total area of 0.2 m^2 recorded 2 impacts of particles having masses larger than 1.5×10^{-8} g. Not a single impact was recorded in the other two mass ranges. The frequency of impacts of particles having masses larger than 1.5×10^{-8} g is thus 9×10^{-5} per m^2 per sec, and is less than 5×10^{-5} per m^2 per sec for particles in the mass ranges of 2×10^{-9} to 6×10^{-9} g and 6×10^{-9} to 1.5×10^{-8} g.

The equipment for meteoric matter research placed on board the automatic interplanetary station launched on October 4, 1959 was turned on for short periods of time for transmission of information. It was adjusted to record particles in the following ranges: 1) particle mass 10^{-9} to 3×10^{-9} g; 2) particle mass 3×10^{-9} to 8×10^{-9} g; 3) particle mass larger than 8×10^{-9} g.

During the time the equipment was in operation from October 4 to 18, impacts by meteoric particles on sensors occupying a total area of 0.1 m^2 were observed.

The total service time of the instrument was 6 hrs, 25 min, 37 sec. During this time 7 impacts of meteoric particles were recorded, including one in the first range, five in the second range, and one in the third. The average number of impacts was accordingly 3×10^{-3} per m^2 per sec, with the breakdown by mass ranges: 2×10^{-3} per m^2 per sec, for particles of mass 3×10^{-9} to 8×10^{-9}, and 4×10^{-4} per m^2 per sec for particles of mass 10^{-9} to 3×10^{-9} and larger than 8×10^{-9} g.

For purposes of comparison, we present below the results of the investigations of meteoric matter by means of the equipment carried by the third artificial earth satellite.[4] An intense increase in the number of impacts was recorded on May 15, 1958. It consisted of from 4 to 11 impacts by particles of mass ranging from 8×10^{-9} g to 2.7×10^{-8} g per m^2 per sec. On following days, the frequency of impact dropped by 3 orders of magnitude, and later by still another order of magnitude, ending up with less than 10^{-4} impact events per m^2 per sec during the last days before the instruments went dead. The reason for the events observed on May 15 is apparently an encounter between the satellite and a meteor shower, since a periodicity close to the precessional period of the satellite was evident in the number of impacts recorded on May 15, and a variation in the number of impacts was also observed as the satellite proceeded along its orbit. However, this interpretation of the phenomenon is not established beyond question because of the excessively high frequency of impacts.[5]

If we compare the results obtained by means of the instrumentation of the third satellite and the three space rockets (Table 1), we may note that the density of meteoric debris in the vicinity of the earth is not constant. It varies in both time and space. One conspicuous fact is that the largest number of recorded "reaction momenta" during the flight of the second and third space rockets refers not to minimum momenta which might be recorded by the instrument in question, but to momenta of large magnitude.

TABLE 1

Research vehicle	Date (day, month, year)	Mass of particles recorded (at $v = 40$ km/sec), g	Number of impacts per m² per sec
Third satellite	15.5.58 16—17.5.58 18—26.5.58	$8 \cdot 10^{-9} - 2,7 \cdot 10^{-8}$	4—11 $5 \cdot 10^{-4}$ $< 10^{-4}$
First space rocket	2.1.59	$2,5 \cdot 10^{-9} - 1,5 \cdot 10^{-8}$ $1,5 \cdot 10^{-8} - 2 \cdot 10^{-7}$ $> 2 \cdot 10^{-7}$	$< 2 \cdot 10^{-3}$ $< 5 \cdot 10^{-4}$ $< 10^{-4}$
Second space rocket	12.9.59	$2 \cdot 10^{-9} - 6 \cdot 10^{-9}$ $6 \cdot 10^{-9} - 1,5 \cdot 10^{-9}$ $> 1,5 \cdot 10^{-8}$	$< 5 \cdot 10^{-5}$ $< 5 \cdot 10^{-5}$ $9 \cdot 10^{-5}$
Third space rocket	4.10.59 18.10.59	$10^{-9} - 3 \cdot 10^{-9}$ $3 \cdot 10^{-9} - 8 \cdot 10^{-9}$ $> 8 \cdot 10^{-9}$	$4 \cdot 10^{-4}$ $2 \cdot 10^{-3}$ $4 \cdot 10^{-4}$

ACKNOWLEDGEMENTS

A sizable team of coworkers were engaged in the research on meteoric particles. The author expresses her profound acknowledgement to L. Z. Rusakov, G. M. Kurtev, A. K. Bektabegov, O. D. Komissarov, A. A. Trukhachev, G. Z. Belyaeva, and L. N. Neugodov, for their active participation in the work.

LITERATURE CITED

1. O. D. Komissarov, T. N. Nazarova, L. N. Neugodov, S. M. Poloskov, and L. Z. Rusakov, Artificial Earth Satellites, Volume 2, Plenum Press, New York, 1960.
2. M. A. Isakovich and N. A. Roi, Artificial Earth Satellites, Volume 2, Plenum Press, New York, 1960.
3. S. A. Baum, S. A. Kaplan, and K. P. Stanyukovich. Vvedenie v kosmicheskuyu gazovuyu dinamiku [Introduction to Cosmic Gas Dynamics], Physics and Mathematics Press, Moscow, 1958.
4. T. N. Nazarova, see Volume 4, p. 402.
5. B. Yu. Levin, Fizicheskaya teoriya metereov i meteornoe veshchestvo v solnechnoi sisteme [Physical Theory of Meteors and Meteoric Matter in the Solar System], U.S.S.R. Academy of Sciences Press, Moscow, 1956.

SOME DIRECTION CONTROL PROBLEMS IN INTERPLANETARY SPACE

B. V. Raushenbakh and E. N. Tokar'

The present paper is devoted to an examination of some problems arising in connection with the guidance of interplanetary missiles. By interplanetary missile we shall mean a device moving beyond the atmosphere limits. The term can be applied to an artificial earth satellite, to a stage of a cosmic rocket, etc. The guidance system of such a missile must permit to change the motion of its center of mass and to bring about any rotation of it around its center of mass, and also to keep the axis of the missile in a given position. Below we shall consider only the second of these problems, namely, rotations of the missile and the preservation of a given orientation.

K. É. Tsiolkovskii proposed two methods for turning a missile around its center: the application of a reaction pull with an action line not passing through the mass center, and the use of reaction wheels. The guidance of a missile with the help of reaction motors can be analyzed quite simply. This is not the case with the guidance by means of reaction wheels.

For the sake of generality we shall suppose that the number of stabilizing wheels in the missile body is n. The extension to an arbitrary number of wheels does not involve any complication in the form of equations of motion. The arrangement of wheels and the orientations of their axes in the missile body will be considered as arbitrary, but fixed in time; we shall also assume that the wheels are statistically and dynamically balanced on their axes; in other words their axes of rotation are principal neutral axes of inertia.

This assumption is evidently correct for wheels with axial symmetry.

EQUATIONS OF MOTION AROUND THE CENTER OF MASS

In order to obtain the equations of motion we shall introduce two coordinate systems having the same origin O in the point where lies the center of mass of the material system consisting of the missile body and of the reaction wheels. In

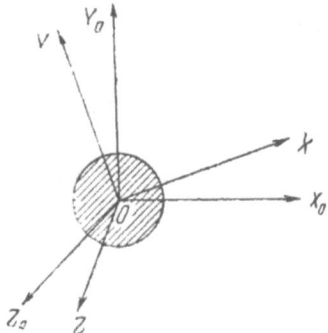

FIG. 1. Relative position of the missile and of the systems of reference $OXYZ$ and $OX_0Y_0Z_0$.

Fig. 1, $OX_0Y_0Z_0$ is the reference system whose axes remain parallel to the axes of some inertial system all the time; $OXYZ$ is a reference system fixed with respect to the body of the missile; the problem of the orientation of the axes OX, OY, and OZ in the missile body will be left unsolved for the moment. The instantaneous angular velocity of the system $OXYZ$ with respect to the system $OX_0Y_0Z_0$ will be denoted by ω; the projections of this angular velocity on the axes OX, OY, and OZ will be denoted by p, q, and r, respectively.

Let us assign a positive direction to the suspension axis of each of the wheels; we shall indicate the orientations of the wheel axes by means of the cosines of the angles between the positive directions of the wheel axes and the positive directions of the OX, OY, and OZ axes. Let the direction cosines of the ith wheel be a_{ix}, a_{iy}, and a_{iz}. The angular velocity of the ith wheel relative to the $OXYZ$ system will be denoted by $\boldsymbol{\omega_i}$; the projection of this vector along the axis of rotation of the wheel will be denoted by ω_i. The projections of the vector ω_i along the axes OX, OY, and OZ will be written, according to the notation chosen above, $\omega_i a_{ix}$, $\omega_i a_{iy}$, $\omega_i a_{iz}$.

The kinetic moment of the material system under consideration in its relative motion around the center of mass, lying in the point O is

$$K = \int_S [rv]\, dm,$$

where **v** is the velocity of the mass element dm relative to the system $OX_0Y_0Z_0$, **r** is the radius vector of the mass element dm considered before; the integration is over the whole system S.

Let us represent the vector **K** as a sum of the kinetic moment of the missile body and of the kinetic moments of reaction wheels where the kinetic moments of the wheels are separated into those determined by the translation velocity of the wheel points in the system $OX_0Y_0Z_0$, and into the kinetic moments determined by the rotation velocity of the wheels relative to the system OXYZ;

$$K = \int_C [rv]\, dm + \sum_{i=1}^{n} \int_{M_i} [rv_{tr}]\, dm + \sum_{i=1}^{n} \int_{M_i} [rv_{rel}]\, dm,$$

where, in the first integral, the integration is carried over the missile body C, and in the other integrals it is carried over the ith wheel M_i.

The velocity of an arbitrary point of the body, **v** (the translation velocity of each point of each wheel being v_{tr}) can be expressed by means of the angular velocity of the system OXYZ according to Euler's formula $v = [\omega r]$. This permits us to group the first n+1 terms.

$$\int_C [rv]\, dm + \sum_{i=1}^{n} \int_{M_i} [rv_{tr}]\, dm = \int_S [r[\omega r]]\, dm,$$

consequently

$$K = \int_S [r[\omega r]]\, dm + \sum_{i=1}^{n} \int_{M_i} [rv_{rel}]\, dm.$$

Let us examine the second term of the expression obtained.

Since the suspension axes of the wheels in the missile body are supposed to coincide with their principal axes of inertia, the kinetic moments of the relative motion of the wheels in the system OXYZ, in the application points are placed in the centers of mass of the wheels, will be $\omega_1 J_1$, $\omega_2 J_2$, ... $\omega_n J_n$. Here ω_1, ω_2 ... ω_n are the angular velocities of the wheels relative to the system OXYZ, introduced above, and J_1, J_2, ... J_n are axial moments of inertia. In the motion of the wheels relative to the OXYZ system the velocities of the centers of mass of the wheels are 0 (the wheels are balanced on their suspension axes in the missile body), which determines the independence of the corresponding kinetic moments upon the center of application. Taking as a general center of application for all wheels the point O, we obtain

$$\int_{M_i} [r\, v_{rel}]\, dm = \omega_i J_i,$$

whence the expression for **K** follows

$$\boldsymbol{K} = \int_S [r\,[\omega r]]\, dm + \sum_{i=1}^{n} \omega_i J_i. \tag{1}$$

Let us represent the kinetic moment of the system so that it will be possible to choose the orientation of the axes OX, OY, and OZ in the missile body so that the expression of the kinetic moment projections along the axes of the system OXYZ will be most simple. The definition of such a system of axes is important not only from the point of view of a simplification in the expression of the equations of motion, but also in order to make the dynamics of the missile to be stabilized better. If we place the planes in which the signals are produced by the sensitive elements of the stabilizing system parallel to the coordinate planes of the system OXYZ, chosen in a suitable way, and if we place the axes of the reaction wheels parallel to the axes of the same system, we can reduce to a minimum the coupling of the missile oscillations around three stabilization axes.

It can be easily seen that the first term in expression (1) is the kinetic moment the system would have if all wheels were stopped, rigidly fixed to the body, and were taken with it as one rigid body. Let us choose as the axes OX, OY, and OZ the principal central axes of inertia of the fictitious rigid body obtained in this way; let its moments of inertia around these axes be A, B and C, respectively. The first term in expression (1) can be written in this case most simply

$$\int_S [r[\omega r]]\, dm = Ap\, e_x + Bq\, e_y + Cr\, e_z,$$

where e_x, e_y, and e_z are the unit vectors parallel to the axes OX, OY, and OZ. Using the expressions for the projections of the vectors ω_i along the axes of the system OXYZ, we can write the second term in expression (1) in the form

$$\sum_{i=1}^n \omega_i J_i = e_x \sum_{i=1}^n \omega_i a_{ix} J_i + e_y \sum_{i=1}^n \omega_i a_{iy} J_i + e_z \sum_{i=1}^n \omega_i a_{iz} J_i.$$

By grouping the terms according to the unit vectors e_x, e_y, and e_z, we get finally for **K**

$$\boldsymbol{K} = (Ap + \sum_{i=1}^n \omega_i a_{ix} J_i)\, e_x + (Bq + \sum_{i=1}^n \omega_i a_{iy} J_i)\, e_y$$

$$+ (Cr + \sum_{i=1}^n \omega_i a_{iz} J_i)\, e_z. \qquad (2)$$

Let us apply the kinetic moment theorem to the motion of the system in question relative to the center of mass; since the point of application of the kinetic moment vector has been chosen in the center of mass O one can write

$$\frac{d\boldsymbol{K}}{dt} = \boldsymbol{M}. \qquad (3)$$

where (dK/dt) is the absolute derivative of the vector **K** with respect to the time, **M** is the principal moment of all external forces acting on the system. By external moments relative to the system we shall mean the moment of external perturbations $\boldsymbol{M_p}$ acting on the missile, and the directing moment of the reaction motors $\boldsymbol{M_r}$ (when together with the reaction wheels the stabilization system of the missile uses as additional devices reaction motors*), and

Calculating the absolute derivative of the vector **K** as the sum of the relative and of the translation velocity of the end of the vector **K** in the system OXYZ, we obtain

$$\frac{d\mathbf{K}}{dt} = \dot{K}_x e_x + \dot{K}_y e_y + \dot{K}_z e_z + [\omega \mathbf{K}].$$

Here $\dot{K}_x$, $\dot{K}_y$, and $\dot{K}_z$ are the time derivatives of the projections of the vector **K** along the axes OX, OY, and OZ

$$\dot{K}_x = A\dot{p} + \sum_{i=1}^{n} \dot{\omega}_i a_{ix} J_i,$$

$$\dot{K}_y = B\dot{q} + \sum_{i=1}^{n} \dot{\omega}_i a_{iy} J_i,$$

$$\dot{K}_z = C\dot{r} + \sum_{i=1}^{n} \dot{\omega}_i a_{iz} J_i.$$

Equating the corresponding projections of the vectors (d**K**/dt) and **M** along the axes of the system OXYZ, we obtain (the summation indices from 1 to n will be left out from now on)

$$A\dot{p} + \Sigma \dot{\omega}_i a_{ix} J_i + q(Cr + \Sigma \omega_i a_{iz} J_i) - r(Bq + \Sigma \omega_i a_{iy} J_i) = M_{px} + M_{rx},$$

$$B\dot{q} + \Sigma \dot{\omega}_i a_{iy} J_i + r(Ap + \Sigma \omega_i a_{ix} J_i) - p(Cr + \Sigma \omega_i a_{iz} J_i) = M_{py} + M_{ry},$$

$$C\dot{r} + \Sigma \dot{\omega}_i a_{iz} J_i + p(Bq + \Sigma \omega_i a_{iy} J_i) - q(Ap + \Sigma \omega_i a_{ix} J_i) = M_{pz} + M_{rz}.$$

Let us transfer the above system as follows:

$$A\dot{p} - \Sigma a_{ix}(a_{ix}\dot{p} + a_{iy}\dot{q} + a_{iz}\dot{r}) J_i + \Sigma a_{ix}(\dot{\omega}_i + a_{ix}\dot{p} + a_{iy}\dot{q} + a_{iz}\dot{r}) + qr(C-B) = M_{px} + M_{rx} - q\Sigma \omega_i a_{iz} J_i + r\Sigma \omega_i a_{iy} J_i, \quad (4)$$

$$B\dot{q} - \Sigma a_{iy}(a_{ix}\dot{p} + a_{iy}\dot{q} + a_{iz}\dot{r}) J_i + \Sigma a_{iy}(\dot{\omega}_i + a_{ix}\dot{p} + a_{iy}\dot{q} + a_{iz}\dot{r}) J_i + rp(A-C) = M_{py} + M_{ry} - r\Sigma \omega_i a_{ix} J_i + p\Sigma \omega_i a_{iz} J_i,$$

*If one takes into account the reaction motors, one should consider the mass of the system under consideration and all the rest of its inertial parameters — moments of inertia, principal axes of inertia, etc. — as variable; however, the mass losses necessary for controlling the position of the missile around its center of mass are small and can be neglected in the calculation.

$$C\dot{r} - \Sigma a_{iz}(a_{ix}\dot{p} + a_{iy}\dot{q} + a_{iz}\dot{r})J_i + \Sigma a_{iz}(\dot{\omega}_i + a_{ix}\dot{p} + a_{iy}\dot{q} + a_{iz}\dot{r})J_i$$
$$+ pq(B-A) = M_{pz} + M_{pz} - p\Sigma\omega_i a_{iy}J_i + q\Sigma\omega_i a_{ix}J_i.$$

Let us explain the meaning of the third terms in the left hand members of these equations. Let us examine the motion of the ith wheel in a coordinate system fixed to the wheel. One of the axes will be chosen as the suspension axis of the wheel with the same orientation as has been chosen before; the origin of the coordinates will be placed in the center of mass of the wheel. The two remaining axes will be arbitrary axes situated in the plane perpendicular to the first axis. Because of the axial symmetry of the wheel its moments of inertia around these two axes will be equal to each other. Let us denote them by J_{ie} (equatorial moment of inertia of the ith wheel).

Since the reference system chosen with respect to the wheel under consideration is a system of principal central axes of inertia, the dynamic equations of Euler can be used as the equations of motion of the wheel.

The projection of the total angular velocity of the wheel along the first coordinate axis (the suspension axis) is given by the relation

$$p' = \omega_i + a_{ix}p + a_{iy}q + a_{iz}r.$$

The projections of the total angular velocity of the wheel along the other two coordinate axes will be denoted by q' and r'. With these notations the first of the dynamic equations of Euler (the equation relating to the projection along the first coordinate axis) is

$$J_i(\dot{\omega}_i + a_{ix}\dot{p} + a_{iy}\dot{q} + a_{iz}\dot{r}) = M'_i.$$

M'_i is the projection of the total moment applied to the wheel along the positive direction of its axis. Evidently M'_i is equal in absolute value to the axial moment on the wheel shaft and has a positive sign if the axial moment acting on the wheel forms with the positive direction of the axis a right hand helix, or negative if the axial moment acts in the opposite direction. The axial moment on the wheel shaft can only be due to the active moment of the motor putting the wheel into motion, M'_{ai}, and by the moment of the resistances on the wheel shaft, M'_{ci}

$$M'_i = M'_{ai} + M'_{ci}.$$

In the preceding formula M'_{ai} and M'_{ci} are algebraic quantities defined exactly as the quantity M'_i. Therefore,

$$J_i(\dot{\omega}_i + a_{ix}\dot{p} + a_{iy}\dot{q} + a_{iz}\dot{r}) = M'_{ai} + M'_{ci}.$$

Multiplying this expression times a_{ix} and summing over i from 1 to n, we obtain an expression for the third term in the first equation of the system (4)

$$\Sigma a_{ix}(\dot{\omega}_i + a_{ix}\dot{p} + a_{iy}\dot{q} + a_{iz}\dot{r})J_i = \Sigma a_{ix}(M'_{ai} + M'_{ci}).$$

Similar expressions can be easily obtained also for the corresponding terms in the two other equations

$$\Sigma a_{iy}(\dot{\omega}_i + a_{ix}\dot{p} + a_{iy}\dot{q} + a_{iz}\dot{r})J_i = \Sigma a_{iy}(M'_{ai} + M'_{ci}),$$
$$\Sigma a_{iz}(\dot{\omega}_i + a_{ix}\dot{p} + a_{iy}\dot{q} + a_{iz}\dot{r})J_i = \Sigma a_{iz}(M'_{ai} + M'_{ci}).$$

By replacing the active moments and the resistant moments on the wheel shaft M'_{ai} and M'_{ci} applied to the corresponding wheels by the moments M_{ai} and M_{ci} directly opposite to them and applied to the missile body, we obtained finally

$$\left.\begin{aligned}&A\dot{p} - \Sigma a_{ix}(a_{ix}\dot{p} + a_{iy}\dot{q} + a_{iz}\dot{r})J_i + qr(C-B) = M_{px} + M_{rx}\\&\quad + \Sigma a_{ix}(M_{ai} + M_{ci}) - q\Sigma\omega_i a_{iz}J_i + r\Sigma\omega_i a_{iy}J_i,\\&B\dot{q} - \Sigma a_{iy}(a_{ix}\dot{p} + a_{iy}\dot{q} + a_{iz}\dot{r})J_i + rp(A-C) = M_{py} + M_{ry}\\&\quad + \Sigma a_{iy}(M_{ai} + M_{ci}) - r\Sigma\omega_i a_{ix}J_i + p\Sigma\omega_i a_{iz}J_i,\\&C\dot{r} - \Sigma a_{iz}(a_{ix}\dot{p} + a_{iy}\dot{q} + a_{iz}\dot{r})J_i + pq(B-A) = M_{pz} + M_{rz}\\&\quad + \Sigma a_{iz}(M_{ai} + M_{ci}) - p\Sigma\omega_i a_{iy}J_i + q\Sigma\omega_i a_{ix}J_i.\end{aligned}\right\} \quad (5)$$

The last two terms in the right hand members of equations (5) are the projections of the gyroscope moments applied to the missile body and determined by the constrained precession of the wheels located in the body; the terms M_{rx}, $\Sigma a_{ix}M_{ai}$, etc., give the projections of the controlling moments produced by the directly acting organs of the stabilization system; these moments must depend upon the angular coordinates of the missile and their derivatives, and may also depend upon the special program of control of the missile position or upon the control signals from the earth.

Let us consider now a particular case of stabilization of the missile by means of three reaction wheels; let us assume that the rotation axes of the first, the second, and the third wheel are respectively parallel to the axes OX, OY, and OZ of the fixed system OXYZ chosen in the missile body as explained before; the positive directions of the rotation axes will coincide with the positive directions of the corresponding axes of the system OXYZ. Then, for the directing cosines, we shall have

$$\begin{Vmatrix} a_{1x} & a_{1y} & a_{1z} \\ a_{2x} & a_{2y} & a_{2z} \\ a_{3x} & a_{3y} & a_{3z} \end{Vmatrix} = \begin{Vmatrix} 1 & 0 & 0 \\ 0 & 1 & 0 \\ 0 & 0 & 1 \end{Vmatrix}.$$

Using the notation

$$\omega_1 = \omega_x, \qquad \omega_2 = \omega_y, \qquad \omega_3 = \omega_z,$$
$$M_{a1} = M_{ax}, \qquad M_{a2} = M_{ay}, \qquad M_{a3} = M_{az},$$
$$M_{c1} = M_{cx}, \qquad M_{c2} = M_{cy}, \qquad M_{c3} = M_{cz},$$

and assuming that the axial moments of inertia of the wheels are the same ($J_1 = J_2 = J_3 = J$), we obtain the equations of motion of the missile to be stabilized for this case:

$$\left. \begin{aligned} (A-J)\dot{p} + (C-B)qr &= M_{px} + M_{rx} + M_{ax} + M_{cx} - \omega_z qJ + \omega_y rJ, \\ (B-J)\dot{q} + (A-C)rp &= M_{py} + M_{ry} + M_{ay} + M_{cy} - \omega_x rJ + \omega_z pJ, \\ (C-J)\dot{r} + (B-A)pq &= M_{pz} + M_{rz} + M_{az} + M_{cz} - \omega_y pJ + \omega_x qJ. \end{aligned} \right\} (6)$$

The obtained equations are similar to the usual dynamic equations of Euler for the missile body. In fact the only difference of the left hand members of these equations from the left hand members of Euler's equations is the presence of the correction J in the first members of the equations; introducing the same corrections in the second members of the equations as follows

$$(C-B)qr = [(C-J)-(B-J)]qr$$

etc., it would not be difficult to obtain a complete similarity of the left hand members of the equations (6) with the left hand members of Euler's equations. The difference would only consist in the fact that instead of the quantities A, B, and C there would appear the differences $(A - J)$, $(B - J)$, $(C - J)$. The right hand members of the equations (6) show a complete similarity with the right hand members of Euler's equations for the missile body: they contain the projections of all moments which are external moments with respect to the missile body. In addition to the moments M_p and M_r the active moments and the resistant moments on the wheel shaft, which are transmitted to the missile body and also the gyroscope moments of the wheels, whose projections $\omega_y rJ$, $\omega_z qJ$, etc., are also contained in the right hand members of the equations derived, are external moments for the missile body.

The observed similarity, however, is apparent, and is due to the circumstance that Euler's equations as well as equations (6) are equations of motion of a system consisting of one or of four rigid bodies, respectively, described with reference to that coordinate system in which the expressions for the projections of the kinetic moments along the coordinate axes are most simple. One should not forget that, unlike the case of Euler's equations, the axes OX, OY, and OZ, to which refer the projections of equations (6), are neither principal nor central axes of inertia with respect to the missile body, and that neither the quantities A, B, and C nor the differences $(A - J)$, $(B - J)$; $(C - J)$ which appear in equations (6) coincide with the moments of inertia of the missile body around these axes.

CONSEQUENCES OF THE LAW OF CONSERVATION OF THE KINETIC MOMENT

In order to relate the nonrotating system $OX_0Y_0Z_0$ to the reference system $OXYZ$ fixed to the missile body, let us introduce three angles δ, ψ, and ϕ (Fig. 2).

The angle δ is defined as the angle between OX', the projection of the axis OX on the plane X_0OY_0 and the axis OX_0. The angle ψ is the angle between OX' and the axis OX. The angle ϕ is the angle between the axis OY and the line of intersection of the planes X_0OY_0 and YOZ. In the case under

Axes of the non-rotating system	Axes of the system fixed to the missile body		
	X	Y	Z
X_0	$\cos\psi\cdot\cos\vartheta$	$-\cos\varphi\cdot\sin\vartheta + \sin\varphi\cdot\sin\psi\cdot\cos\vartheta$	$\cos\varphi\cdot\sin\psi\cdot\cos\vartheta + \sin\varphi\cdot\sin\vartheta$
Y_0	$\cos\psi\cdot\sin\vartheta$	$\cos\varphi\cdot\cos\vartheta + \sin\varphi\cdot\sin\psi\cdot\sin\vartheta$	$\cos\varphi\cdot\sin\psi\cdot\sin\vartheta - \sin\varphi\cdot\cos\vartheta$
Z_0	$-\sin\psi$	$\sin\varphi\cdot\cos\psi$	$\cos\varphi\cdot\cos\psi$

consideration the following sequence of rotations of the missile body for the passage from the position in which the systems OXYZ and $OX_0Y_0Z_0$ coincide, to the given position of the system OXYZ has been chosen: The first rotation around axis OZ_0 by an angle δ, followed by rotation around the new position of the axis OY_0 (OY' in Fig. 2) by an angle ψ, and, finally, in rotation by an angle φ around the axis OX. We report the table of the directing cosines for the axes of the coordinate system fixed to the missile and of the nonrotating one.

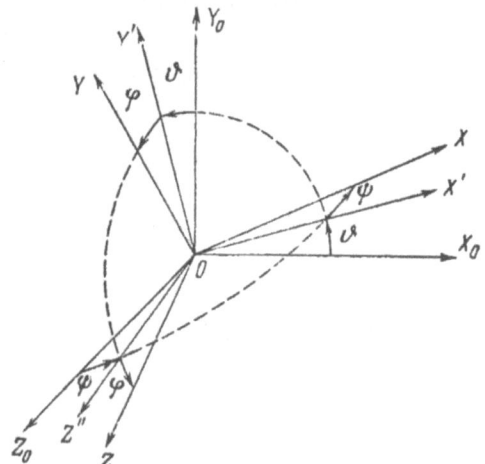

FIG. 2. Sequence of rotations by the angles ϑ, ψ, and φ for carrying the system $OX_0Y_0Z_0$ to the position OXYZ.

Let us express the projections of the angular velocity of the missile body by means of the angles δ, ψ, and φ

$$\left.\begin{aligned} p &= \dot\varphi - \dot\vartheta \sin\psi, \\ q &= \dot\psi \cos\varphi + \dot\vartheta \sin\varphi\cos\psi, \\ r &= \dot\vartheta \cos\varphi\cos\psi - \dot\psi \sin\varphi. \end{aligned}\right\} \qquad (7)$$

The motion of the missile with respect to its center of

mass is expressed by equation (3). If no external moments are applied to the missile (M = 0), (3) leads to the obvious conclusion, known in mechanics as the law of conservation of the kinetic moment

$$K = C,$$

where C is a constant vector.

On the basis of this law, it is possible to derive a number of interesting conclusions in connection with the problem of the rotations of the missile around its center of mass.

Let the projection of the vector K on the axes of the coordinate system which does not rotate in the inertial space be C_{x_0}, C_{y_0}, C_{z_0}. Then its projections in the coordinate system $OXYZ$ will be related to C_{x_0}, C_{y_0}, C_{z_0} through the following expressions

$$\left. \begin{array}{l} C_{x_0} = K_x \cos \psi \cos \vartheta + K_y (\sin \varphi \sin \psi \cos \vartheta - \cos \varphi \sin \vartheta) \\ \qquad + K_z (\cos \varphi \sin \psi \cos \vartheta + \sin \varphi \sin \vartheta), \\ C_{y_0} = K_x \cos \psi \sin \vartheta + K_y (\sin \varphi \sin \psi \sin \vartheta + \cos \varphi \cos \vartheta) \\ \qquad + K_z (\cos \varphi \sin \psi \sin \vartheta - \sin \varphi \cos \vartheta), \\ C_{z_0} = - K_x \sin \psi + K_y \sin \varphi \cos \psi + K_z \cos \varphi \cos \psi. \end{array} \right\} \quad (8)$$

In addition the following evident equality holds:

$$C_{x0}^2 + C_{y0}^2 + C_{z0}^2 = K_x^2 + K_y^2 + K_z^2.$$

The quantities K_x, K_y, and K_z can be easily obtained from formula (2).

Let us consider the missile motion characterized by the fact that the control system located on it maintains some plane fixed with respect to the missile parallel all the time to a given plane in the immobile space. Without any loss in generality we can assume that the axes OX and OY of the mobile system lie in the plane in question, parallel to the plane OX_0Y_0. Then $\psi = \varphi = 0$ for the unperturbed motion, and the equations (8) take the form

$$\left. \begin{array}{l} C_{x_0} = K_x \cos \vartheta - K_y \sin \vartheta, \\ C_{y_0} = K_x \sin \vartheta + K_y \cos \vartheta, \\ C_{z_0} = K_z. \end{array} \right\} \quad (9)$$

In the above relations the angle ϑ can be a given function of time. In fact, let for instance the missile travel along a plane orbit around some celestial body, and let us require of the control system that an axis fixed with respect to the missile and lying in the plane XOY should be always directed toward the center of the celestial body, and that the plane XOY itself should be parallel to the orbit plane. Then, after one revolution, the angle ϑ will change by 2π, since the missile body will have performed one rotation around the axis OZ.

The first two equalities of (9) show that during such a motion around a celestial body the quantities K_x and K_y will be periodic functions of time just as the angle ϑ.

It can be easily seen that

$$K_x = C_{xy} \sin[\vartheta(t) + \theta],$$
$$K_y = C_{xy} \cos[\vartheta(t) + \theta],$$
(10)

where

$$C_{xy} = \sqrt{C_{x0}^2 + C_{y0}^2} \text{ and } \theta = \operatorname{arctg} \frac{C_{x0}}{C_{y0}}.$$

In the unperturbed motion in question here, the angles $\psi = \varphi = 0$, and the angular velocities of the missile body, as can be seen from formulas (7), will be the following

$$p = 0, \ q = 0, \ r = \dot{\vartheta}.$$

In this case, as follows from formula (2), the non-zero projections of the kinetic moment, K_x and K_y, can only depend upon the rotation of the wheels, which had acquired some angular velocities during the time preceding the instant under consideration. This allows us to consider in this case K_x and K_y as the projections of the kinetic moments of the wheels, and, consequently, the number of rotations of the wheels, whose axes are immobile with respect to the missile body must in general change periodically. In the simplest case, when three wheels are situated in the missile, with rotation axes parallel to the axes OX, OY, and OZ, the numbers of rotations changing periodically will refer to the wheels whose axes are parallel to OX and OY. With the notation used above, for the case when the projections of the kinetic moment of the missile body on the axes OX and OY are zero, we obtain

$$K_x = \omega_x J_x \text{ and } K_y = \omega_y J_y, \qquad (11)$$

where J_x and J_y are the moments of inertia on the wheels placed parallel to the axes OX and OY respectively, and ω_x and ω_y are their angular velocities (the index i of ω has been left out since in this case no confusion can arise).

From equations (10) and (11) it follows that ω_x and ω_y are periodic functions of time. After one revolution of the missile around the central body the angular velocity of each wheel reaches a maximum value, goes to zero, changes its sign, and, having reached an absolute value the same as the maximum value, goes back to zero, and finally returns to its initial value. This behavior of the wheels is understandable if one considers that for $r \neq 0$ the angular velocities of the wheels, ω_x and ω_y, give gyroscopic moments different from zero in the first two equations (6); these moments can be eliminated by an appropriate acceleration or deceleration of the wheels. A compensation of the gyroscopic moments is necessary in order to fulfill the condition $p = 0$ and $q = 0$.

It must be noted that the accelerations and decelerations in question inevitably connected to an energy expanse are not at all necessary for a regular motion along the orbit. In fact, if the projections of the vector **K** on the plane of the orbit were equal to zero in the initial moment, i.e. $C_{x_0} = C_{y_0} = 0$, then, as can be seen from expressions (10) and (11), $\omega_x = \omega_y = 0$ during the whole time of motion along the orbit under consideration. Thus, if the energy expanse must be reduced, one must brake the reaction wheels whose axes lie in the plane of the orbit (e.g., by using moments produced by reaction motors).

This braking of the wheels is not always possible, and therefore we shall examine somewhat more accurately the problem of the energy losses in the periodic change in angular velocity of the wheels.

The kinetic energy of the wheels is given by the following expression

$$\frac{1}{2} J_x \omega_x^2 = \frac{1}{2J_x} C_{xy}^2 \sin^2[\vartheta(t)+\theta],$$

$$\frac{1}{2} J_y \omega_0^2 = \frac{1}{2J_y} C_{xy}^2 \cos^2[\vartheta(t)+\theta].$$

By combining these two equations it can be easily seen that if $J_x \neq J_y$ the total kinetic energy of the wheels is a periodic function of time. Consequently, in the general case part of the mechanical energy of the wheels must be periodically transformed into some other form (e.g., into electrical energy) and later must again pass to the wheels as rotational energy. It must be noted that even for $J_x = J_y$ the braking of one wheel and the simultaneous acceleration of the other wheel must always take place by means of the transformation of the kinetic energy of the wheel being slowed down to another form of energy and by the acceleration of the second wheel by some motor producing the motion at the expense of non-mechanical energy. Thus, in the general case the work of the wheels is necessarily connected to a transformation of the energy, which in its turn is connected to additional losses. Let us evaluate these losses.

Let the efficiency of the transformation of mechanical energy into some other form permitting the reverse process be η_1 and the efficiency of the reverse transformation be η_2. The mechanical energy passing to the other form under consideration for a complete braking of the wheel will be equal to $(1/2J)\eta_1 C_{xy}^2$ and the irreversible losses of mechanical energy will be $(1/2J)(1-\eta_1) C_{xy}^2$. When the wheel is started an energy $(1/2J\eta_2) C_{xy}^2$ is expended, and it exceeds the obtained kinetic energy by $[(1/\eta_2)-1](1/2J) C_{xy}^2$. Thus the total energy loss from the braking of the wheels to its start will be

$$\Delta E = \frac{1}{2J}\left(\frac{1}{\eta_2}-\eta_1\right) C_{xy}^2.$$

This equation shows that, for given C_{xy}, η_1, and η_2 the losses can be reduced only by increasing the moment of inertia of the wheel.

An increase in J is connected, however, with an increase in the wheel mass or in its size, or in both simultaneously. Therefore, let us consider the problem of the optimal values of J for two wheels under the condition that the total mass of these wheels is limited. The total energy loss for the two wheels during one acceleration-deceleration cycle will be

$$\Delta E = \frac{1}{2}\left(\frac{1}{J_x}+\frac{1}{J_y}\right)\left(\frac{1}{\eta_2}-\eta_1\right)C_{xy}^2.$$

For a given practical realization of a wheel its mass m can usually be represented approximately as a power of the moment of inertia m = aJ^n where a is some constant, n > 0. The condition n > 0 is quite natural, since when J increases also the wheel mass increases.

The condition that the total mass of the two wheels should be constant is expressed by the equation

$$J_x^n + J_y^n = b, \qquad (12)$$

where b is a constant.

Let us find the minimum of ΔE under condition (12). Using Lagrange's method, we shall seek the extreme of the function Z of the two variables J_x and J_y

$$Z = \frac{1}{2}\left(\frac{1}{J_x}+\frac{1}{J_y}\right)\left(\frac{1}{\eta_2}-\eta_1\right)C_{xy}^2 + \lambda(J_x^n + J_y^n - b),$$

where λ is Lagrange's multiplier. After elementary transformations we find the condition for an extreme, $J_x = J_y$. It can be easily seen that it corresponds to a minimum.

Thus, in order to decrease the energy losses in the control system one must possibly brake the wheels, and, if this is impossible, one must increase their moments of inertia, and take the moments of inertia of the wheels interacting during the motion along the orbit equal to each other.

EXAMPLE OF APPLICATION OF THE OBTAINED RELATIONS TO THE CASE OF SMALL DEVIATIONS OF THE MISSILE FROM A GIVEN POSITION

Let us examine as an example the case of the stabilization of the missile around a given invariable position in space. Let the purpose of the stabilization system be the conservation of this position of the missile, for which the missile coordinate system OXYZ always coincides with some fixed coordinate system $OX_0Y_0Z_0$.

The relative position of the systems OXYZ and $OX_0Y_0Z_0$ will be characterized as before by the angles ϑ, ψ, and φ introduced above. As to the orientations of the wheel axes in the missile body and to the values of the axial moments of inertia of the wheels, we shall follow the same conventions made in deriving equation (6). We shall neglect the value of the axial moment of inertia of the wheels, J, with respect to the moments of inertia A, B, and C. Let us also assume $M_p = M_r = 0$ and let us neglect the values of the gyroscopic moments and of the resistant moments with respect to the values of the controlling moments M_{ax}, M_{ay}, and M_{az}. Let us confine ourselves to small deviations of the missile from the given position. Expression (7) takes in this case the following simple form

$$p = \dot{\varphi}, \quad q = \dot{\psi}, \quad r = \dot{\vartheta}.$$

If we also suppose that the derivatives of the projections of the angular velocity, $\dot{p}$, $\dot{q}$, $\dot{r}$, have the same order of magnitude as the projections p, q, r themselves, the second terms of the left-hand members of equations (6) can be left out as terms of a higher order, and the equations take the form

$$A\ddot{\varphi} = M_{ax},$$
$$B\ddot{\psi} = M_{ay},$$
$$C\ddot{\vartheta} = M_{az}.$$

Let us suppose that the moment acting on the shaft of the reaction wheel placed along the axis OX is a function of the angular velocity produced by it, and, moreover, that it depends (through some control system) upon the angle φ and its derivative $\dot{\varphi}$

$$M_{ax} = -M'_{ax}(\omega_x, \varphi, \dot{\varphi}).$$

If similar equations can be written for M_{ay} and M_{az}, we obtain

$$A\ddot{\varphi} = -M'_{ax}(\omega_x, \varphi, \dot{\varphi}),$$
$$B\ddot{\psi} = -M'_{ay}(\omega_y, \psi, \dot{\psi}),$$
$$C\ddot{\vartheta} = -M'_{az}(\omega_z, \vartheta, \dot{\vartheta}).$$

Hence it follows that the oscillations of the missile around the axes OX, OY, and OZ will be independent of each other.

Let us examine more accurately the first of the above equations. Let us expand the function M'_{ax} in a Taylor series around the point $(\omega_{x_0}, 0, 0)$ corresponding to the unperturbed situation. Retaining the first terms of expansion, we get

$$M'_{ax}(\omega_x, \varphi, \dot{\varphi}) = M'_{ax}(\omega_{x0}, 0, 0) + \frac{\partial M'_{ax}}{\partial \omega}\delta\omega_x + \frac{\partial M'_{ax}}{\partial \varphi}\varphi + \frac{\partial M'_{ax}}{\partial \dot{\varphi}}\dot{\varphi},$$

where the partial derivatives are taken in the point $(\omega_{x_0}, 0, 0)$. The motion of the missile in the plane YOZ will be described by the equation

$$A\ddot{\varphi} + \frac{\partial M'_{ax}}{\partial \dot{\varphi}}\dot{\varphi} + \frac{\partial M'_{ax}}{\partial \varphi}\varphi + \frac{\partial M'_{ax}}{\partial \omega}\delta\omega_x = M_{ax}(\omega_{x0}, 0, 0).$$

In order to eliminate $\delta\omega_x$ let us use the first of the equations (8). For small angles ϑ, ψ, and φ, we can discard the terms of the first and higher orders, and we obtain $C_{x_0} = K_x$ or $C_{x_0} = A\dot{\varphi} + J\omega_x$. A variation around the point under consideration gives

$$\delta\omega_x = -\frac{A}{J}\dot{\varphi}.$$

Consequently, the final equation of motion for the missile in the plane YOZ will have the form

$$A\ddot{\varphi} + \left(\frac{\partial M'_{ax}}{\partial \dot{\varphi}} - \frac{A}{J}\frac{\partial M'_{ax}}{\partial \omega_x}\right)\dot{\varphi} + \frac{\partial M'_{ax}}{\partial \varphi}\varphi = M_{ax}(\omega_{x0}, 0, 0).$$

The member $M_{ax}(\omega_{x_0}, 0, 0)$ of the right side of the equation is the perturbing force determining the statistical error of the system

$$\varphi_{\text{stat}} = \frac{M_{ax}(\omega_{x_0}, 0, 0)}{\frac{\partial M'_{ax}}{\partial \varphi}}.$$

It is interesting to note that the dependence of the moment M_{ax} upon the angular moment of a wheel for $(\partial M'_{ax}/\partial \varphi) < 0$ favors the quenching of the perturbations. For a sufficiently strong dependence of the moment M_{ax} upon the angular velocity ω_x one may not introduce the derivative of the angle in the control law.

DETERMINATION OF THE CONDITIONS FOR VISIBILITY OF SPACE ROCKETS

O. V. Gurko

Observations from the earth of the flight of space ships by optical means may become necessary when interplanetary flights are established. This problem has already come up with the launching of the first Soviet space rockets. In connection with this, the method given below has been evolved for determining the conditions for visibility of space rockets.

The conditions for observation of space rockets will be satisfactory if the observation point is located in the earth's shadow (night at the point), if the rocket (object) is observed at a sufficiently great location angle, and if it possesses the required brightness.

The first and second Soviet space rockets did not have sufficient brightness, due to their relatively small dimensions; this made their observation at significantly great distances difficult, even with the most powerful optical instruments. Therefore in order to improve the visibility conditions for these rockets, devices were installed to permit making them into artificial "comets," at given times, of brightness sufficient for observation.

The selection of the times for producing "comets" for satisfactory observation, and also the selection of observation stations, was made on the principles of the method expounded herein; this method is suitable for determining the conditions for visibility of any space vehicle.

The essence of the method is as follows: we project onto a geographic map of cylindrical projection both the object to be observed and the sun; using transparent material we construct illumination lines (twilight and night) and the lines of location angles for such designated zones; then we determine on the map the positions of these zones with respect to observation stations.

Let us examine several detailed steps in the determination of visibility conditions.

To determine the sun's declination corresponding to the observation date (we consider that the declination remains constant during the period of time in question), we construct on the transparent material a set of zones of equal angles ρ corresponding to the various solar inclination angles. We shall call the zone with $\rho = 72°$ (angle of solar dip $\beta = 18°$) the night zone. The mutual disposition of the equal-angle zone, of the observation station, and the angles used therefor are shown in Fig. 1. The point C, called the middle of

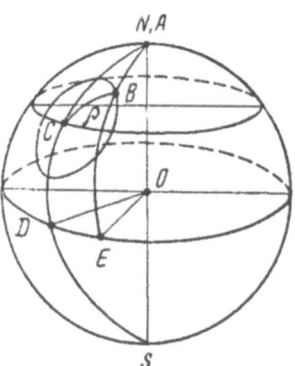

FIG. 1. Relative locations of the middle of the night, C, observation station B (coordinates L, ϕ), and other elements used in the computations. (Arc CD = $-\delta_\odot$; arc AB = 90 deg $- \phi$; arc AC = $90° - \delta_\odot$.

the night, is symmetric with respect to the center of the earth and the projection of the sun on the earth's surface. From this it follows that the middle of the night must be on latitude $\varphi = -\delta_\odot$. The point B is the observation station (its coordinates are L, φ).

Knowing ρ and $\delta_\odot$, we must determine L and φ for the points of the zone (for simplicity, we have assumed point C to be on the Greenwich meridian).

From examination of the spherical triangle ABC

($<$ ACB $= \xi$) we have:

$$\sin \varphi = \cos \rho \cdot \sin \delta_\odot + \sin \rho \cdot \cos \delta_\odot \cos \xi. \qquad (1)$$

Angle ξ is considered the independent variable in this calculation.

To determine the longitudes of the points we seek, we use the equation

$$\sin L = \frac{\sin \rho \cdot \sin \xi}{\sin \varphi}. \qquad (2)$$

Assigning values of angle ξ for a given $\delta_\odot$, we can compute the coordinates of the points of the zone for any ρ and place them on the map or on the transparent material.

For determining visibility zones by this method we used a cylindrical-projection world map such that zones constructed on it did not undergo distortion when the longitude of the middle of the night changed, although distortions were introduced by latitude changes (i.e., by change of observation date). Consequently, zones having centers corresponding to the given value of $\varphi = -\delta_\odot$ could be moved to different regions (if inscribed on the transparent material) without latitude change. This procedure is essential for solution of the visibility problem. In order to solve such a problem with the required accuracy during any time of the year, while the sun's declination changes from $+23.5°$ to $-23.5°$, it is necessary to have a set of "templates" on the transparent material for a whole series of values of $\delta_\odot$ (in intervals of about 3°). It can easily be seen from simple geometrical principles that the "templates" for negative declinations can also be used for positive ones provided they are rotated 180°.

Knowing the date and time of observation, we can determine on the map the position of the middle of the night having a West longitude numerically equal to the Greenwich observation time, and a latitude equal to the sun's declination $\delta_\odot$ with its sign reversed.

Superimposing on middle of the night thus found on the map the middle of the night as found on the "template," we can see if the observation station is located in the zone of night or twilight, and also determine the solar dip angle β.

In order to determine the zone of visibility by location angle we can use the following equation

$$\gamma = \arccos \frac{(R+H)\sin p_1}{\sqrt{R^2 + (R+H)^2 - 2R(R+H)\cos p_1}}, \qquad (3)$$

where H is the height of the object above the earth's surface; p_1 is the central angle between the observation station and the projection of the object on the earth's surface; γ is the location angle; R is the radius of the earth.

Evaluation shows that for H over 100,000 km, the dependence of γ on H can be practically neglected. For values of p_1 corresponding to the selected γ we can use equations (1) (where in place of $\delta_\odot$ we use the latitude of the point of projection of the object onto the earth's surface) and (2) to construct "zones of equal location angles" for heights of 100,000 km; we can also use them for higher altitudes.

It is necessary to note that the change of the geographic latitude for the projection of the object onto the earth's surface, i.e., the center of the zone of equal γ (as for the illuminated zone) leads to distortion of the indicated zone on the map even though other conditions are unchanged. Longitude change does not change the zone configuration. In view of the above, it is also necessary to construct location angles visibility zones for various latitude centers (with a 3° interval, for example) over the range of latitudes within which the anticipated trajectory projections fall.

The location angle zones are constructed in the same way as the visibility zones on transparent material (on "templates"); this permits moving them on the map as necessary. We also need to have on the map the trajectory projections with time marks t' (time elapsed since launch).

Coinciding the center of the "template" zone (having latitude equal to or close to that of the chosen projection of the object on the earth) with the projection on the map of the orbit point corresponding to observation time, we obtain for this given moment a visibility zone from which the object can be observed with location angle equal to or greater than γ. The region of satisfactory observation stations satisfying the visibility conditions must be within the intersecting area of the visibility zone and the location angle zone.

The method described above can be used to solve the problem of choosing the proper time for creating a "comet" in order to permit its observation from given observation stations, the problem of determining the location of observation stations from which the "comet" can be seen if it is created at a given time, the problem of the visibility of a space ship, and other problems concerned with the visibility conditions for objects in space.

Let us investigate the problem of choosing the proper moment for making a "comet." To do this, we must know the date of observation, the Greenwich time of launch t_0 and either the trajectory or else its projection onto the earth's surface. We place on the map several space-rocket trajectories corresponding to various declinations of the moon δ_l (at 3° intervals); this turns out to be sufficient for solving such problems for objects launched on any day.

We shall designate the Greenwich observation time by t_{Gr}, and the time interval from launch to creation of the "comet" by t'. Obviously,

$$t_{Gr} = t' + t_0. \tag{4}$$

To determine t' from (4) it is only necessary to first find t_{Gr} corresponding to night at the observation stations, using the method discussed above. With this objective, we take the center of the "template" having plotted on it the illumination zone for $\delta_\odot$ corresponding to the date of observation, place it on latitude $\varphi = -\delta_\odot$, and move it along the latter until the given observation stations appear within the zone.

The longitude of the center of the zone will then correspond to the Greenwich observation time we seek. Since the launch time is known beforehand, we can determine t' from equation (4). Then the center of the "template" having plotted on it the location angle zone for latitude equal to or greater than the orbit point latitude corresponding to time t' is made to coincide with the projection of this point on the map. The observation stations must fall within the area of intersection

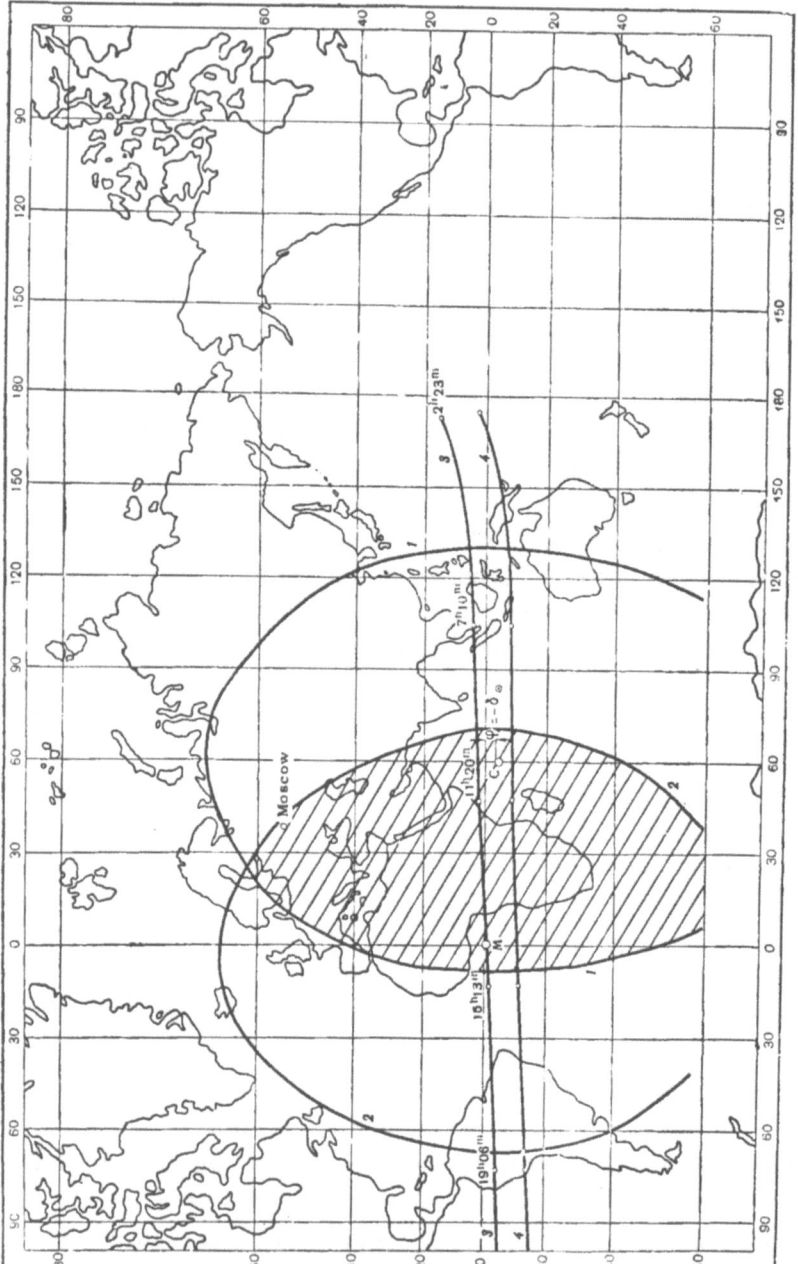

FIG. 2. Arrangement on the surface of the earth of the location angle zone γ = const and the zone of night. C) middle of the night; M) center of the zone γ = const; 1) zone of night; 2) zone of γ = const; 3, 4) trajectories corresponding to $\delta_l = -6°$ and $\delta_l = -18°$; the shaded region shows satisfactory visibility conditions.

VISIBILITY CONDITIONS FOR SPACE ROCKETS

of the visibility and location angle zones (Fig. 2).

In the case where it is not possible to select the proper arrangement of visibility and location angle zones with respect to the observation stations, it is still necessary to consider that for the given trajectory and selected t_0 and $\delta_\odot$, the existing observation stations require information about the conditions of visibility.

If we are given the value of t' and are required to find the stations from which the "comet" can be observed, the solution is carried out in reverse. In order to avoid having to repeat the operations described above for every date and point, it is possible to construct graphs in advance which will permit finding t' and all necessary parameters for visibility from given stations for any trajectory, launch time, and date.

Figure 3 shows such a graph. The abscissa gives the

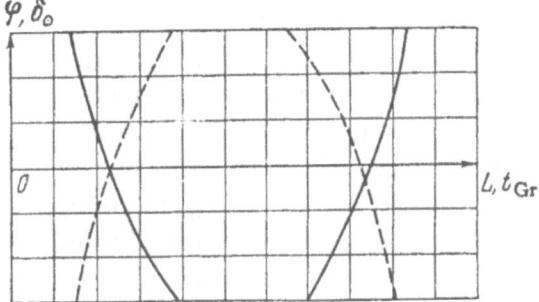

FIG. 3. Graph for determining proper time to create a "comet" (t') and the conditions for its visibility. The solid curve represents the line of γ = const; the dotted curve represents the line of β = const.

longitude L and the Greenwich time t_{Gr}. The ordinate gives the latitude φ (positive sign for north latitude) and the sun's declination $\delta_\odot$. If the projection of the trajectory on the earth's surface with its time marks t' is known, it can be placed on the graph by means of the coordinates L and φ. The curve γ = const indicates the region thereof visible from given stations at an angle less than γ. At the same time, we obtain on the edge of this region the trajectory points corresponding to the observation times $t_{Gr,1} = t'_1 + t_0$ and $t_{Gr,2} = t'_2 + t_0$ (assuming t_0 to be known). Then we determine from the known values of $t_{Gr,1}$ and $t_{Gr,2}$ by means of the curve $\beta = \text{const}$ (for a known $\delta_\odot$) whether it will be nighttime at the station.

Thus we can solve the problem of selecting a portion of the trajectory in which to create the "comet."

Since it is necessary to obtain the visibility parameters with high accuracy, the graphic operations may be replaced with analytic calculations not shown in this paper. In some cases, when the visibility problem requires high accuracy rather than a rapid descriptive solution, it is convenient to use another method briefly described below.

The location angle for a given station depends on the magnitude of the great circle arc ρ_1 connecting the station with the projection of the orbit point under consideration. On the other hand, the illumination of the station depends on the magnitude of the great circle arc ρ, connecting the station with the middle of the night. Thus the problem reduces itself in both cases to finding two great circle arcs, which can be easily determined by solving two spherical triangles:
a) earth's pole-station-projection of orbital point, and
b) earth's pole-station-middle of the night.

This problem is solved with sufficient accuracy on the grid introduced (see Ref. 1). The grid represents the earth's hemisphere with a Postel projection of equal azimuthal spacing (Fig. 4). The horizontal line dividing the hemisphere in

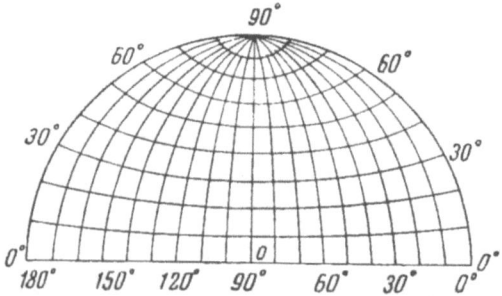

FIG. 4. Terrestrial hemisphere in Postel's projection having equal azimuthal spacing. O is the center; the left hand boundary is the western half of the initial meridian; the right hand boundary is the eastern half of the initial meridian.

half represents the earth's equator; the outside circumference represents the initial meridian.

The quantities ρ_1 and ρ which we seek are found in the following manner. We place tracing material on the grid and mark thereon the center of the sphere. With the aid of the given geodetic coordinates and the grid of meridians and parallels, we mark the terminal points of the arc we seek. The first point is placed on the initial meridian (on the eastern half if this point lies east of the second point, corresponding to a negative longitude difference; on the western half if this point lies west of the second point, corresponding to a positive longitude difference). Rotating the tracing material around the marked center point makes the first point coincide with the "north pole" (90°). The desired quantity is measured from the pole along the initial meridian to its intersection with the parallel of the second point.

For more convenient problem solution one can construct graphs of $\beta = f_1(\rho)$, $\gamma = f_2(\rho_1)$ or else in lieu of the graphs one can adapt the scale lines and work directly on the grid.

The method described (using a map of cylindrical projection) was used to analyze the visibility conditions for Soviet and foreign observation stations with regard to the "comet" created by the first and second Soviet space rockets.

The times of creating these artificial comets were chosen on the basis of the above analysis. In all cases the visibility conditions at the stations corresponded to those calculated; this indicates the correctness and sufficient accuracy of the method described.

LITERATURE

1. A. A. Timofeev and A. V. Lifshits, Tables for Calculating Azimuth and Length of Geodetic Lines of Great Range on the Surface of a Krasov Ellipsoid (in Russian), VTS Press, Moscow, 1956.

ABOUT THE FORMATION OF NO⁺ IN THE UPPER ATMOSPHERE

A. D. Danilov

In references 1, 2, and 3 it was shown that the charge transfer from atomic oxygen ions to nitrogen molecules giving NO^+ ions,

$$O^+ + N_2 \rightarrow NO^+ + N, \tag{1}$$

plays an important role in the upper atmosphere.

The ions formed in this reaction must recombine quite rapidly, disassociating into nitrogen and oxygen.

$$NO^+ + e \rightarrow N + O. \tag{2}$$

In the present work we shall try to examine the above reactions in order to obtain the value and the mechanism for high NO^+-ion concentrations and to compare the results of our calculations with the observed data. The possibility of formation of NO^+ by reactions different from (1), e.g., by the reactions $O^+ + N \rightarrow NO^+$ and $N^+ + O \rightarrow NO^+$, is little probable both because of the slow rate of such reactions and because of the small concentrations of ionized and neutral atomic nitrogen compared with the concentrations of N_2 in the height range under consideration. As to reaction (2), it is most important for the NO^+ elimination process.

If other formation and destruction mechanisms for NO^+ ions are little effective, processes (1) and (2) must balance each other. In other words the rate of formation of NO^+ according to reaction (1) must be equal to the rate of destruction cording to reaction (2):

$$\gamma [N_2] [O^+] = \alpha [NO^+] n_e, \tag{3}$$

where γ is the charge transfer coefficient, α is the coefficient for the dissociative recombination, and n_e is the electron density. From equation (3) we obtain

$$\frac{[NO^+]}{[O^+]} = \frac{\gamma [N_2]}{\alpha n_e} . \qquad (4)$$

Given n_e and N_2, it is easy to obtain from equation (4) the ratio $[NO^+]/[O^+]$ provided the ratio γ/α is known.

Different authors have used different values for the quantities γ and α. Gerjuoy and Biondi[4] used for the rates of the charged transfer reaction and of the dissociative recombination values of 10^{-12} cm$^3 \cdot$ sec^{-1} and 10^{-8} cm$^3 \cdot$ sec^{-1} respectively, which gives for the ratio γ/α the value of 10^{-4}. Bates[1] and V. I. Krasovskii[2] chose the coefficient of charge transfer reaction (1) equal to 10^{-10} cm$^3 \cdot$ sec^{-1}; Bagaryatskii,[5] on the basis of Potter's papers,[3] considered more plausible for this coefficient the value 10^{-8} cm$^3 \cdot$ sec^{-1}. The coefficient of the dissociated recombination was estimated for a long time[6,7] as 10^{-8} to 10^{-7} cm$^3 \cdot$ sec^{-1}. However, more recently the majority[8,9] have taken for this coefficient the value 10^{-6} cm$^3 \cdot$ sec^{-1}. The same value was obtained experimentally by Faire and coworkers[10] for the dissociated recombination of nitrogen.

A comparison of the most reliable value of the charge transfer rate, that which was adopted by Bates[1] and Krasovskii,[2] with the value of dissociative recombination coefficient accepted at present leads to a value of γ/α also equal to 10^{-4}.

According to the preceding remarks a value of $[NO^+]/[O^+]$ has been chosen in the present work for calculating the ratio γ/α which equals 10^{-4}.

The concentration of molecular nitrogen N_2 was calculated from the value of dissociative recombination coefficient and rockets.[11,12] It was also assumed that the rate of concentration of N_2 did not change with the height. This assumption is obviously true even in the presence of a diffusion separation, since by diffusion the concentration of elements light with respect to N_2 increases, but that of heavier elements decreases; therefore the fraction of molecular nitrogen in the total density may change quite negligibly.

The value of n_e for calculating the ratio $[NO^+]/[O^+]$ according to formula (4) was taken from K. I. Gringauz's data.[13] The results of the calculation are presented in Table 1.

TABLE 1

H, km	n_e, 10^5 cm^{-3}	ρ, 10^{-13} g·cm^{-3}	$[N_2]$, 10^9 cm^{-3}	$[N_2]/n_e$	$[NO^+]/[O^+]$	lg $\frac{[NO^+]}{[O^+]}$
120	1.3	200	320	2.5 10^6	2,5 10^2	2.40
140	1.6	32	52	3.4 10^5	34	1.53
160	2.0	9.5	15	7.6 10^4	7.6	0.88
180	3.2	4.4	7.1	2.2 10^4	2.2	0.34
200	4.0	2.7	4.3	1.1 10^4	1.1	0.04
250	13	1.1	1.8	1.4 10^3	1.4 10^{-1}	—0.86
300	20	0.36	0.58	2.9 10^2	2.9 10^{-2}	—1.54
350	16	0.14	0.23	1.4 10^2	1.4 10^{-2}	—1.86
400	14	0.066	0.11	7.9 10	7.9 10^{-3}	—2.10

Figure 1 shows the diagram of the change of the ratio $[NO^+]/[O^+]$ with the height. The solid line is the theoretical line calculated in the present paper according to formula (4). The dots correspond to data obtained by Johnson and co-

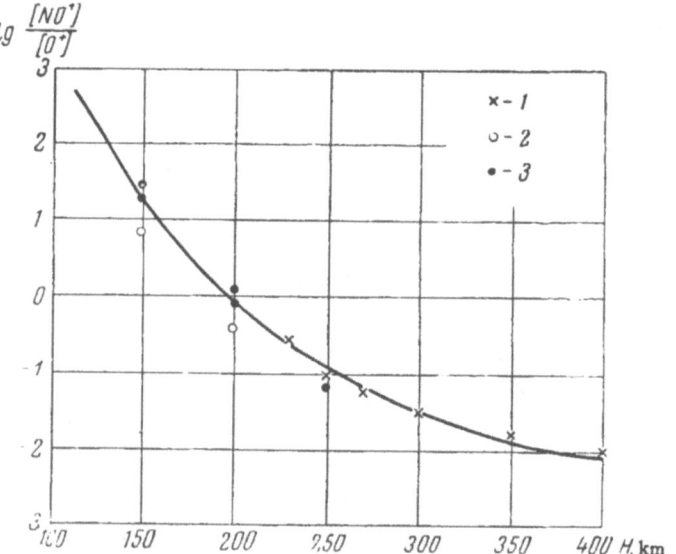

FIG. 1. Behavior with a high $[NO^+]/[O^+]$ ratio. 1, Data from Ref. 14; 2, 3, Data from Ref. 15 (2, day; 3, night); Solid line: result of calculations according to formula (4).

workers[14] with Aerobee; the crosses correspond to data obtained by V. G. Istomin[15] with the help of a mass spectrometer on the third satellite. The satisfactory agreement of the theoretical curve with the experimental data permits us to conclude that the choice of reactions (1) and (2) for studying the problem of the formation and destruction of NO^+ is correct, and that just these reactions are mainly responsible for the formation and destruction of nitrogen oxide ions in the atmosphere above 100 km.

The agreement between the absolute values of calculated and observed values for the ratio $[NO^+]/[O^+]$ leads us to conclude that the value 10^{-4} chosen for the ratio γ/α is correct. Consequently, if the dissociative recombination coefficient is roughly equal to 10^{-6} cm$^3 \cdot$ sec^{-1}, the rate of the charge transfer reaction must be 10^{-10} cm$^3 \cdot$ sec^{-1}.

It is interesting to study the problem of the daily variation of the ratio $[NO^+]/[O^+]$. This ratio, according to formula (4), is inversely proportional to the electron density n_e. As the observations have shown, the electron density changes from day to night. According to Seddon and Jackson's data[16] n_e is an order of magnitude smaller during the night than during the day, in the interval between 100 and 200 km; according to Ya. L. Al'pert[17] the daylight electron density exceeds night-time density by a factor of 4-5 at heights of 200-400 km. Therefore one should expect that night-time values of the ratio $[NO^+]/[O^+]$ must be higher than the daytime values. Table 2 gives the day and night values of the quantity $[NO^+]/[O^+]$ according to Ref. 14. These experimental data suggest that the ratio $[NO^+]/[O^+]$ increases during the night roughly by a factor of 3.

The values obtained for the quantity $[NO^+]/[O^+]$ permit

TABLE 2

H, km	$[NO^+]/[O^+]$			
	Night, 9/20 1956	Night, 2/21 1958	Night, average	Day, 3/23 1958
150	26.9	19.6	23.3	6.9
200	0.8	1.2	1.0	0.36

evaluating the changes in concentration of NO^+ with height. Let us suppose that the number of electrons is equal to the number of positive ions (not taking into account the negative ones)

$$n_e = n_i \qquad (5)$$

If we confine ourselves to considering only the ions O^+, NO^+, and O_2^+, which according to Refs. 14 and 15 are the predominant ions and heights of 100-400 km, equation (5) can be written in the form

$$n_e = [O^+] + [NO^+] + [O_2^+] \qquad (6)$$

or

$$n_e = [O^+]\left(1 + \frac{[NO^+]}{[O^+]} + \frac{[O_2^+]}{[O^+]}\right);$$

consequently

$$[O^+] = \frac{n_e}{1 + \frac{[NO^+]}{[O^+]} + \frac{[O_2^+]}{[O^+]}}. \qquad (7)$$

TABLE 3

H, km	$[NO^+]/[O^+]$	$[O_2^+]/[O^+]$	$1+\frac{[NO^+]}{[O^+]}+\frac{[O_2^+]}{[O^+]}$	n_e, 10^5 cm^{-3}	$[O^+]$, 10^5 cm^{-3}	$[NO^+]$, 10^5 cm^{-3}
140	34	28	63	1.6	0.026	0.89
160	7.6	7.9	16.5	2.0	0.12	0.95
180	2.2	2.5	5.7	3.2	0.55	1.2
200	1.1	0.71	2.8	4.0	1.4	1.5
250	0.14	0.089	1.2	13	11	1.5
300	0.029	0.015	1.0	20	20	0.58
350	0.014	3.3 10^{-3}	1.0	16	16	0.22
400	7.9 10^{-3}	1.0 10^{-3}	1.0	14	14	0.11

The calculation of the value of $[NO^+]/[O^+]$ has been discussed above. The quantity $[O_2^+]/[O^+]$ was taken from Fig. 2, which was constructed from the observed data of Refs. 14 and 15. As a result, using formula (7), we obtained the behavior of the concentration $[O^+]$, and then, using the values of $[O^+]$ and the ratio $[NO^+]/[O^+]$ we found the change of $[NO^+]$ with the

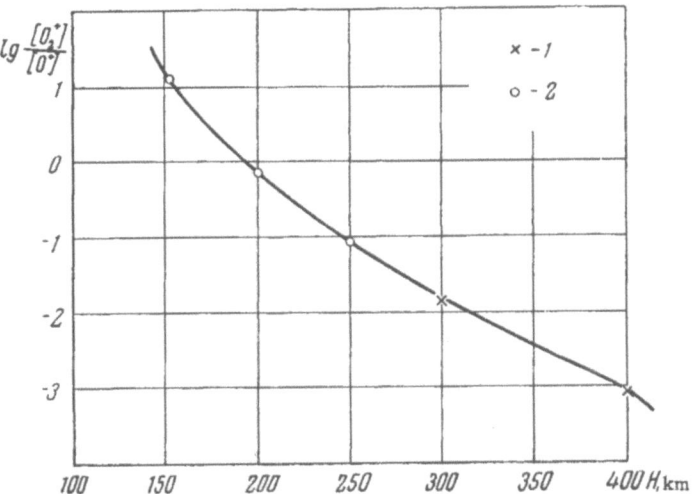

FIG. 2. Ratio $[O_2^+]/[O^+]$ vs. height. 1, data from Ref. 15; 2, data from Ref. 14; the calculation was made on the basis of the curve drawn through the experimental points.

height. The intermediate steps and the results of the calculations are shown in Table 3; the corresponding behavior of $[O^+]$ and $[NO^+]$ is shown in Figs. 3 and 4.

The scanty experimental data do not give the possibility of comparing the theory with experiment. Results obtained by

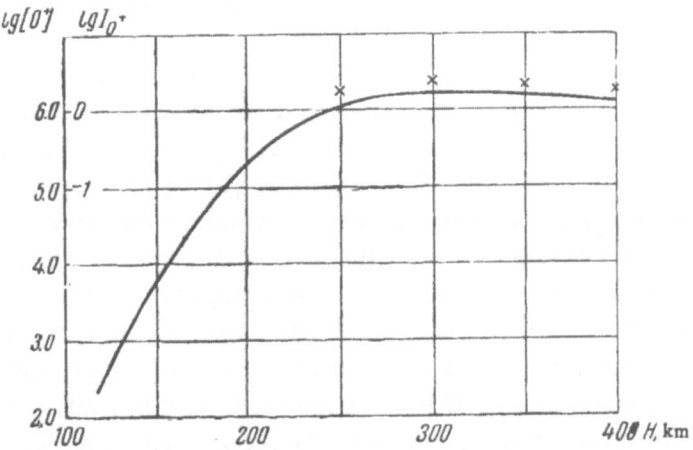

FIG. 3. Calculated dependence of the concentration $[O^+]$ upon the height. The crosses show the values of the ion current of O^+ (in relative units) obtained by V. G. Istomin.

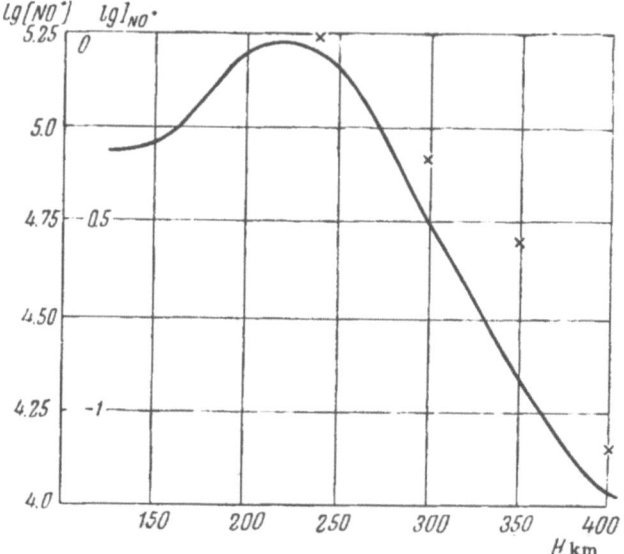

FIG. 4. Calculated dependence of the concentration [NO$^+$] upon the height. The crosses show the values of the ion current of NO$^+$ (in relative units) obtained by V. G. Istomin.

V. G. Istomin (private communication) and also results obtained by Johnson and coworkers[18] permit us only to conclude that there is a very rough agreement between calculated and observed behavior of [O$^+$] and [NO$^+$] with height in the interval 150-400 km.

As can be seen from Fig. 4, the NO$^+$ ions, according to the calculations, are distributed in the atmosphere as a layer with a maximum at about 200-250 km. This can be explained on the basis of formula (4) as follows. [NO$^+$] is proportional to [O$^+$] and to [N$_2$], and is inversely proportional to n_e. The concentration of electrons in the height interval in question changes little, and therefore the most important role in the distribution of [NO$^+$] is played by the way [O$^+$] and [N$_2$] change with the height. Between 100 and 200 km [N$_2$] drops because of the decrease of ρ, but [O$^+$] increases sharply, and consequently the concentration of NO$^+$ increases with the height. Beyond 250 km [O$^+$] remains roughly constant and therefore [NO$^+$] decreases proportionally to the decrease of [N$_2$].

ACKNOWLEDGEMENT

The author expresses his deep gratitude to G. S. Ivanov-Kholodnyi for his constant interest in this work and his participation in the discussion of the results.

LITERATURE

1. D. R. Bates, Proc. Phys. Soc A68, 344, 1955.
2. V. I. Krasovskii, Izv. Akad. Nauk SSSR, Geophys. Ser. 4, 504, 1957.
3. R. F. Potter, J. Chem. Phys. 23, 2462, 1955.
4. E. Gerjuoy and M. A. Biondi, J. Geophys. Res. 58, 295, 1953.
5. B. A. Bagaryatskii, Izv. Akad. Nauk SSSR, Geophys. Ser. 9, 1359, 1959.
6. S. K. Mitra, Upper Atmosphere (Russian Translation), IL, Moscow, 1955, p. 278.
7. E. T. Byram, T. A. Chubb and H. Friedman, Phys. Rev. 98, 1594, 1955.
8. R. B. Bryan, R. B. Holt and O. Oldenberg. Phys. Rev. 106, 83, 1957.
9. T. A. Chubb and H. Friedman, Nature 180 (4584), 501, 1957.
10. A. C. Faire, O. T. Fundingsland, A. U. Aden and K. S. W. Champion, J. Appl. Phys. 29, 928, 1958.
11. V. V. Mikhievich, Artificial Earth Satellites, Volume 2, Plenum Press, New York, 1960.
12. V. V. Mikhievich, B. S. Danilin, A. I. Repnev, and V. A. Sokolov, see Volume 3, p. 119.
13. K. I. Gringauz, Doklady Akad. Nauk SSSR 120 (6), 1234, 1958.
14. C. J. Johnson, E. B. Meadows and J. C. Holms, IGY Rocket Report Series, No. 1, 121, 1958.
15. V. G. Istomin, see Volume 4, p. 411.
16. J. C. Seddon and J. E. Jackson, IGY Rocket Report Series, No. 1, 140, 1958.
17. Ya. L. Al'pert, V. L. Ginzburg, and E. L. Feinberg, Radio-Wave Propagation, Moscow, 1953.
18. C. J. Johnson, J. P. Heppner, J. C. Holms and E. B. Meadows, IGY Rocket Report Series, No. 1, 123, 1958.

OBSERVATION OF SIGNALS FROM THE THIRD SOVIET ARTIFICIAL EARTH SATELLITE, AT CAPE CHELYUSKIN

L. P. Kuperov

Observations of signals originating from the third Soviet artificial earth satellite, received at a frequency of 20,005 kc, were carried out May 16 to June 6, 1958 (13th to 296th passes of the satellite) at Cape Chelyuskin (77°43'N, 104°17'E). The observations were conducted by L. P. Kuperov, O. L. Solovskii, and I. I. Yakubaitis of the staff of the Arctic and Antarctic Scientific Research Institute.

The equipment used comprised a facility for measuring field strength by the comparison method. A class 1 radio receiver with a transmission bandwidth of 1 kc at intermediate frequency and 300 cps at low frequency was employed for the observations. The receiving antenna was an upright whip antenna 2 meters long, with an O. D. of 2 cm. An ÉNO-1 cathode-ray oscillograph (CRO) was placed at the receiver output, with a telephone headset shunt-connected to the CRO input for direct aural monitoring.

A GSS-6 standard signal generator was employed to feed a standardized emf to the receiver input. A dummy antenna consisting of a 51 ohm resistor and a 30 $\mu\mu$f capacitor in series was series-connected to the GSS-6 output.

The entire equipment was mounted on a brass plate sturdily coupled by a short connector to external ground. The shields of the instruments, dummy antenna, and intermediate connectors were coupled by short busses to the brass plate.

Readings of the comparison facility differed by not more than 8% from the readings of the standard comparator at a frequency of 20 Mc. 0.4 μv delivered to the receiver input (via the dummy antenna equivalent) proved sufficient to double the amplitude value of the noise level, with respect to voltage, at a frequency of 20 Mc.

Aural monitoring and measurement of field strength were carried out with the second heterodyne of the receiver on. The frequency of the second heterodyne was set such that the beat tone produced as the GSS-6 intermediate frequency was fed in would correspond to the center frequency of the low-frequency filter bandwidth. This frequency control of the second heterodyne was maintained during the measurements. The signal from the satellite transmitter received at the beat frequency was passed on to the CRO, which reproduced tracings of the signal on the screen. The audio receiver gain was kept at maximum.

The receiver was tuned to 20,005 kc frequency several minutes prior to the moment when signals were expected from the satellite. The preassigned gain for the high-frequency receiver was set and the GSS-6 generator was used to measure the noise level and the level on the oscillograph scale corresponding to $1\mu v$. The values of the CRO scale markings were computed under the assumption that the receiver and oscillograph gain figures did not change during the time the satellite was under observation (about 10-15 min). The receiver was then switched to antenna operation and observations were pursued during the time signals were coming in from the satellite.

The receiver tuning drifted smoothly over a range of ± 2 kc from nominal value during the listening period, since signals from the satellite could acquire a frequency differing from 20,005 kc owing to Doppler effect. When signals came in from the satellite, the observer recorded the time and did a calculation of the field strength on the basis of the CRO scale.

The trajectories of the motion of the satellite's projection over the earth's surface were plotted for each sequence of observations. The plots of the trajectories so obtained made it possible to compute the probability of intercepting the satellite's signals when it was directly over some point on the earth's surface.* A plot of the isolines is shown in Fig. 1. The percentage of cases when radio signals at 20,005 kc frequency

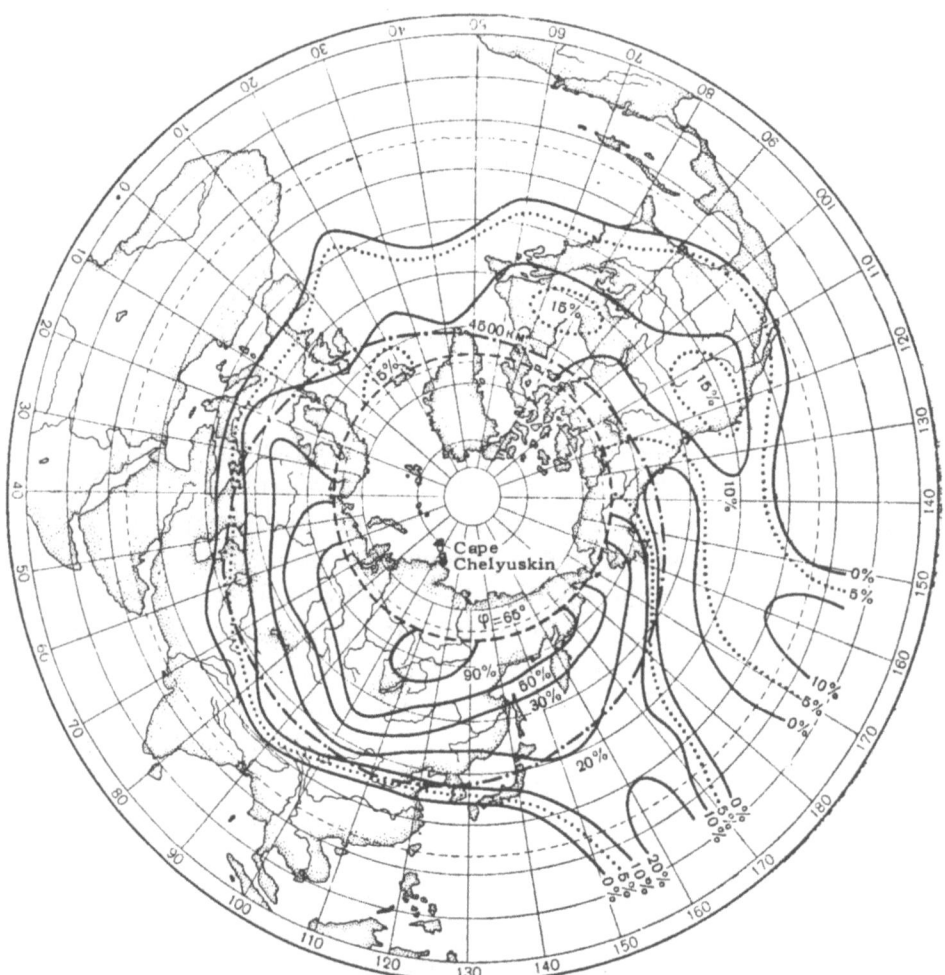

FIG. 1. Plot of isolines of detection probability of satellite signals at Cape Chelyuskin, during the period from May 16 to June 6, 1958.

were detected at Cape Chelyuskin (relative to all cases when the satellite was found at a given point during the time of observation) is shown for each isoline.

*In the period from May 16 to June 6, 1958, the ionosphere along the path of motion of the satellite in the Northern Hemisphere was completely illuminated by the sun. Consequently, the results reported here refer to the sunlit ionosphere.

During the entire period of observations, the maximum horizontal range for detection of satellite signals in cases where the line joining the satellite and Cape Chelyuskin passed completely over dry land was 4500-5000 km. In those cases where the line joining the satellite and Cape Chelyuskin passed to a considerable extent over the sea, the maximum horizontal range was about 8500 km. The horizontal range of radio signals, independent of azimuth, never exceeded 5000 km in those cases where the satellite moved southward from the parallel 65°N. Radio signals from distances in excess of 5000 km were thus detected only when the satellite proceeded from the equator to the parallel 65°N, and the line joining the satellite and Cape Chelyuskin passed over the sea. Figure 1 shows a circle 4500 km in radius with its center situated at Cape Chelyuskin; the zero isoline actually passes near that circle when the satellite passes over dry land, and recedes from the isoline as soon as the greater part of the distance to Cape Chelyuskin is over a water route.

We might draw attention to the fact that, for distances passing over the sea, the zone of reduced values of probability of observation has the same horizontal ranges as the zero isoline for distances traversing dry land, i.e., close to a circle of 4500 km. This supports the assumption that the mechanism responsible for the propagation of the satellite's signals to distances of 5000 km is independent of the nature of the underlying surface, i.e., propagation of radio signals proceeds in that case without skipping and hopping over the earth's surface. For the case of reception of signals from distances of 5000 to 8500 km, there is at most one reflection from the earth's surface in the propagation process. If the point of reflection is on dry land, the radio signal will not arrive at the receiving point on account of heavy losses in the solid crust of the earth.

If we estimate the height of the F_2 layer during the sunlit portion of the day, on the basis of data reported by the ionosphere stations at Moscow and Dikson during May 1958, we find that the satellite was orbiting beneath the F_2 layer when moving toward the parallel 65°N at a distance of 8500 km from the receiving point. When moving away from the parallel 65°N the satellite was at first slightly beneath the F_2 layer (out to distances of 5000 km), and later was constantly above the F_2 layer.

Accordingly, when the satellite was moving toward the parallel 65°N, radio signals reached the receiving point after bouncing once from the F_2 layer (except for cases of line-of-sight reception), when the horizontal range was within 5000 km. From distances greater than 5000 km, for propagation over dry land, radio signals were not detected; the power of the transmitters was apparently inadequate to overcome losses incurred in a single reflection from dry land and a triple traversal of the F_1, E_0, and D layers. This assumption is to be verified in further processing of the observations of satellite signals taken during the period when the ionosphere was in the shadow of the earth, when losses were minimal in the lower-lying layers of the ionosphere.

The maximum horizontal range of 8500 km consists of a maximum distance of 5000 km plus a maximum distance for one "hop" (3500 km) following reflection from the surface of the sea, followed by a reflection from the F_2 layer.

When the satellite was heading away from the parallel 65°N, and radio signals were at the very beginning propagating by skips and hops (except for the case of line-of-sight reception), and later by refraction in the F_2 layer, the maximum distance attained by signals refracted in the F_2 layer was found to be 5000 km. At greater distances, radio signals were not detected either in the case of overland propagation or propagation over a water route, although the satellite was orbiting above the F_2 layer in 80 cases, during the period of observation, at distances greater than 5000 km (in 43% of all cases of observation where the satellite was found to be above the F_2 layer). The reason behind this is apparently the fact that at distances greater than 5000 km, the F_2 layer comes to act as a screen for radio-frequency emissions propagating in the direction of that point on the earth's surface which corresponds to the next "hop" for the arrival of radio waves at the receiving point.

The validity of this assumption is confirmed by an examination of the relationship between the probabilities of refraction and reflection from the F_2 layers at distances out to 5000 km. At closer distances, the probability of refraction (detection of signals at 20 Mc when the satellite is orbiting above the F_2 layer) will exceed the probability of reflection (detection of signals when the satellite is beneath the F_2 layer). As the distance increases, conversely, the probability of reflection will become greater than the probability of refraction. Readings taken over all of the satellite passes, in each of which the satellite was both below and above the F_2 layer at some time during the period when the satellite was outside the range of line-of-sight perception, demonstrated that the probability of refraction was 10% higher than the probability of reflection in the vicinity of the outside boundary of the region of line-of-sight perception (2400 km), while the probability of reflection exceeded the probability of refraction by 23% near the boundary of a circle of radius 5000 km. Near the 5000 km limit, the probability of refraction came to 8%, while the probability of refraction vanished farther out than 5000 km.

In the line-of-sight region, signals were detected with equal probability when the satellite was above, or below the F_2 layer. During the period concerned, the probability of detection of satellite signals in the line-of-sight region was 94%. In all cases where there was no reception of signals from the line-of-sight region (passes Nos. 69, 175, 190), the ionosphere stations at Tiksi Bay [72°N, 129°E]* and Dikson Island [73°N, 83°E] observed, at the corresponding moments of local time, an anomalously high degree of absorption of radio waves, when they performed a vertical probing of the ionosphere.

The sum comprised of the detection probability of signals when the satellite was orbiting outside the range of line-of-sight visibility and above the F_2 layer, of the detection probability of signals when the satellite was outside the line-of-sight region and below the F_2 layer, and of the probability that

---

*Coordinates added by translator.

signals from the line-of-sight region would be inaudible, proved virtually equal to unity for the entire period in question, for the entire region enclosed within a circle of horizontal range 5000 km. This relationship between the detection probabilities of signals, equivalent to the relationship prevailing between the probabilities of refraction in and reflection from the F_2 layer, was also manifested on separate days. The greater the frequency with which signals were picked up from the satellite when it was outside the line-of-sight region and below the F_2 layer, the fewer were the cases of detection of signals, on those days, when the satellite was orbiting above the F_2 layer. Over a three-day period stretching from May 22 to May 24, 1958, while the satellite was above the F_2 layer within the area of a circle of radius 5000 km outside the line-of-sight range, no radio signals were detected. Radio signals were detected in 10 cases out of 14 when the satellite was beneath the F_2 layer. At that time, the state of the ionosphere and of the geomagnetic field was quiescent, the ionization density of the F_2 layer was quite high, the high-frequency field strength (at 16-17 Mc) at remote land-based radio stations was measurably higher than the field strength at low frequencies (8 Mc). Radio signals were accordingly being reflected from the F_2 layer and failed to arrive at the receiving point, when the satellite was orbiting above the F_2 layer.

Conversely, a sharp reduction in the number of cases where signals were detected when the satellite was below the F_2 layer was observed in the period from May 27 to May 31, 1958, and signals were not observed even once on May 29 when the satellite was in that position. On those days, the state of the ionosphere and of the geomagnetic field was intensely disturbed, and the field strength at remote land-based radio stations was found to be highest at low frequencies (8 Mc), rather than at high frequencies. It was therefore concluded that radio signals were passing through the F_2 layer without suffering reflection when the satellite was below the F_2 layer and outside the line-of-sight region, and accordingly escaped detection.

Median values based on peak field strength values (for

TABLE 1

Field Strength of Transmitter on Third Earth Satellite According to Measurements Made at Cape Chelyuskin During the Period from May 16 to June 6, 1958 (28 passes) with Longitude of Ascending Loop Ranging from 10° to 40° E

Location of satellite	Outside line-of-sight region		Inside line-of-sight region		Outside line-of-sight region
	below F_2 layer	below F_2 layer	above F_2 layer	above F_2 layer
Time after equator transit, min	12 13 14 15 16 17 18	19 20 21 22 23	24 25 26 27	28 29 30 31
Median of field strength	0 0.3 0.3 0.3 0.3 0.3 0.2	0.4 0.4 0.5 0.8 0.7	0.5 0.6 0.25 0.3	0.2 0.2 0 0
Peak field strength	0 0.3 0.3 0.3 0.6 0.5 2.0	4.0 1.4 2.5 2.0 2.0	1.4 1.2 1.0 0.6	0.3 0.2 0 0
Number of passes during which observations were performed	19 20 21 23 23 23 23	23 24 26 26 26	26 25 23 22	16 12 11 8

1 min) were computed for the field strength characteristics of the transmitter on the third satellite at 20,005 kc frequency. Only those satellite passes which occurred with the ascending loop between 10° and 40°E were considered in the computations. Those were the cases when the satellite made its closest approach to Cape Chelyuskin. In the computation of median values, cases where the satellite signals faded out were discarded, i.e., the median was based solely on measured quantities. The results of the calculations are tabulated above.

Table 1 shows that the median field strength reached a maximum of 0.8 μv/m 22 min after the satellite passed the equator; the satellite was then in the line-of-sight region and below the F_2 layer. Peak field strength was observed when the satellite entered the line-of-sight region.

Median values of the field strength based on maximum values recorded during each run of observations, for all passes executed by the satellite from May 16 to June 6, 1958 while the satellite was outside the line-of-sight region and either above or below the F_2 layer, are respectively 0.4 and 0.3 μv/m.

CHANGE OF THE ALBEDO OF THE FIRST ARTIFICIAL EARTH SATELLITE AS A RESULT OF THE ACTION OF EXTERNAL FACTORS

I. M. Yatsunskii and O. V. Gurko

A change in the albedo and the nature of the reflection from the surface of a satellite as a result of the action of the atmosphere, meteoritic material, and other physical factors is important in the observation of a satellite, and can also be of scientific interest in its own right. In this connection, an attempt was made to establish the variabilility of the reflecting capability of the surface from observational data on the first artificial earth satellite, which had a mirror surface, and consequently, in the initial period, essentially specular reflection.

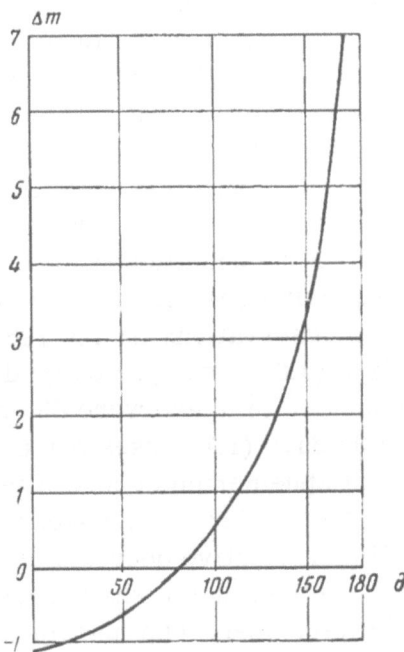

FIG. 1. Dependence of the difference $\Delta m = m_s - m_d$ on the phase angle θ.

This circumstance facilitates the analysis of the available observations.

The method of determining the change of the albedo of the first satellite in flight is based on (1) the known difference of specular from diffuse reflection for a sphere, and (2) the possible decrease of the amount of reflected light in the case of a change of the state of the surface under the action of external factors.

Figure 1 shows the dependence on the phase θ of the difference in stellar magnitudes ($\Delta m = m_s - m_d$) of a sphere with specular and diffuse reflection of light (θ is the angle between the satellite-sun and satellite-observer directions).

It is apparent that with an increase of the angle θ, the difference Δm increases, and for $\theta > 135°$-$140°$, attains 2.5 to 3.0 stellar magnitudes. Consequently, it is impossible to observe a diffusely reflecting sphere in the direction almost opposite the sun, whereas the brightness of a specularly reflecting sphere is independent of the phase.

From October 4 to December 6, 1957, there were more than 200 optical observations of the first artificial earth satellite through type AT-1 astronomical tubes. The phase angle θ and the angle Δ of the sun below the horizon were calculated for all of these observations. Next, the data, consisting of θ, Δ, time of observation, altitude angle β, azimuth A, distance to the satellite S, stellar magnitude m* and the difference Δm, were divided into two groups, corresponding to two periods of time in which the conditions of observation were significantly different. During the first period (from October 9-31), there were 159 observations during 106 passes of the satellite for 23 days (4.6 passes per day); during the second (November 1-December 7), there were 63 observations during 42 passes for 37 days (1.1 passes per day). At the same time, the number of ephemerides transmitted in the first period was 904, i.e., 39 per day; in the second, 1131 (30 per day). The insignificant relative decrease in the number of ephemerides does not explain the sharp drop in the number of observations in November. The fundamental cause of the

---

*The average observed stellar magnitude of the first satellite, launched to a distance of 500 km, was approximately 5 to 5.5.

decrease in the number of observations was the large number of cloudy days in this period.

It is interesting to note that on the basis of 200 analyzed observations, one can make the following conclusions.

1. In the first period, a number of observations were carried out with small altitude angles (20° and less), i.e., with large distances to the satellite. There were observations with large phase angles ($\theta > 130°$).

2. In the second period, there were no observations at large distances and large phase angles.

Table 1 shows data on the number of observations at various β and θ during the two periods under consideration.

Table 2 shows data on β_{av}, θ_{av}, S_{av}, and the height of the satellite over the observational sites on the ascending and descending loops in the two indicated periods. By ascending (direct) loops, we conventionally understand that part of the orbit corresponding to motion of the satellite from southern latitudes to northern; in the case of descending (inverse) loops, this motion takes place from northern latitudes to southern.

TABLE 1

Number of observations in period	Altitude angle β not larger than					Phase angle θ not less than		
	10°	15°	20°	25°	30°	120°	130°	140°
10/9 – 10/31	4	6	14	21	35	20	7	1
11/1 – 12/7	0	1	1	2	3	0	0	0
Δm						1.3	1.8	2.5

TABLE 2

Period of observations	Average altitude angle β_{av}	Average distance to the satellite S_{av}, km	Average phase angle θ_{av}	Average height of flight in the vicinity of the observational site, km	
				Ascending loop	Descending loop
10/9 – 10/31	48°	503	92°	225	400
11/1 – 12/7	56	425	78	220	410

It is clear from the data presented that in October the satellite had not yet lost its properties of specular reflection, for there were observations at phase angles θ up to 140°. Had there been deterioration of the surface, i.e., a sharp decrease of the coefficient of specular reflection, then the brightness of the satellite at $\theta = 140°$ would be weaker by 2.5 stellar magnitudes than with specular reflection, and would be of stellar magnitude 8 on the average. Under these conditions, observations of the satellite with the AT-1 tubes would become almost impossible. In November, as can be seen from Table 1, observations at large phase angles ($\theta > 120°$) are completely absent, although the relative number of cases of large and small phase angles (and similarly for the altitude angles) is practically the same for the two periods, according to the ephemerides distributed.

Hence, one can make a deduction not only regarding a decrease in the albedo of the satellite, but also on a change in the nature of the reflection (diffuse, instead of specular).

A relative increase in the number of failures to observe the satellite with a clear sky during the second period (20% in November as compared with 10% in October) can also serve to confirm these conclusions.

Thus, a statistical analysis of the visual observations of the first artificial earth satellite during two months of flight shows a change of the reflecting properties of the polished aluminum shell of the satellite under the influence of external factors.